Livestock Management

Livestock Management

Editor: Carlos Hassey

CALLISTO REFERENCE

www.callistoreference.com

Callisto Reference,
118-35 Queens Blvd., Suite 400,
Forest Hills, NY 11375, USA

Visit us on the World Wide Web at:
www.callistoreference.com

ISBN: 978-1-63239-780-5 (Hardback)

The publisher's policy is to use permanent paper from mills that operate a sustainable forestry policy. Furthermore, the publisher ensures that the text paper and cover boards used have met acceptable environmental accreditation standards.

Trademark Notice: Registered trademark of products or corporate names are used only for explanation and identification without intent to infringe.

Printed in the United States of America.

Cataloging-in-publication Data

Livestock management / edited by Carlos Hassey.
    p. cm.
Includes bibliographical references and index.
ISBN 978-1-63239-780-5
1. Livestock. 2. Animal culture. 3. Veterinary medicine. 4. Livestock--Diseases. 5. Animal biotechnology.
I. Hassey, Carlos.
SF61 .L58 2017
636--dc23

Table of Contents

Preface...IX

Chapter 1   **Dual Origins of Dairy Cattle Farming – Evidence from a Comprehensive Survey of European Y-Chromosomal Variation**...1
Ceiridwen J. Edwards, Catarina Ginja, Juha Kantanen, Lucía Pérez Pardal, Anne Tresset, Frauke Stock, European Cattle Genetic Diversity Consortium, Luis T. Gama, M. Cecilia T. Penedo, Daniel G. Bradley, Johannes A. Lenstra, Isaäc J. Nijman

Chapter 2   **Where do Livestock Guardian Dogs Go? Movement Patterns of Free-Ranging Maremma Sheepdogs**..14
Linda van Bommel, Chris N. Johnson

Chapter 3   **Priming of Soil Carbon Decomposition in Two Inner Mongolia Grassland Soils following Sheep Dung Addition: A Study using ^{13}C Natural Abundance Approach**..26
Xiuzhi Ma, Per Ambus, Shiping Wang, Yanfen Wang, Chengjie Wang

Chapter 4   **Seroprevalence and Potential Risk Factors for *Brucella* Spp. Infection in Traditional Cattle, Sheep and Goats Reared in Urban, Periurban and Rural Areas of Niger**...35
Abdou Razac Boukary, Claude Saegerman, Emmanuel Abatih, David Fretin, Rianatou Alambédji Bada, Reginald De Deken, Halimatou Adamou Harouna, Alhassane Yenikoye, Eric Thys

Chapter 5   **Comparison of Airway Responses in Sheep of Different Age in Precision-Cut Lung Slices (PCLS)**..47
Verena A. Lambermont, Marco Schlepütz, Constanze Dassow, Peter König, Luc J. Zimmermann, Stefan Uhlig, Boris W. Kramer, Christian Martin

Chapter 6   **Genetic Footprints of Iberian Cattle in America 500 Years after the Arrival of Columbus**..55
Amparo M. Martínez, Luis T. Gama, Javier Cañón, Catarina Ginja, Juan V. Delgado, Susana Dunner, Vincenzo Landi, Inmaculada Martín Burriel, M. Cecilia T. Penedo, Clementina Rodellar, Jose Luis Vega-Pla, Atzel Acosta, Luz A. Álvarez, Esperanza Camacho, Oscar Cortés, Jose R. Marques, Roberto Martínez, Ruben D. Martínez, Lilia Melucci, Guillermo Martínez-Velázquez, Jaime E. Muñoz, Alicia Postiglioni, Jorge Quiroz, Philip Sponenberg, Odalys Uffo, Axel Villalobos, Delsito Zambrano, Pilar Zaragoza

Chapter 7   **Expression of VP7, a Bluetongue Virus Group Specific Antigen by Viral Vectors: Analysis of the Induced Immune Responses and Evaluation of Protective Potential in Sheep**...68
Coraline Bouet-Cararo, Vanessa Contreras, Agathe Caruso, Sokunthea Top, Marion Szelechowski, Corinne Bergeron, Cyril Viarouge, Alexandra Desprat, Anthony Relmy, Jean-Michel Guibert, Eric Dubois, Richard Thiery, Emmanuel Bréard, Stephane Bertagnoli, Jennifer Richardson, Gilles Foucras, Gilles Meyer, Isabelle Schwartz-Cornil, Stephan Zientara, Bernard Klonjkowski

Chapter 8   **Trace Element Distribution in Selected Edible Tissues of Zebu (*Bos indicus*) Cattle Slaughtered at Jimma, SW Ethiopia**..77
Veronique Dermauw, Marta Lopéz Alonso, Luc Duchateau, Gijs Du Laing,
Tadele Tolosa, Ellen Dierenfeld, Marcus Clauss, Geert Paul Jules Janssens

Chapter 9   **Toll-Like Receptor Responses to *Peste des petits ruminants* Virus in Goats and Water Buffalo**...85
Sakthive Dhanasekaran, Moanaro Biswas, Ambothi R. Vignesh, R. Ramya,
Gopal Dhinakar Raj, Krishnaswamy G. Tirumurugaan, Angamuthu Raja,
Ranjit S. Kataria, Satya Parida,Elankumaran Subbiah

Chapter 10   **A Dynamic Spatio-Temporal Model to Investigate the Effect of Cattle Movements on the Spread of Bluetongue BTV-8 in Belgium**................................96
Chellafe Ensoy, Marc Aerts, Sarah Welby, Yves Van der Stede, Christel Faes

Chapter 11   **The Effect of Deltamethrin-treated Net Fencing around Cattle Enclosures on Outdoor-biting Mosquitoes in Kumasi, Ghana**.....................................108
Marta Ferreira Maia, Ayimbire Abonuusum, Lena Maria Lorenz,
Peter-Henning Clausen, Burkhard Bauer, Rolf Garms, Thomas Kruppa

Chapter 12   **Development of a Blocking ELISA Based on a Monoclonal Antibody against a Predominant Epitope in Non-Structural Protein 3B2 of Foot-and-Mouth Disease Virus for Differentiating Infected from Vaccinated Animals**....................114
Yuanfang Fu, Zengjun Lu, Pinghua Li, Yimei Cao, Pu Sun, Meina Tian,
Na Wang, Huifang Bao, Xingwen Bai, Dong Li, Yingli Chen, Zaixin Liu

Chapter 13   **Density of Wild Prey Modulates Lynx Kill Rates on Free-Ranging Domestic Sheep**..122
John Odden, Erlend B. Nilsen, John D. C. Linnell

Chapter 14   **Transmission of *Mycobacterium tuberculosis* between Farmers and Cattle in Central Ethiopia**..130
Gobena Ameni, Konjit Tadesse, Elena Hailu, Yohannes Deresse, Girmay Medhin,
Abraham Aseffa, Glyn Hewinson, Martin Vordermeier, Stefan Berg

Chapter 15   **Diversity and Community Composition of Methanogenic Archaea in the Rumen of Scottish Upland Sheep Assessed by Different Methods**.........................140
Timothy J. Snelling, Buğra Genc, Nest McKain, Mick Watson, Sinéad M. Waters,
Christopher J. Creevey, R. John Wallace

Chapter 16   **The Effect of Different Adjuvants on Immune Parameters and Protection following Vaccination of Sheep with a Larval-Specific Antigen of the Gastrointestinal Nematode, *Haemonchus contortus***....................................149
David Piedra ita, Sarah Preston, Joanna Kemp, Michael de Veer, Jayne Sherrard,
Troy Kraska, Martin Elhay, Els Meeusen

Chapter 17   **Functional Vascular Changes of the Kidney during Pregnancy in Animals**...157
Joris van Drongelen, Rob de Vries, Frederik K. Lotgering, Paul Smits,
Marc E. A. Spaanderman

Chapter 18   **Environmental Drivers of *Culicoides* Phenology: How Important is Species-Specific Variation when Determining Disease Policy?**..172
Kate R. Searle, James Barber, Francesca Stubbins, Karien Labuschagne, Simon Carpenter, Adam Butler, Eric Denison, Christopher Sanders, Philip S. Mellor, Anthony Wilson, Noel Nelson, Simon Gubbins, Bethan V. Purse

Chapter 19   **Antenatal Dexamethasone after Asphyxia Increases Neural Injury in Preterm Fetal Sheep**..185
Miriam E. Koome, Joanne O. Davidson, Paul P. Drury, Sam Mathai, Lindsea C. Booth, Alistair Jan Gunn, Laura Bennet

Chapter 20   **Selenium Supplementation Restores Innate and Humoral Immune Responses in Footrot-Affected Sheep**...193
Jean A. Hall, William R. Vorachek, Whitney C. Stewart, M. Elena Gorman, Wayne D. Mosher, Gene J. Pirelli, Gerd Bobe

Chapter 21   **Fine Epitope Mapping of the Central Immunodominant Region of Nucleoprotein from Crimean-Congo Hemorrhagic Fever Virus (CCHFV)**...................................212
Dongliang Liu, Yang Li, Jing Zhao, Fei Deng, Xiaomei Duan, Chun Kou, Ting Wu, Yijie Li, Yongxing Wang, Ji Ma, Jianhua Yang, Zhihong Hu, Fuchun Zhang, Yujiang Zhang, Surong Sun

Permissions

List of Contributors

Index

Preface

Livestock Management is understood as rearing and raising of domesticated animals in an agricultural setting so that the domesticated animals could be used for production of commodities like food, fiber and labor. This book outlines the processes and applications of livestock management in detail. It unfolds the innovative aspects of animal rearing and farming practices which are closely related to livestock management in an interesting manner. Different approaches, evaluations, methodologies and advanced studies on Livestock Management have been included in this elaborate book. It is an important and invaluable source of information for professionals and researchers in the field.

This book has been a concerted effort by a group of academicians, researchers and scientists, who have contributed their research works for the realization of the book. This book has materialized in the wake of emerging advancements and innovations in this field. Therefore, the need of the hour was to compile all the required researches and disseminate the knowledge to a broad spectrum of people comprising of students, researchers and specialists of the field.

At the end of the preface, I would like to thank the authors for their brilliant chapters and the publisher for guiding us all-through the making of the book till its final stage. Also, I would like to thank my family for providing the support and encouragement throughout my academic career and research projects.

Editor

Dual Origins of Dairy Cattle Farming – Evidence from a Comprehensive Survey of European Y-Chromosomal Variation

Ceiridwen J. Edwards[1,2], Catarina Ginja[3,4], Juha Kantanen[5], Lucía Pérez-Pardal[6], Anne Tresset[7], Frauke Stock[1], European Cattle Genetic Diversity Consortium[¶], Luis T. Gama[4], M. Cecilia T. Penedo[3], Daniel G. Bradley[1], Johannes A. Lenstra[8*], Isaäc J. Nijman[8¤]

1 Smurfit Institute of Genetics, Trinity College Dublin, Dublin, Ireland, 2 Research Laboratory for Archaeology, University of Oxford, Oxford, United Kingdom, 3 Veterinary Genetics Laboratory, University of California Davis, Davis, California, United States of America, 4 Departamento de Genética, Melhoramento Animal e Reprodução, Instituto Nacional dos Recursos Biológicos, Fonte Boa, Vale de Santarém, Portugal, 5 Biotechnology and Food Research, MTT Agrifood Research Finland, Jokioinen, Finland, 6 Area de Genética y Reproducción Animal, SERIDA, Gijón, Spain, 7 Archéozoologie, Archéobotanique, Sociétés, Pratiques et Environnements, CNRS Muséum National d'Histoire Naturelle, Paris, France, 8 Faculty of Veterinary Medicine, Utrecht University, Utrecht, The Netherlands

Abstract

Background: Diversity patterns of livestock species are informative to the history of agriculture and indicate uniqueness of breeds as relevant for conservation. So far, most studies on cattle have focused on mitochondrial and autosomal DNA variation. Previous studies of Y-chromosomal variation, with limited breed panels, identified two *Bos taurus* (taurine) haplogroups (Y1 and Y2; both composed of several haplotypes) and one *Bos indicus* (indicine/zebu) haplogroup (Y3), as well as a strong phylogeographic structuring of paternal lineages.

Methodology and Principal Findings: Haplogroup data were collected for 2087 animals from 138 breeds. For 111 breeds, these were resolved further by genotyping microsatellites *INRA189* (10 alleles) and *BM861* (2 alleles). European cattle carry exclusively taurine haplotypes, with the zebu Y-chromosomes having appreciable frequencies in Southwest Asian populations. Y1 is predominant in northern and north-western Europe, but is also observed in several Iberian breeds, as well as in Southwest Asia. A single Y1 haplotype is predominant in north-central Europe and a single Y2 haplotype in central Europe. In contrast, we found both Y1 and Y2 haplotypes in Britain, the Nordic region and Russia, with the highest Y-chromosomal diversity seen in the Iberian Peninsula.

Conclusions: We propose that the homogeneous Y1 and Y2 regions reflect founder effects associated with the development and expansion of two groups of dairy cattle, the pied or red breeds from the North Sea and Baltic coasts and the spotted, yellow or brown breeds from Switzerland, respectively. The present Y1-Y2 contrast in central Europe coincides with historic, linguistic, religious and cultural boundaries.

Editor: Toomas Kivisild, University of Cambridge, United Kingdom

Funding: CJE was supported by the Enterprise Ireland Basic Research Grants Programme (project numbers SC/1999/409 and SC/2002/510). CG was supported by a grant from Fundação Portuguesa para a Ciência e a Tecnologia (Ref. SFRH/BD/13502/2003). JK was supported by the Academy of Finland and the Finnish Ministry of Agriculture and Forestry (SUNARE-program and 'Russia in flux' program). LP-P was supported by grant BES-2006-13545. Collection of several samples was supported by the Resgen project 98–118 (1999–2002) funded by the European Union. This work was also part-funded by The Wellcome Trust (grant no. 047485/Z/96/Z), and a Eurocores OMLL Programme Grant via CNRS, France. The funders had no role in study design, data collection and analysis, decision to publish, or preparation of the manuscript.

Competing Interests: The authors have declared that no competing interests exist.

* E-mail: J.A.Lenstra@uu.nl

¤ Current address: Genome Biology and Bioinformatics Group, Hubrecht Institute, Utrecht, The Netherlands

¶ For the full list of authors of this consortium, please see the Acknowledgments section.

Introduction

The history of human civilisations has left its footprint in the patterns of genetic variation of livestock species across and within continents [1,2,3,4]. Molecular markers, such as mitochondrial DNA (mtDNA) and autosomal polymorphisms, have been particularly useful in investigating the wild species origin of cattle and the subsequent genetic events that shaped the present pattern of genetic diversity. With regards to cattle in Europe, evidence indicates that: (*i*)

Balkan cattle act as reservoir of high genetic diversity [5]; (*ii*) there is a marked contrast between north and south Europe [6,7,8]; and (*iii*) central European breeds occupy a separate position relative to Mediterranean and northern European cattle [9].

Although the Fertile Crescent is considered the primary centre of taurine cattle domestication, evidence for independent domestication events in other locals is currently debated [10,11,12, 13,14,15]. Ancestral taurine mtDNA lineages have been identified, and confirmed differences seen between northern and southern

European populations of wild cattle (aurochs): the *B. primigenius* haplogroup P was frequent in northern and central Europe [16], while distinct putative auroch matrilines (haplogroups Q and R) were found in modern southern European populations and it appears that these were sporadically introduced into domestic breeds [11,12]. MtDNA sequences also reveal that indicine cattle originated from a different wild aurochs population, *Bos primigenius namadicus*, in the Indus Valley approximately 8,000 years before present [17,18].

Differentiation of paternal lineages via analysis of Y-chromosomal variation adds significantly to what can be inferred from mtDNA and autosomal variation [19]. The absence of interchromosomal recombination outside the pseudoautosomal region (PAR) preserves original arrangements of mutational events, and thus male lineages can be traced both within and among populations. Genetic drift is relatively strong due to the effective population sizes of Y-chromosomes being, at most, 25% of the autosomal effective population size [20]. Effective population size is often reduced further by the relatively high variability of male reproductive success. As a result, the Y-chromosome is a sensitive indicator of recent demographic events, such as population bottlenecks, founder effects and population expansions.

In several herd species, males are more mobile than females and compete for reproduction or, in the case of livestock, are selected on the basis of breeding objectives. So while mtDNA variants stay mostly within the herd, Y-chromosomal variants may reflect the origin of sires as influenced by introgression and upgrading. Indeed, in domestic cattle, a marked difference between the distributions of mitochondrial and Y-chromosomal components has been observed [16,21,22,23,24]. In an initial survey of European breeds, two haplogroups, Y1 and Y2, were found to be dominant in northern and southern Europe respectively [25]. Comparison of cattle Y-chromosome variation over time suggests that the frequency and distribution of these patrilines varied, which could be related with distinct breeding strategies [26]. In European aurochs, Y2 appears to be predominant [27], but so far it is unclear if there was significant introgression from wild bulls into domestic populations.

In the study of human male lineages, the use of Y-specific microsatellites has allowed for refined analyses of the genetic diversity of paternal lineages that can be found within major haplogroups [28,29]. Similarly, in cattle, microsatellite analysis has identified several Y-haplotypes in Portuguese [30], northern and eastern European [31], western-continental, British and Sub-Saharan Africa [14] breeds, as well as in American Creole [22] breeds. Even though different sets of markers were used in these studies, and each only partially covered the diversity pattern of the paternal lineages, they have confirmed that Y-markers exhibit a strong phylogeographic structure in cattle.

Here we report Y-haplotype data on 128 European, one African and nine Asian breeds, incorporating also data from the previous reports. This comprehensive study confirms a clear north-south contrast that is accentuated by a homogeneity of Y1 haplotypes in north-central Europe and of Y2 haplotypes in and around the Alpine region, both regions that are thought to be the origin of highly productive cattle breeds. This genetic boundary correlates with historic differences between northern and southern European cultures and we consider that this may have implications for the investigation of dairy QTL variation.

Results

Cattle haplogroups defined by Y-SNPs

For 238 males from 30 breeds, Y-chromosomal fragments comprising the *ZFY* (1,219 bp and 1,003 bp), *SRY* (2,644 bp) and

DBY (also known as *DDX3Y*; 406 bp) genes were sequenced [32]. These sequences contained five mutational differences when compared to the zebu Y3 sequences [30], and all European animals carried either the Y1 or Y2 taurine haplogroups. Interestingly, comparison of the Y3 sequence with an *SRY* sequence from an Indian Sahiwal zebu (GenBank accession number AY079145) [33] revealed three additional differences downstream of the open-reading frame, indicating zebu-specific Y-chromosomal SNPs. In combination with a SNP in *UTY19* [14,22,25,31], three cosegregating mutations differentiate the taurine Y1 and Y2 haplogroups (**Table S1**). A composite microsatellite in *DBY* [25], with one major allele in both Y1 and Y2 and only present in the Italian Maremmana, was not used for differentiation of haplotypes.

Genotypes of individual SNPs in other animals were combined with Y1-Y2 SNP data and resulted in Y1 or Y2 assignments of 2087 animals from 138 breeds (**Table S1**). The resulting dataset included previously published genotypic information for 1099 individuals from 78 breeds [14,22,31]. The Y3 haplogroup was identified on the basis of microsatellite information as described in the next section. The map of **Figure 1** shows the geographical distribution of Y-haplogroups. The three haplogroups described in cattle (Y1, Y2 and Y3) were detected in Southwest Asia, but only Y3 was present in the two Indian breeds analysed, which is in agreement with their zebu morphology. Y1 was predominant in northern Europe and in a number of Iberian breeds. In contrast, Y2 was dominant in most central, Mediterranean and Iberian breeds, but was also found in several British and Nordic breeds. Although only a single African breed was included in this study, both Y1 and Y2 haplogroups were present.

Several of the breeds that do not confirm this northern Y1 – southern Y2 distribution pattern appear to have been subject to recent introgression from breeds with similar coat colour [34], but carrying the other Y-chromosomal haplogroup. Thus, Y2 was introduced in Dutch Belted (DUB) by crossbreeding with belted Galloway bulls. Danish Red bulls probably introduced Y1 into the Sicilian Modicana (MOD). The presence of Y1 in central dairy breeds, such as Simmental (SIM), Pezzata Rossa Italiana (PRI) and Hinderwald (HIW), is probably explained by crossbreeding with Red Holstein sires. Likewise, crossbreeding with Lowland Pied or English cattle probably accounts for the predominance of Y1 in Russian dairy cattle [35]. Y-chromosomes act as a single haplogroup and are, in general, homogeneous at the Y-chromosome variation level. In most cases, the presence of Y1 and Y2 haplotypes in a given breed can be explained by its recent history [14].

Cattle haplotypes detected through SNPs and microsatellites

The diversity within each haplogroup was further assessed by genotyping *INRA189* [36] and *BM861* [37] Y-specific microsatellite markers. The combination of these data with previously published Y-chromosomal haplotypes [14,22,31] yielded haplotypes for a total of 1472 animals from 111 breeds, a subset of the individuals for which SNP information was available. Haplotype composition and absolute frequencies, as well as unbiased estimates of haplotype diversities with the associated standard deviations (SD) are shown for each breed and geographic group in **Table 1**. We found a total of 19 composite Y-haplotypes. The relationship between current haplotype nomenclatures is summarised in **Table S2**. Locus *INRA189* is the most informative marker with 10 alleles, differentiated among five Y1 and nine Y2 haplotypes. For 21 bulls belonging to the Indian and Southwest Asian breed groups, haplogroup Y3 was identified via the *INRA189*-88 bp allele. Microsatellite marker *BM861* defines one

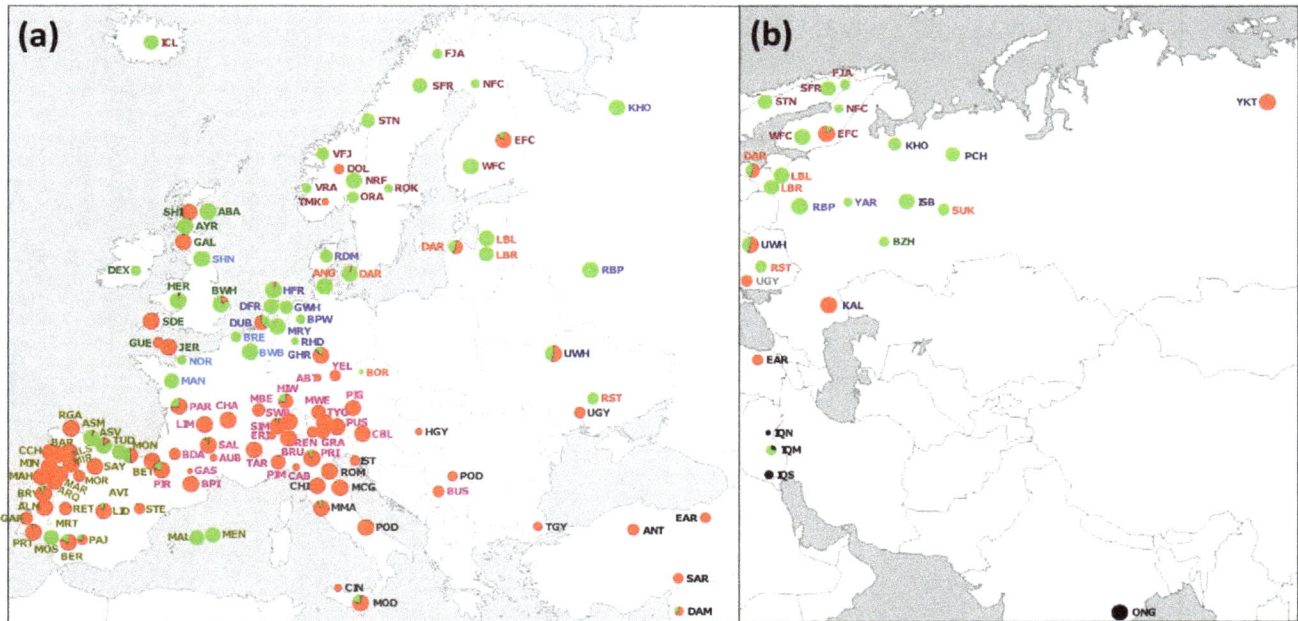

Figure 1. Geographical distribution of Y-haplogroups. (a) Europe, and (b) Eurasia. Green = Y1; red = Y2; black = Y3. Abbreviations of breed names are given in Table 1.

additional Y1 and three Y2 additional haplotypes. Haplotype Y1-98-158 is the most frequent within the Y1 haplogroup and is detected in 82% of the animals from this haplogroup across all geographic breed groups with the exception of the Indian and Podolian. Within the Y2 haplogroup, Y2-102-158 and Y2-104-158 haplotypes account for 62% and 29% of the animals respectively. A map showing the distribution of Y-haplotypes is included in **Figure S1**.

Y-chromosomal diversity within breeds is low (mean of 0.42 ± 0.3) with fixed haplotypes in 65 out of 111 breeds (approximately 59%). Interestingly, the Southwest Asian breed group is the most genetically diverse, with a total of seven haplotypes detected in a limited sample of 24 animals, and an unbiased expected haplotype diversity of 0.57 ± 0.4 (**Table 2**). The Iberian and British groups have intermediate variability, with 11 and five haplotypes observed and diversities of 0.34 ± 0.2 and 0.41 ± 0.3 respectively.

Phylogeography of Y-haplotypes

The phylogenetic relationship among Y-haplotypes was investigated after grouping the breeds into 12 regions on the basis of geography and/or phenotype [34,38]. These breed groups are in agreement with major clusters as defined by autosomal microsatellite (unpublished results; see also **Figure S2**) and SNP data [7]. Haplotype relationships are depicted in **Figure 2** by a median-joining (MJ) network obtained for the complete dataset, as well as by regional MJ networks defined for each of the 12 geographic breed groups. The MJ networks clearly differentiate the indicine Y3 patriline from the taurine Y1 and Y2 haplogroups. The regional networks depict the Y-chromosome diversity found in each region and the relationship among the observed haplotypes defined by the two microsatellites.

AMOVA results are presented in **Table S3** and show a significant ($P < 0.0001$) effect of geographical breed grouping, which accounted for 31% of the total variability. Approximately

51% ($P < 0.0001$) of the Y-chromosome genetic variation was found among breeds within groups and 18% ($P < 0.0001$) within breeds. This was confirmed by the generally high Y-chromosomal F_{ST} values (**Table S4**) of all pairwise breed clusters, which were low mainly for the comparison of the dairy north-western European breed clusters.

In **Figure S2**, Y-chromosomal haplotypes have been indicated in a NeighborNet phylogenetic network of Reynolds' distances based on 30 autosomal microsatellite data. This shows a clear consistency between Y-chromosomal variation and the breed relationships in the autosomal-based phylogeny, with only the British Y2 and Iberian Y1 cattle as major exceptions [as in 14,15].

Discussion

Differentiation of paternal lineages in European cattle

We analysed Y-chromosomal haplotypes in a comprehensive sampling of European cattle. This allowed a differentiation of two SNP-based taurine haplogroups, which were resolved into 18 haplotypes using genotypes from two microsatellite markers. Locus *INRA189* appeared to be, by far, the most informative and, in combination with the other marker and SNPs, showed that paternal lineages strongly depend on geographic origin. A further differentiation was demonstrated by using other microsatellites [14,22,31]. For instance, *BYM1* alleles defined a Y1-98-158 variant in Nordic cattle and a Y2-102-158 variant in Spain [14], while *DYZ1* detected a different Y2-102-158 variant in Southwest Asian breeds [31].

Distribution of zebu and taurine haplotypes

Previous genetic studies have demonstrated a separation of the mitochondrial [39,40,41], Y-chromosomal [25,42,43] and nuclear [21] DNA from taurine and zebu cattle, which supports independent domestications in the Fertile Crescent and Indus valley respectively [44]. In a diverse panel of 2009 European cattle, we exclusively found taurine Y-chromosomal haplotypes.

Table 1. Breed information, including geographical grouping, for the 111 cattle breeds sampled as part of this study, and associated haplotypic data (defined as SNP-*INRA189-BM861*) and diversity values.

Geographic grouping	Breed	Code	Country of origin	Number of Y-chromosomes	Y1-94-158	Y1-96-158	Y1-98-158	Y1-98-160	Y1-100-158	Y1-102-158	Y2-80-158	Y2-90-158	Y2-94-158	Y2-96-158	Y2-98-158	Y2-98-160	Y2-100-158	Y2-102-158	Y2-102-160	Y2-104-158	Y2-104-160	Y2-106-158	Y3-88-156	haplotype diversity (± SD)	total number of haplotypes
India (IND)	Nellore	NEL	Brazil	12																			12	0.000±0.000	1
	Ongole	ONG	India	4																			4	0.000±0.000	1
	Total			**16**																			**16**	**0.000±0.000**	**1**
Africa (AFR)	N'Dama	NDM	Guinea	12			2					9					1							0.247±0.184	3
	Total			**12**			2					9					1							**0.247±0.184**	**3**
Southwest Asia (SWA)	Anatolian Black	ANT	Turkey	5											1		1			2		1		0.300±0.424	4
	Damascus	DAM	Syria	3			1											1				1		0.556±0.416	3
	East Anatolian Red	EAR	Turkey	4														1		2		1		0.278±0.393	3
	Middle Iraqi	IQM	Iraq	4			3															1		0.500±0.000	2
	North Iraqi	IQN	Iraq	1													1							n/a	1
	South Iraqi	IQS	Iraq	3																			3	0.000±0.000	1
	South Anatolian Red	SAR	Turkey	4																2		2		0.222±0.314	2
	Total			**24**			4								1		2	2		6		6	3	**0.574±0.192**	**7**
Podolian (POD)	Chianina	CHI	Italy	20																20				0.000±0.000	1
	Istrian	IST	Croatia	4																4				0.000±0.000	1
	Marchigiana	MCG	Italy	11																11				0.000±0.000	1
	Maremmana	MMA	Italy	19														19						0.000±0.000	1
	Podolica	PODi	Italy	13														9		4				0.154±0.218	2
	Serbian Podolica	PODs	Serbia	4																4				0.000±0.000	1
	Turkish Grey	TGY	Turkey	3																	3			0.000±0.000	1
	Ukrainian Grey	UGY	Ukraine	5								5												0.000±0.000	1

Table 1. Cont.

Geographic grouping	Breed	Code	Country of origin	Number of Y-chromo-somes	Y1-94-158	Y1-96-158	Y1-98-158	Y1-98-160	Y1-100-158	Y1-102-158	Y2-80-158	Y2-90-158	Y2-94-158	Y2-96-158	Y2-98-158	Y2-98-160	Y2-100-158	Y2-102-158	Y2-102-160	Y2-104-158	Y2-104-160	Y2-106-158	Y3-88-156	haplotype diversity (± SD)	total number of haplotypes
			Total	79	5							5						28		43		3		**0.193±0.273**	**4**
Iberian (IBE)	Alentejana	ALN	Portugal	34																34				0.000±0.000	1
	Alistana-Sanabresa	ALS	Spain	12														12						0.000±0.000	1
	Arouquesa	ARQ	Portugal	33														25		8				0.126±0.179	2
	Asturiana de los Valles	ASV	Spain	38	26		6											6						0.256±0.202	3
	Asturiana de Montana	ASM	Spain	19	18													1						0.070±0.050	2
	Avilena Negro Iberica	AVI	Spain	7			1											1				5		0.270±0.214	3
	Barrosã	BAR	Portugal	33														4		29				0.073±0.104	2
	Berrenda	BER	Spain	5														3		2				0.200±0.283	2
	Betizu	BEB	Spain	17															17					0.000±0.000	1
	Brava de Lide	BRV	Portugal	26			2											23		1				0.122±0.091	3
	Cachena	CCH	Portugal	25														1		24				0.027±0.038	2
	Garvonesa	GAR	Portugal	6														6						0.000±0.000	1
	Lidia	LID	Spain	66									1				1	64						0.020±0.028	3
	Mallorquina	MAL	Spain	8			8																	0.000±0.000	1
	Marinhoa	MAH	Portugal	17														17						0.000±0.000	1
	Maronesa	MAR	Portugal	23																23				0.000±0.000	1
	Mertolenga	MRT	Portugal	21			7			1					1			2		10				0.378±0.273	5
	Minhota	MIN	Portugal	28														28						0.000±0.000	1
	Mirandesa	MIR	Portugal	23														23						0.000±0.000	1
	Morucha	MOR	Spain	5														5						0.000±0.000	1
	Mostrenca	MOS	Spain	21			21																	0.000±0.000	1
	Pajuna	PAJ	Spain	4			1											2				1		0.444±0.342	3
	Preta	PRT	Portugal	29			1											1		5		22		0.158±0.178	4
	Retinta	RET	Spain	6														4		2				0.178±0.251	2
	Rubia Gallega	RGA	Spain	44														44						0.000±0.000	1
	Sayaguesa	SAY	Spain	8										1				5		2				0.202±0.286	3

Table 1. Cont.

Geographic grouping	Breed	Code	Country of origin	Number of Y-chromosomes	Y1-94-158	Y1-96-158	Y1-98-158	Y1-98-160	Y1-100-158	Y1-102-158	Y2-80-158	Y2-90-158	Y2-94-158	Y2-96-158	Y2-98-158	Y2-98-160	Y2-100-158	Y2-102-158	Y2-102-160	Y2-104-158	Y2-104-160	Y2-106-158	Y3-88-156	haplotype diversity (± SD)	total number of haplotypes
	Tudanca	TUD	Spain	10		10																		0.000±0.000	1
	Total			**568**	44	57				1			1	1	1		1	277	17	140		28		**0.335±0.244**	**11**
Central (CEN)	Blonde d'Aquitaine	BDA	France	5														3		2				0.200±0.283	2
	Bruna de los Pirineds	BPI	Spain	11														11						0.000±0.000	1
	Busha	BUS	Serbia	5														5						0.000±0.000	1
	Cabannina	CAB	Italy	2														1		1				0.333±0.471	2
	Charolais	CHA	France	31													1	30						0.022±0.030	2
	Limousin	LIM	France	24														24						0.000±0.000	1
	Montbeliard	MBE	France	6														6						0.000±0.000	1
	Parthenaise	PAR	France	15		4												11						0.279±0.198	2
	Piemontese	PIM	Italy	13													2	11						0.094±0.133	2
	Pinzgaur	PGZ	Austria	9														9						0.000±0.000	1
	Pirenaica	PIR	Spain	10														8		2				0.119±0.168	2
	Pustertaler	PUS	Italy	13														13						0.000±0.000	1
	Salers	SAL	France	20														19	1					0.067±0.047	2
	Simmental	SIM	Switzerland	15														15						0.000±0.000	1
	Swiss Brown	SWB	Switzerland	14														14						0.000±0.000	1
	Tarentaise	TAR	France	18														18						0.000±0.000	1
	Tyrolean Grey	TYG	Italy	19													2	17						0.066±0.094	2
	Total			**230**		4											5	215	1	5				**0.056±0.045**	**5**
British (BRT)	Aberdeen Angus	ABA	Scotland	27	25	2																		0.133±0.188	2
	Ayrshire	AYR	Scotland	16	16																			0.000±0.000	1
	British White	BWH	England	21	11	6												1		3				0.252±0.111	4
	Dexter	DEX	Ireland	4	4																			0.000±0.000	1
	Galloway	GAL	Scotland	9														9						0.000±0.000	1
	Hereford	HER	England	21					20									1						0.063±0.045	2
	Highland	HIG	Scotland	10														10						0.000±0.000	1
	Jersey	JER	Jersey	19																19				0.000±0.000	1
	Total			**127**	56	8			20									21		22				**0.413±0.226**	**5**

Table 1. Cont.

Geographic grouping	Breed	Code	Country of origin	Number of Y-chromo-somes	Y1-94-158	Y1-96-158	Y1-98-158	Y1-98-160	Y1-100-158	Y1-102-158	Y2-80-158	Y2-90-158	Y2-94-158	Y2-96-158	Y2-98-158	Y2-98-160	Y2-100-158	Y2-102-158	Y2-102-160	Y2-104-158	Y2-104-160	Y2-106-158	Y3-88-156	haplotype diversity (± SD)	total number of haplotypes
Nordic (NOR)	Blacksided Troender	STN	Norway	7			7																	0.000±0.000	1
	Doela	DOL	Norway	4							4													0.000±0.000	1
	Eastern Finncattle	EFC	Finland	9			1													8				0.148±0.105	2
	Eastern Red Polled	ORA	Norway	5			5																	0.000±0.000	1
	Fjallnara	FNR	Sweden	3			3																	0.000±0.000	1
	Icelandic	ICL	Iceland	8			8																	0.000±0.000	1
	Northern Finncattle	NFC	Finland	3			3																	0.000±0.000	1
	Norwegian (commercial, hybrid)	NRF	Norway	12			12																	0.000±0.000	1
	Swedish Mountain	SFR	Sweden	8			8																	0.000±0.000	1
	Swedish Red Polled	ROK	Sweden	3			3																	0.000±0.000	1
	Telemark	TEL	Norway	2							2													0.000±0.000	1
	Western Finncattle	WFC	Finland	9			9																	0.000±0.000	1
	Western Fjord	VFJ	Norway	6			3		3															0.200±0.283	2
	Western Red Polled	VRA	Norway	3			2		1															0.222±0.314	2
	Total			**82**			**64**		**4**		**6**									**8**				**0.222±0.161**	**4**
Baltic Red (BHR)	Angler	ANG	Germany	10			10																	0.000±0.000	1
	Danish Red	RDM	Denmark	19			18											1						0.070±0.050	2
	Latvian Blue (native)	LBL	Latvia	9			9																	0.000±0.000	1
	Latvian Brown (commercial)	LBR	Latvia	8			8																	0.000±0.000	1
	Latvian Danish Red	DAR	Latvia	7			3											4						0.381±0.269	2
	Suksunskaya	SUK	Russia	5			4											1						0.267±0.189	2
	Ukrainian Red Steppe	RST	Ukraine	5		1	4																	0.133±0.189	2
	Total			**63**		**1**	**56**											**6**						**0.126±0.090**	**3**

Table 1. Cont.

Geographic grouping	Breed	Code	Country of origin	Number of Y-chromosomes	Y1-94-158	Y1-96-158	Y1-86-158	Y1-98-160	Y1-100-158	Y1-102-158	Y2-80-158	Y2-90-158	Y2-94-158	Y2-96-158	Y2-98-158	Y2-98-160	Y2-100-158	Y2-102-158	Y2-102-160	Y2-104-158	Y2-104-160	Y2-106-158	Y3-88-156	haplotype diversity (± SD)	total number of haplotypes
North-West (NWE)	Belgian Blue	BWB	France	21		19	2																	0.060±0.085	2
	Belgian Red	BRE	Belgium	4			4																	0.000±0.000	1
	Normand	NOR	France	46			46																	0.000±0.000	1
	Shorthorn	SHN	Belgium	19			19																	0.000±0.000	1
	Total			**90**		**19**	**71**																	**0.112±0.159**	**2**
Lowland Pied (LLP)	Dutch Belted	DUB	Netherlands	8			3											5						0.357±0.253	2
	Friesian-Dutch	FRH	Netherlands	8			8																	0.000±0.000	1
	German Original Black Pied-West	BPW	Germany	3			3																	0.000±0.000	1
	Groningen Whitehead	GWH	Netherlands	6			6																	0.000±0.000	1
	Holstein Friesian	HFR	Netherlands	65			64									1								0.021±0.015	2
	Jutland (old native)	SJM	Denmark	6			6																	0.000±0.000	1
	Meuse-Rhine-Yssel	MRY	Netherlands	9			9																	0.000±0.000	1
	Red Holstein dual type	RH2	Netherlands	1			1																	n/a	1
	Total			**106**			**100**									**1**		**5**						**0.072±0.039**	**3**
Eastern (EAS)	Bestuzhev	BZH	Russia	4			4																	0.000±0.000	1
	Istobenskaya	ISB	Russia	9			9																	0.000±0.000	1
	Kalmyk	KAL	Russia	12																12				0.000±0.000	1
	Kholomogorskaya	KHO	Russia	6			6																	0.000±0.000	1
	Pechorskaya	PCH	Russia	7			7																	0.000±0.000	1
	Ukrainian Whiteheaded	UWH	Ukraine	11			4		1									6						0.388±0.276	3
	Yakutian cattle	YKT	Siberia	23																22	1			0.029±0.041	2
	Yaroslavskaya	YAR	Russia	3			3																	0.000±0.000	1
	Total			**75**			**33**		**1**									**6**		**34**	**1**			**0.373±0.247**	**5**
Overall	**111 breeds**			**1472**	44	20	448	8	25	1	6	14	1	1	2	1	7	559	18	259	1	36	21	**0.422±0.269**	

Table 2. Haplotypic data for the breeds typed for the Y1-Y2-Y3 SNPs and the two microsatellite loci, *INRA189* and *BM861*.

Geographic grouping	SNPs		Mean number of samples / breed	Microsatellites		Mean number of samples / breed	Total number of haplotypes	Haplotype diversity (± SD)
	Breeds	Samples		Breeds	Samples			
India (IND)	2	16	8.0	2	16	8.0	1	0.000±0.0
Africa (AFR)	1	14	n/a	1	12	n/a	3	0.247±0.2
Southwest Asia (SWA)	7	25	3.6	7	24	3.4	7	0.574±0.4
Podolian (POD)	12	184	15.3	8	79	9.9	4	0.193±0.2
Iberian (IBE)	31	651	21.0	27	568	21.0	11	0.335±0.2
Central (CEN)	30	453	15.1	17	230	13.5	5	0.056±0.1
British (BRT)	10	164	16.4	8	127	15.9	5	0.413±0.3
Nordic (NOR)	14	95	6.8	14	82	5.9	4	0.222±0.2
Baltic Red (BHR)	9	77	8.6	7	63	9.0	3	0.126±0.1
North-West (NWE)	5	126	25.2	4	90	22.5	2	0.112±0.1
Lowland Pied (LLP)	9	208	23.1	8	106	13.3	3	0.072±0.1
Eastern (EAS)	8	74	9.3	8	75	9.4	5	0.373±0.3
Overall	**138**	**2087**	**15.1**	**111**	**1472**	**13.3**	**19**	**0.422±0.3**

This implies that the zebu alleles found for autosomal markers in the Podolian [6,45,46,47], Iberian [6,48,49] and Ukrainian Whitehead [31] breeds came to Europe indirectly, presumably via Anatolia, North Africa or north of the Black Sea. In addition, the presence of zebu Y-chromosomes in Southwest Asia indicate that zebu introgression only took place after the expansion of domestic taurine cattle from Southwest Asia towards Europe [16].

Taurine haplogroups

The divergence of the two haplogroups Y1 and the Y2 without intermediate haplotypes suggests that domestication combined paternal lineages originate from two diverged populations. Remarkably, both haplogroups were found in Southwest Asia. In Europe, the Y1 and Y2 haplogroups found in extant cattle exhibit a clear geographic structure, with Y1 restricted to northern European and Iberian breeds. Initially, this was explained by local aurochs introgression, supposed to carry the Y1 lineage [25]; however, this does not account for the presence of Y1 in Southwest Asia. In addition, a subsequent study [27] exclusively found Y2 haplogroups in European aurochs. We note that this does not exclude that the Y2 lineage was partially contributed by local aurochs, nor that conditions of Neolithic farming may have created opportunities for male introgression from wild animals. However, several analyses of mtDNA have clearly demonstrated that domestic maternal lineages originate from Southwest Asia, with only sporadic female aurochs introgression [11,12,16,50,51].

An obvious possibility is that the current Y-chromosomal haplogroup distribution reflects Neolithic immigration routes. According to archaeological evidence, the dispersal of agriculture in Europe started in Greece around 7,000 BC, moved to southern Italy *circa.* 6,000 BC, and then along a southern route into the western Mediterranean between 5,600 and 5,400 BC, reaching Portugal around 5,300 BC. Migration along the continental route into Poland and Germany occurred between 5,500 and 5,300 BC, reaching north-western France around 5,000 BC; southern Scandinavia, the British Isles and Ireland were reached *circa.* or after 4,000 BC [52,53,54]. It is generally accepted that agriculture spread via these two routes: the *Mediterranean* route and the

Danubian (or *continental*) route. Although Y1 has a clear presence in Iberian cattle, Y2 paternal lineages dominate in the present Mediterranean area. A founder effect in Danubian immigrants could have caused the dominance of Y1 in northern Europe. In this scenario, the presence of Y1 in Iberia would have resulted from movements along the Atlantic seaboard, as documented by the Neolithic archaeological record [55], and the presence of two haplogroups in Britain may indicate a convergence of immigrants of both routes.

Alternatively, colonisation of Britain may have predated the expansion of Y1, which could have arrived in Britain later via the documented import of Dutch sires in the 18th century. This would be in line with analysis of skeletal remains excavated in Sweden, showing that Y1 bulls replaced Y2 bulls during or after the late Middle Ages [26], although Y1 cattle were taken to Iceland by the Vikings *c.* 1,000 AD. However, a Y1 founder effect, long after the introduction of domestic cattle in northern Europe, is the most consistent with the haplotype diversity pattern.

Y1 samples identified in Africa are more likely to be the result of a recent introgression of European cattle rather than the expansion of a genetically heterogeneous sire population domesticated in the Fertile Crescent. Y-specific microsatellite data confirm the existence of a Y2 haplotypic subfamily in African cattle restricted to the African continent [14,15].

Distribution of haplotype diversity

The comprehensive coverage of our study permits a comparison of the diversity of paternal lineages in different regions. Even in a limited sampling, Southwest Asian cattle contain seven out of the 19 taurine haplotypes identified in this study. This supports the theory that the Fertile Crescent was a major centre of cattle domestication and that European cattle are a subset of an initially diverse Southwest Asian domestic population. The high diversity in Spain and Portugal is probably explained by the isolated position of the Iberian Peninsula, by which much of the original diversity has been conserved. However, the increased diversity may also have been influenced by African introgression, which is consistent with mitochondrial and autosomal information

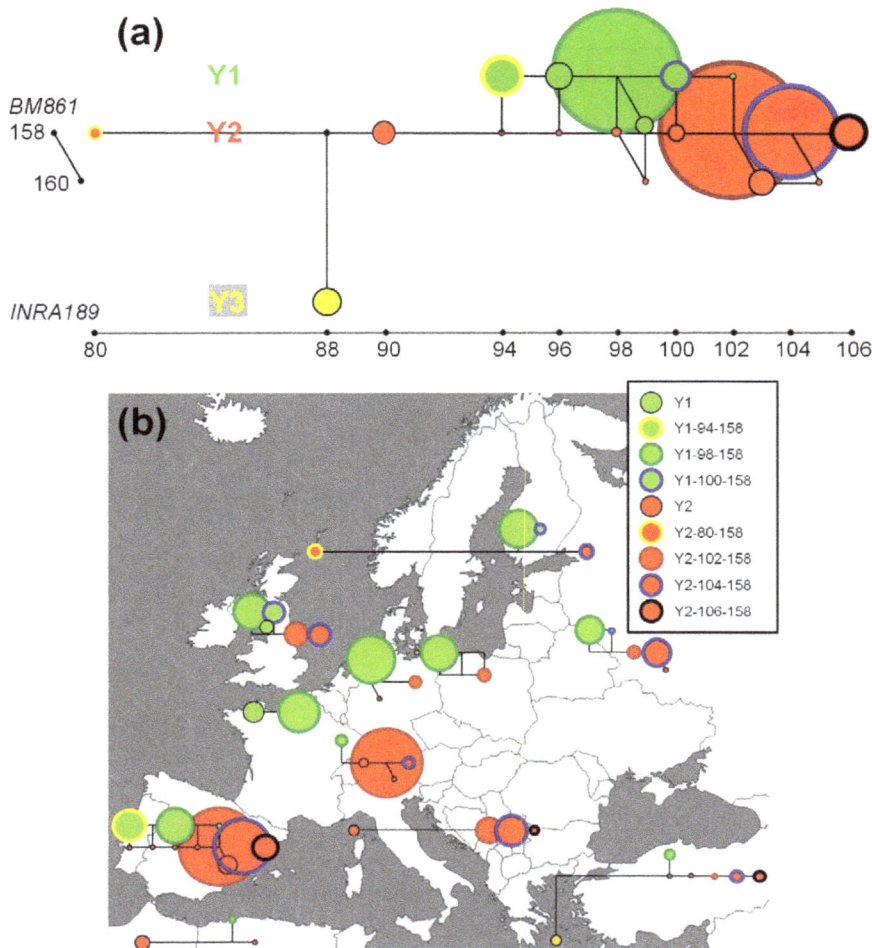

Figure 2. Median-joining networks of Y-chromosome haplotypes using SNPs and two microsatellite markers. The size of each circle represents the number of chromosomes present in each haplotype. (**a**) Haplotype relationships for the complete dataset. (**b**) Regional networks defined for 11 geographic breed groups (India excluded). Haplotypes are indicated with the colouring scheme shown in (**b**).

[6,10,22,56,57]. The diversity in Italian cattle, including the Tuscan breeds with their unusual mtDNA diversity [58], is less than that exhibited in Iberian breeds, with two major Y2 haplotypes. As noted above, the presence of both Y1 and Y2 haplotypes in Britain and the Nordic region may reflect multiple arrivals of cattle, which probably also explains the diversity in Russian cattle.

The diversity in the Mediterranean countries, and in the northern and eastern periphery of Europe, is in striking contrast, with near homogeneity of paternal lineages in north-central and central Europe. The black-pied, red-pied and solid red dairy cattle originating from the Dutch, German and Danish lowlands almost exclusively carry the Y1-98-158 haplotype. We propose that this reflects a founder effect associated with the prehistoric development of dairy farming [59]. Dairy farming by Germanic people in the Roman period has been well documented [Strabo AD23, as translated in 60], while, in the Mediterranean area, cattle were mainly used for draught power and meat consumption [61]. This is consistent with the origin of human lactase tolerance in central Europe [61]. In the 18th century, Rinderpest epidemics led to a replacement of Dutch cattle, with their various coat colours, by black-pied cattle from the Holstein region [34,62]. These became

the ancestors of the highly productive Dutch black-pied cattle, which were exported to several countries. Similarly, the related dairy Red Angler and Danish Red spread to the Baltic countries, Russia and Germany [34,63].

A different haplotype, Y2-102-158 is predominant in southern France, southern Germany and the Alpine region (**Figure 2**). Many of the present Alpine short-headed cattle are derived from, or related to, the mixed-purpose Simmental and Swiss Brown cattle. It is not known when these types of cattle were developed, although archaeological findings indicate an ancient origin of milk consumption in Switzerland [64] and dairy farming in the Roman period has been documented [Pliny the Elder AD77, as translated in 65]. An increase in size of Swiss cattle during the Roman period further suggests importation of Italian cattle [64]. The predominance of a single Y2 haplotype may indicate again a founder effect. Just as the northern-central European dairy breeds, the Alpine breeds also spread to the surrounding regions. Simmentals were exported to several countries and also used for upgrading of local breeds [34]. Swiss Brown has been crossed with German, Italian and Spanish brown mountain breeds, as well as in the Asturian Valley [66]. We propose that historic or recent spreading of popular breeds acts on genetic diversity patterns and, by the selection of proven sires, has a homogenising effect on paternal lineages.

However, southern French beef breeds also carry exclusively the Y2-102-158 haplotype. A close relationship of southern French and Alpine breeds is indicated by autosomal AFLP profiles [9], SNPs [7] and microsatellite genotypes (unpublished results; **Figure S2**). One explanation might be that contacts of Transalpine Gallic people facilitated gene flow, for instance, during a repopulation of Gallia after the devastating Gallic wars. More recently, the Great Famine, from 1315 to 1317, north and west of the Alps may very well have decimated the French cattle and led to the introduction of Alpine cattle. After the Middle Ages, when the size of cattle started to increase, these would have developed into the present beef types.

The divergence of paternal lineages in north-central and central groups of dairy cattle is fully in agreement with their separate positions as indicated by autosomal markers (**Figure S1**; [7,9]). This may constitute evidence for independent developments of dairy cattle in different European regions. The Dutch cattle has been crossed with British breeds [34], but not in the Jersey Island dairy cattle, which obviously developed its high milk production separately. An important implication is that in different regions breeding for milk production selected other gene variants, and that crossbreeding of the different categories of dairy cattle is expected to generate potentially useful allelic combinations.

Apparently, the expansion of the dairy breeds have created, or largely maintained, a sharp genetic contrast of northern and southern Europe, which divides both France and Germany. It may be hypothesised that the northern landscapes, with large flat meadows, are suitable for large-scale farming with specialised dairy cattle (*Niederungsvieh*, lowland cattle), whilst the mixed-purpose or beef cattle (*Höhenvieh*, highland cattle) are better suited to the smaller farms and hilly regions of the south. However, it is also remarkable that in both France and Germany the bovine genetic boundary coincides with historic linguistic and cultural boundaries. In France, the Frankish invasion in the north created the difference between the northern *langue d'oïl* and the southern *langue d'oc*. The German language is still divided into the southern *Hochdeutsch* and northern *Niederdeutsch* dialects, which also correlates with the distribution of the Catholic and Protestant religions. On a larger scale, it is tempting to speculate that the difference between two types of European cattle reflects, and has even reinforced, the traditional and still visible contrast of Roman and Germanic Europe.

We conclude that analysis of paternal lineages contributes to a reconstruction of the history of cattle husbandry. A differentiation of more haplotypes [14,15,22,31], and especially the analysis of historic DNA, may answer the question of the European aurochs contribution, as well as indicating a time depth for the successive genetic events that have created the present pattern of genetic diversity.

Materials and Methods

Sample collection

Details on sample collection and DNA isolation can be found in previous publications from the participating laboratories [14,22,31,67]. Breeds, geographic origins and sample sizes are shown in **Table S1**. This dataset comprises a total of 138 breeds and 2087 individuals. Apart from the 16 zebu (*Bos indicus*) samples from India, all animals were *Bos taurus*. Based on breed histories and geographical origin, a total of 12 geographic breed groups were defined as follows: India (2 breeds), Africa (1), Southwest Asia (7), Podolian (12), Iberian (31), Central (30), British (10), Nordic

(14), Baltic Red (9), North-West (5), Lowland Pied (9) and Eastern (8).

Sequencing and genotyping

Segments of the ZFY, SRY and DBY genes were PCR-amplified and sequenced as previously described [32]. As indicated in **Table S1**, individual SNPs were typed by pyrosequencing [25,30], sequencing [14] or by custom service via the Taqman (Van Haeringen Laboratory, Wageningen, Netherlands) or KASPar (K-Bioscience, Hoddesdon, UK) procedures. Several samples were analysed by more than one method.

The Y-chromosomal microsatellites $BM861$ [37], $INRA189$ [36], $INRA124$ and $INRA126$ [68] were genotyped as described in previous publications [14,30,31,67]. Results from $INRA124$ and $INRA126$ were discounted due to these markers showing amplification in female samples [69]. Allele sizes were standardised via known fixed alleles from several samples or breeds analysed by the different laboratories (Dublin, Jokioinen, Davis and Gijon). Y-haplotypes were defined for each individual by combining data from both SNPs and microsatellite loci (SNP-$INRA189$-$BM861$).

Genotyping of the FAO-recommended autosomal microsatellites has been described previously [5,35,70].

Statistical analyses

Variability at the DNA level was assessed in each breed and within each geographic group by estimating the unbiased expected genetic diversity [71], and by an analysis of molecular variance (AMOVA), using ARLEQUIN version 2.0 [72]. Phylogenetic relationships among haplotypes were analysed by constructing median-joining (MJ) networks for each region using the algorithm of Bandelt et al. [73], as implemented in the NETWORK (Version 4.156; www.fluxus-engineering.com).

Supporting Information

Figure S1 Map showing distribution of SNP Y-haplotypes in: (a) Europe, and (b) Eurasia. Haplotypes are indicated with the colouring scheme shown in (**b**). Abbreviations of breed names are given in Table 1.
(PDF)

Figure S2 NeighborNet graphs of Reynolds' distances, based on 30 autosomal microsatellites. Haplotypes are indicated with the same colouring scheme as for: (**a**) maps in **Figure S1**, and (**b**) maps in **Figure 1**. Abbreviations of breed names are given in Table 1.
(PDF)

Table S1
(DOC)

Table S2
(DOC)

Table S3
(DOC)

Table S4
(DOC)

Acknowledgments

The following members of the European Cattle Genetic Diversity Consortium contributed to the study. **Austria**: *R. Baumung*, BOKU

University, Vienna; **Belgium**: *G. Mommens*, University of Ghent, Merelbeke; **Denmark**: *L.-E. Holm*, Aarhus University, Tjele; *K.B. Withen, B.V Pedersen* and *P. Gravlund*, University of Copenhagen; **France**: *K. Moazami-Goudarzi, M. Gautier* and *D. Laloë*, INRA, Jouy-en-Josas; *A. Oulmouden* and *H. Levéziel*, INRA, Limoges; *P. Taberlet*, Université Joseph Fourier et CNRS, Grenoble; **Germany**: *H. Simianer* and *H. Täubert*, Georg-August-Universität, Göttingen; *G. Erhardt*, Justus-Liebig Universität, Giessen; *I. Medugorac*, Ludwig-Maximilians Universität, Munich; **Iceland**: *E. Eythorsdottir*, Agricultural University of Iceland, Borgarnes; **Italy**: *P. Ajmone Marsan* and *R. Negrini*, Università Cattolica del S. Cuore, Piacenza; *E. Lasagna, V. Landi, F. Panella* and *F. Maria Sarti*, Università degli Studi di Perugia; *M. Longeri, G. Ceriotti* and *M. Zanotti*, Università degli Studi di Milano; *D. Marletta* and *A. Criscione*, Universita degli Studi di Catania; *A. Valentini* and *M.C. Savarese*, Università della Tuscia, Viterbo; *F. Pilla*, Università del Molise, Campobasso; **Latvia**: *Z. Grislis*, Latvia University of Agriculture, Jelgava; **Netherlands**: *D.C.J. van Boxtel*, Utrecht University; *E. Cuppen*, Hubrecht Laboratory, Utrecht; *W. van Haeringen* and *L. van de Goor*, Van Haeringen Laboratory, Wageningen; *M. Felius*, Rotterdam; **Norway**: *I. Olsaker*, Norwegian School of Veterinary Science, Oslo; **Russia**: *Z. Ivanova, R. Popov* and *I. Ammosov*, Yakutian Research Institute of Agricultural Sciences, Yakutsk; *T. Kiselyova*, All-Russian Research Institute for Farm Animals and Breeding, St. Petersburgh-Pushkin; **Serbia**: *M. Ćinkulov*, University of Novi Sad; *S. Stojanovic*, Ministry of Agriculture and Water Management, Belgrade; **Spain**: *S. Dunner*, Universidad Complutense de Madrid; *C. Rodellar*, Veterinary Faculty, Zaragoza; *P.J. Azor, A. Molina* and *E. Rodero*, University of Córdoba; *F. Goyache*, SERIDA, Gijón; **Sweden**: *A. Götherström*, Uppsala University; **Switzerland**: *G. Dolf*, University of Berne; **UK**: *J.L. Williams* and *P. Wiener*, Roslin Institute.

We thank the following persons and institutes for cattle sperm samples: Sue Walters (Shorthorn Society, Kenilworth, UK), J. Dengg (Tierzucht LWK, Salzburg, Austria), Besamungsverein (Neustadt/Aisch, Germany), Barbara Echert (Prüf- und Besamungsstation München Grub, Germany), Dr. Hagena (ZBH eG, Alsfeld, Germany), B. Luntz (LfL Tierzucht, Poing-Grub, Germany), Rinderunion (Herbertingen, Germany), Claudia Schöder (Combibull,Ocholt, Germany), Angelika Weyand (RSH, Schönböken, Germany), ANABIC (Perugia, Italy), Torstein Steine and Nina Hovden Saether (Norway), and Margareta Hård (Sweden). We thank A. Martínez-Martínez and J. V. Delgado (Universidad de Córdoba, Spain) for contributing Mostrenca samples. We also thank Jonna Roito, Tiina Jaakkola and Sari Raiskio (Finland) for DNA extractions, and Anneli Virta (Finland) for skilful technical assistance in A.L.F.™ Express and Megabase runs.

Author Contributions

Conceived and designed the experiments: CJE CG JK LP-P MCTP DGB JAL IJN. Performed the experiments: CJE CG JK LP-P JAL IJN. Analyzed the data: CJE CG JK LP-P AT MCTP DGB JAL. Contributed reagents/materials/analysis tools: FS MCTP DGB JAL. Wrote the paper: CJE CG JAL. Commented on manuscript: LP-P AT LTG.

References

1. Diamond J (2002) Evolution, consequences and future of plant and animal domestication. Nature 418: 700–707.

2. Zeder MA, Emshwiller E, Smith BD, Bradley DG (2006) Documenting domestication: the intersection of genetics and archaeology. Trends Genet 22: 139–155.

3. Driscoll CA, Macdonald DW, O'Brien SJ (2009) From wild animals to domestic pets, an evolutionary view of domestication. Proc Natl Acad Sci U S A 106 Suppl 1: 9971–9978.

4. Groeneveld LF, Lenstra JA, Eding H, Toro MA, Scherf B, et al. (2010) Genetic diversity in farm animals–a review. Anim Genet 41 Suppl 1: 6–31.

5. Medugorac I, Medugorac A, Russ I, Veit-Kensch CE, Taberlet P, et al. (2009) Genetic diversity of European cattle breeds highlights the conservation value of traditional unselected breeds with high effective population size. Mol Ecol 18: 3394–3410.

6. Cymbron T, Freeman AR, Isabel Malheiro M, Vigne JD, Bradley DG (2005) Microsatellite diversity suggests different histories for Mediterranean and Northern European cattle populations. Proc R Soc Lond B Biol Sci 272: 1837–1843.

7. Decker JE, Pires JC, Conant GC, McKay SD, Heaton MP, et al. (2009) Resolving the evolution of extant and extinct ruminants with high-throughput phylogenomics. Proc Natl Acad Sci U S A 106: 18644–18649.

8. Laloë D, Moazami-Goudarzi K, Lenstra JA (2010) Phylogeographic trends of domestic ruminants. Diversity 2: 932–945.

9. Negrini R, Nijman IJ, Milanesi E, Moazami-Goudarzi K, Williams JL, et al. (2007) Differentiation of European cattle by AFLP fingerprinting. Anim Genet 38: 60–66.

10. Beja-Pereira A, Caramelli D, Lalueza-Fox C, Vernesi C, Ferrand N, et al. (2006) The origin of European cattle: evidence from modern and ancient DNA. Proc Natl Acad Sci U S A 103: 8113–8118.

11. Achilli A, Bonfiglio S, Olivieri A, Malusa A, Pala M, et al. (2009) The multifaceted origin of taurine cattle reflected by the mitochondrial genome. PLoS ONE 4: e5753.

12. Achilli A, Olivieri A, Pellecchia M, Uboldi C, Colli L, et al. (2008) Mitochondrial genomes of extinct aurochs survive in domestic cattle. Curr Biol 18: 157–158.

13. Mona S, Catalano G, Lari M, Larson G, Boscato P, et al. Population dynamic of the extinct European aurochs: genetic evidence of a north-south differentiation pattern and no evidence of post-glacial expansion. BMC Evol Biol 10: 83.

14. Perez-Pardal L, Royo LJ, Beja-Pereira A, Curik I, Traore A, et al. (2009) Y-specific microsatellites reveal an African subfamily in taurine (Bos taurus) cattle. Anim Genet 41: 232–241.

15. Perez-Pardal L, Royo LJ, Beja-Pereira A, Chen S, Cantet RJ, et al. (2010) Multiple paternal origins of domestic cattle revealed by Y-specific interspersed multilocus microsatellites. Heredity 105: 511–519.

16. Edwards CJ, Baird JF, MacHugh DE (2007) Taurine and zebu admixture in Near Eastern cattle: a comparison of mitochondrial, autosomal and Y-chromosomal data. Anim Genet 38: 520–524.

17. Troy CS, MacHugh DE, Bailey JF, Magee DA, Loftus RT, et al. (2001) Genetic evidence for Near-Eastern origins of European cattle. Nature 410: 1088–1091.

18. Chen S, Lin BZ, Baig M, Mitra B, Lopes RJ, et al. (2009) Zebu cattle are an exclusive legacy of the South Asia Neolithic. Mol Biol Evol 27(1): 1–6.

19. Underhill PA, Kivisild T (2007) Use of y chromosome and mitochondrial DNA population structure in tracing human migrations. Annu Rev Genet 41: 539–564.

20. Jobling MA, Tyler-Smith C (2003) The human Y chromosome: an evolutionary marker comes of age. Nat Rev Genet 4: 598–612.

21. MacHugh DE, Shriver MD, Loftus RT, Cunningham P, Bradley DG (1997) Microsatellite DNA variation and the evolution, domestication and phylogeography of taurine and zebu cattle (Bos taurus and Bos indicus). Genetics 146: 1071–1086.

22. Ginja C, Penedo MC, Melucci L, Quiroz J, Martinez Lopez OR, et al. (2010) Origins and genetic diversity of New World Creole cattle: inferences from mitochondrial and Y chromosome polymorphisms. Anim Genet 41: 128–141.

23. Bradley DG, Loftus RT, Cunningham P, MacHugh DE (1998) Genetics and domestic cattle origins. Evol Anthrop 6: 79–86.

24. Mohamad K, Olsson M, van Tol HT, Mikko S, Vlamings BH, et al. (2009) On the origin of Indonesian cattle. PLoS ONE 4: e5490.

25. Gotherstrom A, Anderung C, Hellborg L, Elburg R, Smith C, et al. (2005) Cattle domestication in the Near East was followed by hybridization with aurochs bulls in Europe. Proc R Soc Lond B Biol Sci 272: 2345–2350.

26. Svensson E, Gotherstrom A (2008) Temporal fluctuations of Y-chromosomal variation in Bos taurus. Biol Lett 4: 752–754.

27. Bollongino R, Elsner J, Vigne JD, Burger J (2008) Y-SNPs do not indicate hybridisation between European aurochs and domestic cattle. PLoS ONE 3: e3418.

28. Balaresque P, Bowden GR, Adams SM, Leung HY, King TE, et al. A predominantly neolithic origin for European paternal lineages. PLoS Biol 8: e1000285.

29. Underhill PA, Myres NM, Rootsi S, Metspalu M, Zhivotovsky LA, et al. (2010) Separating the post-Glacial coancestry of European and Asian Y chromosomes within haplogroup R1a. Eur J Hum Genet 18: 479–484.

30. Ginja C, Gama LT, Penedo MCT (2009) Y chromosome haplotype analysis in Portuguese cattle breeds using SNPs and STRs. J Hered 100: 148–157.

31. Kantanen J, Edwards CJ, Bradley DG, Viinalass H, Thessler S (2009) Maternal and paternal genealogy of Eurasian taurine cattle (Bos taurus). Heredity 103: 1–12.

32. Nijman IJ, van Boxtel DJ, van Cann LM, Cuppen E, Lenstra JA (2008) Phylogeny of Y-chromosomes from interbreeding bovine species. Cladistics 24: 723–726.

33. Verkaar EL, Nijman IJ, Beeke M, Hanekamp E, Lenstra JA (2004) Maternal and paternal lineages in cross-breeding bovine species. Has wisent a hybrid origin? Mol Biol Evol 21: 1165–1170.

34. Felius M (1995) Cattle breeds - an encyclopedia. Doetinchem, Netherlands: Misset.

35. Li MH, Tapio I, Vilkki J, Ivanova Z, Kiselyova T, et al. (2007) The genetic structure of cattle populations (Bos taurus) in northern Eurasia and the neighbouring Near Eastern regions: implications for breeding strategies and conservation. Mol Ecol 16: 3839–3853.

36. Kappes SM, Keele JW, Stone RT, McGraw RA, Sonstegard TS, et al. (1997) A second-generation linkage map of the bovine genome. Genome Res 7: 235–249.

37. Bishop MD, Kappes SM, Keele JW, Stone RT, Sunden SL, et al. (1994) A genetic linkage map for cattle. Genetics 136: 619–639.

38. Baker CMA, Manwell C (1980) Chemical classification of cattle: 1. Breed groups. Genetics 11: 127–150.
39. Loftus RT, MacHugh DE, Bradley DG, Sharp PM, Cunningham P (1994) Evidence for two independent domestications of cattle. Proc Natl Acad Sci U S A 91: 2757–2761.
40. Loftus RT, MacHugh DE, Ngere LO, Balain DS, Badi AM, et al. (1994) Mitochondrial genetic variation in European, African and Indian cattle populations. Anim Genet 25: 265–271.
41. Bradley DG, MacHugh DE, Cunningham P, Loftus RT (1996) Mitochondrial diversity and the origins of African and European cattle. Proc Natl Acad Sci U S A 93: 5131–5135.
42. Bradley DG, MacHugh DE, Loftus RT, Sow RS, Hoste CH, et al. (1994) Zebu-taurine variation in Y chromosomal DNA: a sensitive assay for genetic introgression in west African trypanotolerant cattle populations. Anim Genet 25: 7–12.
43. Teale AJ, Wambugu J, Gwakisa PS, Stranzinger G, Bradley D, et al. (1995) A polymorphism in randomly amplified DNA that differentiates the Y chromosomes of Bos indicus and Bos taurus. Anim Genet 26: 243–248.
44. Bruford MW, Bradley DG, Luikart G (2003) DNA markers reveal the complexity of livestock domestication. Nat Rev Genet 4: 900–910.
45. Pieragostini E, Scaloni A, Rullo R, Di Luccia A (2000) Identical marker alleles in Podolic cattle (Bos taurus) and Indian zebu (Bos indicus). Comp Biochem Physiol B Biochem Mol Biol 127: 1–9.
46. Ibeagha-Awemu EM, Erhardt G (2006) An evaluation of genetic diversity indices of the Red Bororo and White Fulani cattle breeds with different molecular markers and their implications for current and future improvement options. Trop Anim Health Prod 38: 431–441.
47. Caroli A, Rizzi R, Luhken G, Erhardt G Short communication: milk protein genetic variation and casein haplotype structure in the Original Pinzgauer cattle. J Dairy Sci 93: 1260–1265.
48. Beja-Pereira A, Alexandrino P, Bessa I, Carretero Y, Dunner S, et al. (2003) Genetic characterization of southwestern European bovine breeds: a historical and biogeographical reassessment with a set of 16 microsatellites. J Hered 94: 243–250.
49. Ginja C, Telo Da Gama L, Penedo MC (2010) Analysis of STR markers reveals high genetic structure in Portuguese native cattle. J Hered 101: 201–210.
50. Bollongino R, Edwards CJ, Alt KW, Burger J, Bradley DG (2006) Early history of European domestic cattle as revealed by ancient DNA. Biol Lett 2: 155–159.
51. Stock F, Edwards CJ, Bollongino R, Finlay EK, Burger J, et al. (2009) Cytochrome b sequences of ancient cattle and wild ox support phylogenetic complexity in the ancient and modern bovine populations. Anim Genet 40: 694–700.
52. Waterbolk HT (1968) Food production in prehistoric Europe. Science 162: 1093–1102.
53. Zilhão J (2001) Radiocarbon evidence for maritime pioneer colonization at the origins of farming in west Mediterranean Europe. Proc Natl Acad Sci U S A 98: 14180–14185.
54. Tresset A, Vigne JD (2007) Substitution of species, techniques and symbols at the Mesolithic/Neolithic transition in Western Europe. In: Whittle A, Cummings V, eds. Going Over: the Mesolithic/Neolithic Transition in NW Europe: Proc Brit Acad. pp 189–210.
55. Marchand G, Tresset A (2005) Derniers chasseurs et premiers agriculteurs sur la façade Atlantique de l'Europe. In: Guilaine J, ed. Les Marges, Débitrices ou Créatrices? La Mise en Place du Néolithique et de ses Prolongements à la Périphérie des "Foyers" Classiques. Paris: Errance. pp 255–280.
56. Cymbron T, Loftus RT, Malheiro MI, Bradley DG (1999) Mitochondrial sequence variation suggests an African influence in Portuguese cattle. Proc R Soc Lond B Biol Sci 266: 597–603.
57. Ginja C, Penedo MC, Sobral MF, Matos J, Borges C, et al. (2010) Molecular genetic analysis of a cattle population to reconstitute the extinct Algarvia breed. Genet Sel Evol 42: 18.
58. Pellecchia M, Negrini R, Colli L, Patrini M, Milanesi E, et al. (2007) The mystery of Etruscan origins: novel clues from Bos taurus mitochondrial DNA. Proc R Soc Lond B Biol Sci 274: 1175–1179.
59. Beja-Pereira A, Luikart G, England PR, Bradley DG, Jann OC, et al. (2003) Gene-culture coevolution between cattle milk protein genes and human lactase genes. Nat Genet 35: 311–313.
60. Jones HL (1924) The Loeb Classical Library: The geography of Strabo in eight volumes. London: William Heinemann Ltd.
61. Itan Y, Powell A, Beaumont MA, Burger J, Thomas MG (2009) The origins of lactase persistence in Europe. PLoS Comput Biol 5: e1000491.
62. Faber JA (1962) Cattle plague in the Netherlands during the 18th century. Mededelingen van de Landbouwhogeschool te Wageningen 12: 1–7.
63. Li MH, Kantanen J Genetic structure of Eurasian cattle (Bos taurus) based on microsatellites: clarification for their breed classification. Anim Genet 41: 150–158.
64. Schibler J, Schlumbaum A (2007) History and economic importance of cattle (Bos taurus L.) in Switzerland from Neolithic to Early Middle Ages. Schweiz Arch Tierheilkd 149: 23–29.
65. Healy JF (1991) Pliny the Elder - Natural History. London: Penguin Classics.
66. Martin-Burriel I, Rodellar C, Lenstra JA, Sanz A, Cons C, et al. (2007) Genetic diversity and relationships of endangered Spanish cattle breeds. J Hered 98: 687–691.
67. Edwards CJ, Gaillard C, Bradley DG, MacHugh DE (2000) Y-specific microsatellite polymorphisms in a range of bovid species. Anim Genet 31: 127–130.
68. Vaiman D, Mercier D, Moazami-Goudarzi K, Eggen A, Ciampolini R, et al. (1994) A set of 99 cattle microsatellites: characterization, synteny mapping, and polymorphism. Mamm Genome 5: 288–297.
69. Perez-Pardal L, Royo LJ, Alvarez I, de Leon FA, Fernandez I, et al. (2009) Female segregation patterns of the putative Y-chromosome-specific microsatellite markers INRA124 and INRA126 do not support their use for cattle population studies. Anim Genet 40: 560–564.
70. Econogene (2006) Marker-assisted conservation of European cattle breeds: An evaluation. Anim Genet 37: 475–481.
71. Nei M (1987) Molecular Evolutionary Genetics. New York, NY, USA: Columbia University Press.
72. Schneider S, Roessli D, Excoffier L (2000) Arlequin ver. 2.000: A software for population genetics data analysis. Switzerland: Genetics and Biometry Laboratory, University of Geneva.
73. Bandelt HJ, Forster P, Rohl A (1999) Median-joining networks for inferring intraspecific phylogenies. Genome Biol Evol 16: 37–48.

2

Where Do Livestock Guardian Dogs Go? Movement Patterns of Free-Ranging Maremma Sheepdogs

Linda van Bommel[1,2]*, **Chris N. Johnson**[1]

1 School of Zoology, University of Tasmania, Hobart, Australia, **2** Fenner School of Environment and Society, Australian National University, Canberra, Australia

Abstract

In many parts of the world, livestock guardian dogs (LGDs) are a relatively new and increasingly popular method for controlling the impact of wild predators on livestock. On large grazing properties in Australia, LGDs are often allowed to range freely over large areas, with minimal supervision by their owners. How they behave in this situation is mostly unknown. We fitted free-ranging Maremma sheepdogs with GPS tracking collars on three properties in Victoria, Australia; on two properties, four sheep were also fitted with GPS collars. We investigated how much time the Maremmas spent with their livestock, how far they moved outside the ranges of their stock, and tested whether they use their ranges sequentially, which is an effective way of maintaining a presence over a large area. The 95% kernel isopleth of the Maremmas ranged between 31 and 1161 ha, the 50% kernel isopleth ranged between 4 and 252 ha. Maremmas spent on average 90% of their time in sheep paddocks. Movements away from sheep occurred mostly at night, and were characterised by high-speed travel on relatively straight paths, similar to the change in activity at the edge of their range. Maremmas used different parts of their range sequentially, similar to sheep, and had a distinct early morning and late afternoon peak in activity. Our results show that while free-ranging LGDs spend the majority of their time with livestock, movements away from stock do occur. These movements could be important in allowing the dogs to maintain large territories, and could increase the effectiveness of livestock protection. Allowing LGDs to range freely can therefore be a useful management decision, but property size has to be large enough to accommodate the large areas that the dogs use.

Editor: Cédric Sueur, Institut Pluridisciplinaire Hubert Curien, France

Funding: The research was funded by the Hermon Slade Foundation and the Australian Research Council. Linda van Bommel was supported by an Australian Postgraduate Award. The funders had no role in study design, data collection and analysis, decision to publish, or preparation of the manuscript.

Competing Interests: The authors have declared that no competing interests exist.

* Email: linda.vanbommel@anu.edu.au

Introduction

Livestock Guardian Dogs (LGDs, *Canis familiaris*) are usually dogs of large breeds that are kept with livestock to protect them from predators. Most LGD breeds originated in Europe and Asia, where they have been used for centuries to protect livestock from predators and thieves. They are raised with stock from an early age, and as a result view livestock as their social companions and protect them from threats [1–3]. Experimental and anecdotal evidence shows that these dogs can be effective in protecting a range of livestock species from several types of predators (older studies reviewed in [4], see also [5–9]). In Australia, as in many other parts of the world outside their countries of origin, LGDs are a relatively new, and increasingly popular method to reduce predation on livestock [10,11]. LGDs can be fence-trained so they will remain in the paddock in which their livestock are confined, but they can also be allowed to cross stock fences to roam more freely [11,12]. This free-ranging system is often used on large grazing properties in Australia, where the main threat is posed by feral dogs, domestic dogs, dingoes and their hybrids (*Canis familiaris*, and *Canis dingo*, are hereafter referred to as wild dogs). In these situations, LGDs can potentially roam over large areas. LGDs might be visited by their owners intermittently, sometimes less than once a week [10], and the owners are mostly unaware of their LGDs' movements. Free-ranging could have a significant behavioural function for LGDs,

such as maintaining usage of a territory around the livestock from which wild predators are excluded. On the other hand, roaming by LGDs could potentially leave livestock vulnerable, and cause other problems, such as creating traffic hazards or having negative impacts on wildlife [6,12,13].

In this study, GPS collars were used to study the movements and behaviour of free-ranging Maremma sheepdogs on three properties in Victoria, Australia. On two properties sheep *Ovis aries* were also fitted with GPS collars. We had several specific aims. First, we wanted to know the size of the Maremmas' range to determine how much space these dogs used. We also wanted to know if the Maremmas' behaviour differed with regard to proximity to the core of their range. We measured their behaviour through their speed of movement (m/h) and the straightness versus tortuosity of their movement path. We expected to find relatively low-speed movement in a tortuous path at the core of the range, and high movement speed in a relatively straight line at the edge of their range, reflecting behaviour such as boundary patrolling or seeing off predators at the edge of the range versus resting or attending livestock at the core. Second, we wanted to determine the proportion of time the Maremmas spent with livestock, at what times of the day they were most likely to leave them, and if their behaviour changed when they left the livestock. We expected higher movement speeds in a relatively straight line away from stock versus low movement speeds in a more tortuous path close to stock, similar to the change in behaviour at the edge of the range versus the core. Third, we tested if Maremmas used their range sequentially. Sequential use of a territory indicates that a different part of the range is patrolled each day, and can provide an effective way for an animal to establish a presence over a large area. We also determined their 24-hour activity pattern in order to compare this to the activity patterns of the stock and predators.

Materials and Methods

Ethics statement

All research was carried out in compliance with the Australian Code for the Care and Use of Animals for Scientific Purposes, 7th edition. Ethics approval was obtained from the Animal Ethics Committee of the University of Tasmania (approval number: A0012323).

Study sites

We studied Maremma sheepdogs on three properties in Victoria, Australia. Gillingal Station was situated in eastern Victoria, and Riversdale and Heatherlie were situated in northeast Victoria, approximately 15 km apart. Details of these properties and management of Maremmas and livestock on each are shown in Table 1A and Fig. 1. On all three properties large tracts of native vegetation remained in addition to the pasture used for livestock grazing. Maremmas were free-ranging, and could easily cross stock fences. On Riversdale, the Maremmas could obtain ad-lib dry dog food from a self-feeder in a central location; on Gillingal Station and Heatherlie, the owners visited their dogs daily to feed them.

On all three properties, the main predators of livestock were wild dogs, but smaller predators were also present, including red foxes (*Vulpes vulpes*) and wedge-tailed eagles (*Aquila audax*). Predators had caused large losses of livestock on all three properties before introduction of the Maremmas; after the Maremmas started working, losses greatly decreased (Table 1A). In addition to using Maremmas for predator control, trapping, shooting and baiting of wild dogs still occurred in the vicinity of all properties, and Heatherlie was partly enclosed by an electrified wild-dog exclusion fence.

Data collection

Movements of Maremmas were recorded using Quantum 4000 Enhanced GPS tracking collars (Telemetry Solutions, Concord, USA), which were set to take a location every 30 min, 24 hours a day. This schedule was chosen as the best trade-off between the objective to collect detailed movement data, and the estimated battery life of the collars. The majority of the GPS data was collected in autumn/winter (1 Mar –1 Sept), as the main predators in the area (wild dogs and foxes) have their breeding season in autumn, followed by birth of their young in winter. During this time predators make more incursions into livestock areas, and predation rates are higher than in spring/summer ([14], A.Bowran and M.Fraser pers. comm.). Tracking data were collected during spring/summer only on Riversdale in 2012 and Heatherlie in 2012 (Table 1A). In summer, mean temperatures range between 16.2°C at night and 30°C during the day, with a mean monthly rainfall of 46 mm. In winter, mean temperatures range between 4.9°C at night and 13.3°C during the day, with a mean monthly rainfall of 73.6 mm.

On Gillingal Station and Riversdale, all the Maremmas were collared in each tracking period (Table 1A). On these two properties, all resident Maremmas functioned as one social group. On Heatherlie, five out of seven Maremmas (two females and

three males) were collared in 2011; at that time these dogs formed three distinct social units using sections the property 1–3 km apart. In 2012 one Maremma had died of old age, and all six remaining Maremmas (two females and four males) were collared. Four of these dogs (one female and three males) formed one social group that was predominantly responsible for protecting all livestock over the whole property. One female was old and mostly solitary, and one male suffered extreme social exclusion by the other individuals which severely restricted his movements. This last male was excluded from the analysis. The animals that formed part of a social unit were often found together, but social units also regularly split into sub-groups or individuals for varying lengths of time. All Maremmas were desexed, except for the male on Gillingal Station. This male had been kept sexually intact by his owner for breeding purposes, however, at the time of the research he was no longer used for breeding. Due to his old age his owner did not neuter him, since all other dogs on the property were desexed. On each of Riversdale and Heatherlie, four sheep were fitted with G2C 181 GPS tracking collars (Sirtrack, Hawkes Bay, New Zealand, Table 1A), also programmed to take a location every 30 min, 24 h a day. Maremmas were tracked for an average of 115±11 days (range: 9–174 days) before the collar failed or the battery ran out. On Riversdale the sheep were tracked for 114 days, on Heatherlie for 135 days, at which point the collars were removed because the dog tracking had ceased. The tracking data collected in this study are stored in the Movebank Data Repository [15].

Spatial analysis

Only locations from the GPS collars with an HDOP (Horizontal Dilution of Precision) value of <4 were included in the analysis. This HDOP value was chosen based on a pilot study. In this study, all GPS collars were kept stationary on the lower branches of an apple tree for four days, taking hourly locations. Based on these data, the HDOP value was selected that offered the best balance between filtering out inaccurate locations and data retention. A mean of 1.5% of each dataset was rejected due to an HDOP value that was too high, and the mean HDOP for the remaining locations in each dataset was 1.3. Due to the large size of the datasets that were collected, the loss of a small percentage of accurate locations as a result of applying this filter was not considered a problem [16].

Kernel home ranges and movement analysis. Fixed kernel density distributions [17] were calculated for each dog and sheep for each tracking period to determine home range sizes. Autocorrelation does not affect the accuracy of kernel home range estimates as long as the time interval between successive locations remains relatively constant, and the number of locations is large [18]; accuracy often improves with a shorter time interval despite increased autocorrelation [18]. Therefore, we did not consider autocorrelation to be a problem for our analysis [18–20]. We used an *ad hoc* smoothing parameter designed to prevent under- or over-smoothing, which involved choosing the smallest increment of the reference bandwidth (Href) that results in a 95% home range polygon that was as contiguous as possible, that is, containing no, or the minimum number of, separate activity areas [21–23]. For each dog group, the amount of overlap between the ranges of individuals was calculated. First, this was done on a pair wise basis; for all possible combinations of dog pairs within one group the overlap of home range was determined for the 50% and 95% kernel isopleth range. For each individual, the mean overlap with any of its group mates was calculated. Second, the overlap of home ranges of all Maremmas in the group was determined. Overlap of home ranges could not be calculated for sheep, as the

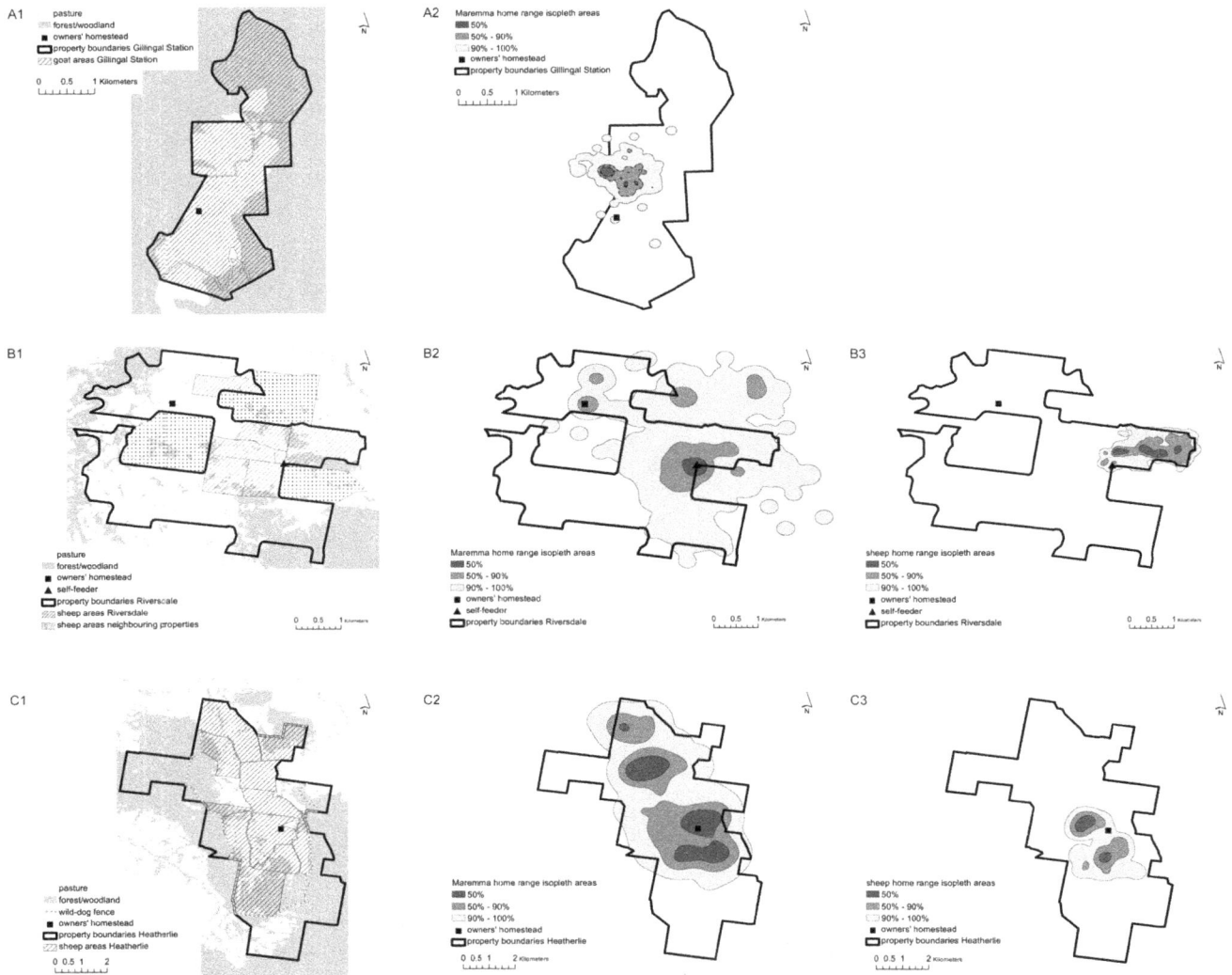

Figure 1. The three research properties used in this study. A. Gillingal Station, B. Riversdale, C. Heatherlie. Image 1 of each property shows the vegetation on the sites, and the areas where the sheep or goats were grazed. In addition, for Heatherlie, the position of the electrified wild-dog fence is indicated. Image 2 of each property shows a representative example of a Maremma home range, depicting the 50%, 50%–90% and 90%–100% kernel isopleth areas. In this figure, tracking data was used from a Maremma in 2010 on Riversdale and in 2011 on Heatherlie, to coincide with sheep home ranges depicted in image 3. Image 3 of Riversdale and Heatherlie shows a representative example of a sheep home range, depicting the 50%, 50%–90% and 90%–100% kernel isopleth areas.

sample size of collared individuals in one paddock was too low in most cases.

To investigate whether behaviour changed depending on location within the home range, kernel isopleths (i.e. probability contours) were calculated for each 10% increase in kernel density. Mean movement speed (m/hour) was calculated for each individual within each isopleth. In addition, path tortuosity was used to investigate how convoluted or straight the movement path was, depending on the location in the home range. For each dataset, the total movement path was divided into separate paths in each of three areas in the home range: within the 50% isopleth, between the 50% and 90% isopleths and outside of the 90% isopleth (figure 1). A path started when the animal crossed into an isopleth area, and ended when it crossed into the next area. Tortuosity was calculated as L/R^2, where L is the path length, and R is the net displacement (the distance in a straight line between

start point and end point). We choose R^2 as opposed to R, as R^2 commonly increases linearly with path length [24]. Tortuosity may be scale-dependent [24], so we measured tortuosity for path segments in four lengths: 0.1, 0.25, 0.5 and 1 km. For each scale of analysis, we partitioned each separate path into sequential segments of length (L). This led to a large number of path segments, and therefore tortuosity measurements, within each isopleth area for each individual at each scale of analysis. In order to get a representative measurement of tortuosity for each isopleth area at each scale of analysis for each individual, the median of the tortuosity values of all path segments in the corresponding dataset was calculated. The mean could not be used due to the presence of a small number of outliers with extremely high values for tortuosity in most datasets.

Association with livestock. The presence of a Maremma in a paddock containing livestock does not necessarily indicate that

Table 1. Details of the three research properties.

Property	Gillingal Station	Riversdale	Heatherlie
1A. Management, Losses and Tracking periods			
Property management			
Location	Eastern Victoria	North-eastern Victoria	North-eastern Victoria
Topography	Hilly, elevation 250–450 m a.s.l.	Hilly, elevation 200–800 m a.s.l.	Hilly, elevation 200–900 m a.s.l.
Surroundings	State forest on all sides	Grazing properties, pine plantations, native vegetation	Grazing properties, pine plantations, native vegetation
Size (ha)	800	1,214	2,428
Livestock	3,000 Boer and wild-caught feral goats; cattle	1,500 Merino sheep; cattle	6,000 – 8,000 Merino and crossbred sheep; cattle
Number of Maremmas	3; 1M, 2F*	4; 3M, 1F*	7 (2011); 3F, 4M* 6 (2012); 2F, 4M*
Stock guarded by Maremmas	Boer goats	Merino sheep	Merino and crossbred sheep
Stock management	Gates left open; stock roamed the whole property	728 ha was used for sheep. 3-4 paddocks in the area were used at one time, depending on the season	Rotational grazing. Number of paddocks in use, and the size of sheep flock in those paddocks changed frequently.
Losses to predators#			
Before Maremmas	Up to 34 adults in one night	100 adult sheep and 100% of lambs lost yearly	150-200 adult sheep and 40% of lams lost yearly
Purchase of Maremmas	1999	2006	2009
With Maremmas	No adults lost, small percentage of goat kids lost yearly	No adults lost, 70% lamb survival	No adults lost, 90% lamb survival
GPS tracking periods			
Estimated age of Maremmas for the majority of time they were tracked	8, 11, 12 yrs	4, 4, 5, 5 yrs	2, 3, 3, 4, 4, 10, 11 yrs
Maremmas	June – August 2009	March – August 2010 April – September 2011 January – August 2012	April – September 2011 June – December 2012
Sheep	-	March – August 2010	May – September 2011
1B. Kernel home range sizes			
Maremmas			
95% isopleth (ha)	31.0±3.9	2010 286.3±81.4 2011 211.5±66.1 2012 summer 148.0±55.4 2012 winter 185.2±91.3	2011 197.0±31.6 2012 summer 915.5±154.1 2012 winter 1160.9±163.4
50% isopleth (ha)	4.1±0.4	2010 20.4±4.3 2011 19.3±5.1 2012 summer 14.8±4.9	2011 55.2±9.2 2012 summer 173.2±45.9 2012 winter 252.1±40.6

Table 1. Cont.

Property	Gillingal Station	Riversdale	Heatherlie
		2012 winter	
Sheep			
95% isopleth (ha)		69.2±14.9	94.4±3.2
		13.8±8.0	
50% isopleth (ha)		10.8±3.1	15.9±5.7

* M – male, F – female.
\# Information on losses of livestock before and after obtaining LGDs has been sourced from records kept by the farmers themselves.

the dog is closely associating with the livestock at that time. However, it does indicate that the Maremma is within a certain distance of its stock, and, in addition, that the dog is choosing to utilise the same area as the livestock. Therefore, the percentage of locations of the Maremmas that fell within sheep or goat paddocks was used as a measure to determine the association between the dogs and livestock.

To test for differences in behaviour with and without livestock, mean movement speed (m/hour), and path tortuosity were calculated for each of the two categories for each Maremma. Tortuosity was calculated as for the kernel isopleth areas, with, in this case, the total movement path of each dataset divided into separate paths in two areas; in sheep paddocks and outside sheep paddocks. A path started and ended when the animal crossed from one area into the other. If a Maremma spent more than 2 hours in close proximity to the owners' homestead, that part of the movement path was excluded from the analysis, as the motivation for those particular movements was presumably unrelated to animal-guarding activities. The separate movement paths were analysed at the four length scales described above for the isopleth areas, and median tortuosity values of the segments were calculated at each scale of analysis for each area (in or out of sheep paddocks) for each individual. Maremmas that were located in livestock paddocks 100% of the time were excluded from this analysis.

To determine which time of the day Maremmas were most likely to leave their livestock, for each category (with and without livestock), the percentage of the total number of locations within that category was calculated per hour in a 24 hour period. Then for each hour, the percentage with livestock was subtracted from the percentage without livestock.

Sequential use of territory. Minimum convex polygons (MCPs) were calculated for the area used each day (from midnight to midnight) per individual per tracking period, following the method of Demma and Mech [25]. MCPs were considered the best method for this analysis, as the objective was to determine the extent of the daily area of use, without regard to the density distribution of the locations. The main MCP biases were minimised by using only locations with an HDOP of <4.0, using an adequate number of GPS locations for the calculation of each MCP (48 locations per 24 hour period), and the lack of significant geographic barriers to movement [26]. The wild dog exclusion fence on Heatherlie did not influence the calculations, as no GPS locations ever occurred outside of fence, and lines from the MCPs never crossed the fence.

The average percentage of overlap of the MCP of consecutive days was determined per individual. Sequential use was defined as a daily overlap of <50%. To test for recurring patterns of re-use of specific areas of the home range, we determined the percentage of overlap during periods of 20 consecutive days for all datasets. To do this, we first determined the percentage of overlap between the first day in the series and each of the 19 consecutive days. We then took the second day as the initial day and repeated the process with the next 19 consecutive days, etc. Data were averaged across day number per individual.

24-hour activity pattern. The mean movement speed per hour was calculated for each individual Maremma and sheep, and compared across the 24 hour daily cycle.

All spatial analyses were done using ArcGis 10 [27] and custom written code in R [28].

Statistical analysis

Averages are given ± SE. Repeated measures ANOVA was used to test for statistical differences in variables for categories of

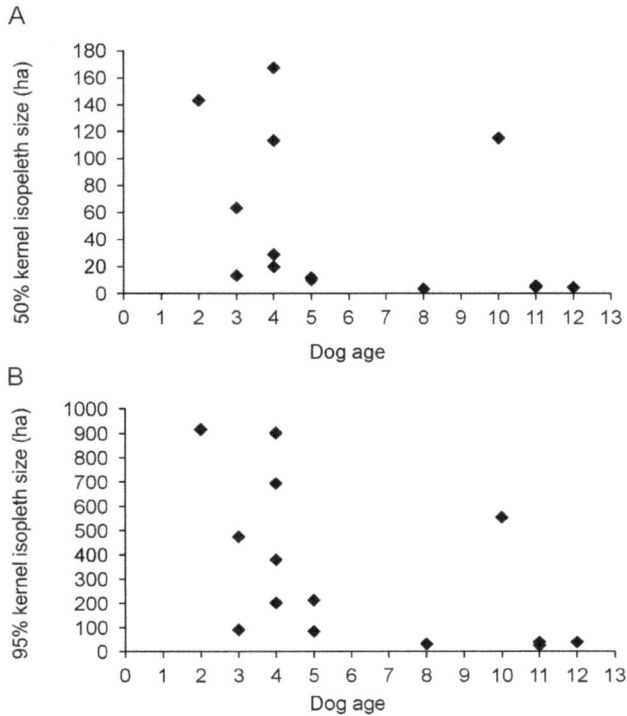

Figure 2. Relationship between age of the Maremma and size of the kernel isopleth home range (14 Maremmas in total). A. 50% kernel isopleth core area, and B. 95% kernel isopleth range. Each point represents the mean size of the range of an individual dog.

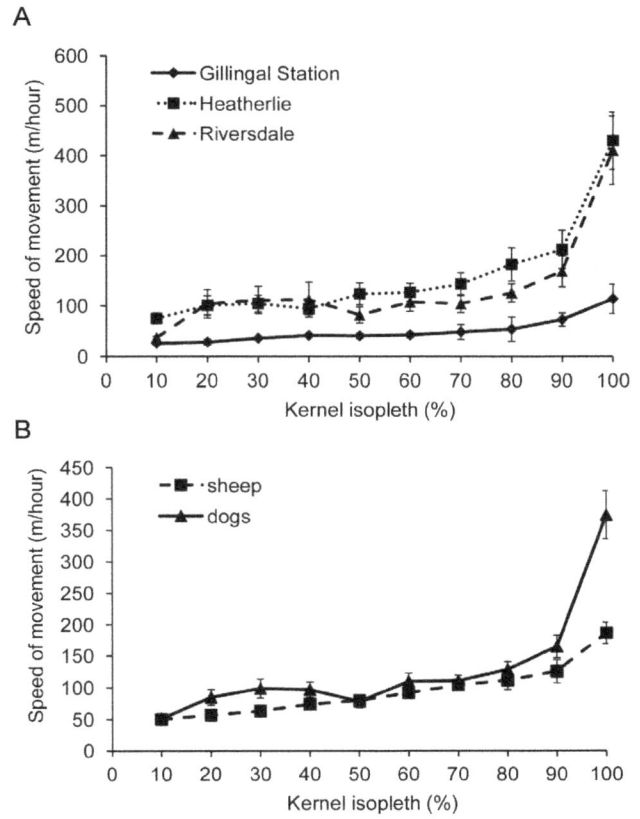

Figure 3. Mean speed of movement. (A) Relative to location in the kernel isopleths of Maremmas on the three properties (14 Maremmas in total). Movement speed increased significantly towards the edge of the home range ($F_{9, 117} = 46.2$, $P < 0.01$), the difference between the three properties is not significant. (B) Relative to location in the kernel isopleth of Maremmas compared to sheep when they were collared at the same time (9 Maremmas and 8 sheep). Maremmas had higher speed of movement than sheep, and a greater increase in movement speed towards the edge of their home range ($F_{1, 10} = 6.07$, $P < 0.05$).

analysis within each species, for the effect of age of the dog, differences between the different properties, and to test for differences between species. To test for differences between species, data was only used from the time period in which both species were collared simultaneously. The main part of the analysis focuses on data from autumn/winter only. On Riversdale and Heatherlie in 2012, variables were calculated for each individual dog for summer and winter separately, and tested for significant differences using repeated measures ANOVA. All tests used were two sided with a 95% confidence level. All statistical analysis were done using R Statistical Software [28].

Results

Home ranges

Kernel home range sizes at the 95% and 50% isopleths are shown in Table 1B. Home range sizes of Maremmas varied significantly with dog age ($F_{1, 10} = 5.57$, $P < 0.05$); older dogs had smaller ranges (Fig. 2). As a result, the Maremmas on Gillingal Station had the smallest ranges, followed by Riversdale and then Heatherlie (Fig. 1, Table 1B), but after accounting for age, the difference between the properties was not significant ($F_{2, 10} = 2.81$, $P = 0.10$). For sheep, no significant size differences were found between the home ranges on Riversdale and Heatherlie ($F_{1, 6} = 2.86$, $P = 0.14$). On Gillingal Station the Maremmas' home ranges covered approximately a quarter of the property, focussing on the area that was mainly used by the purebred Boer goats (personal observation). The 50% core of the Maremmas' ranges centred on the areas where these goats camped at night. During all tracking periods on Riversdale, and the 2012 tracking period on Heatherlie, the Maremmas' entire home range (100% kernel

isopleth) encompassed all the sheep-grazing paddocks, except for one paddock on Heatherlie, and extended up to two kilometres beyond them. On Riversdale, this range included sheep paddocks of neighbouring properties. The 50% core of the dog's range centred on the self-feeder on Riversdale, and encompassed parts of the main paddocks that contained sheep during the tracking period. On Heatherlie, the 50% core of the home range centred on the main parts of the main sheep grazing paddocks. In 2011 on Heatherlie, the home ranges were smaller than in 2012, due to the separation of the three dog groups in 2011. Entire home ranges (100% kernel isopleth) still encompassed all sheep paddocks on the side of the property on which the dogs were working, and extended up to one km beyond them. The 50% core of the dogs' ranges centred on the area used by the sheep to camp at night. On both Riversdale and Heatherlie, Maremmas reduced visits to paddocks when sheep were removed, and increased visits once they were present again. Maremma home ranges were significantly larger than those of the sheep they were guarding ($F_{1, 14} = 9.17$, $P < 0.01$) (Table 1B). Summer and winter ranges on Riversdale and Heatherlie in 2012 were not significantly different from each other ($F_{1, 17} = 1.40$, $P = 0.25$).

The overlap of home ranges of any two Maremmas belonging to the same social group was 76.2%±3.7% for the 50% isopleth range, and 61.9%±3.4% for the 95% isopleth range ($F_{1, 13} = 8.07$, P<0.05). Overlap of ranges of all Maremmas within one social group was 38.4%±6.8% for the 50% isopleth range, and 54.8%±6.6% for the 95% isopleth range ($F_{1, 13} = 9.92$, P<0.01). The amount of overlap between pair wise home range comparisons was not significantly influenced by dog age ($F_{1, 11} = 1.30$, P = 0.30), however, with increasing dog age the amount of overlap between ranges of all Maremmas within a social group increased ($F_{1, 11} = 7.15$, P<0.05). Property did not significantly influence overlap of either pair wise home range comparisons ($F_{1, 11} = 0.52$, P = 0.49), or of the whole group ($F_{1, 11} = 0.03$, P = 0.85). There were no significant differences in amount of overlap of pair wise home ranges between summer and winter on Riversdale and

Heatherlie 2012 ($F_{1, 15} = 2.87$, P = 0.11), although there was a trend for the overlap of ranges of the whole group to be larger in winter ($F_{1, 15} = 3.43$, P = 0.08).

Activity relative to home range location

Movement speed increased significantly with increasing distance from the centre of the home range for both Maremmas ($F_{9, 117} = 46.2$, P<0.01), and sheep ($F_{9, 63} = 19.16$, P<0.01). For Maremmas, overall movement speed decreased significantly with increasing age of the dog ($F_{1, 10} = 9.43$, P<0.05). This led to Maremmas on Gillingal Station having the lowest movement speed, followed by Riversdale and Heatherlie (Fig. 3), however, after accounting for the effect of age these differences between properties were not significant ($F_{2,10} = 1.53$, P = 0.26). No differences in movement speed for sheep were found between Riversdale and Heatherlie ($F_{1,6} = 3.82$, P = 0.1). Maremmas had higher speed of movement than sheep, and a greater increase in movement speed towards the edge of their home range ($F_{1, 10} = 6.07$, P<0.05) (Fig. 3). Maremmas' movement speed was on average 24.4±5.0 m/h higher in winter than in summer in all isopleth areas on both properties ($F_{1, 79} = 14.77$, P<0.01).

For Maremmas, tortuosity values decreased significantly towards the edge of the home range at all scales of analysis (0.1 km: $F_{2, 26} = 4.67$, P<0.05, 0.25 km: $F_{2, 26} = 16.58$, P<0.01, 0.5 km: $F_{2, 26} = 16.17$, P<0.01, 1 km: $F_{2, 18} = 16.44$, P<0.01) (Fig. 4). Tortuosity values in the 90%–100% and the 50%–90% kernel isopleth areas were lower than in the 50% kernel isopleth area, corresponding to straighter movement paths. In most cases the tortuosity value in the 90%–100% kernel isopleth area was also lower than in the 50%–90% area. For sheep this decrease was also found, but was significant only at three scales of analysis (0.1 km: $F_{2, 14} = 2.82$, P = 0.09, 0.25 km: $F_{2, 14} = 4.37$, P<0.05, 0.5 km: $F_{2, 14} = 5.48$, P<0.05, 1 km: $F_{2, 12} = 8.42$, P<0.01). Tortuosity values increased significantly with age for Maremmas at three scales of analysis (0.1 km: $F_{1, 10} = 30.02$, P<0.01, 0.25 km: $F_{1,10} = 33.09$, P<0.01, 0.5 km: $F_{1, 10} = 37.20$, P<0.01). At the 1 km scale of analysis there was no effect of age ($F_{1, 7} = 1.92$, P = 0.21), but 4 of the older dogs (>8 years) could not be included in this analysis as their movement paths within each isopleth area were never long enough to calculate a tortuosity value. After accounting for the effect of age, there were significant differences

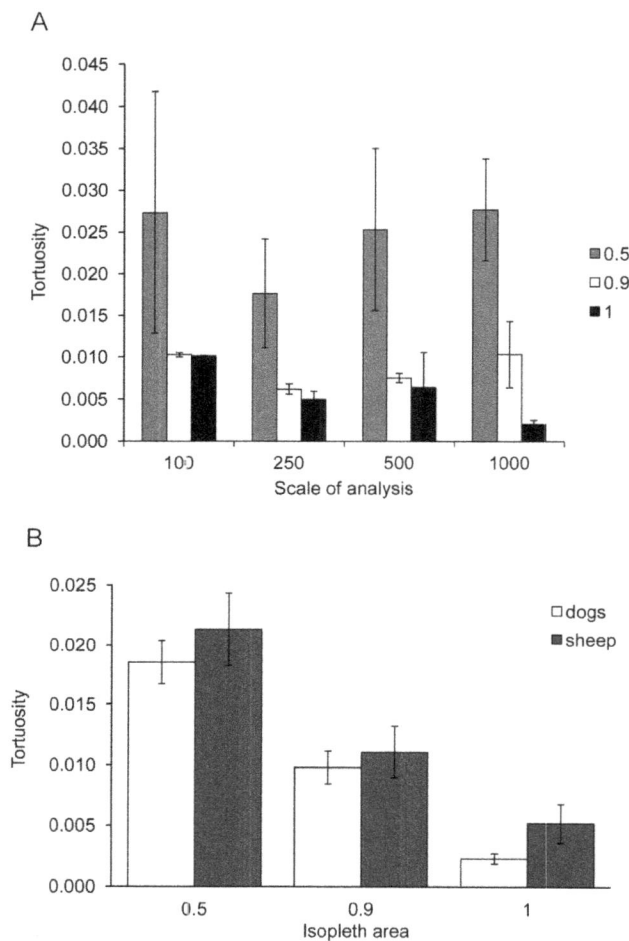

Figure 4. Mean median tortuosity values. A. Maremmas on the three properties within each kernel isopleth area for all scales of analysis (14 Maremmas in total). Tortuosity values decreased significantly towards the edge of the home range at all scales of analysis (0.1 km: $F_{2, 26} = 4.67$, P<0.05, 0.25 km: $F_{2, 26} = 16.58$, P<0.01, 0.5 km: $F_{2, 26} = 16.17$, P<0.01, 1 km: $F_{2, 18} = 16.44$, P<0.01). B. Maremmas compared to sheep when they were collared at the same time for the 1km scale of analysis (9 Maremmas and 8 sheep). The other scales of analysis showed a similar trend, the differences between sheep and Maremmas were not significant. Median values of tortuosity for individual dogs within each kernel isopleth area were only included in the overall calculation if the overall number of line segments used to calculate that median was equal to or greater than 10.

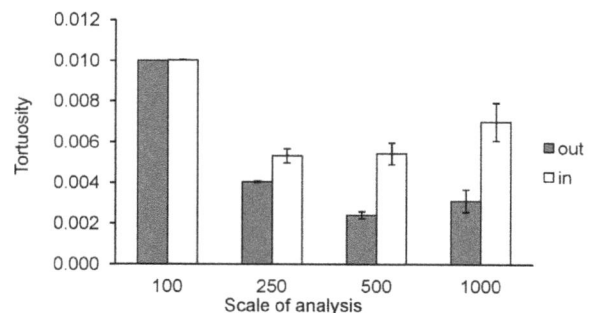

Figure 5. Median tortuosity values of movement paths of Maremmas inside and outside of livestock areas at four different scales of analysis (10 Maremmas in total). Inside livestock areas tortuosity values were significantly higher for all four scales of analysis (0.1 km: $F_{1, 9} = 11.25$, P<0.01; 0.25 km: $F_{1, 9} = 14.08$, P< 0.01; 0.5 km: $F_{1, 9} = 40.71$, P<0.01 and 1 km: $F_{1, 9} = 31.96$, P<0.01). Median values of tortuosity for individual dogs within each livestock area were only included in the overall calculation if the number of line segments that was used to calculate that median was equal to or greater than 10.

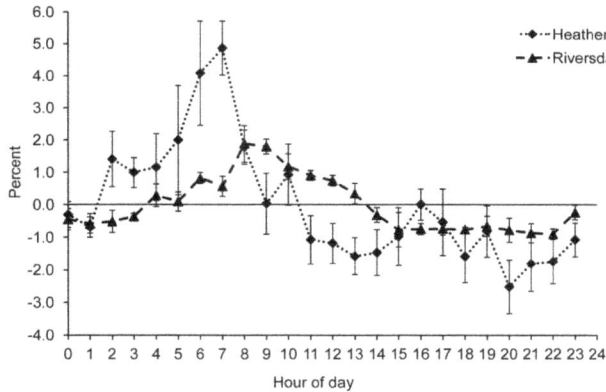

Figure 6. The time of the day at which Maremmas are more likely to leave their livestock (10 Maremmas in total). When the graph is positive, it indicates the times of the day that Maremmas are more likely to leave stock, when the graph is negative, it indicates the times of the day that Maremmas are less likely to leave stock. This number is calculated as follows: the percentage of the total number of locations for each category (with and without livestock) that fell within that hour of the day was calculated for each dog. Then for each hour, the percentage with livestock was subtracted from the percentage without livestock.

between properties in tortuosity values (0.1 km: $F_{2, 10} = 22.82$, $P<0.01$, 0.25 km: $F_{2, 10} = 8.40$, $P<0.01$, 0.5 km: $F_{2, 10} = 5.21$, $P<0.05$, 1 km: $F_{2, 7} = 1.92$, $P = 0.21$); the Maremmas on Gillingal Station had the highest values followed by Riversdale and then Heatherlie. There were no significant differences between Riversdale and Heatherlie in tortuosity values for sheep (0.1 km: $F_{1, 6} = 0.49$, $P = 0.51$, 0.25 km: $F_{1, 6} = 0.23$, $P = 0.65$, 0.5 km: $F_{1, 6} = 0.31$, $P = 0.60$, 1 km: $F_{1, 6} = 0.41$, $P = 0.55$). Tortuosity for sheep was higher than for Maremmas in all kernel isopleth areas at all scales of analyses (Fig. 4), but these differences were not significant (0.1 km: $F_{1, 14} = 2.18$, $P = 0.16$, 0.25 km: $F_{1, 14} = 1.26$, $P = 0.28$, 0.5 km: $F_{1, 14} = 0.15$, $P = 0.70$, 1 km: $F_{1, 11} = 3.05$, $P = 0.11$). No significant differences were found in tortuosity values for Maremmas between summer and winter on Riversdale and Heatherlie in 2012 (0.1 km: $F_{1, 23} = 0.45$, $P = 0.45$, 0.25 km: $F_{1, 23} = 0.12$, $P = 0.74$, 0.5 km: $F_{1, 20} = 0.58$, $P = 0.46$, 1 km: $F_{1, 20} = 0.11$, $P = 0.75$).

Association with livestock

Maremmas spent significantly more time in than outside of livestock paddocks ($F_{1,13} = 288.2$, $P<0.01$), spending an average of $91.3\% \pm 2.4\%$ of their time in livestock areas. Dog age did not significantly influence the time spent with livestock ($F_{1,10} = 0.01$, $P = 0.92$). On Gillingal station they spent 100% of their time in livestock areas, on Riversdale this was $85.6\% \pm 4.8\%$ and on Heatherlie this was $90.9\% \pm 2.5\%$; these differences were not significant ($F_{2,10} = 1.02$, $P = 0.40$). There were no significant difference in the proportion of time spent with sheep between summer and winter for Riversdale and Heatherlie in 2012 ($F_{1,17} = 0$, $P = 1$).

Outside sheep paddocks, Maremmas travelled faster than inside sheep areas: 300.1 m/hour ±42.2 m/hour vs. 192.5 m/hour ±18.7 m/hour ($F_{1, 9} = 13.57$, $P<0.01$). There was a trend for overall movement speed to decrease with Maremma age ($F_{1, 6} = 4.98$, $P = 0.06$), and Maremmas on Heatherlie travelled faster than the Maremmas on Riversdale ($F_{2, 6} = 9.81$, $P<0.05$). Speed of movement in sheep paddocks was not significantly different

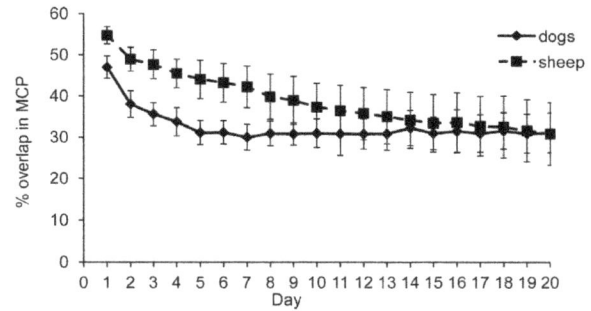

Figure 7. The percentage of overlap in MCP's in consecutive 20-day periods for Maremmas compared to sheep, the difference between the two species is not significant (9 Maremmas and 8 sheep). Both species were collared during the same time period. First the overlap between the first day of the series and each of the 19 consecutive days was calculated, after which the process was repeated by using the second day as the initial day, and calculating the overlap between that day and each of 19 consecutive days.

between summer and winter for Riversdale 2012 and Heatherlie 2012 ($F_{1,6} = 0.50$, $P = 0.50$). Speed of movement outside sheep paddocks was on average 61.5 ± 28.4 m/h higher in winter than in summer ($F_{1, 6} = 9.27$, $P<0.05$).

Outside sheep paddocks, path tortuosity values for Maremmas were lower than inside sheep paddocks, indicating that movement paths were straighter (Fig. 5). This difference was significant for all four scales of analysis (0.1 km: $F_{1, 9} = 11.25$, $P<0.01$; 0.25 km: $F_{1, 9} = 14.08$, $P<0.01$; 0.5 km: $F_{1, 9} = 40.71$, $P<0.01$ and 1 km: $F_{1, 9} = 31.96$, $P<0.01$). Tortuosity values were significantly higher with increasing age of the dog at two scales of analysis (0.1 km: $F_{1, 7} = 12.31$, $P<0.01$; 0.25 km: $F_{1, 7} = 5.29$, $P = 0.05$; 0.5 km: $F_{1, 7} = 2.11$, $P = 0.19$; 1 km: $F_{1, 7} = 3.02$, $P = 0.13$). After accounting for dog age, tortuosity values in and out of sheep paddocks did not differ significantly between properties (0.1 km: $F_{1, 7} = 0.01$, $P = 0.95$; 0.25 km: $F_{1, 7} = 0.21$, $P = 0.66$; 0.5 km: $F_{1, 7} = 0.01$, $P = 0.92$; 1 km: $F_{1, 7} = 0.09$, $P = 0.77$), and were not significantly different between summer and winter for Riversdale 2012 and Heatherlie 2012 (0.1 km: $F_{1, 15} = 0.79$, $P = 0.39$; 0.25 km: $F_{1, 13} = 0.35$, $P = 0.57$; 0.5 km: $F_{1, 13} = 0.48$, $P = 0.50$; 1 km: $F_{1, 13} = 0.38$, $P = 0.55$).

Maremmas were more likely to leave livestock in the early hours of the day (Fig. 6). There was no significant difference in this pattern between dogs of different ages ($F_{1, 7} = 1.30$, $P = 0.29$),

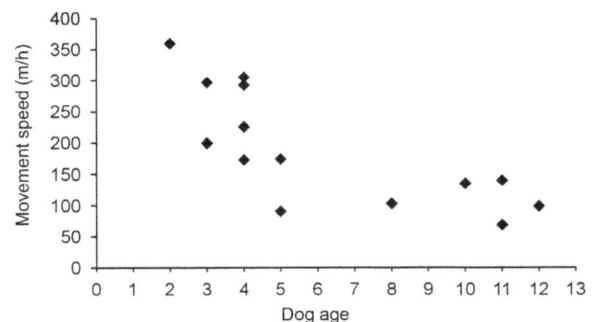

Figure 8. Relationship between age of the Maremma and movement speed (m/h). Each point represents the mean movement speed of an individual dog (14 Maremmas in total).

between properties ($F_{1, 7} = 0.04$, $P = 0.86$), nor between summer and winter for Riversdale and Heatherlie in 2012 ($F_{1, 167} = 0$, $P = 1$).

Sequential use of territory

For Maremmas, the average daily percentage overlap of the MCP of two consecutive days was $45.8\% \pm 2.4\%$. The amount of overlap in MCPs of consecutive days increased with the age of the dog ($F_{2, 10} = 23.40$, $P < 0.01$). After accounting for age of the dogs, there was a trend for the Maremmas on the different properties to have different MCP overlap values; on Gillingal Station daily overlap was $57.0\% \pm 4.2\%$, on Riversdale it was $38.2\% \pm 1.3\%$, and on Heatherlie it was $46.0\% \pm 2.4\%$ ($F_{2, 10} = 3.67$, $P = 0.06$). Average daily percentage overlap of the MCP did not significantly differ between summer and winter for Riversdale and Heatherlie in 2012 ($F_{1,7} = 0.82$, $P = 0.40$). For sheep the average was significantly higher than for dogs: $54.8\% \pm 2.1\%$ ($F_{1, 18} = 10.29$, $P < 0.01$). The sheep on Riversdale had on average 11.0% more daily overlap of MCP's than the sheep on Heatherlie ($F_{1, 6} = 113.1$, $P < 0.01$).

No consistent pattern of re-use of particular parts of the home range was found for either Maremmas or sheep over 20 day periods (Fig. 7). For both species the percentage of overlap between MCPs of each initial day with each of the following 19

days decreased with increasing number of days (Maremmas: $F_{19, 228} = 9.49$, $P < 0.01$, sheep: $F_{19, 113} = 14.57$, $P < 0.01$) (Fig. 7). For Maremmas, older dogs had significantly higher percentages of overlap in MCPs of each of the 19 consecutive days than younger dogs ($F_{1, 9} = 6.22$, $P < 0.05$). After accounting for age, there were no significant differences between properties ($F_{2, 9} = 1.48$, $P = 0.28$). The sheep on Riversdale on average had $35.8\% \pm 1.76\%$ more overlap in the MCPs of each of the 19 consecutive days than the sheep on Heatherlie ($F_{1, 6} = 196.4$, $P < 0.01$), due to the highly regular paddock rotation schedule on Heatherlie. There were no significant differences in the pattern of overlap of MCPs between Maremmas and sheep ($F_{1,14} = 0.54$, $P = 0.48$). However, if the sheep on Heatherlie are excluded, sheep have on average $20.0\% \pm 0.5\%$ more overlap in MCPs in each of the 19 consecutive days than Maremmas ($F_{1, 10} = 14.20$, $P < 0.01$). For Maremmas, the decrease in overlap of MCPs was also seen in summer and winter on Riversdale and Heatherlie in 2012, but there was a trend for summer to be different from winter ($F_{1, 139} = 3.64$, $P = 0.06$). This was mainly caused by the percentage overlap between day one and each consecutive day being higher in the last three days in summer compared to winter.

24-hour activity patterns

Average hourly movement speed of Maremmas significantly decreased with increasing dog age ($F_{1, 10} = 28.26$, $P < 0.01$) (Fig. 8). There were also significant differences between the properties ($F_{2, 10} = 5.47$, $P < 0.05$), being lowest on Gillingal Station (91.5 ± 11.4 m/hour), followed by Riversdale (165.4 ± 27.9 m/hour) and Heatherlie (245.9 ± 33.7 m/hour). In winter, Maremmas travelled on average 35.3 m/h more than in summer ($F_{1,7} = 10.84$, $P < 0.05$). Sheep on Riversdale had a significantly higher average movement speed than sheep on Heatherlie ($F_{1, 6} = 7.20$, $P < 0.05$); 150.6 ± 8.1 m/hour on Riversdale versus 115.7 ± 10.7 m/hour on Heatherlie. Maremmas had a significantly higher average movement speed than the sheep they were guarding ($F_{1, 14} = 7.93$, $P < 0.05$).

Figure 9 shows the 24-hour activity pattern of both Maremmas and sheep. Both species had a distinct early morning peak in activity after sunrise and a late afternoon peak in activity before sunset. The activity pattern was significantly different between the two species ($F_{1, 14} = 7.49$, $P < 0.05$). The morning activity peak began earlier for Maremmas than for sheep, and activity peaked later in the afternoon. In addition, Maremmas maintained a higher level of activity at night than sheep. The pattern of activity in a 24 hour period was very similar between different aged dogs, and between properties for the Maremmas, but with increasing age the mean activity per hour over the whole 24-hour period for dogs decreased ($F_{1, 10} = 28.26$, $P < 0.01$), and there were differences between the different properties ($F_{2, 10} = 5.47$, $P < 0.05$). The Maremmas on Gillingal station were least active, followed by Riversdale and then Heatherlie (Fig. 9). There was a significant difference in activity pattern between sheep on Riversdale and Heatherlie ($F_{1, 6} = 7.20$, $P < 0.05$), mainly due to a higher morning activity peak of the sheep on Heatherlie.

The 24-hour activity pattern was significantly different between summer and winter for the dogs on Riversdale and Heatherlie in 2012 ($F_{1,191} = 43.28$, $P < 0.05$). The timing of the morning and afternoon activity peaks shifted as sunrise and sunset times changed, and in addition, in winter the dogs had a slightly overall higher activity level than in summer.

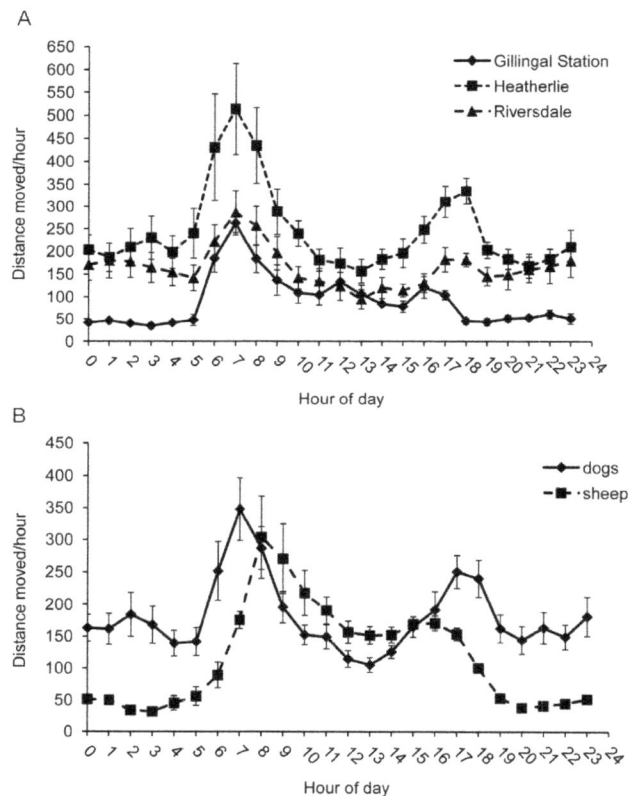

Figure 9. Average 24-hour activity patterns. (A) Maremmas on the three research properties (14 Maremmas in total). The overall level of activity was significantly different between the three properties $F_{2, 10} = 5.47$, $P < 0.05$. (B) Maremmas compared to sheep when both were collared during the same time period (9 Maremmas and 8 sheep). The activity pattern was significantly different between the two species ($F_{1, 14} = 7.49$, $P < 0.05$). The Y-axis represents the average distance moved per hour in a 24 hour period.

Discussion

Unsupervised free-ranging Maremmas spent the majority of their time in livestock areas, apparently choosing to remain in proximity to their livestock most of the time. However, movements away from livestock did occur, mainly at night. These movements were generally characterised by high-speed travel on relatively straight paths. These high-speed movements were related to the finding that movements become faster towards the edge of the range boundary, because the Maremmas' ranges encompassed the livestock areas, and therefore leaving the livestock paddocks entailed moving towards the range edge. Travel between livestock paddocks, and between livestock paddocks and the owners' homestead, accounted for a number of straight high-speed movements. In other cases, fast movements away from livestock could represent responses to predator incursions or perceived predator threats, or simply exploratory behaviour. Alternatively, the movements could be related to patrolling territorial boundaries. No published research has as yet provided clear evidence that LGDs are territorial, and might use territorial exclusion of predators to protect livestock. However, a number of studies have recorded the use of territorial signalling by LGDs, such as scent-marking, regular barking and boundary patrolling [29–33].

Movements of sheep also became faster and less tortuous towards range edges, but less so than for Maremmas. Overall, sheep movements were always more tortuous than dogs'. This probably reflects different motivations for movement between the two species. In the case of the sheep, water was plentiful on both properties, and movements were mainly governed by grazing and resting patterns. In the case of the dogs, food was provided by their owners and water was easily obtained. The movements in the core of their range in livestock areas probably reflect activities such as resting, play and following livestock. Movements towards the edge of their range could be motivated by a number of factors (as explained above).

In pair wise comparisons, Maremmas belonging to the same social group had a relatively high degree of overlap of 50% and 95% isopleth ranges, indicating that any two dogs within the same group share a large portion of their home range. The amount of overlap of the ranges of all members of the same group was lower, especially in the 50% isopleth. This indicates that members of the same social group, while sharing a large part of their home range with the other dogs in the group, do all occupy slightly different areas, especially at the core of their range. Therefore, the group as a whole occupies a larger area than each individual dog on their own, which could potentially increase their effectiveness for livestock protection. The relatively low percentage of overlap between areas used by the same dog on consecutive days points to sequential use of the home range by the Maremmas. In sheep, the amount of overlap between areas used on consecutive days was higher than for dogs, but only slightly. In sheep this behaviour is probably caused by their grazing patterns. Sequential use of the home range would probably optimise foraging for sheep, as the quality of areas recently grazed has declined and new areas offer a better alternative [34]. If the Maremmas consistently move with their sheep, this could explain a large part of the sequential use of the range by the dogs. In addition, for the dogs, traversing a different area of the range each day probably facilitates maintaining a presence throughout the whole range.

The activity patterns of Maremmas show that they are most active in the early morning and late afternoon, and maintain a relatively high level of activity during the night. This activity pattern matches that of the main predators in the areas where the dogs were working [35–38]. Compared to the Maremmas, the peaks of sheep activity were later in the morning and earlier in the afternoon, and sheep were relatively inactive all during the night. Consistent daily sheep activity patterns have also been documented in other studies [39–41]. Late start of grazing and earlier bedding down seems to be related to season [41], and is probably the result of this study taking place mostly in autumn and winter.

Dog age significantly influenced the general level of activity of the Maremmas, with Maremmas displaying less activity with increasing age. Older dogs moved less per hour, and their movement paths tended to be more tortuous. This decrease in activity likely leads to smaller home ranges, and therefore the decrease in home range size with increasing age of the dog that was observed in this study. This probably means that as a LGD gets older, the size of the area in which it can effectively protect livestock decreases. One way to counter this effect would be to have older LGDs working together with younger dogs. The older LGDs can teach the younger dogs the job [11], and as the younger dogs maintain a higher level of activity and have a larger home range, they still protect the livestock over the larger area that the older LGDs used to occupy.

Other than a higher overall level of activity in winter, and small differences in the patterns of re-use of territory, no differences were found between movements of Maremmas in summer and winter. The higher activity of the Maremmas in winter could be related to the breeding season of wild dogs and foxes, which can lead to an increase in incursions in livestock areas [14]. This would require a higher level of vigilance from the Maremmas, which could lead to higher activity. Other than that, it seems that seasonality has little effect on Maremma movements, other than a slight adjustment to different sunrise/sunset times. This is probably due to two main factors. First, food was provided by their owners, so foraging was not necessary, and water was plentiful on all properties throughout the year. Second, most Maremmas were desexed, and mating and breeding related behaviours did not occur in different seasons.

Proper management of pups, and later of adult dogs, is crucial for LGDs to develop the behaviour which causes them to choose to spend the majority of their time with livestock. Pups need to be thoroughly socialised to the species they need to protect later in life; without proper bonding to livestock as a pup, a mature LGD is unlikely to voluntarily remain with livestock unsupervised for any length of time [2,42,43]. In a free-range system, some movements away from livestock are always likely, no matter how well-bonded the dog is to its stock. However, van Bommel & Johnson [10] found that the management style of LGDs (free-ranging vs. restricted in movements, i.e. fence trained or not) did not influence how well LGDs were able to protect their livestock. On large Australian properties where wild dogs or dingoes are the main predator, free-ranging of LGDs is often considered by livestock managers to be the most effective management strategy, as LGDs are able to provide each other with backup in case of a predator attack [10,44]. This would not be possible if LGD movements were restricted. Territoriality could explain why LGDs can still be effective in protecting stock even though they do not spend all their time with them. If LGDs set up a territory around the livestock, and predators are deterred by those territorial boundaries, then the LGDs continuous presence with livestock is not required to keep the stock safe. On the contrary, some movement away from stock is actually necessary for the LGDs to patrol and maintain territorial boundaries. More research is needed to investigate if or to what extend LGDs work through territorial exclusion of predators.

When producers are faced with the management decision of whether to have their LGDs free-ranging or to restrict their movements, the potential size of the range that the LGDs will use should be taken into consideration, and compared to the size of the property that the dog is meant to work on. We found that 95% kernel isopleth ranges of LGDs measured up to 1161 ha. The size of the range is likely influenced by many variables, and will sometimes be smaller, but sometimes also larger. One factor that is potentially important in determining the size of the dogs' range is the combined size of the ranges of the different groups of livestock they are guarding. However, regardless of the size of the area that the livestock use, LGDs are likely to have a larger range than the stock. In this study the boundaries of the Maremmas' ranges extended up to 2 km beyond the boundaries of the ranges of the stock, and in the case of the Maremmas on Riversdale, the dogs even chose to include the sheep on a neighbouring property in their range. These factors should all be taken into consideration when deciding if a LGD can operate in a free-range system. If a property is smaller than the potential size of the range of the dogs, free-ranging might not be the best management decision, and the choice can be made for example to fence train the LGD, or to upgrade boundary fences to prevent the dog leaving the property. Roaming by dogs outside their property boundaries could lead to problems such as traffic accidents, concerned neighbours, or conflicts with local councils, as in most Australian states dogs are required by law to be contained within their owners' property if not accompanied by a human. However, if the property can accommodate the range that the LDGs will use, free-ranging can be an effective management system, allowing the LGDs to maintain a presence, and deter predators, throughout and around the livestock areas.

Acknowledgments

We thank the owners of Gillingal Station, Riversdale and Heatherlie for generously allowing us to use their Maremmas and livestock in this research, and for allowing free access to their properties. We also thank Bob Forester and Dale Richardson for help with the analysis of the data.

Author Contributions

Conceived and designed the experiments: LVB CNJ. Performed the experiments: LVB. Analyzed the data: LVB. Contributed reagents/materials/analysis tools: LVB CNJ. Wrote the paper: LVB CNJ.

References

1. Coppinger L, Coppinger R (2007) Dogs for Herding and Guarding Livestock. In: Granding T, editor. Livestock handling and transport. 3rd ed. Oxford, England: CABI International. pp. 199–213.
2. Coppinger R, Coppinger L (2001) Dogs: A New Understanding of Canine Origin, Behaviour and Evolution. New York: University of Chicago Press.
3. Coppinger R, Lorenz J, Coppinger L (1983) Introducing Livestock Guarding Dogs to Sheep and Goat Producers. In: Decker DJ, editor. Proceedings of the 1st Eastern Wildlife Damage Control Conference. Ithaca, New York: Cornell University, Ithaca, NY. pp. 129–132.
4. Rigg R (2001) Livestock guarding dogs: their current use world wide. IUCN/SSC Canid Specialist Group.
5. Rigg R, Findo S, Wechselberger M, Gorman ML, Sillero-Zubiri C, et al. (2011) Mitigating carnivore–livestock conflict in Europe: lessons from Slovakia. Oryx 45: 272–280.
6. Gehring T, VerCauteren K, Provost M, Cellar A (2010) Utility of livestock-protection dogs for deterring wildlife from cattle farms. Wildlife Research 37: 715–721.
7. Hansen I, Staaland T, Ringsø A (2002) Patrolling with Livestock Guard Dogs: A Potential Method to Reduce Predation on Sheep. Acta Agriculturae Scandinavica, Section A – Animal Sciences 52: 43–48.
8. Marker LL, Dickman AJ, Macdonald DW (2005) Perceived Effectiveness of Livestock-Guarding Dogs Placed on Namibian Farms. Rangeland Ecology & Management 58: 329–336.
9. Otstavel T, Vuoric K, Simsd D, Valrosa A, Vainioe O, et al. (2009) The first experience of livestock guarding dogs preventing large carnivore damages in Finland. Estonian Journal of Ecology 58: 216–224.
10. van Bommel L, Johnson CN (2012) Good dog! Using livestock guardian dogs to protect livestock from predators in Australia's extensive grazing systems. Wildlife Research 39: 220–229.
11. van Bommel L (2010) Guardian Dogs: Best Practice Manual for the use of Livestock Guardian Dogs. Invasive Animals CRC, Canberra.
12. Gehring TM, VerCauteren KC, Cellar AC (2011) Good fences make good neighbors: implementation of electric fencing for establishing effective livestock-protection dogs. Human-Wildlife Interactions 5: 106–111.
13. Gingold G, Yom Tov Y, Kronfeld Schor N, Geffen E (2009) Effect of guard dogs on the behavior and reproduction of gazelles in cattle enclosures on the Golan Heights. Animal Conservation 12: 155–162.
14. Fleming PJS, Korn TJ (1989) Predation of Livestock by Wild Dogs in Eastern New South Wales. Australian Rangeland Journal 11: 61–66.
15. van Bommel L, Johnson CN (2014) Movements of free-ranging Maremma sheepdogs. Data from: Where do livestock guardian dogs go? Movement patterns of free-ranging Maremma sheepdogs. Movebank Data Repository (movebank.org). DOI 10.5441/001/1.pv048q7v.
16. Recio MR, Mathieu R, Denys P, Sirguey P, Seddon PJ (2011) Lightweight GPS-tags, one giant leap for wildlife tracking? An assessment approach. PLoS One 6: e28225.
17. Worton BJ (1989) Kernel Methods for Estimating the Utilization Distribution in Home-Range Studies. Ecology 70: 164–168.
18. De Solla S, Bondurianksy R, Brooks R (1999) Eliminating autocorrelation reduces biological relevance of home range estimates. Journal of Animal Ecology 68: 221–234.
19. Fieberg J (2007) Kernel Density Estimators of Home Range: Smoothing and the Autocorrelation Red Herring. Ecology 88: 1059–1066.
20. Reynolds TD, Laundre JW (1990) Time Intervals for Estimating Pronghorn and Coyote Home Ranges and Daily Movements. Journal of Wildlife Management 54: 316–322.
21. Berger K, Gese E (2007) Does interference competition with wolves limit the distribution and abundance of coyotes? Journal of Animal Ecology 76: 1075–1085.
22. Kie J, Matthiopoulos J, Fieberg J, Powell R, Cagnacci F, et al. (2010) The home-range concept: are traditional estimators still relevant with modern telemetry technology? Philosophical Transactions of the Royal Society B: Biological Sciences 365: 2221–2231.
23. Jacques C, Jenks J, Klaver R (2009) Seasonal movements and home-range use by female pronghorns in sagebrush-steppe communities of western South Dakota. Journal of Mammalogy 90: 433–441.
24. Turchin P (1998) Quantitative analysis of movement: measuring and modeling population redistribution of animals and plants: Sinauer Associates, Sunderland, Massachusetts, USA.
25. Demma DJ, Mech LD (2009) Wolf use of summer territory in northeastern Minnesota. The Journal of Wildlife Management 73: 380–384.
26. White GC, Garrott RA (1990) Analysis of wildlife radio-tracking data: Academic press.
27. ESRI (2011) ArcGIS Desktop: Release 10. Redlands, CA: Environmental Systems Research Institute.
28. R Development Core Team (2008) R: A language and environment for statistical computing. R Foundation for Statistical Computing, Vienna, Austria. Available: http://www.R-project.org.
29. Linhart SB, Sterner RT, Carrigan TC, Henne DR (1979) Komondor Guard Dogs Reduce Sheep Losses to Coyotes: A Preliminary Evaluation. Journal of Range Management 32: 238–241.
30. Black HL, Green JS (1985) Navajo Use of Mixed-breed Dogs for Management of Predators. Journal of Range Management 38: 11–15.
31. Hansen I, Smith ME (1999) Livestock-guarding dogs in Norway Part II: Different working regimes. Journal of Range Management 52: 312–316.
32. Green JS, Woodruff RA (1983) The Use of Three Breeds of Dog to Protect Rangeland Sheep from Predators. Applied Animal Ethology 11: 141–161.
33. McGrew JC, Blakesley CS (1982) How Komondor Dogs Reduce Sheep Losses to Coyotes. Journal of Range Management 35: 693–696.
34. Bailey DW, Gross JE, Laca EA, Rittenhouse LR, Coughenour MB, et al. (2006) Mechanisms that result in large herbivore grazing distribution patterns. Journal of Range Management Archives 49: 386–400.
35. Brook LA, Johnson CN, Ritchie EG (2012) Effects of predator control on behaviour of an apex predator and indirect consequences for mesopredator suppression. Journal of applied ecology 49: 1278–1286.
36. Thomson PC (1992) The Behavioural Ecology of Dingoes in North-western Australia. II. Activity Patterns, Breeding Season and Pup Rearing. Wildlife Research 19: 519–530.
37. Ables ED (1969) Activity studies of red foxes in southern Wisconsin. The Journal of Wildlife Management: 145–153.
38. Phillips M, Catling P (1991) Home range and activity patterns of red foxes in Nadgee Nature Reserve. Wildlife Research 18: 677–686.
39. Warren JT, Mysterud I (1991) Summer habitat use and activity patterns of domestic sheep on coniferous forest range in southern Norway. Journal of Range Management 44: 2–6.
40. Squires V (1974) Grazing distribution and activity patterns of Merino sheep on a saltbush community in south-east Australia. Applied Animal Ethology 1: 17–30.

41. Bowns JE (1971) Sheep behavior under unherded conditions on mountain summer ranges. Journal of Range Management: 105–109.

42. Sims DE, Dawydiak O (2004) Livestock Protection Dogs: Selection, Care, and Training: Alpine Blue Ribbon Books, Loveland, CO, USA.

43. Green JS, Woodruff RA (1990) ADC Guarding Dog Program Update: A Focus on Managing Dogs. In: Davies LR, Marsh RE, editors. Proceedings of the 14th Vertebrate Pest Conference. Sacramento, California: University of California, Davis, CA. pp. 233–236.

44. van Bommel L (2010) Guardian Dogs: Best Practice Manual for the use of Livestock Guardian Dogs. Invasive Animals CRC, Canberra.

3

Priming of Soil Carbon Decomposition in Two Inner Mongolia Grassland Soils following Sheep Dung Addition: A Study Using ^{13}C Natural Abundance Approach

Xiuzhi Ma[1], Per Ambus[2], Shiping Wang[3]*, Yanfen Wang[4], Chengjie Wang[5]

1 College of Forestry, Inner Mongolia Agricultural University, Huhhot, China, **2** Department of Chemical and Biochemical Engineering, Technical University of Denmark, Lyngby, Denmark, **3** Laboratory of Alpine Ecology and Biodiversity, Institute of Tibetan Plateau Research, Chinese Academy of Sciences, Beijing, China, **4** Department of Life College, University of Chinese Academy of Sciences, Beijing, China, **5** College of Ecology and Environmental Science, Inner Mongolia Agricultural University, Huhhot, China

Abstract

To investigate the effect of sheep dung on soil carbon (C) sequestration, a 152 days incubation experiment was conducted with soils from two different Inner Mongolian grasslands, *i.e.* a *Leymus chinensis* dominated grassland representing the climax community (2.1% organic matter content) and a heavily degraded *Artemisia frigida* dominated community (1.3% organic matter content). Dung was collected from sheep either fed on *L. chinensis* (C$_3$ plant with δ^{13}C = −26.8‰; dung δ^{13}C = −26.2‰) or *Cleistogenes squarrosa* (C$_4$ plant with δ^{13}C = −14.6‰; dung δ^{13}C = −15.7‰). Fresh C$_3$ and C$_4$ sheep dung was mixed with the two grassland soils and incubated under controlled conditions for analysis of ^{13}C-CO$_2$ emissions. Soil samples were taken at days 17, 43, 86, 127 and 152 after sheep dung addition to detect the δ^{13}C signal in soil and dung components. Analysis revealed that 16.9% and 16.6% of the sheep dung C had decomposed, of which 3.5% and 2.8% was sequestrated in the soils of *L. chinensis* and *A. frigida* grasslands, respectively, while the remaining decomposed sheep dung was emitted as CO$_2$. The cumulative amounts of C respired from dung treated soils during 152 days were 7–8 times higher than in the un-amended controls. In both grassland soils, ca. 60% of the evolved CO$_2$ originated from the decomposing sheep dung and 40% from the native soil C. Priming effects of soil C decomposition were observed in both soils, *i.e.* 1.4 g and 1.6 g additional soil C kg^{-1} dry soil had been emitted as CO$_2$ for the *L. chinensis* and *A. frigida* soils, respectively. Hence, the net C losses from *L. chinensis* and *A. frigida* soils were 0.6 g and 0.9 g C kg^{-1} soil, which was 2.6% and 7.0% of the total C in *L. chinensis* and *A. frigida* grasslands soils, respectively. Our results suggest that grazing of degraded Inner Mongolian pastures may cause a net soil C loss due to the positive priming effect, thereby accelerating soil deterioration.

Editor: Dafeng Hui, Tennessee State University, United States of America

Funding: The work was supported by the National Basic Research Program (2010CB833502) and Chinese National Natural Science Foundation (97711001, 31160117, 31260119). The funders had no role in study design, data collection and analysis, decision to publish, or preparation of the manuscript.

Competing Interests: The authors have declared that no competing interests exist.

* E-mail: wangship2008@yahoo.cn

Introduction

The availability of soil organic carbon (C) for microbial decomposition is crucial for many processes within the C cycle, and assessment of soil dynamics is of great concern in terms of climate change and soil fertility [1]. Animal dung returned to soil can constitute important source of C, and maintain long-term soil fertility in grassland ecosystems [2–4]. However, dung application can also potentially increase soil respiration [5–8]. Studies have shown that the addition of easily degradable C to soil may stimulate microbial activity to such an extent that the turnover of soil organic matter (SOM) is accelerated temporarily, an effect that is frequently called the priming effect (PE) [3,9–12]. When a positive PE occurs, the addition of material such as animal slurry to soils may not result in a net C sequestration, but rather a net C loss [9]. The intensity, direction, and extent of PE depends on several parameters, including the amount and quality of added C,

soil microbial activity and community structure [13–15], soil pH [16] and aggregate size [17].

Distinct signatures in ^{13}C content between native soil C and 'new' introduced labile C compounds added in the form of animal slurry or manure enables quantification of the interaction in C turnover between different C pools [3,18–20]. By this means, researchers have shown that incorporated slurry-C was lost twice as fast as the native soil C in two soils with different C contents. Slurry incorporation induced a PE, which was most pronounced in the soil with the highest C content [18]. Following the application of slurries with different particle sizes to a grassland soil, significant increases of soil CO$_2$ effluxes (by 2–8 times) were observed in all slurry fractions and the highest was found in the smaller slurry particles [3]. In a study with additions of different substrate quality combinations and C-13 characteristics, Kuzyakov & Bol (2004) have identified three distinct C sources for soil CO$_2$ emissions and observed that addition of labile C (sugar) lead to changes in SOM

[21]. The ^{13}C natural abundance trace technique has also been applied in the two-phase model of CO_2 emission after dairy or pig slurry application, the first phase (0–48 h) dominated by the incorporation of labile slurry C from the liquid phase, while beyond 48 h slurry-derived C was mainly from less mobile particulate C [12,18,22–23]. However, whereas previous studies mainly focused on the decomposition of cattle dung or slurry [3,5,6,24] or pig slurry [25], less information is available concerning decomposition of sheep faeces [26].

Inner Mongolia's grasslands in Northern China are representative of large areas of the Eurasian steppe belt [27]. Sheep is the primary livestock in Inner Mongolia's grassland and large amount of sheep dung is applied as fertilizer except for cooking energy [28]. More than 70% of the natural Inner Mongolian grassland area is extensively degenerated as a consequence of increases in livestock numbers and the change of farming systems during the last three decades [28]. Knowledge regarding the importance of sheep dung excretion for soil C cycling in this ecosystem remains sparse. Addition of artificial sheep excreta to Inner Mongolian steppe in autumn did not impact soil microbial biomass C, but microbial activity significantly increased [29–30]. None of these studies, however, have achieved detailed information about the fates of sheep dung derived C in the Inner Mongolian grassland soils, and to what extent heavy grazing by sheep, and thus deposition of dung C, will affect the overall soil C balance through increased sequestration or losses due to priming.

We hypothesized that sheep dung additions to Inner Mongolian grassland soils i) cause a positive priming effect on soil C turnover, and ii) lead to differentiated net C loss in soils with contrasting SOM content. The effects of interaction between soil type and sheep dung addition on soil respiration and soil C sequestration were investigated using the ^{13}C natural abundance technique through a five-month incubation experiment.

The objectives of the study were (1) to assess the input of sheep dung-derived C to the soil C-pool; (2) to examine the extent of SOM priming due to application of dung C; and (3) to quantify the net C pool changes in soils subject to intensive sheep dung application.

Materials and Methods

Ethics Statement

There was no activities involved the endangered or protected species in this study, and Sheep (vertebrates) were involved in this study, so the permission to use sheep in this study was granted from local authority (Dr. Yongfei Bai, Director of Inner Mongolia Grassland Research Station) before we took the soil sample in the field.

Field site details

Soil incubated in the study was taken from the Inner Mongolia Grassland Ecosystem Research Station, Chinese Ecosystem Research Network (IMGERS, 116°42′E, 43°38′N). This region is part of the temperate semiarid steppe belt of Eurasia. The mean annual temperature is slightly above zero (0.8°C) with a January mean of −21°C (absolute minimum −42°C) and a July mean of 19°C (absolute maximum 39°C). Mean annual precipitation is 330.3 mm but fluctuates greatly among years, and most rainfall events occur in July and August (both means for years 1982–2007; IMGERS weather data). The annual frost-free period generally lasts 90–110 days [31].

Field sampling, dung collection and preparation

The soil was collected from two grassland sites. One site was characterized by *Leymus chinensis* vegetation (C_3 plant), which is a dominant species of the climax grassland community in the Inner Mongolian steppe [32]. Under moderate grazing, *L. chinensis* vegetation is replaced by *Cleistogenes squarrosa* (C_4 plant), which in turn is replaced finally by *Artemisia frigida* (C_3 plant) vegetation under heavy grazing [33–34]. About 87% of plants in this geographical region possess a C_3 photosynthetic pathway [33], and consequently the ^{13}C isotopic signature in soil organic carbon reflects a C_3 dominated community (Table 1). Further details on the site conditions are given by Chen & Wang [32].

Soils were collected from 0–10 cm depth in the two grasslands. Five 10×10 m^2 plots situated 2 m apart were sampled randomly using a 6 cm diameter auger to achieve ca. 10 kg of soil from each plot. Soil was composited into one bulk sample, vegetation and coarse roots were removed by hand, and the soil was sieved to pass a 2 mm mesh and stored at <5°C under field moist conditions until it was used.

Sheep dung was collected from twenty Mongolian sheep (two-year old), housed in metabolic cages with the approval of the Chinese Experimental Animal Committee of the Chinese Academy of Sciences and the owner of sheep. Dung collection did not influence sheep feeding but limited their activities freely in successive ten days by the metabolic cages. The sheep were divided randomly into two groups (ten for each group), and one group was fed on *L. chinensis*, while another group was fed on *C. squarrosa*. After 5 days trial to allow for equilibration of the ^{13}C content in the digestive tract, dung was collected twice per day in plastic bags which were attached to the tails of the sheep. Dung samples were collected during the following five consecutive days. The dung was kept frozen (−20°C) until used. The difference of the C_3 and C_4 sheep dung are given in Table 2. There was no significant different of C, N content and C/N ratios for C_3 and C_4 sheep dung.

Experimental setup

The incubation experiment was conducted over a 5-month period. Before the start of incubation, soil was mixed thoroughly and the soil moisture was adjusted with demineralized water to 40% of water-holding capacity (WHC). The sheep dung was thawed and homogenized by hand before being mixed into the soil. The incubation experiment included six treatments, i.e. the full combination of the two soils and three applications of sheep dung (C_3, C_4 and no addition). Dung was added as 60 g fresh weight portions mixed thoroughly with 1 kg of soil (air-dried equivalent). The soil-dung mixtures were transferred to 2 l Kilner jars in triplicate that were gently tapped on the lab bench to compress the soil. Two sets of jars were prepared, one for gas sampling and one for soil sampling. To minimize water losses from the soils, jars were covered with perforated Parafilm that was only removed 30 min before gas sampling events. The jars were incubated at 20±1°C in a controlled temperature cabinet throughout the 152 days of the experiment. Water content was held constant by regular watering to weight.

Headspace sampling for analysis of CO_2 flux and δ^{13} of CO_2

Samples for CO_2 efflux and d^{13}C isotopic analysis were collected 16 times on days 1, 2, 4, 6, 9, 14, 19, 24, 41, 55, 71, 83, 100, 121, 137 and 152 after sheep dung amendment. The Kilner jars were sealed gas tight by lids equipped with a rubber septum to allow headspace gas to be sampled by syringe and

Table 1. Physical and chemical characteristics of the two grassland soils used in the incubation study.

Soil	L. chinensis **soil**	A. frigida **soil**
Dominant vegetation	Leymus chinensis, Stipa grandis	Artemisia frigida, Cleistogenes squarrosa
Soil type	Dark chestnut	Light chestnut
Soil texture	Silty loam	Sandy
Soil organic matter (SOM) (%)	2.1±0.06	1.3±0.05
Soil total N content (g kg^{-1})	1.9±0.1	1.3±0.1
Soil microbial biomass C (mg/kg^{-1})	390±13	251±17
pH (water: soil = 2.5:1)	6.3±0.02	6.6±0.04
δ^{13} value (‰ vs VPDB)	−22.2±0.1	−22.4±0.8

Numbers are mean ± SE of n = 5 replicate bulk soil samples.

needle. At each gas sampling event, the headspace was sealed for 30 min and four 10-ml headspace samples were collected every 10 min. The sampling involved a three-step procedure. First, the headspace gas was mixed with a 20-ml sampling syringe several times. Second, a 10-ml gas sample was extracted from the headspace and 5 ml used to pressurize an evacuated 2-ml crimp-sealed vial for the ^{13}C isotopic analysis of CO_2. The residual 5 ml gas sample was analyzed immediately for CO_2 concentration by gas chromatography using a HP 6890 GC equipped with a Chromosorb 101 column (30°C), He carrier gas and Thermal Conductivity Detection. Gas fluxes were calculated from the change in CO_2 concentration inside the Kilner jars over the 30 min enclosure period. The relationship between CO_2 concentrations vs. time was significantly linear ($R^2 = 0.93$). Flux rates were thus calculated from the slope of the linear regression lines and expressed as mg C kg^{-1} soil (DM) day^{-1}.

The ^{13}C of CO_2 stored in the 2-ml pressurized vials was determined within 1 week. We used a PreCon (Thermo Scientific, Bremen, Germany) trace gas preparation-concentration unit coupled in continuous flow mode to a Delta PLUS isotope ratio mass spectrometer (IRMS, Thermo Scientific). As laboratory standard we used commercial CO_2 which had been calibrated against certified $^{13}CO_2$ standards (Messer Griesheim, Krefeld, Germany). Samples of the certified standards were also included in each batch of analysis. Results relating to ^{13}C characteristics are reported as ‰ vs. VPDB [35].

Soil sampling and analyses

Soil samples were collected from the jars at days 17, 43, 86, 127 and 152 after dung addition using a 2 cm diameter polyethylene pipe (15 cm long). The subsamples were mixed and the visible small sheep dung was sought out. The larger sheep dung particles were carefully removed by tweezers and the smaller fractions of sheep residues in the soil was absorbed by electrostatic effect, which produced by a polyethylene bottle rubbing against a piece of fabric [36].

Soil samples were weighed in Ag-foil capsules, arranged on a microtiter plate, wetted with water to approximately field capacity, and placed in a desiccator containing a beaker with concentrated (12M) HCl. The carbonates are released as CO_2 by the acid treatment in 6 to 8 h. The soil samples are then dried at 60°C prior to isotope determination [37]. The soil was then finely ground by a ball mill, and a ca. 30 mg subsample was weighed into a tin combustion cup for determination of total carbon and ^{13}C:^{12}C ratio following flash combustion on an elemental analyzer (EA 1110, CE Instruments, Milan, Italy) coupled in continuous flow mode to the IRMS.

Calculations

In this study, we assumed that the C_3 and C_4 sheep dung materials go through the same transformation and transport processes, so we can differentiate the carbon source both in soil and CO_2 efflux based on the different isotope value.

We calculated the percentage of dung-derived C in relation to the total soil C according to equation (1):

$$D = \frac{\delta_4 S - \delta_3 S}{\delta_4 d - \delta_3 d} \times 100 \tag{1}$$

where $\delta_4 S$ and $\delta_3 S$ are the δ^{13}C isotope values of soils amended with C_4 or C_3 dung at the time of sampling, and $\delta_4 d$ and $\delta_3 d$ are the δ^{13}C isotope values of the original C_4 and C_3 dung prior to amendment [18,38]. The difference in δ^{13}C between the C_3 and C_4 dung in our incubation was 10.5‰ (Table 2).

Table 2. Characteristics of sheep dung used in the incubation study. Dung was collected from sheep either fed on L. chinensis (C_3 dung) or C. squarrosa (C_4 dung).

Dung type	Dry matter	Organic C	Total N	C/N	Dry matter	Total C	δ^{13} value
	(% w:w)	(% of DM)			(Per 60 g fresh material) [a]		(‰ vs VPDB)
C_3 dung	86.1±0.4	43.9±0.3[b]	1.4±0.08	31.3±0.1	51.7±0.4	22.7±0.15	−26.2±0.04
C_4 dung	85.1±0.6	44.4±0.2	1.29±0.03	34.4±0.1	51.1±0.2	22.7±0.18	−15.7±0.06

[a]The content based on 60g fresh dung portions as applied in the experiment.
[b]Numbers are mean ± SE of n = 3 replicates.

The fractions of dung-derived C incorporated in the soil at the sampling time in relation to the total dung C applied was calculated with equation (2):

$$P = \frac{D \times SDW \times SC \times 100}{DC} \qquad (2)$$

where SDW is soil dry weight, SC is the content of soil organic carbon and DC is the amount of dung C added into soil at the beginning of the incubation.

The difference in $\delta^{13}C$ values between the respired CO_2 from the C_4 and C_3 dung treatments was used to quantify the proportions of dung versus soil-derived CO_2-C respired from the soil. The dung-derived CO_2 was calculated with equation (3):

$$R = \frac{\delta_4 AS - \delta_3 AS}{\delta_4 d - \delta_3 d} \times 100 \times CO_{2C4} \qquad (3)$$

where $\delta_4 AS$-$\delta_3 AS$ is the difference in $\delta^{13}C$ values of CO_2 emitted from C_4 dung and C_3 dung treatments, and CO_{2C4} is the flux of CO_2 in the C_4 dung treatment. The approach assumes that any fractionation in ^{13}C vs. ^{12}C during respiration of sheep dung C and soil C is similar for the C_3 and C_4 dung.

Statistical analyses

Repeated Measures Define Factors of General Linear Model (SPSS 13.0, SPSS Inc. Chicago, Illinois, USA) was used to assess the impacts of treatment, sampling day, soil type and their interactions on the $\delta^{13}C$ values of soils and CO_2 emission, the CO_2 respired efflux, dung-derived carbon, soil-derived carbon and the primed carbon effect. The sampling day was treated as within-subject variables, and soil type and the treatment were used as a between-subject variable. For each observation of CO_2 emission, cumulative CO_2 emission, $\delta^{13}C$ of CO_2 emission, dung-derived carbon and soil-derived carbon, the significance of differences between treatments was assessed by two-way ANOVA and Least Significance Difference (LSD).

Results

Soil carbon derived from sheep dung

Independent of soil type, soil $\delta^{13}C$ did not differ (p>0.05) between the control soil and the C_3 dung amended soil throughout the 152 days period (Table 3). In the C_4 dung soils, $\delta^{13}C$ significantly exceeded the C_3 and control treatments from day 127 onwards in the *L. chinensis* soil, and from day 43 onwards in the *A. frigida* soil. The temporal incorporation of C_4 sheep dung increased soil $\delta^{13}C$ values by 0.76‰ from day 17 to day 152 in the *L. chinensis* soil, and by 0.46‰ in the *A. frigida* soil (Table 3).

An increasing amount of dung-derived C appeared in the soil C fractions for both soil types, except for the *L. chinensis* soil at day 86. After 152 days of incubation, about 3.8% and 4.9% of total organic soil C was derived from the applied dung in the *L. chinensis* and *A. frigida* soils, respectively. This was equivalent to 3.5% and 2.8% of the total applied sheep dung C (Table 3).

Daily and cumulative CO_2 fluxes

The addition of sheep dung to the soil significantly increased CO_2 flux throughout the experiment in both soils when compared with the control ($P<0.05$; Fig. 1 A, B). There was no difference in CO_2 emission between the C_3 dung and C_4 dung treatments for the two soils, except on days 24 and days 100 in the *L. chinensis* soil, where C_3 dung amended soil emitted most CO_2. The two control soils used in the study showed almost uniform CO_2 emission

patterns, maintaining a constant rate (average 3.2 mg C kg^{-1} soil day^{-1}) except for the initial increase in CO_2 flux on days 1–3 which probably resulted from the soil wetting event. A two-phase pattern of soil CO_2 emission was found in the incubation study for both soils. The first phase was observed during 0–55 days after sheep dung amendment, during which CO_2 fluxes decreased to 26% and 43% of the initial CO_2 emission in *L. chinensis* and *A. frigida* soil, respectively. Then a second peak of CO_2 occurred after 55 days, decreasing again after ca. 100 days (*L. chinensis* soil) and 71 days (*A. frigida* soil) (Fig. 1 A, B). There was no interactive effect between soil type and dung treatment during the incubation.

The cumulative CO_2 losses from sheep dung amended soils were ca. 7–8 times higher than in the control soils after 152 days ($P<0.01$, Fig. 1 C, D). However, there was no difference in total CO_2 emission between the C_3 and C_4 sheep dung treatments ($P>0.05$, Fig. 1 C, D).

Isotopic characteristics of emitted CO_2

Slightly higher $\delta^{13}C$ values of CO_2 were found in all the soils included in the two control soils at the beginning of the experiment. These increased towards a peak value at day 6, and then decreased to a minimum asymptotic value in all treatments at day 14 (Fig. 2). Generally, there was no difference in ^{13}C-CO_2 between the C_3 dung treatment and control, except at days 55 and 121 for *A. frigida* soil ($P>0.05$). However, the $\delta^{13}C$ value of CO_2 from the C_4 dung treated soils was significantly higher than that in the control soil in most occasions ($P<0.05$). For the C_4 dung, C_3 dung, and control treatments, respectively, average $\delta^{13}C$ of CO_2 emissions during 152 days incubation were −14.6‰, −20.7‰, and −24.3‰ for *L. chinensis* soil, and −15.2‰, −20.5‰, and −24.0‰ for *A. frigida* soil (Fig. 2).

CO_2 emission sources and primed CO_2 emission in dung amended soil

The simple method of estimating the contribution of dung-derived C in respired CO_2 is only valid when the respiration rates from the C_3 and C_4 dung treated soils are the same[38], as was the case in the current study. Two peaks of dung-derived C were observed in both soils (Fig. 3). As a proportion of the total CO_2-C respired from the dung treated *L. chinensis* soil, during the first 24 days of the experiment, the dung-derived C increased from 11.5% (day 2) to 90.9% (day 24), and from the dung treated *A. frigida* soil increased from 17.9% (day 2) to 91.0% (day 24) (Fig. 3). The cumulative amount of CO_2-C respired from the C_4 dung treatment was 5.11 and 5.33 g C kg^{-1} for *L. chinensis* soil and *A. frigida* soil, respectively (Appendix S1). The amount of dung-derived C recovered in respiration was nearly 1.5 times that from soil-derived C in both soils, *i.e.* 59.7% and 58.9% of the evolved CO_2 originated from the decomposing dung in the *L. chinensis* soil and *A. frigida* soil, respectively (Fig. 3).

More CO_2 emissions in C_4 dung treated soil than control soil in this study was ascribed to the priming process. The occurrence of positive priming was observed in both soils, with a more pronounced priming effect in *A. frigida* soil than in *L. chinensis* soil (Fig. 4). Thus, compared with the control soils, an additional 1.34 g C (*L. chinensis*) and 1.55 g C (*A. frigida*) was emitted as CO_2 after sheep dung was applied (Appendix S1 and Table 4).

Net carbon budget

Over the 152 days incubation period, 3.5% and 2.8% (*i.e.* 0.79 g C and 0.64 g C) of the amended sheep dung was recovered in the *L. chinensis* soil and *A. frigida* soil carbon fractions, respectively (Table 3 and Table 4), and 13.4% and 13.8% of the

Table 3. The dynamics of $\delta^{13}C$ (Mean $\pm$ SE) in control soils and in soil treated with C_3 and C_4 dung, and the percent of dung-derived C incorporated in the soil in relation to soil C (D) and applied dung C (P).

Days after addition	Control	δ^{13} C (‰ vs VPDB) C_3 dung soil	C_4 dung soil	D (% of soil C)	P (% of applied dung C)
L. chinensis dominated soil					
17	-22.3 ± 0.1	-22.6 ± 0.1	-22.5 ± 0.02	1.4 ± 0.1	1.3 ± 0.1
43	-22.2 ± 0.1	-22.2 ± 0.1	-22.0 ± 0.1	1.1 ± 0.2	1.2 ± 0.1
86	-22.1 ± 0.01	-20.0 ± 0.04	-22.0 ± 0.1	0.5 ± 0.1	0.5 ± 0.01
127	-22.0 ± 0.1^a	-22.1 ± 0.1^a	-21.8 ± 0.1^b	3.4 ± 0.2	3.2 ± 0.2
152	-22.2 ± 0.1^a	-22.1 ± 0.3^a	-21.7 ± 0.1^b	3.8 ± 0.4	3.5 ± 0.2
A. frigida dominated soil					
17	-22.3 ± 0.03	-22.2 ± 0.1	-22.2 ± 0.03	0.03 ± 0.1	0.02 ± 0.00
43	-22.5 ± 0.1	-22.4 ± 0.1	-22.3 ± 0.04	1.0 ± 0.1	0.6 ± 0.01
86	-22.5 ± 1.5^a	-22.3 ± 0.1^a	-22.1 ± 0.1^b	2.1 ± 0.1	1.2 ± 0.04
127	-22.4 ± 0.1^a	-22.3 ± 0.1^a	-21.9 ± 0.2^b	4.0 ± 0.3	2.3 ± 0.1
152	-22.2 ± 0.03^a	-22.3 ± 0.01^a	-21.8 ± 0.02^b	4.9 ± 0.3	2.8 ± 0.2

Data are shown for each sampling during the 152 days incubation.
Different superscript letters represent statistical significance at $P<0.05$ at the same sampling time among treatments.

amended sheep dung was emitted as CO_2 in the *L. chinensis* soil and *A. frigida* soil carbon fractions, respectively.

A priming effect was found in both soils during the incubation period. Considering the apparent supply of dung C to the soil C component, in comparison with the primed CO_2 loss from dung treated soil, the net soil C loss was nearly two times higher from the low C *A. frigida* soil (0.91 g C kg^{-1} soil) compared to the high C

L. chinensis soil (0.55 g C kg^{-1} soil). As a result, 2.6% and 7.0% of soil C was lost due to the application of sheep dung from the *L. chinensis* and *A. frigida* soils, respectively (Table 4).

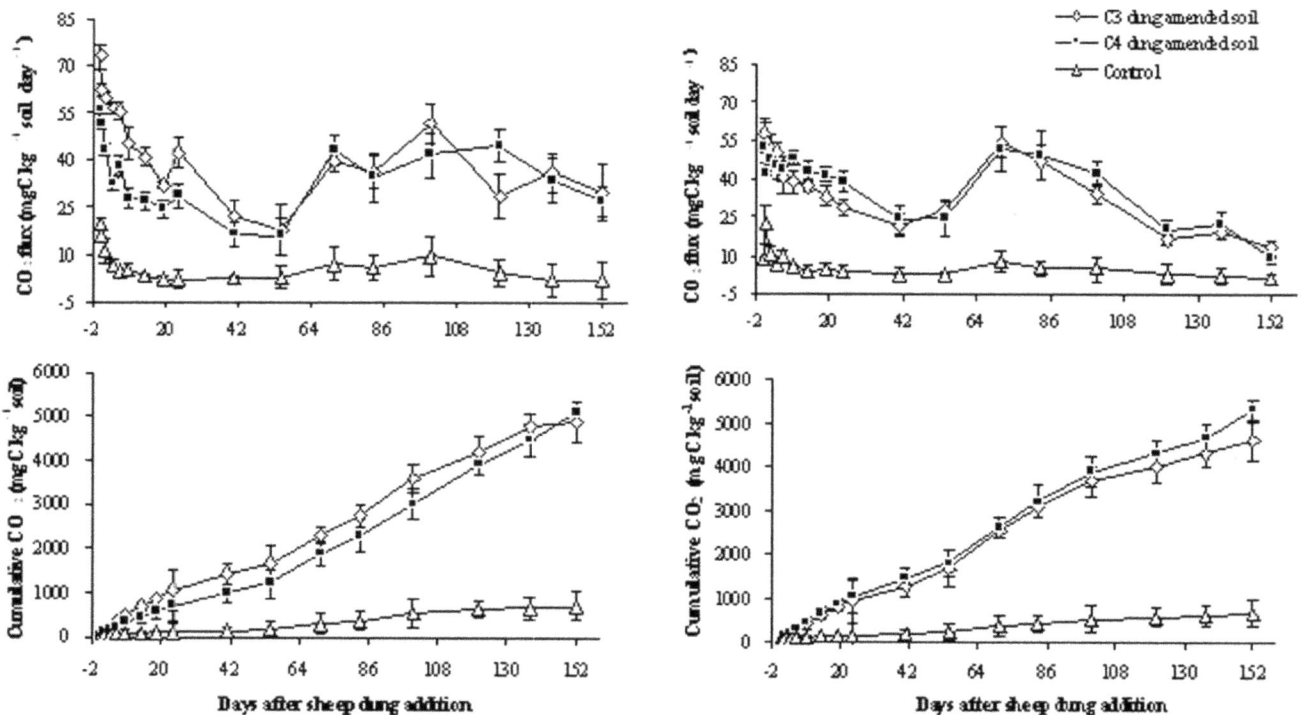

Figure 1. Temporal dynamics of CO_2 fluxes (mg C kg^{-1} day^{-1}) (A and B), and cumulative CO_2-C loss (mg C kg^{-1} day^{-1}) (C and D) during 152 days of incubation of *L. chinensis* and *A. frigida* soils amended with C_3 and C_4 dung. Values are the mean (n = 3) $\pm$ SE (bars).

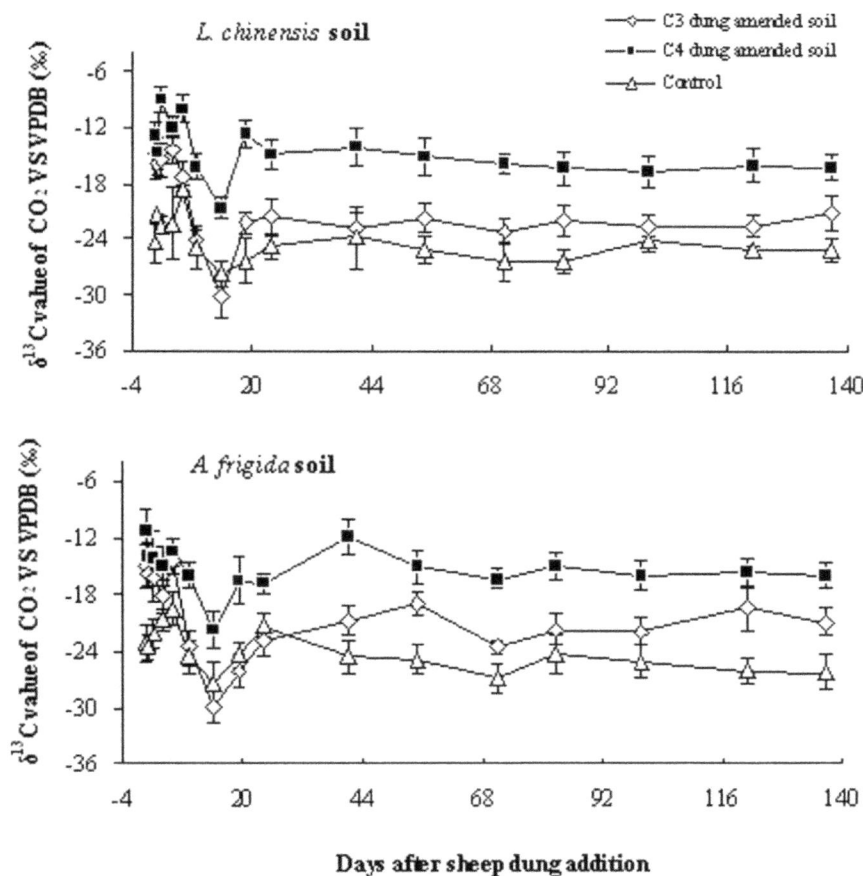

Figure 2. Temporal dynamics of $\delta^{13}C$ (‰ vs VPDB) signatures in CO_2 emitted during 152 days following sheep dung addition to *L. chinensis* and *A. frigida* soil (n = 3).

Discussion

Dung-derived C in the soil

The $\delta^{13}C$ signatures differed by 10.5‰ between the C_4 and C_3 sheep dung in our study, which makes it possible to differentiate the soil- and dung-derived C in bulk samples as well as in respired CO_2 based on the ^{13}C natural abundance characteristics. Distinct differences in $\delta^{13}C$ signatures between C_4 and C_3 dung treated (or control) soils emerged 43 days after dung incorporation into *L. chinensis* dominated soil, and 86 days in *A. frigida* dominated soil, suggesting differentiated time lags in the apparent transformation of dung C to the soil C component. However, the limited temporal resolution of soil sampling impeded a detailed identification of the exact temporal dynamics. In contrast, low but significant CO_2 emissions derived from dung C was observed initially in the incubations, which indicates that dung decomposition commenced immediately, but a transfer to the soil C component was not apparent until after several weeks of incubation.

Bol et al. [38] observed that after a 150 days field experiment, 12.6% of applied cattle dung C was retained in a grassland top soil C component. Another field experiment in a temperate grassland showed that a maximum of 60% of cow dung C was retained in the soil after 56 days, declining to around 20% after 372 days [39]. In our study, only 2.8–3.5% of sheep dung C appeared in the soil C component after 152 days of incubation at 20°C. Probably the relatively low dung water content (*i.e.* 14.5%) in our study caused constrained decomposition compared with other studies

(e.g. 84% water content in the study by Bol et al. [38]. Slurry generally decomposes faster than dung due to its liquid nature and missing of various dissolved compounds [40]. The characteristics of the dung, such as its C/N ratio, is also an important factor which affected the decomposition of excreta. The C/N ratios in our experiment (31.3 for C_3 dung and 34.4 for C_4 dung) were much higher than the 0.7–10.9 for Bertora et al. [2], and high C/N excreta might be prone to slow mineralization compared to low C/N excreta [2,41–42].

CO_2 fluxes

An apparent two-phase pattern of CO_2 emission was observed in the current study, which was attributed to the two-phase pattern of sheep dung decomposed in both soil.

During the whole period of the incubation (152 days), in both soils, ca 40% of the total CO_2 was released from dung treated soil itself, while 60% released from the decomposed dung (Fig. 3). The two-stage decomposition patterns have also been observed in other studies on dung decomposition, and it is proposed that in the first stage CO_2 emission is due to the decomposition of labile C from soil and easily degradable dung fractions, while in the second phase, the decomposers attack more recalcitrant material (3, 18, 37). The first phase of decomposition, as indicated by the increased CO_2 efflux, lasted for ca. 6 days, which is longer compared with the 24–48 h duration observed in other studies, suggesting that the labile fraction of sheep dung C is more

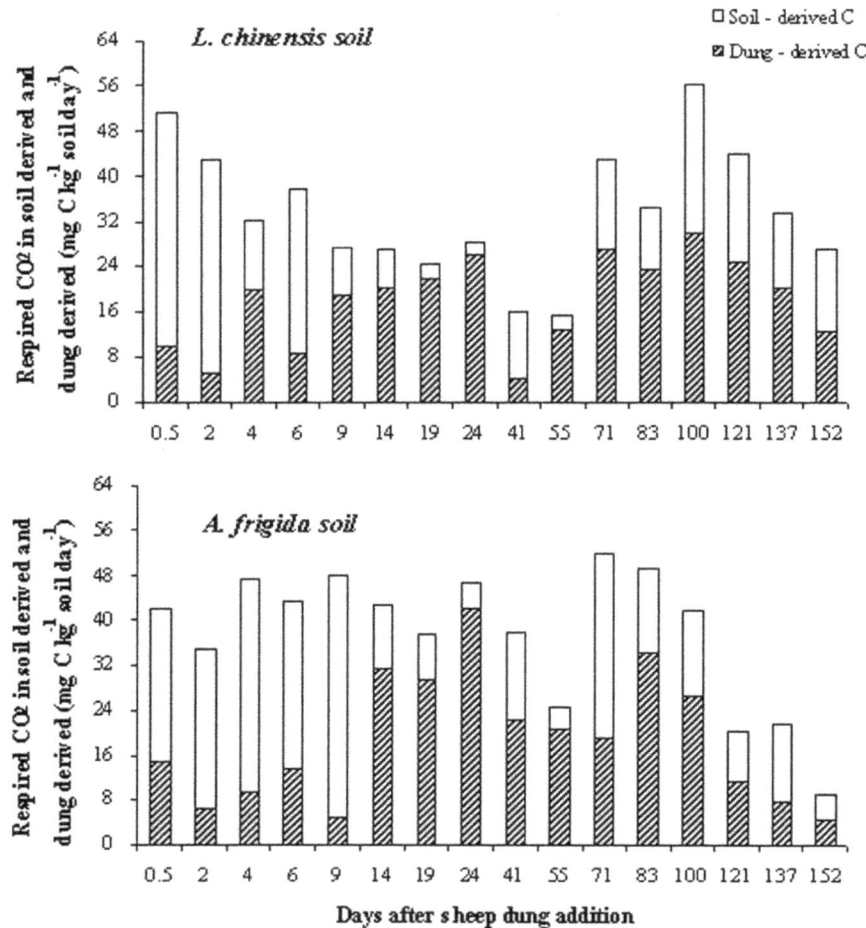

Figure 3. The relative contribution of dung-derived CO₂-C and soil-derived CO₂-C during 152 days incubation calculated from the δ¹³ C signature of CO₂ after sheep dung addition (n = 3).

recalcitrant and less available compared to other excreta such as wet cattle dung [24,29].

Isotope characteristics of emitted CO_2

The $\delta^{13}C$ signal of emitted CO_2 from both C_4 incorporated soils were significantly higher than the C_3 and the control soil in our experiment, which was likely due to the incorporation and the microbial turnover of polymeric biologic cell wall materials from C_4 dung into C_3 grassland soils of C_4 dung C [3]. Similar phenomena have been observed in other studies within the first

hours [25–26]. For the control soil, in general the $\delta^{13}C$ of emitted CO_2 is slightly higher than the $\delta^{13}C$ of soil undergoing decomposition. For example, Fangueiro et al. [3] reported that the $\delta^{13}C$ of CO_2 emitted from untreated soil was on average 5.4‰ higher than the value for the SOM undergoing decomposition. Angers et al. [25] reported 5.0‰ higher $\delta^{13}C$ of emitted CO_2 than $\delta^{13}C$ in the soil itself. At the same time, slightly higher $\delta^{13}C$ of emitted CO_2 in the C_4 dung treated soil (average $-14.5‰$) than the sheep dung itself ($-15.7‰$) was found in our incubation, the interaction effect of sheep dung and soil and the isotope

Table 4. Fates of soil and sheep dung carbon after 152 days incubation.

Soil type	Organic C of soil	Fate of sheep dung C		Soil CO₂ emission (g C kg dry soil⁻¹)			Net soil C loss	Total soil C loss
		Recovered in soil	Emitted as CO₂	Soil derived C	Control	Priming		
L. chinensis	21.0±0.06	0.79±0.01	3.05±0.04	2.06±0.01	0.72±0.02	1.34±0.03	0.55±0.01[a]	2.6±0.2[b]
A. frigida	13.0±0.05	0.64±0.01	3.14±0.03	2.19±0.01	0.64±0.01	1.55±0.01	0.91±0.02	7.0±0.3

[a]Net soil C loss was given by the value of sheep dung C sequestrated in the soil during 152 days subtracted from the primed soil CO₂.
[b]Soil C loss (%) is the percentage total CO₂-C loss compared to soil total organic C content.

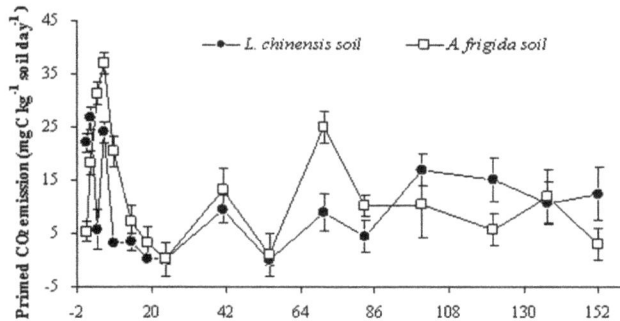

Figure 4. Primed CO_2 emission (mg C kg^{-1} day^{-1}) during 152 days of incubation from sheep dung amended *L. chinensis* and *A. frigida* soils. The numbers indicate soil C derived CO_2 emitted in excess to the control soil (n = 3).

fractionation associated with the microbial turnover maybe was the possible reason.

Soil priming effects after sheep dung addition

A priming effect (PE) is defined as a short-term change in the turnover of soil organic matter caused by the addition of labile organic C to the soil [43]. Here, we determined the priming effect as the excess emissions of soil C derived CO_2 in dung treated soil compared to control soils. Primed CO_2 emissions were observed throughout the 152 days incubation in both grassland soils. Specifically, 1.3 g and 1.6 g of excess soil C kg^{-1} was emitted as CO_2 when dung was added to the *L. chinensis* soil and *A. frigida* soil, respectively, corresponding to 6.2% and 11.9% of total soil C. Such a priming effect is in the upper range of primed C losses of 2.3%–8.9% observed in previous studies [44–45].

Priming of soil C decomposition is believed to occur during relatively short-term periods upon addition of labile substrates to soils, and always occurs only in the early stage of substrate addition after which it rapidly ceases [9,10,18,46]. The rapid decrease in the priming effect was likely caused by the depletion of easily decomposable substances added with more complex materials [23]. However, primed soil C decomposition was apparent in our study throughout the entire 152 days period, although with a quantitative variation during the experiment, which may be related to the high C/N ratio and slow decomposition rate characteristics of sheep dung. The priming effect depends not only on the decomposability of the various carbon pools in the environment, but also on the state of the microorganisms [47–48], and involves not only one mechanism but rather a succession of processes partly connected with succession of microbial communities and functions [13]. Further research is needed to test the fundamental processes and mechanisms involved in the priming effect of soil organic matter decomposition due to grazing in Inner Mongolian grasslands.

The loss of sheep dung C via CO_2 was the same in the two grassland soils in our experiment, which contrasts with results by Bol et al. [18] who reported that more slurry-derived C was respired from a C-rich soil compared to a C-poor soil during 0–9 days after slurry incorporation. The authors of that study

attributed this to the more pronounced enhancement of basal soil respiration in C-rich soil compared to the C-poor soil. Furthermore, after labile organic C addition, energy limitation in C-poor soil may have ceased, which subsequently facilitated more activation of soil microorganisms, and more enzymes were produced that were capable of SOM degradation [17,43].

Conclusion and implications for grassland management

The addition of C_4 sheep dung to a C_3 grassland soil enabled us to successfully trace the fate of dung-derived C in the soil and calculate the soil organic C budget for two soils from the Inner Mongolian steppe. Although sheep dung provided an additional organic carbon source for the grassland soils, a large part was emitted as CO_2 to the atmosphere. After sheep dung addition, a positive priming effect of soil C decomposition was observed in both a high-C *L. chinensis* soil and a low-C *A. frigida* soil. Therefore, the balance of soil organic carbon storage was negative when sheep dung was mixed into the soil. This effect was more pronounced for the degraded community of *A. frigida* soils, from which more soil C was lost compared with the climax community of *L. chinensis* soils. This finding is contrary to the conventional conception of carbon storage in temperate grassland, which predicts that the livestock excreta applied to grassland soils could return essential nutrients for plant growth and increase fertility and SOM contents.

The results suggest that intensive grazing management in the temperate steppe should be avoided. In Inner Mongolian grasslands, the succession from *L. chinensis* dominated communities to *A. frigida* dominated communities will result in decreased plant productivity and soil carbon inputs because of the decreased litter and root nutrient return [34]. Bearing in mind that complex plant-soil interactions which exist under field conditions have not been considered in our study, and assuming that the present conclusion can be extrapolated to field conditions, a further acceleration of the decreasing soil C pool in degraded grasslands may occur. Under actual grazing conditions, sheep dung may be mixed merely within the very top grassland soils, suggesting that the current calculations based on well-mixed soil and dung may overestimate soil C losses. Future work is thus needed to examine whether the priming effect of sheep dung amendments observed in this study can be extended to field conditions.

Supporting Information

Appendix S1
(DOC)

Acknowledgments

The authors gratefully acknowledge thank Andreas Wilkes for correction of English and grammar. We also acknowledge Dr Mette Sustmann Carter for her skilled laboratory and technical assistance.

Author Contributions

Conceived and designed the experiments: XZM SPW YFW. Performed the experiments: XZM PA. Analyzed the data: XZM. Contributed reagents/materials/analysis tools: XZM CJW. Wrote the paper: XZM.

References

1. Pausch J, Kuzyakov Y (2011) Soil organic carbon decomposition from recently added and older sources estimated by δ^{13}C values of CO_2 and organic matter. Soil Biol Biochem 55: 40–47.

2. Bertora C, Alluvione F, Zavattaro L, Groenigen JWV, Velthof G, et al. (2008) Pig slurry treatment modifies slurry composition, N_2O and CO_2 emissions after soil incorporation. Soil Biol Biochem 40: 1999–2006.

3. Fangueiroa D, Chadwick D, Dixon L, Bol R (2007) Quantification of priming and CO_2 emission sources following the application of different slurry particle size fractions to a grassland soil. Soil Biol Biochem 39: 2608–2620.

4. Pineiro G, Paruelo JM, Oesterheld M (2006) Potential long-term impacts of livestock introduction on carbon and nitrogen cycling in grasslands of southern South America. Global Change Biol 12: 1267–1284.

5. Dungait JA, Bol R, Evershed RP (2005) Quantification of dung carbon incorporation in a temperate grassland following spring application using bulk stable isotope determinations. Isot Environ Healt S 41: 3–11.

6. Fangueiro D, Chadwick D, Dixon L, Grilo J, Walter N, et al. (2010) Short term N_2O, CH_4 and CO_2 production from soil sampled at different depths and amended with a fine sized slurry fraction. Chemosphere 81: 100–108.

7. Lin XW, Wang SP, Ma XZ, Xu GP, Luo CY, et al. (2009) Fluxes of CO_2, CH_4, and N_2O in an alpine meadow affected by yak excreta on the Qinghai-Tibetan plateau during summer grazing periods. Soil Biol Biochem 41: 718–725.

8. Ma XZ, Wang SP, Wang YF, Jiang GM, Nyren P (2006) Short-term effects of sheep excreta on carbon dioxide, nitrous oxide and methane fluxes in typical grassland of Inner Mongolia. New Zeal J Agr Res 49: 285–297.

9. Kuzyakov Y, Friedel JK, Stahra K (2000) Review of mechanisms and quantification of priming effects. Soil Biol Biochem 32: 1485–1498.

10. Luo Y, Durenkamp M, De Nobili M, Lin Q, Brookes PC (2011) Short term soil priming effects and the mineralisation of biochar following its incorporation to soils of different pH. Soil Biol Biochem 43: 2304–2314.

11. Ohm H, Hame U, Marschner B (2007) Priming effects in soil size fractions of a podzol Bs horizon after addition of fructose and alanine. J Plant Nutr Soil Sc 170: 551–559.

12. Rochette P, Angers DA, Chantignya MH, Bertranda N, Cote D (2004) Carbon Dioxide and Nitrous Oxide Emissions following Fall and Spring Applications of Pig Slurry to an Agricultural Soil. Soil Sci Sci AM J 68: 1410–1420.

13. Blagodatskaya E, Kuzyakov Y (2008) Mechanisms of real and apparent priming effects and their dependence on soil microbial biomass and community structure: critical review. Biol Fert Soils 45: 115–131.

14. Hamer U, Marschner B (2005) Priming effects in different soil types induced by fructose, alanine, oxalic acid and catechol additions. Soil Biol Biochem 37: 445–454.

15. Shen J, Bartha R (1996) The priming effect of substrate addition in soil-based biodegradation tests. Appl Environ Microb 62: 1428–1430.

16. Pineiro G, Paruelo JM, Oesterheld M (2006) Potential long-term impacts of livestock introduction on carbon and nitrogen cycling in grasslands of southern South America. Global Change Biol 12: 1267–1284.

17. Ohm H, Hame U, Marschner B (2007) Priming effects in soil size fractions of a podzol Bs horizon after addition of fructose and alanine. J Plant Nutr Soil Sc 170: 551–559.

18. Bol R, Moering J, Kuzyakov Y, Amelung W (2003) Quantification of priming and CO_2 respiration sources following slurry C incorporation in two grassland soils with different C content. Rapid Commun Mass Sp 17: 2585–2590.

19. Blagodatskaya E, Yuyukina T, Blagodatsky S, Kuzyakov Y (2011) Turnover of soil organic matter and of microbial biomass under C_3–C_4 vegetation change: Consideration of ^{13}C fractionation and preferential substrate utilization. Soil Biol Biochem 43: 159–166.

20. Kuzyakov Y, Bol R (2004) Using natural ^{13}C abundance to differentiate between three CO_2 sources during incubation of a grassland soil amended with slurry and sugar. J Plant Nutr Soil Sc 167: 669–677.

21. Kuzyakov Y, Bol R (2006) Sources and mechanisms of priming effect induced in two grassland soils amended with slurry and sugar. Soil Biol Biochem 38: 747–758.

22. Chantigny MH, Rochette P, Angers DA (2001) Short-term C and N dynamics in a soil amended with pig slurry and barley straw: a field experiment. Can J Soil Sci 81: 131–137.

23. Glaser B, Bol R, Preedy N, McTiernan K, Clark M, et al. (2001) Tracing slurry-derived carbon and nitrogen in a temperate grassland using $\delta^{13}C$ and $\delta^{15}N$ natural abundance. J Plant Nutr Soil Sc 164: 467–474.

24. Lovell RD, Jarvis SC (1996) Effect of cattle dung on soil microbial biomass C and N in a permanent pasture soil. Soil Biol Biochem 28: 291–299.

25. Angers DA, Rochettea P, Chantignya MH, Lapierre H (2007) Use of ^{13}C abundance to study short-term pig slurry decomposition in the field. Soil Biol Biochem 39: 1234–1237.

26. Kristiansen SM, Brandt M, Hansen EM, Magid J, Christensen BT (2004) ^{13}C signature of CO_2 evolved from incubated maize residues and maize-derived sheep faeces. Soil Biol Biochem 36: 99–105.

27. Bai YF, Han XG, Wu JG, Chen ZZ, Li LH (2004) Ecosystem stability and compensatory effects in the Inner Mongolia grassland. Nature 43: 181–184.

28. Tong C, Wu J, Yong S, Yan J, Yong W (2004) A landscape scale assessment of steppe degradation in the Xilin River Basin, Inner Mongolia, China. J Arid Environ 59: 133–149.

29. Liu ZK, Wang SP, Han JG, Wang YF, Chen ZZ (2004) Changes of soil chemical properties in sheep urine patches in Inner Mongolia steppe. Chinese Journal of Applied Ecology 15(12), 2255–2260.

30. Ma XZ, Wang SP, Jiang GM, Haneklaus S, Schnug E, et al. (2007) Short-term effect of targeted placements of sheep excrement on grassland in Inner Mongolia on soil and plant parameters. Commun Soil Sci Plan 38: 1589–1604.

31. Chen WW, Wolf B, Zheng XH, Yao ZS, Butterbach-Bahl K, et al. (2011) Annual methane uptake by temperate semiarid steppes as regulated by stocking rates, aboveground plant biomass and topsoil air permeability. Global Change Biol 17: 2803–2816.

32. Chen ZZ, Wang SP (2000) Chen, Z.Z, Wang, S.P., 2000. Plant response to different grazing intensity in Inner Mongolia grasslands. In: Typical Steppe Ecosystem of China, Science Press, Beijing. pp. 13–21.

33. Chen L, Michalk DL, Millar GD (2002) The ecology and growth patterns of Cleistogenes species in degraded grassland of eastern Inner Mongolia, China. J Appl Ecol 39: 584–594.

34. Wang SP, Li YH, Wang YF, Han YH (1998) The succession of Artemisia frigida rangeland and multivariation analysis under different stocking rates in Inner Mongolia. Acta Agrestia Sinica 6: 299–305.

35. Formanek P, Ambus P (2004) Assessing the use of $\delta^{13}C$ natural abundance in separation of root and microbial respiration in a Danish beech (Fagus sylvatica L.) forest. Rapid Commun Mass Sp 18: 897–902.

36. Bao SD (2000) Agrochemical Analysis of Soil, Fourth ed. Agriculture Press of China, Beijing.

37. Harris D, Horwa WR, van Kessel C (2001) Acid fumigation of soils to remove carbonates prior to total organic carbon or carbon-13 isotope analysis. Soil Sci Soc Am J 65: 1853–1856.

38. Bol R, Kandeler E, Amelung W, Glaser B, Marx M, et al. (2003). Short-term effects of dairy slurry amendment on carbon sequestration and enzyme activities in a temperate grassland. Soil Biol Biochem 35: 1411–1421.

39. Dungait JAJ, Bol R, Bull ID, Evershed RP (2009) Tracking the fate of dung-derived carbohydrates in a temperate grassland soil using compound-specific stable isotope analysis. Org Geochem 40: 1210–1218.

40. Haynes RJ, Williams PH (1993) Nutrient cycling and soil fertility in the grazed pasture ecosystem. Adv Agron 49: 119–199.

41. Lupwayi NZ (1999) Leucaena hedgerow intercropping and manure application in the Ethiopian highlands I. Decomposition and nutrient release. Biol Fert Soils 28: 182–195.

42. Wang SP, Li YH (1997) The influence of different stocking rates and grazing periods on the chemical components in feces of grazing sheep and relationship among the fecal components. Chinese Journal of Animal Nutrition 9: 49–56.

43. Kuzyakov Y (2002) Review: Factors affecting rhizosphere priming effects. J Plant Nutr Soil Sc 165: 382–396.

44. Dendooven L, Bonhomme E, Merckx R, Vlassak K (1998) N dynamics and sources of N_2O production following pig slurry application to a loamy soil. Biol Fert Soils 26: 224–228.

45. Flessa H, Beese F (2000) Laboratory estimates of trace gas emissions following surface application and injection of cattle slurry. J Environ Qual 29: 262–268.

46. Kuzyakov Y, Yilmaz G, Stahr K (1999) Decomposition of plant residues of Lolium perenne in soils and induced priming effects under different land use. Agribiological Research 52: 25–34.

47. Kuzyakov Y, Bol R (2005) Three sources of CO_2 efflux from soil partitioned by ^{13}C natural abundance in an incubation study. Rapid Commun Mass Sp 19: 1417–1423.

48. Kuzyakov Y (2010) Priming effects: Interactions between living and dead organic matter. Soil Biol Biochem 42: 1363–1371.

Seroprevalence and Potential Risk Factors for *Brucella* Spp. Infection in Traditional Cattle, Sheep and Goats Reared in Urban, Periurban and Rural Areas of Niger

Abdou Razac Boukary[1,2,3,7], **Claude Saegerman**[2], **Emmanuel Abatih**[3], **David Fretin**[4], **Rianatou Alambédji Bada**[5], **Reginald De Deken**[3], **Halimatou Adamou Harouna**[6], **Alhassane Yenikoye**[7], **Eric Thys**[3]*

1 Department of Livestock promotion and Management of Natural Resources, ONG Karkara, Niamey, Niger, 2 Department of Infectious and Parasitic Diseases, Research Unit in Epidemiology and Risk Analysis applied to the Veterinary Sciences (UREAR), University of Liege, Liege, Belgium, 3 Department of Biomedical Sciences, Unit of Biostatistics and Epidemiology, Institute of Tropical Medicine, Antwerp, Belgium, 4 Department of Bacteriology and Immunology, Veterinary and Agro-chemical Research Centre, Uccle, Belgium, 5 Service of Microbiology-Immunology-Infectious Pathology, Interstate School of Veterinary Sciences and Medicine, Dakar, Senegal, 6 Ministry of Livestock, Direction of Animal Health, Maradi, Niger, 7 Faculty of Agronomy, University of Niamey, Niamey, Niger

Abstract

Introduction: In Niamey, Niger, interactions within the interface between animals, humans and the environment induce a potential risk of brucellosis transmission between animals and from animals to humans. Currently, little is known about the transmission of *Brucella* in this context.

Results: 5,192 animals from 681 herds were included in the study. Serum samples and hygroma fluids were collected. A household survey enabled to identify the risk factors for transmission of brucellosis. The true adjusted herd-level prevalence of brucellosis ranged between 11.2% and 17.2% and the true adjusted animal-population level prevalence was 1.3% (95% CI: 0.9–1.8%) based on indirect ELISA test for *Brucella* antibodies. Animals aged of 1–4 years were found to be more susceptible than animals less than 1 year old (Odds ratio [OR] of 2.7; 95% CI: 1.43–5.28). For cattle, the odds of brucellosis seropositivity were higher in rural compared to the periurban areas (OR of 2.8; 95% CI: 1.48–5.17) whereas for small ruminants the risk of seropositivity appeared to be higher in urban compared to periurban areas (OR of 5.5; 95% CI: 1.48–20.38). At herd level, the risk of transmission was increased by transhumance (OR of 5.4; 95% CI: 2.84–10.41), the occurrence of abortions (OR of 3.0; 95% CI: 1.40–6.41), and for herds having more than 50 animals (OR of 11.0; 95% CI: 3.75–32.46). *Brucella abortus* biovar 3 was isolated from the hygromas.

Conclusion: brucellosis in Niger is a serious problem among cattle especially in the rural areas around Niamey and among sheep in the urban areas of Niamey. The seroprevalence varies across strata and animal species with important risk factors including herd size, abortion and transhumance at herd level and age at animal population level. For effective control of brucellosis, an integrated approach seems appropriate involving all stakeholders working in public and animal health.

Editor: Qijing Zhang, Iowa State University, United States of America

Funding: The authors wish to thank the Belgian Directorate for Development (DGD) for funding this study and VAR-Belgium for the contribution in the laboratory tests. DGD had no role in study design, data collection and analysis, decision to publish, or preparation of the manuscript. VAR had a role in the analysis and preparation of the manuscript.

Competing Interests: The authors have declared that no competing interests exist.

* E-mail: ethys@itg.be

Introduction

Worldwide, brucellosis remains an important disease in humans, domestic and wild animals [1]. It is an infectious disease caused by bacteria of the genus *Brucella* which comprises eight species ranked according to their pathogenicity and host preferences. Six of the eight species can be isolated from terrestrial mammals: *B. abortus*, *B. melitensis*, *B. suis*, *B. canis*, *B. ovis et B. neotomae* [2]. The disease is endemic in sub-Saharan Africa (SSA), with significant effects on economic and social conditions of people in this region [3]. Indeed, brucellosis has an important impact on the health and productivity of livestock greatly reducing their economic value [4]. The epidemiology of brucellosis in SSA is complex and the prevalence varies across geographic regions and

livestock systems [5]. The disease incidence is influenced by management factors, herd size, population density, type of animal breed and biological features such as herd immunity [6,7,8,9,10,11]. In West Africa, the rates of infection vary greatly from one country to another, within a country and production systems [12,13,14,15,16]. It is generally accepted that the prevalence of brucellosis is much higher in the pastoral grazing systems than the urban and periurban systems where herd sizes are smaller [5,10,17,18,19].

In Niger, brucellosis was first reported in 1953 in humans [20], but it was not until 1970 that the first preliminary serological studies were conducted to assess the prevalence of the disease in animals [21]. There are few data on human brucellosis in West

Africa, particularly in Niger [5,6,7,24]. Gidel et al. [21] showed seroprevalence rates ranging from 1% to 17% in humans in pastoral areas of Côte d'Ivoire, Niger and Burkina Faso. According to the same authors, the prevalence of the disease in 1974 was 0.5% in the city of Niamey [21]. Since then, very little research has been conducted in order to assess the magnitude of, and risk factors for the disease transmission within different production systems. Later, investigations in pastoral livestock systems of the country in 1986 by Akakpo et al. [12], Akakpo and Bornarel [22], and in 1991 by Bloch and Diallo [13] have confirmed the presence of brucellosis in cattle with apparent prevalence rates ranging between 1.4% and 30.9%.

The increased demand for animal products following the growth of the urban population and the depletion of food resources in pastoral areas due to climate change is forcing livestock keepers and their animals to move to the peripheral cities [23]. This has led to the development of a dynamic and complex livestock production system in the urban and suburban regions of Niamey city [24]. Breeders are in most cases installed on unhealthy and unmanaged land without adequate infrastructure to conduct their activities [25]. Dietary habits of Niger population such as consumption of unpasteurized dairy products, close contact with infected herds and with contaminated environmental sources could be major risk factors for the spread of *Brucella* infections among humans [24,25,39,43]. The contribution of these and other factors to the epidemiology of brucellosis in livestock production systems in Niger is not yet known.

The aim of this study was to determine the prevalence of *Brucella* infection, using indirect Enzyme-linked Immunosorbent assay (iELISA) in cattle, goats and sheep in the urban, periurban and surrounding rural areas of Niger and to identify risk factors for infection both in human and livestock populations. In addition, we used some hygroma fluid to identify a field circulating strain of *Brucella*.

Materials and Methods

2.1. Ethics statement

This study involves a questionnaire based survey of farmers as well as blood sampling from their animals. The study protocol was assessed and approved by the Niger National Advisory Committee on Ethics with reference number 010/2009/CCNE and by the Ministry of Agriculture and Livestock of the Republic of Niger with reference number 00109 on 28 January 2010. Participants provided their verbal informed consent for animal blood sampling as well for the related survey questions, according to the Niger procedures at the time of the study. Collection of blood samples was carried out by professional veterinarians adhering to the regulations and guidelines on animal husbandry. In each village, a meeting was held with the community members to explain the purpose of the study. Farmers were not forced to participate in the survey and animal blood sampling. Name, region and village of the farmers were registered. Paper questionnaires were encoded and recorded in Excel and names were replaced by their coded versions for analysis. Paper questionnaires were stored in Niger.

2.2. The study area

The study zone was composed of three strata in accordance with the classification established by Boukary et al. [24]: the urban (Ur), the periurban (Pu) of Niamey and the rural areas (Ru) (**Fig. 1**).

The urban area was formed by the Urban Community of Niamey (UCN) located along the Niger River in the western part of the country, between 2° 10' and 2° 14' longitude E and 13° 33'

and 13° 36' latitude N and covered an area of about 12,500 ha with nearly one million inhabitants.

The periurban area covers a ray ranging from 5 to 25 km around the capital. It is populated by the long-established resident population and a population of immigrants composed mainly of Fulani herders. The installation of the latter was promoted by the development of the dairy industry and the increase in demand for milk in the capital [24]. They occupy makeshift homes generally subjected to inadequate measures of sanitation and hygiene. Their animal breeding strategy consists in keeping only lactating females and genitor males in the sites. The renewal of the animals is done from the main transhumant herd located mostly in rural areas of Balleyara and Torodi [24].

For the rural area, the community of Balleyara located about 110 km northeast of the capital and the community of Torodi located 80 km southwest of Niamey at the border with Burkina Faso and Benin were considered as are the main rural poles which supply the city of Niamey with cattle, small ruminants and animal products.

Studying the interactions between rural, urban and surrounding rural areas through various exchange relationships between people and their herds seem very interesting in understanding the mechanisms of transmission of zoonotic diseases such as brucellosis and justifies the inclusion of this rural strata in the present study.

2.3. Study design and data collection

The study took place between December 2007 and October 2008 within the three strata previously defined and was conducted in two phases. First, a cross-sectional household survey was carried out and secondly, blood sampling and hygroma fluid collection were performed on animals belonging to herds led by the households surveyed. These samples were used for laboratory analysis.

2.3.1 The cross-sectional household survey. Since the study area was divided into three strata; urban, peri-urban and surrounding rural areas of Niamey, the first step was to identify the number of sampling sites. A total of 45 sampling sites were randomly selected from a roster of 375 sites identified within the three strata. In each study stratum, the approximate number of herds (which belonged to different sites) was listed with the assistance of local veterinary officers and farmers' leaders. The total number of herds to be included in this study was calculated using an expected herd level seroprevalence "p" of 14.2% [22], a confidence level of 95%, desired absolute precision (d) of 0.05 and using the following formula $n = (1.96)^2 p(1-p)/d^2$ [26]. This yielded a total of 187 herds to be sampled from each strata. However, since herds turn to be similar within sites, a correction factor of magnitude 2 [27] was applied to account for the clustering of herds within sites. In addition, contingencies were adjusted for by adding another 25% of herds leading to a total of 234 herds to be sampled from each strata. The sampling of the herds within sites was based on a proportionate sampling scheme since the total number of herds within each site was available. All animals present at the time the herd was visited were sampled.

In this study, herd means all animals reared within the household surveyed (i.e., ecosystem) and it was regarded as the primary sampling unit according to the study area. So, there were as many herds as households surveyed.

The questionnaire used in the face-to-face interview with the head of the selected households included questions related to risk factors for transmission of brucellosis both in animals and humans. At the animal level, information was collected on species (goats, sheep, cattle), age (in years), and gender (male or female). At the herd level, the factors included: herd size (number of animals in

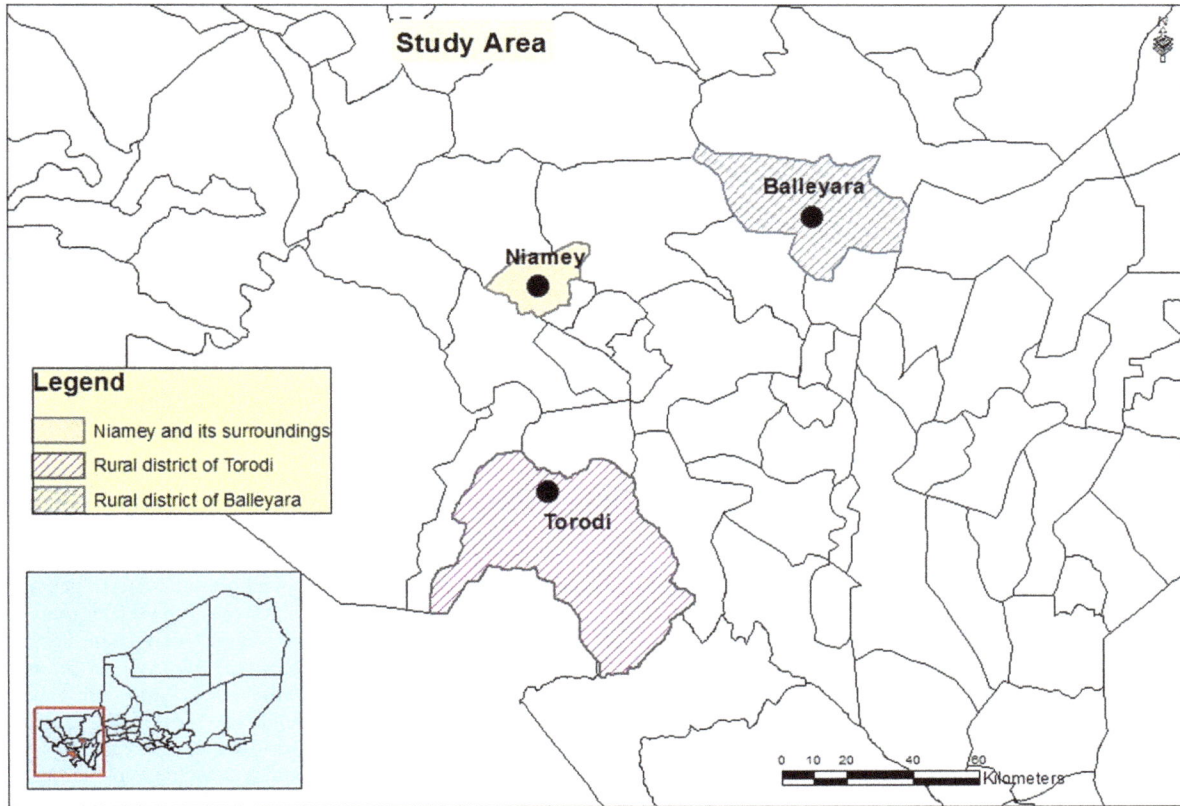

Figure 1. Location of the study areas in Niger.

the household), occurrence of abortion (yes/no), whereas relative to the household, the factors were: practices related to livestock (acquisition modes of the animals by the household, method of rearing animals, handling of newly arrival animals, fate of dead animals or aborted foetuses), and the social status of the household (native of the locality or migrant). The full questionnaire in French is available as supporting information document (see questionnaire S1).

2.3.2. Blood sample collection and testing.   Five thousand one hundred and ninety-two (5,192) serum samples and sixteen (16) hygroma fluid samples were collected from animals (Table 1). The blood sample collection was made during the face-to-face interviews with the head of the household.

The collected samples were stored in a deep freezer ($-20°C$) at the National Reference Laboratory for AIDS and Tuberculosis (NRL-HIV/TB) of Niamey (Niger), until they could be analysed at the National Reference Centre for Brucellosis, Veterinary and Agrochemical Research Centre (CODA-CERVA) in Belgium. All assays except MLVA performed at CODA-CERVA are accredited (ISO 17025).

2.3.3. Serological testing.   For procedural reasons, our samples were sent to Belgium 2 years after collection. Serological tests were conducted between September 2009 and February 2010. An indirect ELISA described previously by Limet et al. [28] was used. The antigenic use in this test is a purification of the lipopolyssacharide of *Brucella abortus* W99. Briefly, 50 µl of serum dilutions (1:50 in buffer consisting of 0.1 M glycine, 0.17 M sodium chloride, 50 mM EDTA, 0.1% (volume) Tween 80, and distilled water, pH 9.2) were added to the wells in duplicate. The plates were incubated for 1 h at room temperature. Binding of

antibodies was detected using a protein-G peroxydase conjugate (Biorad, Belgium). The conjugate was incubated for 1 h at room temperature. Citrate–phosphate buffer containing 0.4% o-phenyl-enediamine and 2 mM H2O2 was used to visualize the peroxydase activity. The difference in optical densities (OD) at A 490 and 630 nm was read on a Bio Kinetics Reader EL-340 (Biotek Instruments, Vermont, USA). Negative control serum and dilution buffer was added in duplicates on each plate as controls. This ELISA fulfils the requirement laid down in the OIE Manual of Standards for Diagnostic Tests and Vaccines [1].

Table 1. Total number of herds surveyed and animals tested in the urban (Ur), peri-urban (Pu) and rural areas (Ru) of Niger.

Variable	Ur	Pu	Ru	Total
Data on herds surveyed				
- *Number of sites indentified*	19	131	225	375
- *Number of sites selected*	9	13	23	45
- *Number of herds (households interviewed)*	239	215	227	681
Data on animals tested				
- *Cattle*	973	1,473	724	3,170
- *Sheep*	216	320	650	1,186
- *Goats*	106	150	583	839
Total number of animals tested	1,295	1,943	1957	5,195

2.3.4. Bacteriological testing. Directly after 15 minutes centrifugation at 3000 rpm, isolation of *Brucella* sp. from Hygroma was performed according to the technique described by Alton et al [29] and Bankole [30]. Isolate of *Brucella* were typed by classical method and molecular method (MLVA). A 15 locus VNTR typing was carried out according to Le Flèche et al. [31]. The 15 loci have been classified into two panels, panel 1 (eight minisatellite loci) and panel 2 (seven microsatellite loci) (Table 2). The profile obtained from the MVLA was compared to other strain profiles using MVLA Public Databases (MLVAbank 2012).

2.4. Statistical analysis

2.4.1. Determination of the true prevalence of brucellosis. The estimation of the true prevalence (TP) of brucellosis at the animal population level was done using the formula proposed by Rogan and Gladen [32]:

$$TP = (AP + Sp - 1)/(Se + Sp - 1)$$

where AP is the apparent prevalence; Se is the sensitivity and Sp is the specificity.

Because no prior data were available for Niger, the specificity (Sp) and sensitivity (Se) of the iELISA were the values of the study carried out on traditional livestock farming systems in Ivory Coast by Thys et al. [14]. The values of Se and Sp for the iELISA and their 95% confidence intervals based on this study were as follows:

$$Se : 0.9639(95\%CI : 0.9272 - 0.9984)$$

$$Sp : 0.9861(95\%CI : 0.9600 - 0.9989)$$

Table 2. Loci of the Variable Number Tandem Repeats analysis (VNTR) used in the study (according to [31]).

Panel	Reference VNTR [a]	Name of marker [b]
1	BRU1322_134bp_408bp_3u	Bruce06
	BRU1134_18bp_348bp_4u	Bruce08
	BRU211_63bp_257bp_3u	Bruce11
	BRU73_15bp_392bp_13u	Bruce12
	BRU424_125bp_539bp_4u	Bruce42
	BRU379_12bp_182bp_2u	Bruce43
	BRU233_18bp_151bp_3u	Bruce45
	BRU2066_40bp_273bp_3u	Bruce55
2	BRU1543_8bp_152bp_2u	Bruce04
	BRU1250_8bp_158bp_5u	Bruce07
	BRU588_8bp_156bp_7u	Bruce09
	BRU548_8bp_152bp_3u	Bruce16
	BRU339_8bp_146bp_5u	Bruce18
	BRU329_8bp_148bp_6u	Bruce21
	BRU15C5_8bp_151bp_6u	Bruce30

Legend;
[a]reference VNTR: naming nomenclature includes repeat unit size, PCR product size in strain 16 M, corresponding repeat copy number,
[b]common name of the marker.

A herd was considered positive if at least one animal tested positive for *Brucella* antibodies by the iELISA test within the herd. The animal population and herd-level AP were estimated using an intercept-only random effects logistic regression model with site and herd as random effects to account for the survey design characteristics of the study. Treating herd and site for the animal population level analysis and site as a random effect in the herd-level analysis as random effects accounted for the clustering of animals within herds and the clustering of herds within sites respectively. In addition, it accounted for differences in number of animals within herds and number of herds within sites. For the different sub-populations such as male or female cattle, sheep and goat within each stratum, the AP and TP were computed using random effects logistic regression models. Normal 95% confidence intervals were computed for both the AP and the TP.

2.4.2. Risk factor analysis. The risk factor analysis was performed at the animal and herd/household levels. It should be noted that the data on the herd are combined and processed together with data on the household. In what follows, the 'herd level' denotes the analysis of factors collected at the household and herd-level. In addition, the animal population level analysis was done separately for cattle and small ruminants. Initially, a univariate analysis was performed using a univariate random effects logistic regression model at the animal population level as well as at the herd level. The animal population level model used as response, the brucellosis status of the animals and each animal level risk factor or indicator variable in turn as explanatory variables whereas the herd level model used as response, the herd level brucellosis status and corresponding herd level risk/indicator factors as covariates.

For the animal population level analysis, herd and site location were used as random effects to account for potential clustering of animals within herds (dependence of results from the same herd) and clustering of herds within sites whereas site location was used as a random effect for the herd level analysis to account for the effects of clustering of herds within sampling sites. At the animal population level and at the herd level, the variable representing the three strata was forced into the model to account for variations in prevalence across strata.

Variables with a p-value<0.10 in the univariate analysis were further evaluated in a multivariable random effects logistic regression analysis. A manual forward stepwise selection approach was applied to choose the final model. In the first step of this approach, univariate models were built for each covariate. The best univariate model was selected based on the AIC values (the smaller the better). The remaining variables were then added each in turn to the best univariate model to form two-variable models. The best two-variable model was selected as that with the smallest AIC among the two-variable models. This procedure was repeated until the addition of one more variable failed to improve the model fit; in other words once the AIC started to increase or remained constant. The model with the smallest AIC was considered to be the most appropriate model for the data.

The effects of confounding were investigated by observing the change in the estimated odds ratios of the variables that remain in the model once a non-significant variable is removed. When the removal of a non-significant variable led to a change of more than 25% of any parameter estimate, that variable was considered a confounder and was not removed from the model.

Multicollinearity was assessed among the independent variables using the Cramer's phi prime statistic which expressed the strength of the association between two categorical covariates. Values >0.7 were indicative of co-linearity and in this case only the variable most significantly associated with the response was kept in the model [35].

All two-way interaction terms of the variables remaining in the final model were assessed for significance based on the likelihood ratio test comparing the model with the desired interaction term and the corresponding model with no interaction terms.

The intra-class correlation coefficient (ICC), which is a measure of the degree of clustering of animals belonging to the same herd or of herds belonging to the same site, was computed. In random effects logistic regression models, the individual level variance σ^2 on the logit scale is usually assumed to be fixed to $\pi^2/3$ [34]. The variability attributed to differences between herds was given by:

$$ICC_{Herd} = \sigma^2_{INT:Herd}/(\sigma^2_{INT:Herd} + \pi^2/3)$$

whereas that between sites was computed as:

$$ICC_{Site} = \sigma^2_{INT:Site}/(\sigma^2_{INT:Site} + \pi^2/3)$$

If the ICC is low or zero in either case, it implies that the animals within herds or the herds within sites are independent (there is no clustering) of each other and therefore random effects should not be included in the analysis. On the contrary an ICC close to 1 implies that there is high between herds or between site heterogeneity implying the clustering of individual animals within herds or clustering of herds within sites respectively [35].

The models were built using the xtmelogit function in STATA, version 12.1, software (SataCorp LP, College station, Texas). Model selection was done using Laplacian approximation whereas parameter estimates from the final model were obtained using Adaptive Gaussian Quadrature [36]. The robustness of the final model was assessed by increasing the number of Quadrature (integration) points and monitoring changes in parameter estimates [37].

Results

3.1. Herd structure

A total of 5,195 animals composed of 3,170 cattle, 1,186 sheep and 839 goats were sampled. These animals belonged to 681 herds which in turn were nested within the 45 sites (9 in the urban region, 13 in the peri-urban region and 23 in the rural area). The number of herds reduced from 702 to 681 because of incomplete information for 21 of the sampled herds. Regardless of region of origin it was found that the herds were mixed and included the three species: cattle, sheep and goats. However, in the urban and peri-urban areas, cattle were the most numerous. They were respectively 72% and 78% of the herds which showed that these farms were mainly oriented to dairy production. In rural pastoral areas, herds were more balanced with 38% of cattle, 33% sheep and 29% goats (**Table 1**). The different cattle breeds included: Azawak, Mbororo, Djelli, Goudali and crossbred and were found to be widely distributed across strata. For small ruminants, the common breeds for sheep were Oudah (Bali Bali) and Ara Ara whereas for goats the common breed was Sahel. All were also found to vary widely across strata.

3.2. *Brucella* seroprevalence and potential risk factors

3.2.1. Brucella seroprevalence results. Of the 5,195 sera examined, 2.6% tested positive for iELISA (137/5195, 95% CI: 2.2–3.1%). The estimated overall animal population-level true

prevalence in the study population was 1.3% (95% CI: 1.1–3.4) (**Table 3**). The prevalence of brucellosis was highly variable among the animal species considered.

Brucellosis prevalence varied according to strata. In cattle, it was significantly higher in rural areas with a true prevalence (TP) of 4.6% (95% CI: 3.1–6.2) against 2.0% and 1.8% in urban and peri-urban areas respectively. For small ruminants, the prevalence of brucellosis also varied across strata even though differences were not statistically significant i.e. the 95% confidence intervals overlapped. In sheep, the overall true prevalence of brucellosis was 3.6% (95% CI: 1.1–6.1) in urban areas where it is higher than in periurban and rural areas (**Table 3**).

At the herd level, the estimation of the true prevalence of brucellosis across the three strata indicated that 91 out of 681 herds investigated (13.7%) were found to be maintaining infected animals (Table 3). The true herd-level prevalence (THP) of brucellosis ranged between 11.2% and 17.2% according to the area in consideration.

3.2.2. Potential risk/indicator factors associated with sero-prevalence of brucellosis based on univariate random effects logistic regression analysis. The results of the univariate analysis which was based on random effects models correcting for animal level clustering indicated that at the animal population level, age was significantly associated with brucellosis seropositivity for cattle (P<0.05) (**Table 4**). In general, it was observed that the prevalence of brucellosis was significantly higher in older animals compared to young animals since their confidence intervals do not overlap. Animals aged between 1 and 4 years appeared more at risk than young animals and animals older than 4 years.

Among small ruminants, the effects of gender could not be evaluated using the random effect logistic regression model, because there were no positive cases among males. However, a univariate analysis was performed using Firth's logistic regression analysis. Firth's logistic regression analysis was used in place of the traditional exact logistic regression analysis to overcome the computational limitations and convergence issues caused by the sparseness (separation) of the data. The method uses penalized maximum likelihood (PML), which is carried out iteratively until model convergence to estimate the associated odds ratios, standard errors, and 95% confidence intervals [38]. The results indicated that gender was not significantly associated with brucellosis seropositivity among small ruminants but since the p-value was <0.10 it was considered as a potential risk factor to be included in the multivariable analysis (**Table 4**).

The univariate random effects logistic regression analysis with a random effect for site and a fixed effect for strata, revealed that the herd level risk factors: herd composition, transhumance, abortion in the herd, acquisition of animals, handling of newly arrived animals, herd size and origin of herds, all appeared to be highly significantly associated with the herd level brucellosis sero-positivity (P<0.05) (**Table 5**).

3.3. Multiple random effects logistic regression model

The results of the multivariable random effects logistic regression analysis at the animal-population level indicated that for cattle, the variables representing strata and age were important risk factors whereas for small ruminants, only the variable representing strata was found to be important (**Table 6**). On the other hand, out of the 8 potential risk factors initially considered in the multiple random effects logistic regression model only transhumance, abortion in the herd and herd size) were included in the final herd level model (**Table 7**). None of the two-way interaction terms were statistically significant (p>0.05). No

Table 3. Apparent prevalence (AP) and estimated true prevalence (TP) of brucellosis at the individual animal and herd levels.

Species	Gender	Urban						Periurban						Rural					
		N	Positive	AP	95% CI	TP	95% CI	N	Positives	AP	95% CI	TP	95% CI	N	Positive	AP	95% CI	TP	95% CI
Cattle	Female	908	26	2.0	1.1–2.9	0.6	0.1–3.4	1337	32	1.8	1.1–2.6	0.5	0.1–2.9	649	36	4.4	2.8–5.9	3.1	1.8–7.3
	Male	65	1	1.0	0.0–3.3	0	–	136	3	1.5	0–3.6	0.1	0–4.1	75	5	5.3	0.2–10.4	4.1	0.0–14.9
	Total	973	27	2.0	1.1–2.9	0.6	0.1–3.4	1473	35	1.8	1.2–2.5	0.5	0.1–2.9	724	41	4.6	3.1–6.2	3.4	2.1–7.5
Sheep	Female	206	10	4.1	1.4–6.8	2.8	0.6–9.0	274	3	0.8	0.0–1.9	–	–	575	17	2.5	1.2–3.7	1.1	0.3–4.6
	Male	10	0	–	–	–	–	46	0	–	–	–	–	75	0	–	–	–	–
	Total	216	10	3.6	1.1–6.1	2.3	0.3–8.1	320	3	0.6	0.0–1.5	–	–	650	17	2.1	1.0–3.2	0.8	0.1–3.9
Goats	Female	95	1	0.8	0.0–2.7	–	–	133	0	–	–	–	–	535	3	0.4	0.0–1.7	–	–
	Male	11	0	–	–	–	–	17	0	–	–	–	–	48	0	–	–	–	–
	Total	106	1	0.7	0.0–2.2	–	–	150	0	–	–	–	–	583	3	0.4	0.0–0.9	–	–
Herd	Total	227	42	13.5	9.1–18.0	12.8	8.4–22.3	215	27	12.0	7.7–16.4	11.2	7.0–20.6	239	33	17.8	12.9–22.6	17.2	12.5–27.4

Legend: * According to the formula proposed by Rogan and Gladen [30].

evidence of confounding was present and the estimated Cramer's phi prime statistic values were all less than 0.7 indicating no important correlations between the independent variables. Increasing the number of quadrature points had no influence on the estimated fixed effects and the variance component parameters indicating that the models were robust.

The variance components of the final model for cattle indicated that the ICC for herd was $ICC_{HERD} = 0.27$ and for small ruminants $ICC_{HERD} = 0.07$. The substantial ICC for cattle implies that there is considerable between-herd heterogeneity and thus clustering of animals within herd whereas for small ruminants the low ICC implies that the animals within herd are independent (there is no clustering). The ICC for the herd-level data was 0.34 suggesting that there is considerable clustering of herds within sites. The considerable cattle-level clustering and herd-level clustering demonstrates the potential for herd-level and site-level interventions to influence brucellosis seropositivity.

From the final model for cattle (**Table 6**), it can be seen that the odds of brucellosis sero-positivity were significantly higher in rural areas as compared to periurban areas with an OR of 2.8. In addition, for cattle between 1 and 4 years old the odds of brucellosis seropositivity were 2.7 times higher compared to those that are 1 year old. For small ruminants, the odds of brucellosis seropositivity were significantly higher in urban areas as compared to periurban areas with an OR of 5.5.

At herd level, the final multivariable model (**Table 7**) yielded that for households that reported the presence of abortions in the herd, the odds of seropositivity were 3 times higher as compared to households which did not report the occurrence of abortions. Also for herds that reported the practice of transhumance, the odds of sero-positivity were 5.4 times higher compared to those that did not practice transhumance. Finally, for herds with more than 50 animals, the odds of brucellosis seropositivity were 11 times higher compared to herds with less than 10 animals.

3.4. Strain typing and identification

Of the 16 hygroma samples collected and cultured, only one was positive after 3 days of incubation and showed round (1–2 mm diameter), convex colonies with entire edges and smooth shiny surfaces. Colonies required CO2 for growth, produced H_2S and grew in the presence of basic fuchsin, thionin and safranin. The determination of biotype was based on the results of four tests: hydrogen sulphide production, agglutination by monospecific anti-A and anti-M sera, growth in the presence of dyes, and carbon dioxide requirement. The profile of this isolate was classified as *B. abortus* biovar 3, according to the Corbel and Brinley-Morgan [50] classification. The number of tandem repeats for each locus is shown in **Table 8**. Considering only the first panel, this profile appeared to be related to *B. abortus* biovar 3 reference strain Tulya and dromedary strain BCCN 93_26 from Uganda (Le Flèche_2006). This type is also close to *B. abortus* biovar 3 strain BCCN 93_26 from Sudan (Le Flèche_2006), *B. abortus* biovar 3 strain 11-KEBa2, 14-KEBa2 and 15-KEBa2 from Kenya (Muendo_2011) and *B. abortus* biovar 3 reference strain Tulya (Ferreira_2012). The relationship between these strains and our isolate is shown in **Fig. 2**.

Discussion

The study confirms that brucellosis is present in Niger and that herd level seroprevalence varied by abortion status of the herd, herd size and method of rearing animals.

Due to lack of unbiased data and standardized method to estimate the seroprevalence in Niger and across the West African

Table 4. Potential risk/indicator factors associated with individual animal-level brucellosis seropositivity among 5195 animals nested within 681 herds.

Variable	Number tested (Positive)	% Positive (95% CI)	Odds ratio (95% CI)	P-value
Cattle				
Strata				0.003
Periurban	1473 (35)	2.4(1.7–3.3)	1 (Ref.)	
Urban	973 (27)	2.8(1.8–4.0)	1.3 (0.54–2.67)	
Rural	724 (41)	5.7(4.1–7.6)	2.8 (1.37–5.60)	
Age (years)				<0.001
≤1	912 (16)	1.8 (1.0–2.8)	1 (Ref.)	
>1 and <4	1307 (61)	4.7(3.6–6.0)	3.7 (1.87–7.17)	
≥4	951 (26)	2.7(1.8–4.0)	1.7 (0.83–3.68)	
Gender				0.944
Bull	276 (9)	3.3 (1.5–6.1)	1 (Ref.)	
Cow	2894 (94)	3.2 (2.6,4.0)	1.1 (1.53–2.36)	
Small ruminants (sheep and goats)				
Strata				0.018
Peri-urban	470 (3)	0.6 (0.1–1.9)	1 (Ref.)	
Urban	322 (11)	3.4 (1.7–6.0)	5.4 (1.41–20.88)	
Rural	1233 (20)	1.6 (1.0–2.5)	2.4 (0.68–8.56)	
Age (years)				0.161
≤1	318 (4)	1.3 (0.3–3.2)	1 (Ref.)	
>1 and <4	723 (12)	1.7 (0.9–2.9)	1.3 (0.55–3.14)	
≥4	984 (18)	1.8 (1.1–2.9)	2.1 (0.79–5.69)	
Gender				0.026
Male	207 (0)	0.0 (0–1.8)	1 (Ref.)	
Female	1818 (34)	1.9 (1.3–2.6)	8.0 (0.94–131.35)[exact]	

Exact: estimates based on Firth's logistic regression model; Ref: reference group.

sub-region, comparing our findings with those from other studies should be made with caution. The apparent prevalence in our study was low compared with that obtained in other studies conducted in Niger. Indeed, using the Rose Bengal Test (RBT), Akakpo et al. [12] found an AP rate of brucellosis of 27.7% in the Kirkissoye ranch not far from Niamey, while Bloch and Diallo [13] reported an AP rate ranging from 3.7% to 9.5%. Using RBT, Boukary et al. [39] reported an AP rate of brucellosis comprised between 2.4 and 5% in smallholder dairy cattle herds in the urban and periurban areas of Niamey. The difference in prevalence between our study and the previous ones may be partly explained by the methodology used in the study protocol. In fact, in some studies, the lack of sampling frames or their imperfection does not allow to achieve representative sampling [40]. Another important issue is the difference in sensitivity and specificity of serological tests used for screening. This factor contributes to the variability of results among researchers [5,6,41,42]. The reported high prevalence in the other studies might be due to false-positive serum reactions [10]. The RBT used for screening individual animals at national-local-level is cheap, rapid and highly sensitive [1]. However, its specificity is low because the smooth lipopolysaccharides of the *Brucella* antigen can cross-react with antibodies produced by closely related Gram-negative bacteria such as *Yersinia enterocolitica* O:9, *Escherichia coli* O:157, *Salmonella spp.*, and *Sternotrophomonas maltophilia* as well as antibodies produced by *B. abortus* S19 vaccine [41,42].

The fact that the risk of transmission of brucellosis in animals at the population and herd level varied significantly depending on the strata is in agreement with the findings of several authors who demonstrated variations in the prevalence of brucellosis related to the production systems [5,7,15,16,18]. In cattle, we found that that the risk of brucellosis seropositivity was higher in rural areas compared to periurban and urban areas. The reason for the higher prevalence in the rural areas was probably due to the fact that in this area, free animal movement is common [24,39]. It is now well documented that the dynamics and frequent migration of pastoral herds might increase the chance of coming into contact with other potentially infected herds and exposure to geographically limited or seasonally abundant diseases [7,11,12,16,43]. Considering the contagious nature of *Brucella* species, sharing grazing land and drinking water facilitate transmission of the disease [8,9,10]. Another factor that may explain the high sensitivity of cattle to *Brucella* spp. in rural areas is linked to the herd composition. We also observed in that area that the herds are equally mixed, while in the urban and periurban areas cattle are more abundant than sheep and goats. Although the factor "herd composition" was not retained in our final model, our results based on a univariate random effects model showed that the risk of contamination increases sharply in mixed herds where the odds of brucellosis seropositivity was 8.9 times higher compared to pure cattle herds. This is in accordance with Holt et al. [33] and Megersa et al. [11].

Table 5. Potential risk factors associated with herd level seroprevalence of brucellosis based on a univariate random effects model with strata forced in as a fixed effect and site as a random effect.

Variable code	Level	Odds ratio (95% C.I)	P-value
Herd Composition	Animal species that occur within the herd belonging to the herd surveyed		<0.001
	1: Cattle	1(Ref.)	
	2: Cattle + Sheep or Goat	4.8(1.20–19.46)	
	3: Sheep or Goat	3.3(0.92–12.00)	
	4: Cattle + Sheep + Goat	8.9(2.58–30.90)	
Herd size	Number of animals owned by the herd		<0.001
	1:< = 10	1(Ref.)	
	2:>10 and < = 50	3.3(1.27–8.40)	
	3:>50	27.9(9.9–78.7)	
Abortion	Presence of females who aborted among the animals belonging to the herd surveyed		<0.001
	1: No	1 (Ref.)	
	2: Yes	4.5(2.23–8.95)	
Acquiring animals	Acquisition modes of the animals by the herd		0.025
	1: Heritage	1 (Ref.)	
	2: Fostering	1.2(0.34–4.69)	
	3: Purchase	1.7(0.80–3.72)	
	4: Mix	2.7(1.32–5.65)	
Transhumance	Method of rearing animals of sedentary type (not migratory : No) or nomadic (transhumant : Yes)		<0.001
	1: No	1(Ref.)	
	2: Yes	9.1(5.06–16.30)	
Handling	Handling of newly arrival animals (mixed with other animals or quarantined)		0.022
	1: Quarantined	1 (Ref.)	
	2: Mixed	1.8(1.08–2.85)	
Native	Origin of the herd surveyed : native of the locality (Yes) or migrant (No)		<0.001
	1: Yes	1(Ref.)	
	2: No	4.3(2.15–8.64)	

Sero-prevalence: Having or not at least one animal testing positive by Elisa-test within the herd (1 or 0). Strata: Stratum in which the investigations took place (Urban, Periurban, Rural). Site: Means the village, hamlet or the district selected for the study within the different strata. Herd: Herd surveyed within the different sites. Ref.: reference group.

Considering the strata, the odds of testing positive to brucellosis in cattle were significantly higher in the rural areas than in urban ones. This can be explained by the difference in management. The Fulani of the periurban area of Niamey developed a new strategy for dairy production which involves keeping only animals in production (dairy cows in early stage of gestation or in lactation) the rest of the herd being kept in rural areas [24]. The low prevalence of brucellosis in cattle in periurban results from this strategy as only apparently healthy animals are selected for milk production [43]. This is in agreement with our observation that the seroprevalence of brucellosis increased with the incidence of abortions. Indeed, the odds of seropositivity were 3.0 times higher in the herds where the presence of abortions was reported as compared to those which did not report the occurrence of abortions. This is in accordance with several authors who found that the prevalence of brucellosis within herds is positively correlated with the incidence of abortions in females [6,44,45].

Contrary to what we observed in cattle, the risk of infection with brucellosis in small ruminants was much higher in urban compared to rural and periurban areas. Indeed, the odds of

brucellosis seropositivity were 5.4 times higher in urban compared to periurban areas for small ruminants. Difference in management can also explain this, as small ruminants play a very important economic role in urban areas. For many households, keeping sheep and goats is a way of saving money [25]. Males are kept separately where they are fed with forage complemented and with kitchen waste. Their market value is much higher than that of females and they are usually sold when there is a need for cash or are slaughtered during religious ceremonies [23]. This explains the low number of males in the samples used in our study and also their low susceptibility to brucellosis infection. Unlike rural areas where herds are usually mixed, urban flocks are in most cases separated from cattle. Ewes and she-goats of the different flocks are typically collected by a shepherd who brings them to the pasture [24,25]. These specific conditions of raising small ruminants in urban areas promote aggregation of animals within neighborhoods, pastures and water points, favouring the transmission of the disease [11,43,45].

Transhumance in Niger is much more pronounced in pastoral areas where large herds have to run long distances searching for

Table 6. Final model of animal population level risk factors associated with brucellosis seropositivity among cattle and small ruminants.

Variable code	Level	Odds ratio (95% C.I)	P-value
	Cattle		
Strata	Periurban	1(Ref.)	
	Urban	1.4(0.73–2.62)	0.323
	Rural	2.8(1.48–5.17)	0.003
Age (years)			
	≤1	1(Ref.)	
	>1 and <4	2.7(1.43–5.28)	0.002
	≥4	1.2(0.59–2.60)	0.527
Random effects		SE(95% CI)	
Herd level variance1.20		0.45(0.57–2.50)	
	Small ruminants		
Strata	Periurban	1 (Ref.)	
	Urban	5.5(1.48–20.38)	0.011
	Rural	(0.70–8.50)	0.161
Random effects		SE(95% CI)	
Herd level variance0.26		0.42(0.01–6.13)	

Legend: Ref.: reference group.

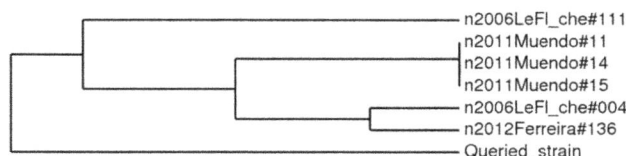

Figure 2. Clustering analysis of a field strain of *Brucella abortus* **3 from Niger (Queried_Strain) with field and reference strains in the** *Brucella* **multiple loci variable number tandem repeats analysis (MLVA) database (MVLABANK, 2012) using panels 1 and 2.** The data are given in columns from left to right: year of isolation and 'alias'.

herds compared to those that did not practice transhumance. Similarly, our results showed that herd size was linked to *Brucella* seropositivity (P<0.001) and that the risk of contamination was much higher in larger herds compared to those with a limited number of animals. These results corroborated those of several other authors [6,10,40,46]. In Niger, the history of migration is closely linked to that of transhumance. Under pressure from repeated drought and deterioration of their livelihoods, pastoralists tend increasingly to become sedentary [25]. These people usually are installed on marginal lands where sanitation and hygiene infrastructures are generally lacking [24]. The absence of veterinary services brings these migrants to assist themselves pregnant or aborted females [24]. This will expose them to a higher risk of dissemination and transmission of the disease.

In our study, the prevalence of brucellosis in cattle was highly correlated with the age of the animals. Indeed, for cattle between 1 and 4 years old, odds of brucellosis seropositivity were 3.7 times higher compared to those that are 1 year old or younger. That higher seropositivity of animals between 1 and 4 years old could be explained due to the increase in exposure [12,18,47,48,49]. Indeed

pasture and water points [12,21,24,25]. We observed that the risk of contracting the disease increases significantly in herds with high mobility. The odds of seropositivity were 9.1 times higher in these

Table 7. Final model of herd-level risk factors associated with brucellosis sero-positivity among 681 herds which nested within 45 sites.

Variable code	Level	Odds ratio (95% C.I)	P-value
Strata	Stratum in which the investigations took place		
	Periurban	1(Ref.)	
	Urban	1.5(0.37–6.25)	0.334
	Rural	1.8(0.55–5.70)	0.557
Herd size	Number of animals in the herd		
	<=10	1 (Ref.)	
	>10 and <=50	1.9(0.71,5.15)	0.199
	>50	11.0(3.75,32.46)	<0.001
Abortion	Presence of females who aborted among the animals belonging to the herd surveyed		
	No	1(Ref.)	
	Yes	3.0(1.40–6.41)	0.005
Transhumance	Method of rearing animals of sedentary type (not migratory : No) or nomadic (transhumant : Yes)		
	No	1(Ref.)	
	Yes	5.4(2.84–10.41)	<0.001
Random effects		**SE**	**95% CI**
Site level variance	1.69	0.68	(0.77–3.72)

Legend: Ref.: reference group.

Table 8. The Multiple Loci Variable Number Tandem Repeats analysis (MLVA) profiles showing number of variable tandem repeats (VTR) for a *B. abortus* biovar 3 isolate from Niger (Queried Strain) and its closest MLVA neighbour profile.

	Strain	REF Tulya	BCCN 93–26	11-KEBa2	14-KEBa2	15-KEBa2	REF Tulya	Queried Strain
	Host	human	dromedary	cattle	cattle	cattle	cattle	cattle
	Publication	Le Flèche 2006	Le Flèche 2006	Muendo 2011	Muendo 2011	Muendo 2011	Ferreira 2012	This study
	Country	Uganda	Sudan	Kenya	Kenya	Kenya	-	Niger
VTR	bruce06	3	3	3	3	3	3	3
	bruce08	5	5	5	5	5	5	5
Panel 1	bruce11	4	4	4	4	4	4	3
	bruce12	11	11	11	11	11	11	11
	bruce42	2	2	2	2	2	2	2
	bruce43	2	2	2	2	2	2	2
	bruce45	3	3	3	3	3	3	3
	bruce55	3	3	3	3	3	3	3
VTR	bruce18	8	6	7	7	7	8	8
	bruce19	40	40	40	40	40	42	21
Panel 2	bruce21	8	8	8	8	8	8	8
	bruce04	6	6	6	6	6	6	6
	bruce07	5	8	5	5	5	5	2
	bruce09	3	3	3	3	3	3	3
	bruce16	11	7	12	12	12	11	12
	bruce30	5	7	5	5	5	5	7

these animals are more mobile and therefore more exposed to infection by *Brucella* within the transhumant herds. Animals less than one year old are generally kept in the household together with lactating females.

At the household level, our results showed that mixing of newly arrived animals into the herd is highly correlated with brucellosis seropositivity. Females infected with *Brucella spp.* excrete high concentrations of the organism in their milk, placental membranes and aborted foetuses [51]. Therefore, there is a high risk of transmission of the pathogen between animals and from animals to humans through direct contact with contaminated material such as foetal membranes, aborted foetuses and other animal products. According to an investigation conducted in the periurban and rural areas of Niger by Boukary et al. [43], it seems that due to the lack of veterinary services, farmers assist in delivering cows without gloves or masks, which puts them at high risk of infection with *Brucella*.

Our study aimed also to identify strains of *Brucella spp* circulating in Niger. Out of the 16 hygroma samples, one sample was found positive after culture and *Brucella abortus* biovar 3 was isolated. We were not able to isolate *Brucella* from the remaining 15 samples analyzed, although they were collected from animals tested positive for iELISA. This can be explained by the fact these samples would probably not contain enough germs to allow their isolation. Another reason is that the shelf life and transport conditions of samples may have a negative effect on the survival of *Brucella*. Indeed our hygroma samples were collected between December 2007 and October 2008. They were kept during 2 years at $-20°$ C prior to shipment to Belgium, where they were analyzed. Possible electric power outages during storage, thawing of samples during transport and handling may have affected the

quality of the hygroma liquid. The difficulty of isolating *Brucella* from hygroma fluids under similar conditions to ours has already been mentioned by Bankole et al. [30].

Our *Brucella* isolate shows the same characteristics as those already isolated in Niger by Akakpo et al. [12]. In fact, strains of *Brucella abortus* isolated in Africa are known to grow slowly, to be sometimes negative on the oxidase test and to have a specific oxidative pattern [53]. This finding is similar to the results obtained by several authors in West and Central Africa who reported the presence of *B. abortus* biovar 3 or intermediate 3/6 [12,22,30,40,46]. So, *Brucella abortus* biovar 3 is very common in Africa. Considering the panels of MLVA profile, our isolate profile appeared to be related to *B. abortus* biovar 3 strains isolated in Uganda and Sudan [31] and those isolated in Kenya [54]. The profile also appeared to share some similarities with *B. abortus* biovar 3 reference strain Tulya isolated by Ferreira et al. [55].

In conclusion, the present study confirms the existence of *Brucella* in cattle, sheep and goats from the three studied strata. It highlights the presence of *Brucella abortus* biovar 3 and stresses that age, practice of transhumance, herd size and occurrence of abortions are risk factors for the spread of the disease within animals. These risk factors are related to the complexity of interactions that exist within and between the different production systems and the different practices observed in urban, periurban and rural areas.

At present, there is no officially coordinate program control of brucellosis in Niger. The role played by the disease in limiting livestock production and its economic impact on the livestock industry in Niger has not yet been evaluated. Attitudes of communities have to be defined regarding the brucellosis, the feasibility and the acceptability of potential measures. Measures including selective vaccination programme in herds with high prevalence combined with the slaughtering of known infected animals *(test and slaughter)* in herds with low infection rates as well as testing animals newly introduced into the herd can be considered [52,56]. For effective control of this disease in the context of sub-Saharan Africa, an integrated approach should be promoted that takes into account the relationship between humans, animals and environment. A multisectorial framework involving physicians, veterinarians, and all the stakeholders working in public and animal health in the context of a "One Health" approach is recommended.

Supporting Information

Questionnaire S1   Questionnaire for the cross-sectional household survey on animal husbandry practices and eating habits among rural livestock keepers.
(DOC)

Acknowledgments

The authors thank the staff of VAR-Belgium for their contribution in the laboratory tests. They thank also the NGO Karkara, the association of farmers Gajel Sudu-Baba, the Central Laboratory of livestock in Niamey, municipal services of Torodi, Balleyara and Niamey and VSF-Belgium for their contribution to the experiments in the field.

Author Contributions

Conceived and designed the experiments: ARB CS ET. Performed the experiments: ARB DF. Analyzed the data: ARB CS EA ET. Contributed reagents/materials/analysis tools: ARB EA DF CS. Wrote the paper: ARB CS EA DF RAB RDD AHH AY ET.

References

1. OIE (2009) Bovine brucellosis. Manual of Diagnostic Tests and Vaccines for Terrestrial Animals, Paris, France, pp. 1–35. Available: http://www.oie.int/fileadmin/home/eng/health_standards/tahm/2.04.03_bovine_brucell.pdf. Accessed 2013 Jan 11.
2. Halling SM, Young EJ (1994) Brucella. In : Hui Y.H., Gorham J.R., Murrell K.D., Cliver D.O., (Eds). Foodborne Disease Handbook – Disease caused by bacteria. New York: Marcel Dekker, INC. pp 63–69.
3. FAO (2009) FAOSTAT, Food and Agricultural Organization Statistic Division. http://faostat.fao.org/site/573/DesktopDefault.aspx?PageID = 573#ancor. Accessed 2011 Aug 31.
4. Ly C (2007) Santé animale et pauvreté en Afrique. In : Ahmadou Aly Mbaye, David Roland-Holst, Joachim Otte, (Eds), Agriculture, élevage et pauvreté en Afrique de l'Ouest. Rome : CREA-FAO.pp 71–85.
5. Mangen MJ, Otte J, Pfeiffer D, Chilonda P (2002) Bovine brucellosis in Sub-saharan Africa: Estimation of ser-prevalence and impact on meat and milk offtake potential. Rome: FAO.58 p.
6. McDermott JJ, Arimi SM (2002) Brucellosis in sub-Saharan Africa: epidemiology, control and impact. Vet Microbiol 90: 111–134.
7. Acha P, Szyfres B (2005) Zoonoses et maladies transmissibles à l'homme et aux animaux. 3ème edition. Paris: OIE. 693 p.
8. Muma JB, Godfroid J, Samui KL, Skjerve E (2007) The role of *Brucella* infection in abortions among traditional cattle reared in proximity to wildlife on the Kafue flats of Zambia. Rev. sci tech Off Int Epiz 26: 721–730.
9. Mekonnen H, Kalayou S, Kyule M (2010) Serological survey of bovine brucellosis in Barka and Arado breeds (*Bos indicus*) of Western Tigray, Ethiopia. Prev Vet Med 94: 28–35.
10. Makita K, Fèvre ME, Waiswa C, Eisler M, Thrusfield M, et al. (2011) Herd prevalence of bovine brucellosis and analysis of risk factors in cattle in urban and peri-urban areas of the Kampala economic zone, Uganda. BMC Veterinary Research 7: 60.
11. Megersa B, Biffa D, Abunna F, Regassa A, Godfroid J, et al. (2011) Seroprevalence of brucellosis and its contribution to abortion in cattle, camel, and goat kept under pastoral management in Borana, Ethiopia. Trop Anim Health Prod 43: 651–656.
12. Akakpo AJ, Saley M, Bornarel P, Sarradin P (1986) Epidémiologie de la brucellose bovine en Afrique tropicale II: Analyse sérologique et identification des deux premières souches de Brucella abortus biotype 3 au Niger. Rev Elev Méd vét Pays trop 39 :175–179.
13. Bloch N, Diallo I (1991) Enquête sérologique et allergologique sur les bovins du Niger. Rev. Elev Med Vet Pays Trop 44: 117–122.
14. Thys E, Yahaya MA, Walravens K, Baudoux C, Bagayoko I, et al. (2005) Etude de la prévalence de la brucellose bovine en zone forestière de la Côte d'Ivoire. Rev. Elev Med Vet Pays Trop 58: 205–209.
15. Cadmus SIB, Ijabone IF, Oputa HE, Adesoken HK, Stack JA (2006) Serological survey of brucellosis in livestock animals and workers in Ibadan Nigeria, African Journal of Biomedical Research 9: 163–168.
16. Sanogo M, Cissé B, Ouattara M, Walravens K, Praet N, et al. (2008) Prévalence réelle de la brucellose bovine dans le centre de la Côte d'Ivoire. Rev. Elev Med Vet Pays Trop 61: 147–151.
17. Kang'Ethe EK, Ekuttan CE, Kimani VN, Kiragu MW (2007) Investigations into the prevalence of bovine brucellosis and the risk factors that predispose humans to infection among urban dairy and non-dairy farming households in Dagoretti division, Nairobi, Kenya. East African Medical Journal 84: 96–99.
18. Chimana HM, Muma JB, Samui KL, Hangombe BM, Munyeme M et al. (2010) A comparative study of seroprevalence of brucellosis in commer-cial and small-scale mixed dairy-beef cattle enterprises of Lusaka province and Chibombo district, Zambia. Trop Anim Health Prod 42: 1541–1545.
19. Matope G, Bhebhe E, Muma JB, Lund A, Skjerve E (2010) Herd-level factors for Brucella seropositivity in cattle reared in smallholder dairy farms of Zimbabwe. Prev Vet Med 94: 213–221.
20. Merle F (1953) Apparition de la fièvre de Malte au Niger. Bull. Soc. Path. Exot.46: 211–214.
21. Gidel R, Albert JP, Mao GL, Retif M (1974) La brucellose en Afrique occidentale et son incidence sur la santé publique. Résultats de dix enquêtes épidémiologiques effectuées en Côte-d'Ivoire, Haute-Volta et Niger de 1970 à 1973. Rev. Elev Med Vet Pays Trop 27: 403–418.
22. Akakpo AJ, Bornarel P (1987) Epidémiologie des brucelloses animales en Afrique tropicale: Enquêtes clinique, sérologique et bactériologique. Rev. sci. tech. Off. Int. Epiz. 6, 981–4181027.
23. Thys E, Schiere H, Van Huylenbroeck G, Mfoukou-Ntsakala A, Oueadraogo M, et al. (2006) Three approaches for the integrated assessment of urban household livestock production systems: Cases from Sub-Saharan Africa. Outlook on Agriculture,35: 7–18.
24. Boukary AR, Chaïbou M, Marichatou H, Vias G (2007) Caractérisation des systèmes de production laitière et analyse des stratégies de valorisation du lait en milieu rural et périurbain au Niger : cas de la communauté urbaine de Niamey et de la commune rurale de Filingué. Rev. Elev Med Vet Pays Trop 60: 113–120.
25. Marichatou H., Kore H., Motcho HK, Vias G (2005) Synthèse bibliographique sur les filières laitières au Niger. p.40. Available:http://www.repol.info/IMG/pdf/Synthese_biblio_du_Niger.pdf. Accessed 2013 Jan 11.
26. Thrusfield M (2005) Surveys. In Thrusfield M (ed.) *Veterinary epidemiology*. Blackwell, Oxford: 228–242 32.
27. Wagner B, Salman MD (2004) Strategies for two-stage sampling designs for estimating herd-level prevalence. Prev. vet. Med. 66, 1–17.
28. Limet JN, Kerkhofs P, Wijffels R, Dekeyser P (1988) Le diagnostic sérologique de la brucellose bovine par ELISA. Ann Méd Vét 132: 565–575
29. Alton GG, Jones LM, Angus RD, Verger GM (1988) Techniques for the Brucellosis Laboratory. INRA. pp 29–61, 82–85
30. Bankole AA, Saegerman C, Berkvens D, Fretin D, Geerts S, et al. (2010) Phenotypic and genotypic characterisation of Brucella strains isolated from cattle in the Gambia. Veterinary Record 166: 753–756.
31. Le Flèche P, Jacques I, Grayon M, Dahouk SA, Bouchon P, et al. (2006) Evaluation and selection of tandem repeat loci for a Brucella MLVA typing assay. BMC Microbiol 6: 9.
32. Rogan W, Gladen B (1978) Estimating prevalence from the results of a screening test. Am J Epidemiol 107: 71–76.
33. Holt HR, Eltholth MM, Hegazy MY, El-Tras WF, Tayel AA, et al. (2011) *Brucella* spp. infection in large ruminants in an endemic area of Egypt: cross-sectional study investigating seroprevalence, risk factors and livestock owner's knowledge, attitudes and practices (KAPs). BMC Public Health 11: 341. doi:10.1186/1471-2458-11-341.
34. Snijders TAB, Bosker RJ (1999) Multilevel analysis: an introduction to basic and advanced multilevel modeling. London: SAGE publications. 266 p.
35. Solorio-Rivera JL, Segura-Correa JC, Sánchez-Gil LG (2007) Seroprevalence of and risk factors for brucellosis of goats in herds of Michoacan, Mexico. Prev Vet Med. 82: 282–290.
36. Twisk JWR (2003) Applied Longitudinal Data Analysis for Epidemiology: A practical guide. Cambridge: Cambridge University Press. 320 p.
37. Frankena K, Somers JG, Schouten WG, van Stek JV, Metz JH, et al. (2009) The effect of digital lesions and floor type on locomotion score in Dutch dairy cows. Prev Vet Med 88: 150–157.

38. Heinze G, Schemper M (2002) A solution to the problem of separation in logistic regression. Stat Med. 30:21(16), 2409–19.

39. Boukary AR, Thys E, Abatih E, Gamatie D, Ango I, et al. (2011) Bovine Tuberculosis Prevalence Survey on Cattle in the Rural Livestock System of Torodi (Niger). Plos One 6(9), e24629.

40. Domenech J. Corbel M, Thomas E, Lucet P (1983) La brucellose bovine en Afrique centrale: VI. Identification et typage des souches isolées au Tchad et au Cameroun. Rev Elev Méd vét Pays trop 36: 19–25.

41. Nielsen K (2002) Diagnosis of brucellosis by serology. Vet Microbiol 90: 447–459.

42. Saegerman C, De Waele L, Gilson D, Godfroid J, Thiange P, et al. (2004) Field evaluation of three serum ELISA using monoclonal antibodies or protein G as peroxidase conjugate for the diagnosis of bovine brucellosis. Vet Microbiol 100: 91–105.

43. Boukary AR, Saegerman C, Rigouts L, Matthys F, Berkvens D, et al. (2010) Preliminary results of the study on zoonotic brucellosis and tuberculosis in Niamey. In: Globalization of Tropical Animal Diseases and Public Health Concerns; proceedings of 13th AITVM 2010 International Conference, 23–26 August 2010, Bangkok, Thailand. [Bangkok]: [Chulalongkorn University; Utrecht: Association of Institutions for Tropical Veterinary Medicine (AITVM)]; 2010. pp. 22–24.

44. Schelling E, Diguimbaye C, Daoud S, Nicolet J, Boerlin P, et al. (2003) Brucellosis and Q-fever seroprevalences of nomadic pastoralists and their livestock in Chad. Prev Vet Med 61: 279–293.

45. Ibrahim N, Belihu K, Lobago F, Bekena M (2010) Sero-prevalence of bovine brucellosis and risk factors in Jimma zone of Oromia Region, South-western Ethiopia. Trop Anim. Health Prod 42, 35–40.

46. Sanogo M, Abatih E, Thys E, Fretin D, Berkvens D, et al. (2012) Risk factors associated with brucellosis seropositivity among cattle in the central savannah-forest area of Ivory Coast. Prev Vet Med. 107: 51–56.

47. Turkson P, Boadu D (1992) Epidemiology of bovine brucellosis in the Coastal Savanna zone of Ghana. Acta Tropica, 52: 39–43.

48. Musa MT, Shigidi MTA (2001) Brucellosis in Camels in Intensive Animal Breeding Areas of Sudan. Implications in Abortion and Early-Life Infections. Revue Élev Méd vét Pays trop 54: 11–15.

49. Faye B, Castel V, Lesnoff M, Rutabinda D, Dhalwa J (2005) Tuberculosis and brucellosis prevalence survey on dairy cattle in Mbarara milk basin (Uganda). Prev Vet Med 67: 267–281.

50. Corbel MJ, Brinley-Morgan WJ (1984) Genus *Brucella*, Meyer and Shaw 1920. 173AL. In: Krieg N R., Holt J G., editors. Bergey's Manual of Systematic Bacteriology.Vol. 1. Williams & Wilkins. pp 377–388.

51. Corbel M (1988) Brucellosis. In: Laing J. editor Fertility and Infertility in Veterinary Practice. 4th edition. ELBS, Bailliere Tindall: pp.190–221.

52. Saegerman C, Berkvens D, Godfroid J, Walravens K (2010) Chapter 77: Bovine brucellosis. In: Infectious and Parasitic Disease of Livestock. Lavoisier et Commonwealth Agricultural Bureau – International (ed.), France, 971–1001.

53. Verger JM, Grayon M, Doutre MP, Sagna F (1979) *Brucella abortus* d'origine bovine au Sénégal : identification et typage. Rev Elev Méd vét Pays trop 32, 25–32.

54. Muendo EN, Mbatha PM, Macharia J, Abdoel TH, Janszen PV, et al. (2012) Infection of cattle in Kenya with Brucella abortus biovar 3 and Brucella melitensis biovar 1 genotypes. Trop Anim Health Prod. 44, 17–20.

55. Ferreira AC, Chambel L, Tenreiro T, Cardoso R, Flor L, et al. (2012) MLVA16 typing of Portuguese human and animal Brucella melitensis and Brucella abortus isolates. PLoS One. 7(8): e42514. doi: 10.1371/journal.pone.0042514.

56. Benkirane A. (2001) Surveillance épidémiologique et prophylaxie de la brucellose des ruminants: l'exemple de la région Afrique du nord et Proche-Orient. Rev. Sci. Tech. 20, 757–767.

Comparison of Airway Responses in Sheep of Different Age in Precision-Cut Lung Slices (PCLS)

Verena A. Lambermont[1][9], **Marco Schlepütz**[2][9], **Constanze Dassow**[2], **Peter König**[3], **Luc J. Zimmermann**[1], **Stefan Uhlig**[2], **Boris W. Kramer**[1], **Christian Martin**[2]*

1 Department of Pediatrics, Maastricht University Medical Center, Maastricht, the Netherlands, **2** Institute of Pharmacology and Toxicology, University Hospital Aachen, Aachen, Germany, **3** Institute of Anatomy, University of Lübeck, Airway Research Center North (ARCN), Member of the German Center for Lung Research (DZL), Lübeck, Germany

Abstract

Background: Animal models should display important characteristics of the human disease. Sheep have been considered particularly useful to study allergic airway responses to common natural antigens causing human asthma. A rationale of this study was to establish a model of ovine precision-cut lung slices (PCLS) for the *in vitro* measurement of airway responses in newborn and adult animals. We hypothesized that differences in airway reactivity in sheep are present at different ages.

Methods: Lambs were delivered spontaneously at term (147d) and adult sheep lived till 18 months. Viability of PCLS was confirmed by the MTT-test. To study airway provocations cumulative concentration-response curves were performed with different allergic response mediators and biogenic amines. In addition, electric field stimulation, passive sensitization with house dust mite (HDM) and mast cells staining were evaluated.

Results: PCLS from sheep were viable for at least three days. PCLS of newborn and adult sheep responded equally strong to methacholine and endothelin-1. The responses to serotonin, leukotriene D_4 and U46619 differed with age. No airway contraction was evoked by histamine, except after cimetidine pretreatment. In response to EFS, airways in PCLS from adult and newborn sheep strongly contracted and these contractions were atropine sensitive. Passive sensitization with HDM evoked a weak early allergic response in PCLS from adult and newborn sheep, which notably was prolonged in airways from adult sheep. Only few mast cells were found in the lungs of non-sensitized sheep at both ages.

Conclusion: PCLS from sheep lungs represent a useful tool to study pharmacological airway responses for at least three days. Sheep seem well suited to study mechanisms of cholinergic airway contraction. The notable differences between newborn and adult sheep demonstrate the importance of age in such studies.

Editor: Ali Önder Yildirim, Institute of Lung Biology and Disease (iLBD), Helmholtz Zentrum München, Germany

Funding: The authors have no funding or support to report.

Competing Interests: The authors have declared that no competing interests exist.

* Email: chmartin@ukaachen.de

[9] These authors contributed equally to this work.

Introduction

The prevalence of asthma increased during the last decades [1,2] and may be related to Western lifestyle factors [2,3]. However, the causal reasons and underlying mechanisms are not well understood. Several studies have shown that in most cases of persistent asthma, the initial asthma-like symptoms occur during the first years of life [4,5]. It has been suggested that the child's environment plays an important role to develop asthma later in life. Children exposed to a farm environment had less asthma and atopy than children grown-up in a non-farming setting [6–9]. The mechanism associated with this protective effect is unknown. It is suggested that the child's immune system may be stimulated along a Th1 pathway by early exposure to increased concentrations of bacterial components present in stables such as endotoxin (LPS) [10]. This theory is known as the hygiene hypothesis [11]. For some of these children asthma symptoms seem to remit with time,

but many children develop asthmatic symptoms which persist throughout their life and are associated with more severe symptoms ending in the loss of lung function. About 15% of the wheezing infants develop persistent wheezing and clinical asthma later in life [12].

Animal models of asthma should display the pathology of the human disease and have carefully to be selected. Several studies indicate for instance that the innervation of the lung differs considerably between and within species [13,14]. Moreover, rodent airways do not or only weakly respond to leukotrienes [15], mediators that readily cause bronchoconstriction in humans [16] and guinea pigs [17]. Sheep show a greater resemblance to humans concerning lung development compared to rodents. Rodents and guinea pigs undergo the alveolar phase of lung development postnatally whereas sheep and humans undergo this phase in the uterus [18–20]. In addition sheep can, like rodents, be sensitized to house dust mite (HDM) antigen which is a common human antigen in asthma, and have allergen-specific IgE responses and acute eosinophil responses to allergen challenge [21,22]. Therefore, sheep have been considered particularly useful as models to study allergic airway responses to common human natural antigens.

Airway responses can be visualized by precision-cut lung slices (PCLS), which are viable lung tissue slices of uniform thickness ($\approx$250 µm). PCLS can easily be prepared from different species and are already established for many species including rat, mouse, guinea pig, non-human primates and humans [14,17,23–26]. PCLS represent a highly useful model to study bronchial and pulmonary vascular responses by videomicroscopy [23,27]. The responses in pulmonary vessels strips of newborn and adult sheep have shown interesting differences in reactivity [28]. The diameter of the pulmonary vessels increased from the newborn to the adult animals and the maximum velocity of shortening was in newborns much higher than in adult sheep. The responses of airways in newborn and adult sheep have not yet been investigated. We hypothesize that differences in airway smooth muscle reactivity in sheep are present at different ages.

In the present study the bronchoconstriction of newborn and adult sheep was studied in PCLS. A rationale of this study was to establish a model of ovine PCLS for the measurement of airway responses to early allergic response mediators and to allergens (after passive sensitization) in newborn and in adult animals.

Materials and Methods

Animal model

All animal procedures were approved by the Animal Ethics Committee of the University of Maastricht, The Netherlands. The newborns (n = 7) were delivered spontaneously at term (147 ± 2 days GA). The lambs were euthanized directly after surgical delivery with an intra-venous injection of pentobarbital. Adult sheep (n = 5) were euthanized at ~18 months with pentobarbital.

Preparation of PCLS

PCLS were prepared from adult or newborn sheep lung as described in previous studies [14,17,23–25] with some modifications. Briefly, the lungs were filled via the lobular bronchus with 1.5% low-melting-point agarose solution and put onto ice until the agarose had solidified. Tissue cores (1 cm in diameter) with a penetrating airway were punched out. Those cylinders were then cut perpendicular to the airway by means of a Krumdieck tissue slicer (Alabama Research and Development, Munford, AL, USA) into approximately 250 µm thin PCLS. PCLS were transferred into a 10 cm cell culture dish and incubated under cell culture conditions (37°C, 5% CO_2 atmosphere) in minimal essential medium (MEM) that was frequently changed during the next 4 hours (h) and incubated over night. The medium exchange supported the removal of tissue released mediators as well as the wash out of agarose from airways. For measurements, only slices with airways free of agarose, with beating cilia and an intact and relaxed airway smooth muscle layer were used. We only studied PCLS with comparable airway size in parallel experiments to reduce interslice variations. Electric field stimulation (EFS) on PCLS and passive sensitization studies were performed within 24 h after preparation. All other physiological measurements were conducted within 48 h after preparation.

Viability of PCLS

Viability of PCLS over three day incubation was confirmed by intracellular reduction of a tetrazolium dye to its according purple formazan (MTT-test) and constriction responses. PCLS were transferred into cavities of a standard 24-well plate (1 PCLS/well) and incubated with 900 µL MEM + 100 µL 3–(4,5–dimethylthiazol–2–yl)–2,5–diphenyl tetrazolium bromide solution (MTT, 7 mg/mL) for 15 min. The supernatant was discarded and the formazan was dissolved by incubation of PCLS with 200 µL formic acid/propanol (5%/95%) solution for 20 min. 100 µL of the purple supernatant were taken and transferred to 96-well plates to measure the extinction at 550 nm (Tecan GENios Microplate Reader). All reactions were carried out at room temperature and in the dark. For negative control measurements PCLS were digested with 0.2% (v/v) Triton–X100 (300 µL, 20 min, 37°C) before performing the MTT-assay.

Videomicroscopy

If not otherwise stated, PCLS were kept in cavities of standard 24-well plates and were immersed in 1 mL MEM during the experiment. The plate was then mounted on the stage of an inverted Leica DMIL microscope (Leica Microsystems, Wetzlar, Germany). The airways were imaged and digitized by videomicroscopy (SensiCam 365KL digital camera, Visitron Systems, Munich, Germany; Optimas 6.5 software, Optimas, Bothell, WA, USA). The airway area before any provocation was defined as 100%-initial airway area [%-IAA].

Airway provocations by early allergic response mediators and biogenic amines

Cumulative concentration-response curves were performed with methacholine (10^{-10} M–10^{-4} M; 5 min/conc.; pictures every 5 s), serotonin (10^{-10} M–10^{-4} M; 5 min/conc.; pictures every 5 s), histamine (10^{-9} M–10^{-4} M; 5 min/conc.; pictures every 5 s), endothelin-1 (10^{-12} M–10^{-6} M; 10 min/conc.; pictures every 10 s), leukotriene D4 (LTD$_4$, 10^{-12} M–10^{-6} M; 10 min/conc.; pictures each 5 s) and the thromboxane A$_2$ analogue U46619 (10^{-10} M–10^{-5} M; 10 min/conc.; pictures every 5 s). To study a potential effect of the histamine H$_2$-receptor on bronchoconstriction, PCLS were pre-incubated with 10 µM cimetidine for 15 min prior to performing the histamine concentration-response curve.

Electric field stimulation (EFS)

As described before [29], EFS of PCLS was carried out in standard 12-well plates at a reaction volume of 1 mL standard MEM. The PCLS were placed in between two platinum electrodes of 12 mm distance and were mounted by a Teflon ring. The electric field was applied by a Hugo Sachs Electronics Stimulator II (Hugo Sachs Electronics, March Hugstetten, Germany). The electric stimuli were defined by a frequency of 50 Hz, pulse duration of 1 ms, a current amplitude of 200 mA, a train width of 2.5 s and a train rhythm of 60 s. Each train lasted 3.3 min. After the first control stimulation the muscarinic antagonist atropine (10 µM) was added and incubated for 15 min prior to the second stimulation. In an additional set of EFS experiments, frequency response curves were conducted on PCLS from adult and newborn sheep. In this, the frequency was steadily increased from 0.4 Hz–100 Hz, while the pulse duration and current amplitude were kept constant at 1 ms and 200 mA, respectively. Each frequency was applied once for 2.5 s and after a pause of one minute the next frequency was applied. The airway behavior of PCLS upon stimulation was monitored by videomicroscopy. Pictures were taken every 2.5 s. Airway area before the first stimulation was defined as 100% initial airway area (IAA).

Passive sensitization

Passive sensitization was performed by incubation with 1% serum from house dust mite (HDM)−sensitized sheep overnight. Medium was replaced by fresh serum−free MEM directly before provocation with 5000 Units of HDM from ALK−SHERAX (Wedel, Germany), which is normally used for intracutaneous testing of atopy. Airway responses were followed by videomicroscopy for 20 min. Pictures were taken every 10 s and pruned for clarity to each fifth data point.

Mast cell staining

Mast cells were stained with toluidine blue or with alcian blue and safranin. Briefly, PCLS were fixed with formaldehyde 4% (Roti−Histofix 4%, Carl Roth GMBH & Co. KG, Karlsruhe, Germany), dehydrated and embedded in paraffin. Then, 5 μm thick sections were cut, deparaffinized, rehydrated and incubated in 1% toluidin blue for 30 min. After rinsing with water, sections were dehydrated and coverslipped. Alternatively after rehydration, sections were incubated in 0.05% Alcian Blue in 0.02 M acetate buffer, pH 5.8 with 0.2 M MgCl for 4 h. After rinsing in water, sections were incubated for 5 min in 0.25% safranin 0.02 M acetate buffer, pH 5.0. After a final rinsing step, sections were dehydrated and coverslipped.

Data analysis

Data are shown as means ± standard error of the mean (SEM). Concentration− response curves were fitted by non-linear regression (4−parameter logistic equation). To analyze differences between curves, each parameter of the 4−parameter logistic equation was separately compared by the Extra sum-of-squares F-test. The 4-parameter logistic equation is $[Y = \text{Bottom} + (\text{Top-Bottom})/(1+10^{\wedge}((\text{LogEC50-X})*\text{HillSlope})]$, with the bottom being the maximal response (i.e. fit values aim towards the minimum in the concentration response curves), the top being the initial situation (i.e. fit values aim towards 100%-IAA), the logEC50 being calculated to the half maximal response and the slope initially kept variable (i.e. unconstrained from a standard value of 1). A shared concentration-response curve was plotted for adult and newborn sheep, if no difference was found in any parameter of the 4−parameter logistic equation. Minimal airway areas in EFS before and after atropine treatment were compared by Student's t−test (due to homogeneity of variance). Mixed model analysis considering changes in airway area dependent on time and treatment was performed on time courses in the passive sensitization experiments. P-values <0.05 were considered significant. The statistical analysis was performed by either GraphPad Prism 5 (GraphPad Software, La Jolla, CA) or SAS 9.1 (SAS Institute Inc., Cary, NC).

Results

Viability of PCLS

More than 50 slices were obtained from one lung lobe of either newborn or adult sheep. The viability of sheep PCLS was demonstrated by the MTT test and PCLS were viable for at least three days (figure 1). In figure 1A, high extinctions in this test indicate intact cellular reduction systems, which correlate to viability. PCLS of adult sheep had an extinction OD of 0.2 stable over 3 days compared to deterged control PCLS with an extinction OD of 0.05. PCLS of newborn sheep had also a robust extinction OD of up to 0.3 over 3 days. Functionality of airway smooth muscle contraction was shown by the application of 10^{-4} M methacholine, which evoked a strong bronchoconstriction in sheep PCLS (figure 1B and C).

Airway provocations by allergic response mediators or biogenic amines in adult and newborn sheep

In sheep PCLS all of the mediators of the early allergic response in humans, except for histamine, evoked a marked bronchoconstriction (figure 2A−F). In general, concentration-response curves are characterized by their efficacy and potency. Shared concentration-response curves in PCLS from adult and newborn sheep were found for methacholine and endothelin-1 with half-maximal responses (log EC_{50} [M]) of -7.0 ± 0.1 and -7.7 ± 0.1, respectively (figure 2A and 2D). Age related differences in potency were found for serotonin and LTD_4. For serotonin the $logEC_{50}$ [M] values were -6.3 ± 0.4 for newborn and -6.9 ± 0.2 for adult sheep (figure 2B), whereas for LTD_4 $logEC_{50}$ values were smaller for newborn than for adult sheep (-9.7 ± 0.4 [M] vs. -8.0 ± 0.6 [M]) (figure 2E). Differences in efficacy were found for U46619, as airways of adult sheep responded markedly to U46619 ($logEC_{50}$ [M] $= -5.9\pm1.0$; max contraction to $64.6\pm8.4\%$−IAA at 10^{-5} M), whereas airways of newborn sheep were mostly unresponsive (max. contraction to $93.4\pm6.5\%$−IAA at 10^{-5} M) (figure 2F).

Airway provocation by histamine with or without cimetidine

Since in the preceding experiments no airway contraction was observed by histamine, PCLS were pretreated with or without cimetidine, which blocks the H_2−receptor. Again, airways in control PCLS, i.e. without cimetidine, did not contract in adult sheep (figure 3). In contrast, stimulation of airways in PCLS with histamine after cimetidine pretreatment resulted in a marked concentration-dependent contraction ($logEC_{50} = -6.3$) (figure 3). No data were available for PCLS from newborn sheep.

Intrinsic activation of bronchoconstriction in sheep PCLS

In addition to the preceding exogenous activation of bronchoconstriction in PCLS by exogenous application of mediators, we studied whether an intrinsic activation of airway response is possible in sheep PCLS. In this regard, two approaches were chosen: EFS to study neutrally-induced bronchoconstriction and passive sensitization to evoke an early allergic response by mast cell degranulation after antigen challenge.

In response to EFS, PCLS from both adult and newborn sheep, strongly contracted to $43.7\pm13.9\%$ IAA and $34.3\pm6.5\%$ IAA (figure 4), respectively. This neurally−induced response was potently blocked by atropine resulting in an airway contraction to only $84.0\pm7.6\%$ IAA for PCLS from adult sheep and 78.6 ± 6.8 for PCLS from newborn sheep.. Moreover, to distinguish between a difference in sensitivity of PCLS from newborn and adult sheep to EFS, frequency response curves were conducted. However, a shared curve was found (figure 4 C), indicating equal sensitivity of PCLS from newborn and adult sheep to neural stimulation.

With respect to passive sensitization, provocation with 5000 U of HDM evoked a weak but significant early allergic response in PCLS from adult and newborn sheep, that had been incubated with serum from HDM sensitized sheep (figure 5A and 5B). Comparing the HDM-induced airway contractions in PCLS from adult and newborn sheep, a significant difference was found in the respective kinetics. The response to HDM was more sustained in adult than in newborn sheep. In order to study these responses, we stained for mast cells in the lung tissue since we consider them essential for the response. In accordance to the weak response, very few mast cells were found in PCLS from sheep at both ages as shown in figure 5C and 5D. We found on average less than one mast cell per 5 μm thick section.

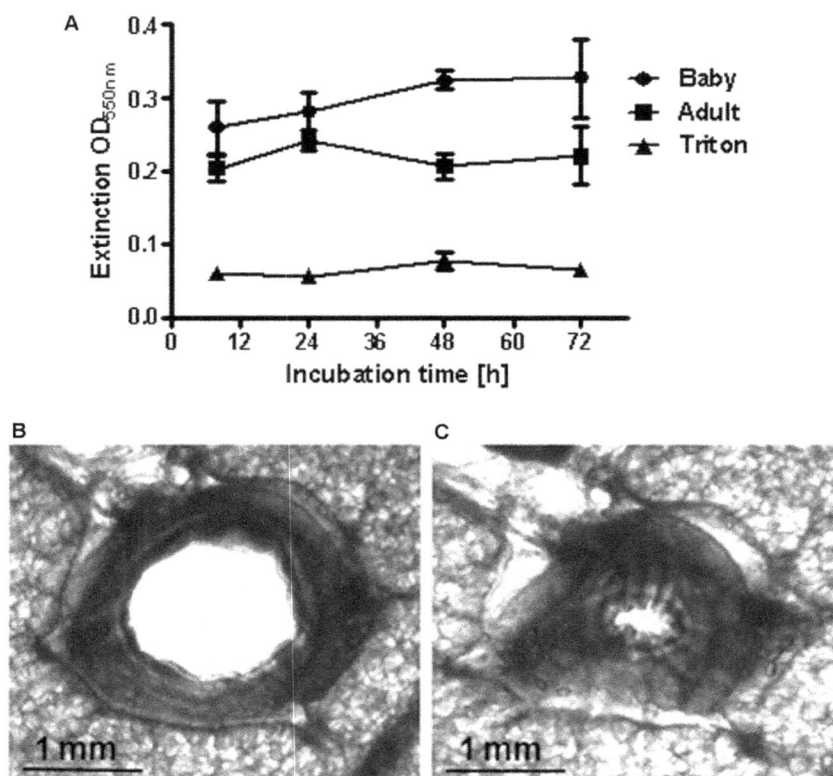

Figure 1. Viability of sheep PCLS. A) Viability was followed by the MTT test. High extinctions indicate high viability. Data are shown as mean ± SEM, n = 3 PCLS from 3 animals per group. Exemplary photographs (in black and white) of a PCLS before (**B**) and after (**C**) provocation with 10^{-4} M metacholine.

Discussion

To our knowledge this is the first time, the airway reactivity of newborn and adult sheep was compared in PCLS. Many of the responses were similar in newborn and adult sheep, such as the bronchoconstriction to EFS, and to the mediators methacholine, histamine and endothelin-1. On the other hand different responses in newborn and adult sheep were found for the eicosanoids LTD_4 and thromboxane as well as the biogenic amine serotonin. The response to allergen after passive sensitization was comparable in the extent of contraction, but prolonged in adult sheep.

PCLS are viable, can readily be obtained from different species and offer a great opportunity to study airway responses under cell culture conditions [14]. The microarchitecture of lung tissue, i.e. airways and vessels embedded in the parenchyma, remains intact and permit physiological measurements [23–26]. Even nerves remain functional and can be specifically activated [29]. In the past, PCLS from rat, mouse, guinea pig, non-human primate and human lung have been investigated [14,26]. Allergic responses in sheep have been suggested to be similar to humans [18,19,21,22], but the response of airways to early allergic response mediators in newborn compared to adult sheep had not been investigated yet. Recurrent wheezing is common in young infants and toddlers [30]. For some of these children asthma symptoms seem to remit with time, but many children develop asthmatic symptoms which persist throughout their life [12,31]. Therefore, it is important to compare the allergic responses in newborn to adult lung tissue.

Given the importance of small airways for asthma and COPD [24,25,32,33], the possibility to study these airways in PCLS is of clinical relevance. Small airways are defined by their inner diameter being less than 2 mm in human (airway generations >11) and less than 780 μm in rat airways [25,34]. Since sheep are comparable in weight to humans, lung/airway sizes and generations should correlate to the human structure. In this context of note, the current experiments were exclusively performed on small airways, since they were less than 2 mm in diameter. Therefore, our results reflect the situation only for the peripheral lung reducing interslice variations since airway responses can differ between large and small airways [23,29].

In the present study PCLS from sheep were viable for at least three days, not different from the viability of human PCLS [24], making it possible to study the effects of cytokines, growth factors and maybe even remodeling processes [35,36]. In addition, recent work by Behrsing and colleagues [37] has demonstrated that the process of slicing itself elicits a massive cytokine response, confounding early results of cytokine measurements in response to PCLS challenge. Moreover, in contrast to our study, they also used transcriptionally active agents like humulin, hydrocortisone and retinoic acid in their incubation medium. In the current study, we have performed frequent medium exchanges during the first 24 h after slicing to minimize the effect of endogenously released mediators. Nonetheless in general, the impact of mechanical damage conducted to create the PCLS should be noticed when cytokines and mediators are studied in PCLS.

PCLS of newborn and adult sheep responded very similar to methacholine and endothelin-1, mediators also effective in humans [17]. However, the response to serotonin, the thromboxane analogue U46619 and LTD_4 depended on the age of the sheep. In newborn sheep the response to serotonin and U46619

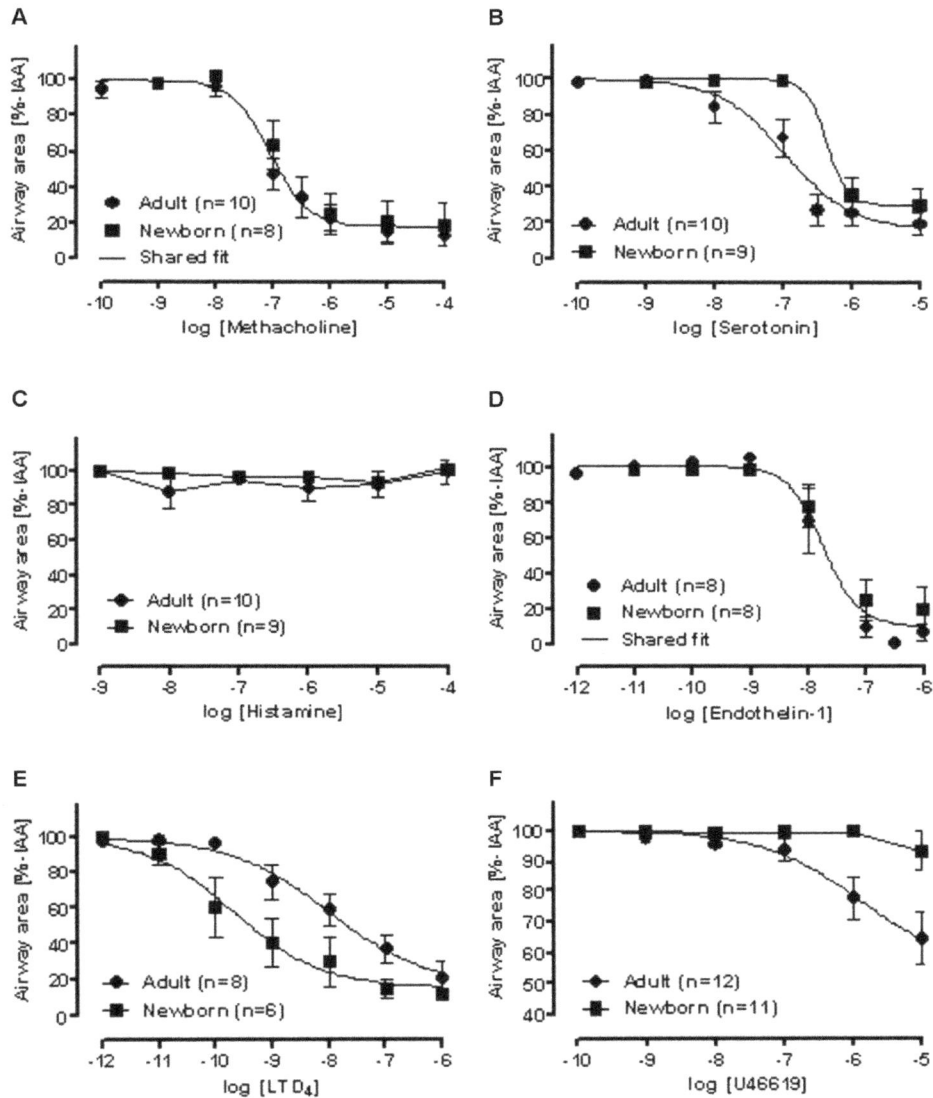

Figure 2. Concentration-response curves for common mediators of early allergic response in PCLS of adult (•) and newborn sheep (■). A) methacholine, **B**) serotonin, **C**) histamine, **D**) endothelin-1, **E**) leukotriene D_4 (LTD_4), and **F**) U46619. n = number of PCLS, whereby a total of 7 newborn and 5 adult sheep were examined. Data are shown as mean ± SEM. A shared concentration-response curve was plotted for adult and newborn sheep, if no difference was found in any parameter of the 4−parameter logistic equation.

was weaker than in adult sheep, whereas the response to LTD_4 was stronger in newborn sheep. This is an important difference compared to rodent airways that do not or at best very weakly respond to LTD_4 [15]. Leukotrienes contribute to airway obstruction in many human asthmatics [29]. In this regard sheep PCLS, in particular newborn, may appear as a suitable animal model to study bronchoconstriction. From pharmacological and clinical point of view these results are very interesting. The strong bronchial response of newborn sheep to leukotrienes suggests that young animals will be more sensitive to the effects of montelukast, which is a leukotriene-receptor antagonist and used to prevent the wheezing and shortness of breath caused by asthma. On the other hand, in the present study it is demonstrated that adult sheep have a stronger bronchial response to serotonin. Therefore, adult sheep may be more sensitive to the effects of ketanserin, which is a highly selective antagonist for contractile serotonin 5-HT_{2A} receptors.

Our studies show a lifetime dependent response pattern to different antagonist, which suggest specific disease treatment with different antagonist depending on the age of the sheep.

Compared to human PCLS, there are also differences: sheep airways reacted to serotonin rather than to histamine, which is the opposite to what is observed in human and guinea pigs PCLS, in which histamine (EC50 2.7 mM) is effective but serotonin is not [17,24,25]. The reason for these species differences is not clear. It was suggested that it may, at least in part, be related to age differences in tissue as shown in young healthy guinea pigs versus middle-aged adults [17,24,25]. However, the present study demonstrated that the bronchoconstriction in both newborn and adult sheep was the same and that age differences may not play such a crucial role if one compares responses from animal tissue with that from human tissue.

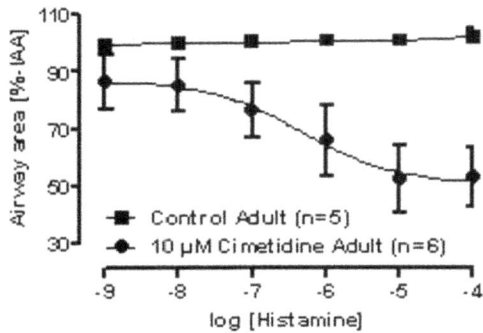

Figure 3. H₂−receptor counteracts histamine-induced bronchoconstriction in adult sheep PCLS. PCLS from adult sheep were either pre-incubated with the H_2-receptor antagonist cimetidine before histamine provocation (●) or kept untreated (control, ■). Data are shown as mean ± SEM, n = number of PCLS of separate sheep. The concentration response curves were fitted by the 4−parameter logistic equation.

Importantly, however, the lack of responsiveness to histamine was not due to the lack of H_1-receptors. This was revealed by experiments with the H_2-receptor antagonist cimetidine (figure 3), which has also minor affinity for H_1-, H_3- and H_4-receptors [38,39]. Here, we consider effects of H_3 and H_4 activities unlikely, because H_3 is predominantly expressed in brain tissue and H_4 is attributed to immune cell responses [40,41]. Moreover, in literature there are no reports about H_3 and H_4 expression in sheep lung. Since no analysis of the different histamine receptor expression was performed in lung tissue in the present study the discussion about H_3 and H_4 remains speculative and is a clear limitation of this study. Normally, in the lungs histamine acts on H_1- and H_2-receptors and its effect is therefore balanced by the H_1-mediated contraction and the H_2-mediated relaxation of airway smooth muscles [42–45]. The observation that histamine contracts tracheal and bronchial airways, but relaxes smaller bronchi and bronchioles [46], indicates that there is a shift towards predominance of H_2−receptors towards the peripheral airways. Thus, the present finding that histamine was able to contract ovine airways only after blockade of H_2-receptors, may reflect the fact that the PCLS were derived from the lung periphery. In addition, the expression of histamine H_2-receptors in the lungs may also differ depending on the type of lung sample, previous sensitization or infection, frequently leading to down-regulation of H_2-receptors [47–49]. Therefore, H_2-receptor deficiency in the airways may, at least in part, explain the commonly observed airway hyperreactivity to histamine in asthmatics.

In the present study, the early allergic response in passively sensitized sheep PCLS was weak. One explanation might be that after degranulation mast cells mainly release histamine, which – as seen before – is ineffective unless H_2-receptors are blocked. Alternatively, this finding can be explained by the small number of mast cells found in the airways of PCLS from non-sensitized sheep. It was demonstrated that exposure to HDM (active sensitization) doubled the number of mast cells mostly in alveolar septa and in airway walls [22,50]. Notably, these responses required at least 16 weekly HDM challenges [50], while in the present study the sheep lungs were unchallenged and PCLS were only passively sensitized with serum from actively sensitized sheep. This might explain the small numbers of mast cells in the present study. Another interesting observation in the allergen-induced bronchoconstriction after passive sensitization was the prolonged airway

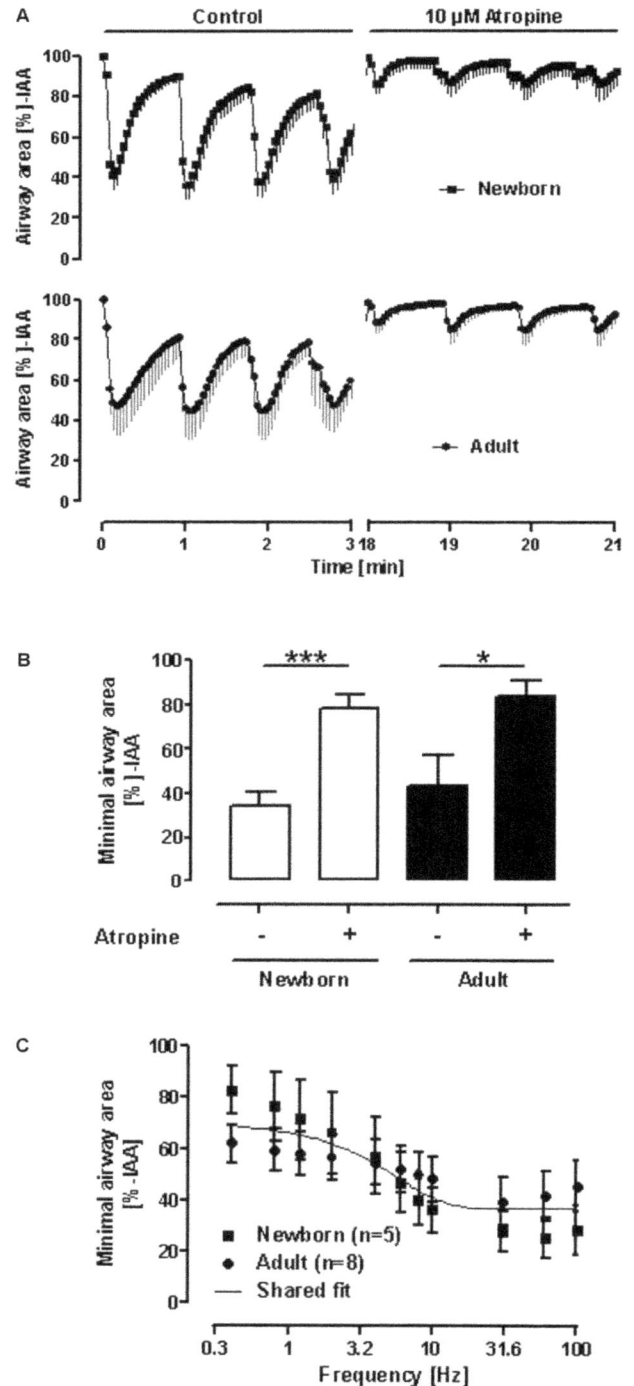

Figure 4. Electric field stimulation (EFS) of PCLS from newborn and adult sheep. A) Course of airway area changes during repeated EFS in absence or presence of atropine. **B)** Statistical analysis on minimal airway area during EFS as obtained in **A**. Data are shown as mean ± SEM, n = 7 for newborn sheep and n = 4 for adult sheep, whereby each PCLS was taken from an independent sheep; * $p < 0.05$; *** $p < 0.001$ in Student's t−test. **C)** Frequency-response curves of EFS-induced airway contractions. Data are shown as mean ± SEM, n = 5 PCLS from three newborn sheep and n = 8 PCLS from eight adult sheep. A shared frequency-response curve was plotted for adult and newborn sheep, since there was no difference in any parameter of the 4−parameter logistic equation assuming that the top is equal to 100%-IAA and EF_{50} is larger than zero Hz.

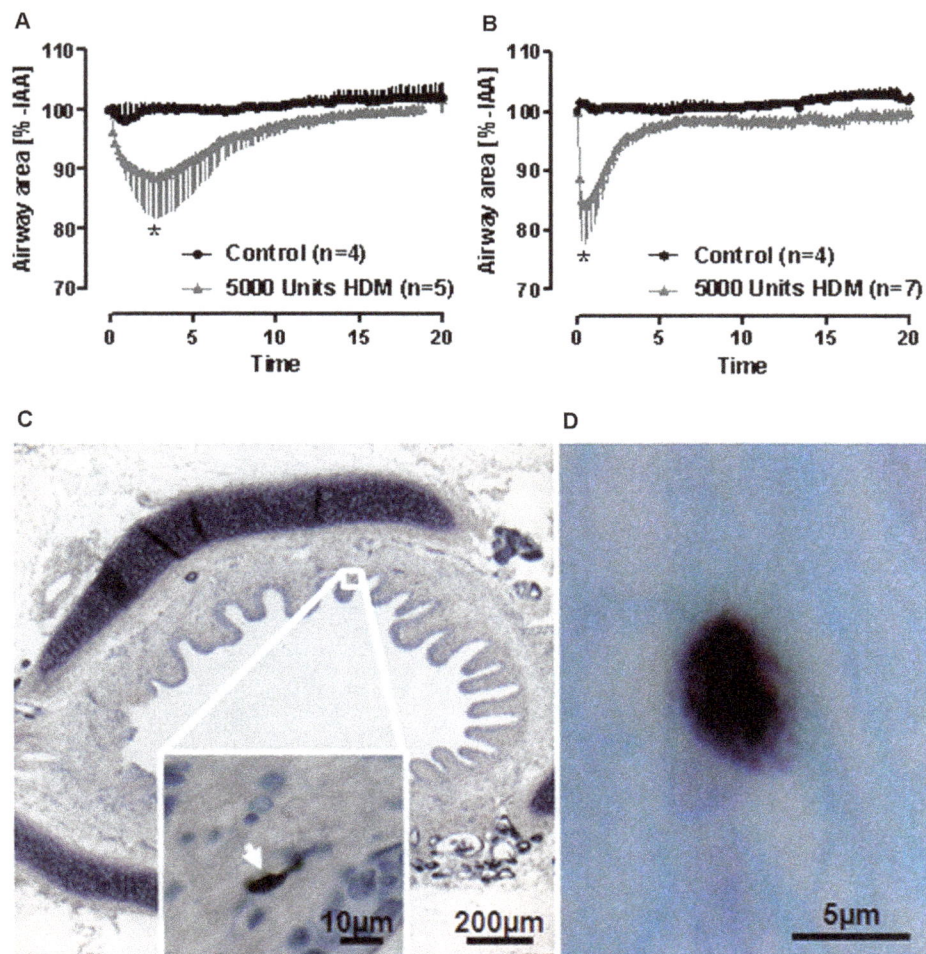

Figure 5. Early allergic response in sheep PCLS after passive sensitization. PCLS from newborn (**A**) or adult (**B**) sheep were passively sensitized with serum from actively sensitized sheep against HDM and provoked with 5000 U HDM. Data are shown as mean ± SEM, n as indicated on PCLS from different sheep. Time courses were statistically compared by mixed model analysis; *p<0.05. If one compares HDM groups in A and B, courses are also significant different, which points out differences in the kinetics of allergen-induced bronchoconstriction in adult and newborn sheep. **C, D**) Less than one mast cell per 5 μm thick section were found in PCLS from sheep. PCLS were stained for mast cells with toluidine blue (**C, arrow**) or with alcian blue and safranin (**D**).

contraction in PCLS from adult sheep compared to newborn sheep. To our knowledge this is the first study examining the early allergic response of adult and newborn animals in parallel. Reasons, such as a sustained mediator release, diminished degradation of broncho-constrictors or different receptor density in adult sheep, are therefore speculative. Nonetheless, this example confirms, that extreme caution is mandated selecting the appropriate animal model, when studying childhood asthma or asthma in adults as responses may differ tremendously.

We further demonstrated that ovine PCLS respond to electric field stimulation. Airway responses to electric field stimulation differ largely between species [14]. PCLS from rats and marmoset airways contract by about 20%, whereas guinea pigs, sheep and humans contract at maximum by about 40–60% [14]. The present study adds that the maximum contraction (>50%) in both newborn and adult sheep did not differ. Moreover, since the sensitivity of sheep PCLS from newborn and adult sheep did not differ, one may speculate that the neural network in sheep lung is widely established by birth. The shared frequency-response curve is also in line with the exogenous application of methacholine, in

which also a shared curve, i.e. same sensitivity to the agonist, was found. These findings are rounded off by the high atropine-sensitivity of EFS-induced airway contractions pointing out that lung innervation in sheep is mostly cholinergic. Hence, airway responses in sheep PCLS are a reasonable proxy for human airways, if cholinergic airway contractions are intended to study.

In our model we focused on the acute response to exogenously added inflammatory mediators and passive HDM sensitization to demonstrate that ovine PCLS are a reliable tool for in vitro measurement of airway responses. It was beyond the scope of this study to examine the role of cytokine and mediator secretion in response to a physiologically relevant stimulus or active sensitiza-tion. In future research it will be relevant to compare acute versus chronic responses after HDM sensitization and also to compare HDM to sensitization with other allergens. In conclusion, PCLS from sheep lungs represent a useful tool to study pharmacological airway responses for at least three days. Sheep seem well suited to study mechanisms of cholinergic airway contraction. Their airway pharmacology differs in some respects to that observed in humans. Bronchoconstriction is similar in newborn and adult sheep, except for the lipid mediators (LTD4, U46619) and serotonin. Early allergic response is weak which is probably based on a small number of mast cells.

Author Contributions

Conceived and designed the experiments: LJZ SU BWK CM. Performed the experiments: VAL MS CD PK. Analyzed the data: VAL MS CD PK LJZ. Contributed reagents/materials/analysis tools: LJZ SU BWK CM. Wrote the paper: VAL MS.

References

1. Manning PJ, Goodman P, O'Sullivan A, Clancy L (2007) Rising prevalence of asthma but declining wheeze in teenagers (1995–2003): ISAAC protocol. Ir Med J 100: 614–615.
2. Beasley R, Crane J, Lai CK, Pearce N (2000) Prevalence and etiology of asthma. J Allergy Clin Immunol 105: S466–472.
3. Britton J (2003) Parasites, allergy, and asthma. Am J Respir Crit Care Med 168: 266–267.
4. Martinez FD, Wright AL, Taussig LM, Holberg CJ, Halonen M, et al. (1995) Asthma and wheezing in the first six years of life. The Group Health Medical Associates. N Engl J Med 332: 133–138.
5. Barbee RA, Dodge R, Lebowitz ML, Burrows B (1985) The epidemiology of asthma. Chest 87: 21S–25S.
6. Von Ehrenstein OS, Von Mutius E, Illi S, Baumann L, Bohm O, et al. (2000) Reduced risk of hay fever and asthma among children of farmers. Clin Exp Allergy 30: 187–193.
7. Riedler J, Eder W, Oberfeld G, Schreuer M (2000) Austrian children living on a farm have less hay fever, asthma and allergic sensitization. Clin Exp Allergy 30: 194–200.
8. Braun-Fahrlander C, Gassner M, Grize L, Neu U, Sennhauser FH, et al. (1999) Prevalence of hay fever and allergic sensitization in farmer's children and their peers living in the same rural community. SCARPOL team. Swiss Study on Childhood Allergy and Respiratory Symptoms with Respect to Air Pollution. Clin Exp Allergy 29: 28–34.
9. Riedler J, Braun-Fahrlander C, Eder W, Schreuer M, Waser M, et al. (2001) Exposure to farming in early life and development of asthma and allergy: a cross-sectional survey. Lancet 358: 1129–1133.
10. von Mutius E, Braun-Fahrlander C, Schierl R, Riedler J, Ehlermann S, et al. (2000) Exposure to endotoxin or other bacterial components might protect against the development of atopy. Clin Exp Allergy 30: 1230–1234.
11. Renz H, Blumer N, Virna S, Sel S, Garn H (2006) The immunological basis of the hygiene hypothesis. Chem Immunol Allergy 91: 30–48.
12. Guilbert T, Krawiec M (2003) Natural history of asthma. Pediatr Clin North Am 50: 523–538.
13. Abraham WM (2000) Animal models of asthma. In: Busse WW, Holgate ST, editors. Asthma and rhinitis. Oxford; Malden, MA, USA: Blackwell Science. pp. 1205–1227.
14. Schleputz M, Rieg AD, Seehase S, Spillner J, Perez-Bouza A, et al. (2012) Neurally Mediated Airway Constriction in Human and Other Species: A Comparative Study Using Precision-Cut Lung Slices (PCLS). PloS one 7: e47344.
15. Held HD, Martin C, Uhlig S (1999) Characterization of airway and vascular responses in murine lungs. Br J Pharmacol 126: 1191–1199.
16. Dahlen SE, Hedqvist P, Hammarstrom S, Samuelsson B (1980) Leukotrienes are potent constrictors of human bronchi. Nature 288: 484–486.
17. Ressmeyer AR, Larsson AK, Vollmer E, Dahlen SE, Uhlig S, et al. (2006) Characterisation of guinea pig precision-cut lung slices: comparison with human tissues. Eur Respir J 28: 603–611.
18. Kramer BW (2011) Chorioamnionitis - new ideas from experimental models. Neonatology 99: 320–325.
19. Pringle KC (1986) Human fetal lung development and related animal models. Clin Obstet Gynecol 29: 502–513.
20. Wolfs TG, Jellema RK, Turrisi G, Becucci E, Buonocore G, et al. (2012) Inflammation-induced immune suppression of the fetus: a potential link between chorioamnionitis and postnatal early onset sepsis. J Matern Fetal Neonatal Med 25 Suppl 1: 8–11.
21. Bischof RJ, Snibson K, Shaw R, Meeusen EN (2003) Induction of allergic inflammation in the lungs of sensitized sheep after local challenge with house dust mite. Clin Exp Allergy 33: 367–375.
22. Snibson KJ, Bischof RJ, Slocombe RF, Meeusen EN (2005) Airway remodelling and inflammation in sheep lungs after chronic airway challenge with house dust mite. Clin Exp Allergy 35: 146–152.
23. Martin C, Uhlig S, Ullrich V (1996) Videomicroscopy of methacholine-induced contraction of individual airways in precision-cut lung slices. Eur Respir J 9: 2479–2487.
24. Wohlsen A, Martin C, Vollmer E, Branscheid D, Magnussen H, et al. (2003) The early allergic response in small airways of human precision-cut lung slices. Eur Respir J 21: 1024–1032.
25. Wohlsen A, Uhlig S, Martin C (2001) Immediate allergic response in small airways. Am J Respir Crit Care Med 163: 1462–1469.
26. Seehase S, Schleputz M, Switalla S, Matz-Rensing K, Kaup FJ, et al. (2011) Bronchoconstriction in nonhuman primates: a species comparison. J Appl Physiol 111: 791–798.
27. Rieg AD, Rossaint R, Uhlig S, Martin C (2011) Cardiovascular agents affect the tone of pulmonary arteries and veins in precision-cut lung slices. PLoS one 6: e29698.
28. Belik J, Halayko A, Rao K, Stephens N (1991) Pulmonary vascular smooth muscle: biochemical and mechanical developmental changes. J Appl Physiol 71: 1129–1135.
29. Schleputz M, Uhlig S, Martin C (2011) Electric field stimulation of precision-cut lung slices. J Appl Physiol (1985) 110: 545–554.
30. Robison RG, Singh AM (2012) Chapter 11: the infant and toddler with wheezing. Allergy Asthma Proc 33 Suppl 1: S36–38.
31. Been JV, Lugtenberg MJ, Smets E, van Schayck CP, Kramer BW, et al. (2014) Preterm birth and childhood wheezing disorders: a systematic review and meta-analysis. PLoS Med 11: e1001596.
32. Sturton RG, Trifilieff A, Nicholson AG, Barnes PJ (2008) Pharmacological characterization of indacaterol, a novel once daily inhaled 2 adrenoceptor agonist, on small airways in human and rat precision-cut lung slices. J Pharmacol Exp Ther 324: 270–275.
33. Sturton G, Persson C, Barnes PJ (2008) Small airways: an important but neglected target in the treatment of obstructive airway diseases. Trends Pharmacol Sci 29: 340–345.
34. Yeh HC, Schum GM, Duggan MT (1979) Anatomic models of the tracheobronchial and pulmonary regions of the rat. Anat Rec 195: 483–492.
35. Switalla S, Lauenstein L, Prenzler F, Knothe S, Forster C, et al. (2010) Natural innate cytokine response to immunomodulators and adjuvants in human precision-cut lung slices. Toxicol Appl Pharmacol.
36. Kasper M, Seidel D, Knels L, Morishima N, Neisser A, et al. (2004) Early signs of lung fibrosis after in vitro treatment of rat lung slices with CdCl2 and TGF-beta1. Histochem Cell Biol 121: 131–140.
37. Behrsing HP, Furniss MJ, Davis M, Tomaszewski JE, Parchment RE (2013) In vitro exposure of precision-cut lung slices to 2-(4-amino-3-methylphenyl)-5-fluorobenzothiazole lysylamide dihydrochloride (NSC 710305, Phortress) increases inflammatory cytokine content and tissue damage. Toxicol Sci 131: 470–479.
38. Wade L, Bielory L, Rudner S (2012) Ophthalmic antihistamines and H1-H4 receptors. Curr Opin Allergy Clin Immunol 12: 510–516.
39. Bielory L, Ghafoor S (2005) Histamine receptors and the conjunctiva. Curr Opin Allergy Clin Immunol 5: 437–440.
40. Zampeli E, Tiligada E (2009) The role of histamine H4 receptor in immune and inflammatory disorders. Br J Pharmacol 157: 24–33.
41. Gantner F, Sakai K, Tusche MW, Cruikshank WW, Center DM, et al. (2002) Histamine h(4) and h(2) receptors control histamine-induced interleukin-16 release from human CD8(+) T cells. J Pharmacol Exp Ther 303: 300–307.
42. Bongers G, de Esch I, Leurs R (2010) Molecular pharmacology of the four histamine receptors. Adv Exp Med Biol 709: 11–19.
43. Okamoto T, Iwata S, Ohnuma K, Dang NH, Morimoto C (2009) Histamine H1-receptor antagonists with immunomodulating activities: potential use for modulating T helper type 1 (Th1)/Th2 cytokine imbalance and inflammatory responses in allergic diseases. Clin Exp Immunol 157: 27–34.
44. Chand N, Eyre P (1975) Classification and biological distribution of histamine receptor sub-types. Agents Actions 5: 277–295.
45. Eyre P, Chand N (1982) Histamine receptor mechanisms of the lung. In: Ganellin CR, Parsons ME, editors. Pharmacology of histamine receptors. Bristol; Boston: PSG. pp. 298–332.
46. Yre P (1969) The pharmacology of sheep tracheobronchial muscle: a relaxant effect of histamine on the isolated bronchi. Br J Pharmacol 36: 409–417.
47. Chand N (1980) Distribution and classification of airway histamine receptors: the physiological significance of histamine H2-receptors. Adv Pharmacol Chemother 17: 103–131.
48. Chand N (1980) Is airway hyperactivity in asthma due to histamine H2-receptor deficiency? Med Hypotheses 6: 1105–1112.
49. Foreman JC (1991) Histamine H2 Receptors and Lung Function. In: Arrang JM, Uvnäs B, editors. Histamine and histamine antagonists. Berlin; New York: Springer Berlin Heidelberg. pp. 285–304.
50. Van der Velden J, Barker D, Barcham G, Koumoundouros E, Snibson K (2012) Increased mast cell density and airway responses to allergic and non-allergic stimuli in a sheep model of chronic asthma. PLoS One 7: e37161.

Genetic Footprints of Iberian Cattle in America 500 Years after the Arrival of Columbus

Amparo M. Martínez[1], Luis T. Gama[2,3], Javier Cañón[4], Catarina Ginja[5], Juan V. Delgado[1], Susana Dunner[4], Vincenzo Landi[1], Inmaculada Martín-Burriel[6], M. Cecilia T. Penedo[7], Clementina Rodellar[6], Jose Luis Vega-Pla[8]*, Atzel Acosta[9], Luz A. Álvarez[10], Esperanza Camacho[11], Oscar Cortés[4], Jose R. Marques[12], Roberto Martínez[13], Ruben D. Martínez[14], Lilia Melucci[15,16], Guillermo Martínez-Velázquez[17], Jaime E. Muñoz[10], Alicia Postiglioni[18], Jorge Quiroz[17], Philip Sponenberg[19], Odalys Uffo[9], Axel Villalobos[20], Delsito Zambrano[21], Pilar Zaragoza[6]

1 Departamento de Genética, Universidad de Córdoba, Córdoba, Spain, 2 L-INIA, Instituto Nacional dos Recursos Biológicos, Fonte Boa, Vale de Santarém, Portugal, 3 CIISA – Faculdade de Medicina Veterinária, Universidade Técnica de Lisboa, Lisboa, Portugal, 4 Departamento de Producción Animal, Facultad de Veterinaria, Universidad Complutense de Madrid, Madrid, Spain, 5 Centre for Environmental Biology, Faculty of Sciences, University of Lisbon & Molecular Biology Group, Instituto Nacional de Recursos Biológicos, INIA, Lisbon, Portugal, 6 Laboratorio de Genética Bioquímica, Facultad de Veterinaria, Universidad de Zaragoza, Zaragoza, Spain, 7 Veterinary Genetics Laboratory, University of California Davis, Davis, California, United States of America, 8 Laboratorio de Investigación Aplicada, Cría Caballar de las Fuerzas Armadas, Córdoba, Spain, 9 Centro Nacional de Sanidad Agropecuaria, San José de las Lajas, La Habana, Cuba, 10 Universidad Nacional de Colombia, Sede Palmira, Valle del Cauca, Colombia, 11 IFAPA, Centro Alameda del Obispo, Córdoba, Spain, 12 EMBRAPA Amazônia Oriental, Belém, Pará, Brazil, 13 Centro Multidisciplinario de Investigaciones Tecnológicas, Dirección General de Investigación Científica y Tecnológica, Universidad Nacional de Asunción, San Lorenzo, Paraguay, 14 Genética Animal, Facultad de Ciencias Agrarias, Universidad Nacional de Lomas de Zamora, Lomas de Zamora, Argentina, 15 Facultad Ciencias Agrarias, Universidad Nacional de Mar del Plata, Balcarce, Argentina, 16 Estación Experimental Agropecuaria Balcarce, Instituto Nacional de Tecnología Agropecuaria, Balcarce, Argentina, 17 Instituto Nacional de Investigaciones Forestales, Agrícolas y Pecuarias, Coyoacán, México, 18 Área Genética, Departamento de Genética y Mejora Animal, Facultad de Veterinaria, Universidad de la República, Montevideo, Uruguay, 19 Virginia-Maryland Regional College of Veterinary Medicine, Virginia Tech, Blacksburg, Virginia, United States of America, 20 Instituto de Investigación Agropecuaria, Estación Experimental El Ejido, Los Santos, Panamá, 21 Universidad Técnica Estatal de Quevedo, Quevedo, Ecuador

Abstract

Background: American Creole cattle presumably descend from animals imported from the Iberian Peninsula during the period of colonization and settlement, through different migration routes, and may have also suffered the influence of cattle directly imported from Africa. The introduction of European cattle, which began in the 18th century, and later of Zebu from India, has threatened the survival of Creole populations, some of which have nearly disappeared or were admixed with exotic breeds. Assessment of the genetic status of Creole cattle is essential for the establishment of conservation programs of these historical resources.

Methodology/Principal Findings: We sampled 27 Creole populations, 39 Iberian, 9 European and 6 Zebu breeds. We used microsatellite markers to assess the origins of Creole cattle, and to investigate the influence of different breeds on their genetic make-up. The major ancestral contributions are from breeds of southern Spain and Portugal, in agreement with the historical ports of departure of ships sailing towards the Western Hemisphere. This Iberian contribution to Creoles may also include some African influence, given the influential role that African cattle have had in the development of Iberian breeds, but the possibility of a direct influence on Creoles of African cattle imported to America can not be discarded. In addition to the Iberian influence, the admixture with other European breeds was minor. The Creoles from tropical areas, especially those from the Caribbean, show clear signs of admixture with Zebu.

Conclusions/Significance: Nearly five centuries since cattle were first brought to the Americas, Creoles still show a strong and predominant signature of their Iberian ancestors. Creole breeds differ widely from each other, both in genetic structure and influences from other breeds. Efforts are needed to avoid their extinction or further genetic erosion, which would compromise centuries of selective adaptation to a wide range of environmental conditions.

Editor: Sergios-Orestis Kolokotronis, Fordham University, United States of America

Funding: This work was partially funded by the Instituto Nacional de Investigación y Tecnología Agraria y Alimentaria (INIA) PET2007-01-C07-04, PET2007-05-C03-03, RZ01-002-C2-1, RZ01-002-C2-2, RZ2004-00009, RZ 2004-00022-00-00, RZ2006-00003-C02-01, RZ 2006-00003-C02-02, RZ 2006-00007-C03-03, RZ2008-00005-C02-02, RZ2008-00006-C02-02 and RZ 2008-00008-00-00 projects. C. Ginja has received funding from the European Union Seventh Framework Programme (FP7/2007–2013) under grant agreement n°PCOFUND-GA-2009-246542 and from the Foundation for Science and Technology of Portugal (DFRH/WIIA/15/2011). This work was also partially funded by the Veterinary Genetics Laboratory, University of California, Davis (USA) and the Diputación de Córdoba (Spain). The funders had no role in study design, data collection and analysis, decision to publish, or preparation of the manuscript.

Competing Interests: The authors have declared that no competing interests exist. The Empresa Brasileira de Pesquisa Agropecuária (EMBRAPA) (Brazilian Enterprise for Agricultural Research) is a state-owned company affiliated with the Brazilian Ministry of Agriculture, which is devoted to pure and applied research on agriculture. EMBRAPA conducts agricultural research on many topics including animal agriculture and crops. EMBRAPA (www.embrapa.br) is not a commercial company therefore the authors think that Jose R. Marques has no conflict of interest.

* E-mail: jvegpla@oc.mde.es

Introduction

"That many breeds of cattle have originated through variation, independently of descent from distinct species, we may infer from what we see in South America, where the genus Bos was not endemic, and where the cattle which now exist in such vast numbers are the descendants of a few imported from Spain and Portugal."

Charles Darwin, in The Variation of Animals and Plants Under Domestication, 1868

Columbus's trip to the Americas was one of the most important events in the history of humanity, as it produced major social and economic changes on both sides of the Atlantic. The Pre-Columbian American civilizations were predominantly agriculturalist but few were livestock keepers. The only domesticated species in the Americas were the dog, turkey, guinea pig and two Andean camelids [1]. One of the major impacts of Columbus's trip was the exchange of plant and animal genetic resources among continents, which revolutionized the way of life and food habits of populations in both Europe and the Americas [2].

Livestock species were brought from the Iberian Peninsula to the Americas since the late 15th century, starting with the second trip of Columbus, which departed from the Spanish city of Cádiz in 1493. In this trip, which had a re-supply in the Canary Islands, Columbus brought horses, cattle, sheep, goats and pigs to the Americas for the first time [3]. Afterwards, many other conquerors and settlers followed, and cattle brought from the Iberian Peninsula, and possibly directly from Africa at a later stage, spread throughout the Americas, adapting to a wide range of environmental conditions and giving origin to the populations currently known as Creole cattle [4]. After nearly 300 years of expansion of Creole cattle in the American continents, and with the development of more intensive production and breeding systems, several other European breeds were introduced into the Americas in the 19th century [5]. By the end of the 19th century, Indian cattle breeds, of the Zebu or *Bos indicus* type, were also introduced and quickly disseminated throughout the Americas, where they were extensively crossed with local populations, especially in tropical regions [6].

For over three centuries, Creole cattle were used as a source of draught power, food and leather, playing a key role in the settlement of human populations and the development of agriculture throughout the Western Hemisphere [7]. However, the successive introduction of different cattle breeds starting in the 19th century resulted in the progressive replacement of many Creole populations, which have completely disappeared in several regions or were displaced to marginal areas, where they still subsist nowadays [8]. Even though these extant populations present high levels of genetic diversity [9] and result from several centuries of adaptation to local environments, it is not clear how much of the ancestral Iberian founder contributions have been retained, or if the successive waves of other cattle introduced over the years have replaced the original contribution of Iberian stock.

The study of genetic diversity within and across breeds provides insight into population structure and relationships, and is essential for the development of conservation and breeding programs. Microsatellite genetic markers have been extensively used to assess between- and within-breed genetic diversity and inbreeding levels, introgression from other genetic groups, genetic differentiation and population structure [10–14]. The phylogeny of cattle has also been investigated with other types of genetic markers, including mtDNA [15], the non-recombining region of the Y chromosome

[16] and single nucleotide polymorphisms [17–19] The insight on breed development and introgression provided by the different types of genetic markers is complementary, with neutral genetic markers such as microsatellites essentially reflecting the consequences of genetic drift, founder effects and population admixture. This is particularly important in the case of Creoles where founder effects and genetic drift must have been dramatic considering that the total number of Iberian cattle brought to the Americas was probably less than 1000 [4].

Knowing the genetic history of Creole cattle in the Americas should provide a better understanding of livestock gene flow during the period of discovery and settlement by Iberian colonizers, and the influence that may have resulted from the later introductions of cattle from other European origins and of Zebus from India that begun in the 19th century. In addition, the assessment of genetic diversity and structure of Creole cattle populations is crucial for the development of appropriate management programs aimed at their recognition, conservation and genetic improvement.

The objective of this study was to use neutral genetic markers to retrospectively assess the origins and evolutionary trajectories of American Creole cattle, and investigate the influence that Iberian, European and Zebu breeds may have had on their genetic make-up. The influence of African cattle to the Creole breeds is also discussed, particularly the indirect contribution mediated by their Iberian counterparts. Using a subset of 81 cattle breeds sampled in Europe and the Americas, we show that the majority of the Creole breeds still maintain distinct genetic signatures of Iberian cattle, but some have been admixed with cattle from other geographic regions, mostly of the Zebu type in tropical regions and British and Continental breeds in other parts of the Americas.

Results

Genetic Diversity and Breed Differentiation

A set of 19 microsatellite markers was used to analyze samples of the 81 cattle breeds included in this study (Table S1), which represented the Creole (27 breeds), Iberian (39 breeds), British (5 breeds), Continental European (4 breeds) and Zebu (6 breeds) groups, with the geographical distribution shown in Figure 1.

The microsatellite markers used allowed the detection of a mean number of 6.78 ± 1.88 alleles/locus per breed and 11.93 ± 3.52 alleles/locus per breed group, with global observed and expected heterozygosities of 0.688 ± 0.018 and 0.711 ± 0.025, respectively (Table S2). Taken together, Creole cattle showed the highest mean (14.21 ± 3.74) and effective (4.08 ± 0.57) number of alleles, allelic richness (4.69 ± 0.51), and observed and expected heterozygosities (0.719 ± 0.004 and 0.805 ± 0.014, respectively), when compared with the other breed groups (Table 1).

The average F-statistics and their 95% confidence intervals obtained with 10,000 bootstraps over loci were $f = 0.0326$ (0.0231–0.0451), $F = 0.1360$ (0.1250–0.1479) and $<\text{theta}> = 0.1069$ (0.0977–0.1170). The group means for within-breed deficit in heterozygosity were highest for the Spanish and Zebu breeds (nearly 0.048), and lowest for the Continental European breeds (-0.002 ± 0.026). The Portuguese Mertolenga and Brava, the Spanish Negra Andaluza and the Mexican Criollo Poblano had the highest within-breed F_{IS}, with estimates close to 0.11 (Table S2).

Genetic distances among breed pairs, estimated by $<\text{theta}>$ values, ranged from 0.01 to 0.33 (results not shown for individual breeds), with a mean distance of Creoles relative to other breed groups as follows: 0.016 for Spanish, 0.018 for Portuguese, 0.023 for Continental European, 0.033 for British and 0.095 for Zebu

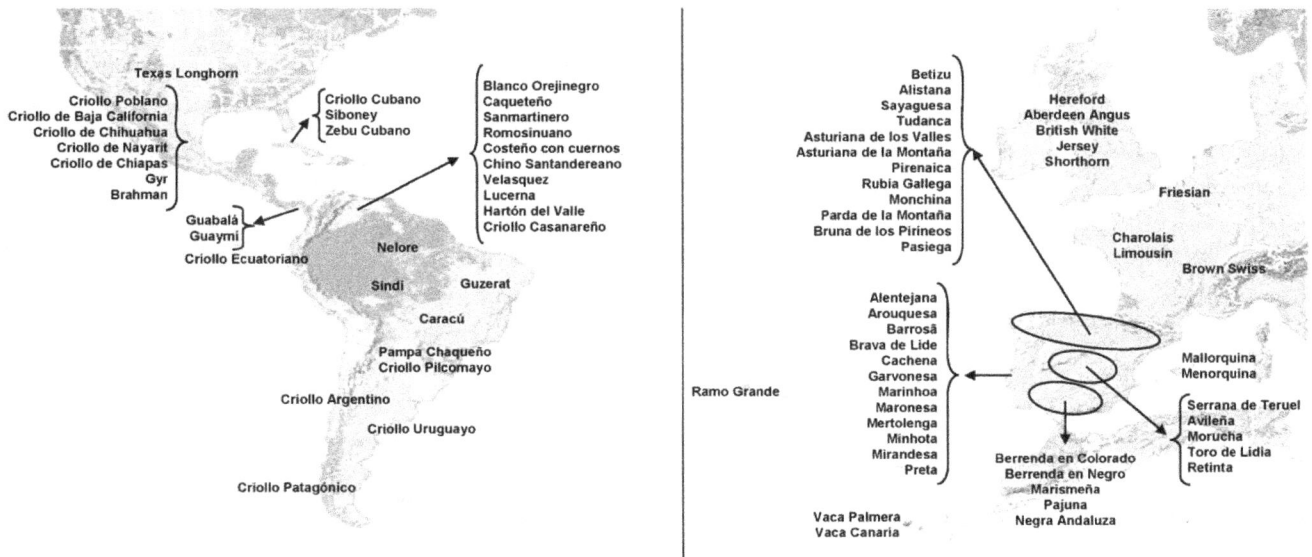

Figure 1. Geographic distribution of the 81 cattle breeds from America and Europe.

breeds (Table 2). The estimated number of migrants, i.e., the number of individuals exchanged between populations per generation that would balance the diversifying effect of genetic drift, was highest for the Creole, Spanish and Portuguese pairs, while the Zebu had the lowest number of migrants relative to all the other groups.

The results from the Factorial Correspondence Analysis (Figure 2) indicate that the first three FCA axes explain <25% of the variability. The first axis accounts for about 16% of the variability and separates *Bos indicus* from the remaining breeds. The second component, which accounts only for 5% of the variability, essentially separates the Iberian and the European breeds, while the third component accounts for a small 4% of the

total variability and allows for some splitting among breeds in the same genetic or geographical group. The Creoles occupy a more central position in the graph and, depending on the breed considered, they have a closer proximity to the Iberian, European or Indicine clusters, reflecting the influence that these groups have had in their genetic make-up.

The AMOVA results indicated that the highest percentage of variation among groups (11.4%, P<0.001) was found when breeds deriving from *B. taurus* and *B. indicus* were compared (results not shown). When the genetic differentiation of Creoles relative to Iberian, European and Zebu breeds was considered, the largest amount of variability was found between Creole and Zebu

Table 1. Genetic variability estimated for different groups of cattle breeds.

Breed Group	N	Am ± SD	Ae	Ar	Ho ± SD	He ± SD
Creole	907	14.21±3.74	4.08	4.69	0.719±0.004	0.805±0.014
Spanish	1,199	12.53±3.39	3.85	4.46	0.677±0.003	0.777±0.018
Portuguese	675	10.74±3.59	3.62	4.27	0.677±0.004	0.749±0.025
British	200	8.89±2.21	3.26	4.41	0.653±0.008	0.754±0.015
Continental European	184	9.89±3.45	3.95	4.13	0.720±0.008	0.760±0.020
Zebu	168	11.32±3.16	3.31	4.41	0.654±0.009	0.735±0.026
Mean		*11.93±3.52*	*3.68±0.34*	*4.40±0.19*	*0.683±0.030*	*0.763±0.025*

Number of individuals sampled (N), mean number of alleles (Am), effective number of alleles (Ae), allelic richness (Ar), observed (H$_o$) and expected (H$_e$) heterozygosities and their standard deviations (SD). Groups of breeds: *CREOLE*: Criollo Argentino (CARG), Criollo Patagónico (PAT), Caracú (CAR), Blanco Orejinegro (BON), Caqueteño (CAQ), Criollo Casanareño (CC), Chino Santandereano (CH), Costeño con Cuernos (CCC), Hartón del Valle (HV), Lucerna (LUC), Romosinuano (RMS), Sanmartinero (SM), Velasquez (VEL), Cubano (CUB), Siboney (SIB), Criollo Ecuatoriano (EC), Criollo de Baja California (CBC), Criollo de Chiapas (CHI), Criollo de Chihuahua (CHU), Criollo de Nayarit (CNY), Criollo Poblano (CPO), Guabalá (GUA), Guaymí (GY), Pampa Chaqueño (PA), Criollo Pilcomayo (PIL), Criollo Uruguayo (CUR) and Texas Longhorn (TLH); *SPANISH*: Alistana (ALS), Asturiana de las Montañas (ASM), Asturiana de los Valles (ASV), Avileña (AVI), Berrenda en Colorado (BC), Berrenda en Negro (BN), Betizu (BET), Bruna de los Pirineos (BRP), Mallorquina (MALL), Menorquina (MEN), Monchina (MON), Morucha (MOR), Marismeña (MAR), Negra Andaluza (NAN), Pajuna (PAJ), Parda de Montaña (PM), Pasiega (PAS), Pirenaica (PIRM), Retinta (RET), Rubia Gallega (RGA), Sayaguesa (SAY), Serrana de Teruel (STE), Toro de Lidia (TL), Tudanca (TUD), Vaca Canaria (VCA) and Vaca Palmera (PAL); *PORTUGUESE*: Alentejana (ALT), Arouquesa (ARO), Barrosã (BARR), Brava de Lide (BRAV), Cachena (CACH), Garvonesa (GARV), Marinhoa (MARI), Maronesa (MARO), Mertolenga (MERT), Minhota (MINH), Mirandesa (MIRA), Preta (PRET) and Ramo Grande (RG); *BRITISH*: Aberdeen Angus (AA), British White (BWC), Hereford (HER), Jersey (JER), Shorthorn (SH); *CONTINENTAL EUROPEAN*: Charolais (CHAR), Friesian (FRI), Limousin (LIM), Brown Swiss (BSW); *ZEBU*: Brahman (BRH), Gyr (GYR), Guzerat (GUZ), Nelore (NEL), Sindi (SIN), Zebu Cubano (CUZ).

Table 2. Genetic distances among breed groups.

	CRE	SP	PT	BR	EU	ZEB
CRE		0.016	0.018	0.033	0.023	0.095
SP	15.80	–	0.013	0.036	0.020	0.135
PT	13.78	18.51	–	0.041	0.027	0.143
BR	7.43	6.66	5.79	–	0.032	0.159
EU	10.43	11.96	9.12	7.57	–	0.156
ZEB	2.38	1.60	1.50	1.33	1.35	–

Genetic distances estimated by Weir and Cockerham <theta> (above diagonal) and corresponding number of migrants (below diagonal). Breed groups: CRE – Creole; SP – Span sh; PT – Portuguese; BR – British; EU – Continental European; ZEB –Zebu. See Table 1 for the definition of breeds included in each group.

populations (9.15%, P<0.001), and the lowest between the Creole and Iberian breeds (1.09%, P<0.001).

Population Genetic Structure

The Neighbor-net built with the Reynolds distances (Figure 3) supports the existence of two major clusters, corresponding to *B. indicus* and *B. taurus* breeds, with several Creole breeds grouped in the *B. indicus* cluster, which is interpreted as a sign of Zebu influence in their genetic make-up. These included the Creoles from Cuba and Ecuador, the Pilcomayo from Paraguay, the Criollo de Chiapas from Mexico and some Creole breeds from Colombia (Chino Santandereano, Caqueteño and Criollo Casanareño). Among the Creoles showing a residual zebu influence, the Texas Longhorn and the majority of the Mexican Creoles were closely clustered at the centre of the dendrogram, displaying a common origin with the Spanish Marismeña. Another Creole cluster, made-up by the Romosinuano and Costeño con Cuernos from Colombia and, to a lesser extent, the two breeds from Panama, showed a common origin with the breeds from the Canary Islands and the Portuguese Mertolenga. The Creoles from Argentina and Uruguay and the Caracu from Brasil formed an independent cluster at the center of the dendrogram, with a weak relationship with British breeds. On the other hand, the Pampa Chaqueño from Paraguay and the Harton del Valle, Lucerna and Blanco Orejinegro from Colombia showed a clear influence of British breeds.

Among Iberian breeds, several different clusters could be identified, such that nearly all Portuguese breeds grouped together, with the major exception of the Mirandesa, which clustered with breeds with a close geographic distribution, both in Portugal and Spain. Another cluster corresponded to the breeds from the Balearic Islands, which grouped with a few breeds from northern Spain, while the majority of the Spanish breeds clustered together. A distinct cluster corresponded to the breeds from the Canary Islands, which also included the Portuguese Mertolenga. Two Spanish breeds were isolated from the remaining clusters, i.e., the Marismeña and the Berrenda en Negro. The remaining Iberian breeds (Minhota and Ramo Grande from Portugal, Bruna de los Pirineos, Serrana de Teruel and Parda de Montaña from Spain) were close to Continental European breeds, indicating some admixture with these breeds.

The Bayesian clustering model-based method [20] allowed for assessment of the genetic structure and admixture among breeds. When the number of ancestral populations varied from K = 2 to 81, the largest change in the log of the likelihood function (ΔK) was when K = 71 (Figure S1).

The results for K = 2 (Figure 4) indicate a clear separation between *B. indicus* and *B. taurus* breeds. Moreover, these results confirm the admixture of Zebu with some of the Creole breeds, especially Siboney, Criollo Cubano, Criollo Ecuatoriano, Pilcomayo, Casanareño and Velasquez, while other breeds, such as the Creoles from Argentina and Uruguay, and the Romosinuano, Sanmartinero and Blanco Orejinegro from Colombia, show minor signs of Zebu admixture.

When three ancestral populations were inferred, the breeds from Northern Spain and the Portuguese Mirandesa and Marinhoa, separated from the remaining *B. taurus* breeds, whereas the other breeds from Portugal and Southern Spain remained clustered with the Creole breeds. As the number of inferred ancestral populations increased, admixture among breeds became more apparent, but some Creole breeds, such as the two Argentinean and the Uruguayan Creoles, Caracú from Brazil, Texas Longhorn, Creoles of Baja California and Poblano from Mexico, and the Romosinuano and Costeño con Cuernos from Colombia, remained very homogeneous at K = 8. For the 81 cattle breeds analysed, the most likely number of inferred ancestral populations was K = 71 (Figure S2), as assessed by the method of Evanno et al. (2005). The computed individual membership coefficients resulted in about 60–70% of the individuals classified within their source ancestral population, assuming a threshold of q>0.8. The Zebu breeds Brahman, Guzerat, Gyr, Nelore and Sindi grouped together in the same cluster with values of q around 0.700 while Cuban Zebu grouped in the same cluster with Criollo Cubano. The Mexican Creoles, with the exception of the Criollo de Chiapas, clustered together, in the same way that the Creoles from Colombia Chino Santandereano, Velasquez, Casanare and Caqueteño formed a unique cluster, although with low q values (Table S3).

Ancestral Genetic Contributions to Creole Cattle

The estimated genetic contributions of each potential ancestral breed group (Iberian, British, Continental European and Zebu) to Creole cattle are shown in Figure 5 and Table S4, as computed by the likelihood estimation of admixture proportions developed by Wang [21], and implemented by the LEADMIX software. The admixture estimates indicate that, for the Creole cattle considered as a single group, Iberian cattle contributed nearly 62% to the genetic pool, Zebu breeds contributed about 17% and Continental European and British breeds about 10% each.

The Neighbor-net indicated the existence of various Creole clusters, which is also supported by the analysis carried-out with STRUCTURE. These clusters likely reflect different contributions from the ancestral genetic groups to the current genetic pool of Creoles. Therefore, a similar analysis of estimated genetic contributions was carried out with LEADMIX for each of the five identified Creole clusters, as shown in Figure 5. These analyses revealed clear differences among the five clusters in the relative contributions of the four parental genetic groups. The Creoles from Panama, Mexico, United States and some Colombian breeds (Clusters 1 and 2) showed the strongest Iberian influence, with nearly 70 to 80% of the genetic pool contributed by Iberian breeds, with the remaining contributions corresponding to Continental and Zebu breeds, in about equal proportions. The Creoles from the southern region of the Americas (Cluster 3) had an important influence of about 60% from Iberian breeds, but also showed influence from British cattle. Cluster 4, which corresponds to Creole breeds widely dispersed in tropical areas, showed an important contribution from Zebu breeds, even though the Iberian contribution was still predominant. The Paraguayan and Colombian breeds included in Cluster 5 show a major influence of

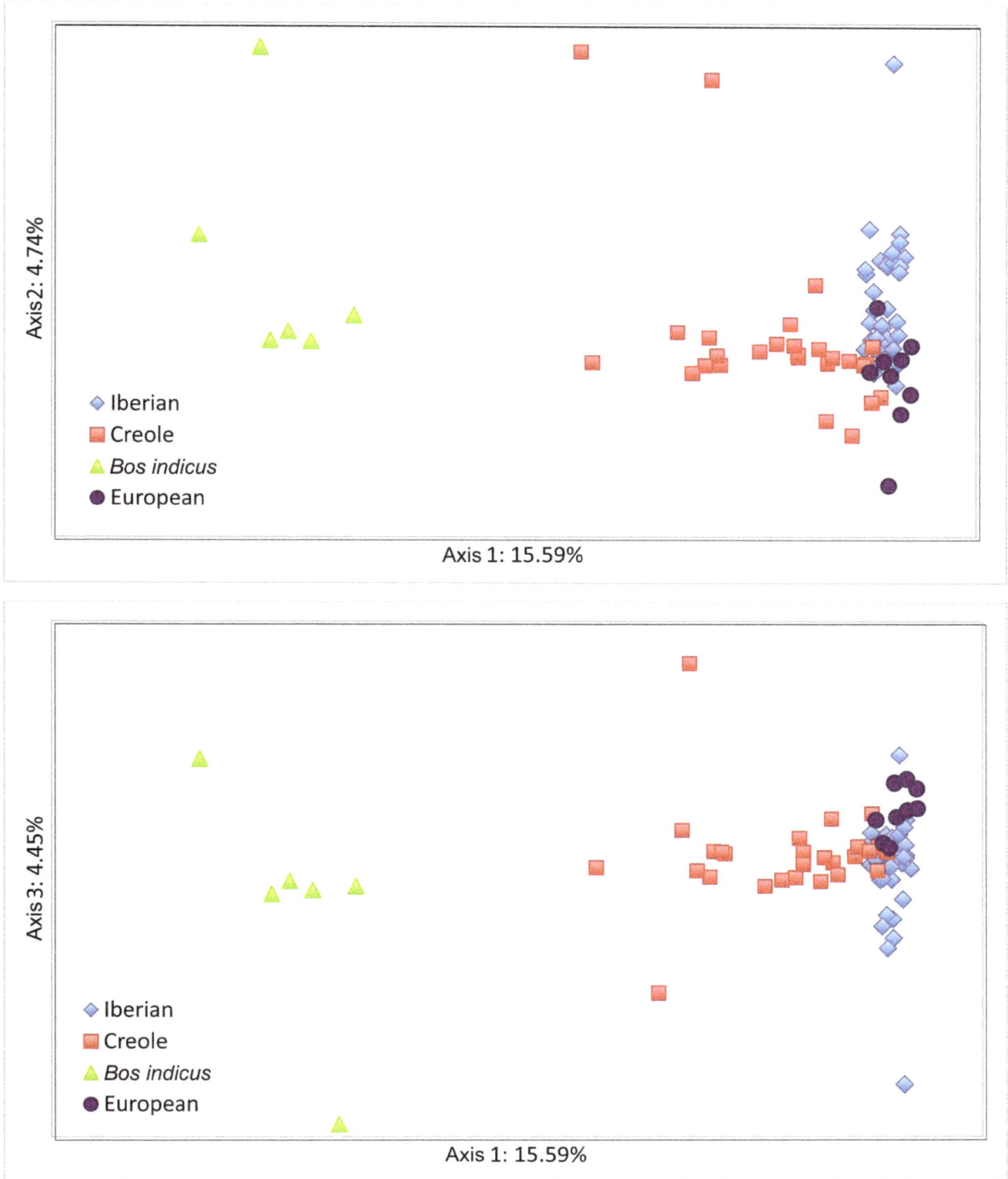

Figure 2. Graphical representation of the three first axes from the factorial correspondence analysis of the 81 cattle breeds from America and Europe.

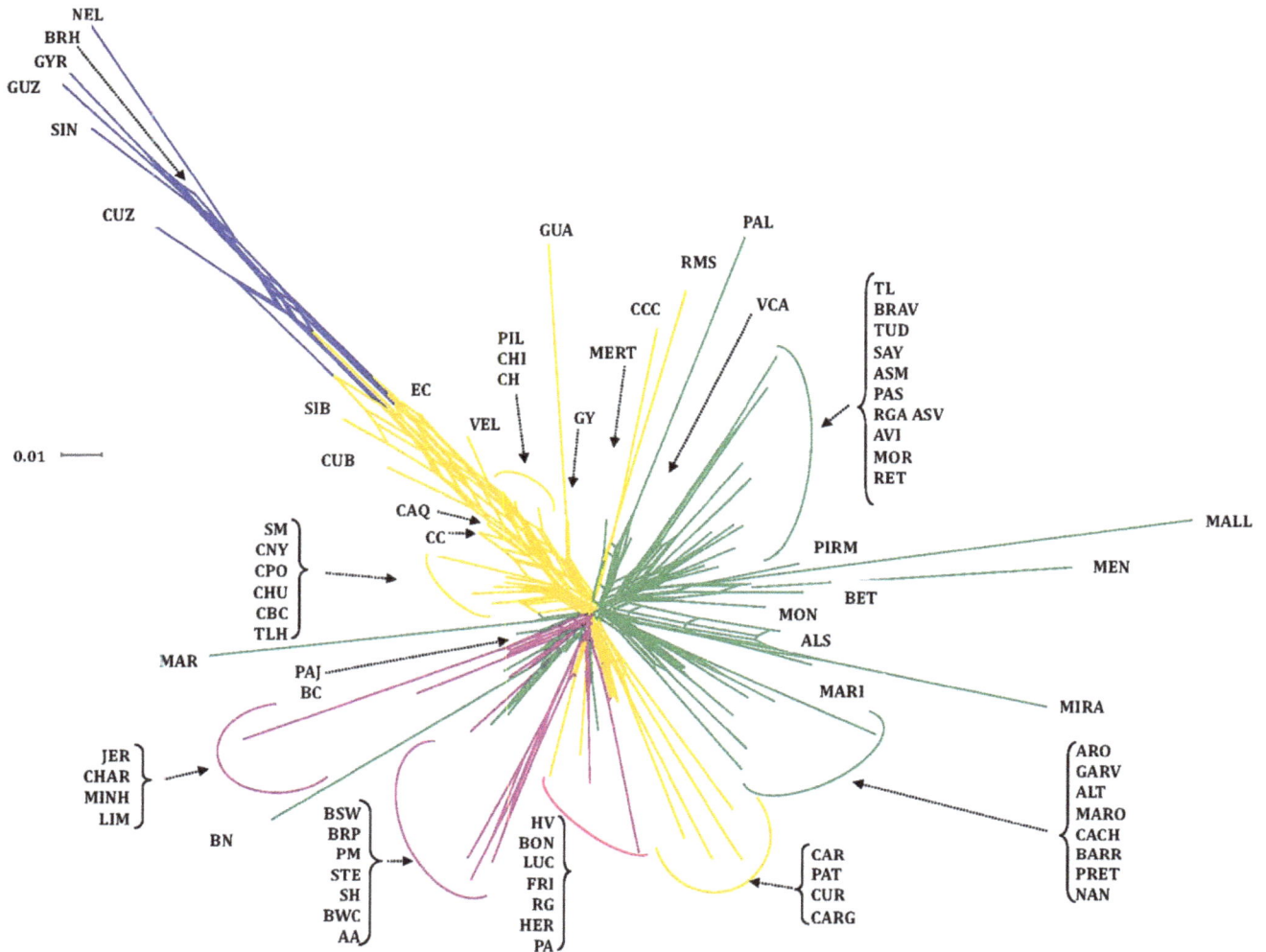

Figure 3. Neighbor-net dendrogram constructed from the Reynolds genetic distances among 81 cattle breeds. Yellow: Creole; Green: Iberian; Pink: British and Continental European; Blue: Indian Zebu. *SPANISH.* Betizu (BET), Toro de Lidia (TL), Menorquina (MEN), Alistana (ALS), Sayaguesa (SAY), Tudanca (TUD), Asturiana de los Valles (ASV), Asturiana de las Montañas (ASM), Retinta (RET), Morucha (MOR), Avileña (AVI), Pirenaica (PIRM), Rubia Gallega (RGA), Mallorquina (MALL), Monchina (MON), Serrana de Teruel (STE), Parda de Montaña (PM), Bruna de los Pirineos (BRP), Pasiega (PAS), Berrenda en Colorado (BC), Berrenda en Negro (BN), Marismeña (MAR), Pajuna (PAJ), Negra Andaluza (NAN), Vaca Canaria (VCA), Vaca Palmera (PAL); *PORTUGUESE.* Alentejana (ALT), Arouquesa (ARO), Barrosã (BARR), Brava de Lide (BRAV), Cachena (CACH), Garvonesa (GARV), Marinhoa (MARI), Maronesa (MARO), Mertolenga (MERT), Minhota (MINH), Mirandesa (MIRA), Preta (PRET), Ramo Grande (RG); *CREOLE.* Guabalá (GUA), Guaymí (GY), Texas Longhorn (TLH), Criollo Poblano (CPO), Criollo de Baja California (CBC), Criollo de Chihuahua (CHU), Criollo de Nayarit (CNY), Criollo de Chiapas (CHI), Blanco Orejinegro (BON), Caqueteño (CAQ), Sanmartinero (SM), Romosinuano (RMS), Costeño con Cuernos (CCC), Chino Santandereano (CH), Velasquez (VEL), Lucerna (LUC), Hartón del Valle (HV), Criollo Casanareño (CC), Criollo Ecuatoriano (EC), Criollo Uruguayo (CUR), Pampa Chaqueño (PA), Criollo Pilcomayo (PIL), Criollo Argentino (CARG), Criollo Patagónico (PAT), Caracú (CAR), Cubano (CUB), Siboney (SIB); *ZEBU.* Gyr (GYR), Brahman (BRH), Sindi (SIN), Guzerat (GUZ), Nelore (NEL), Zebu Cubano (CUZ); Other *EUROPEAN.* Friesian (FRI), Hereford (HER), Brown Swiss (BSW), Aberdeen Angus (AA), British White (BWC), Charolais (CHAR), Jersey (JER), Limousin (LIM), Shorthorn (SH).

British and Continental breeds, with a smaller but still detectable contribution of Iberian cattle.

Discussion

The genetic relationships between Creole cattle and their presumed ancestral sources remain largely unexplored. Estimates of genetic diversity and population structure have been previously reported for some Creole populations [9,22–26], for Iberian cattle [13,27–30], Zebu breeds [31,32] and European cattle [10,33]. Moreover, mtDNA and Y-chromosome markers were used to investigate the origins of Creole cattle [34–38], but their genetic relationship with other cattle breeds which could have influenced them remained unclear.

Our study combines several data sets that cover a wide range of Creole, European and Indicine cattle populations, thus providing a more comprehensive insight about the genetic influences that Creole breeds received since the arrival of the first Iberian cattle in the American continents in the late 1400's.

Of the total genetic variability, nearly 11% is explained by breed differences, which is slightly higher than what has been reported for other cattle breeds around the world, generally in the range of 7 to 9% [10,28,38]. This could be justified by the inclusion in this study of cattle breeds representing the two well differentiated phylogenetic groups of *B. indicus* and *B. taurus* [39].

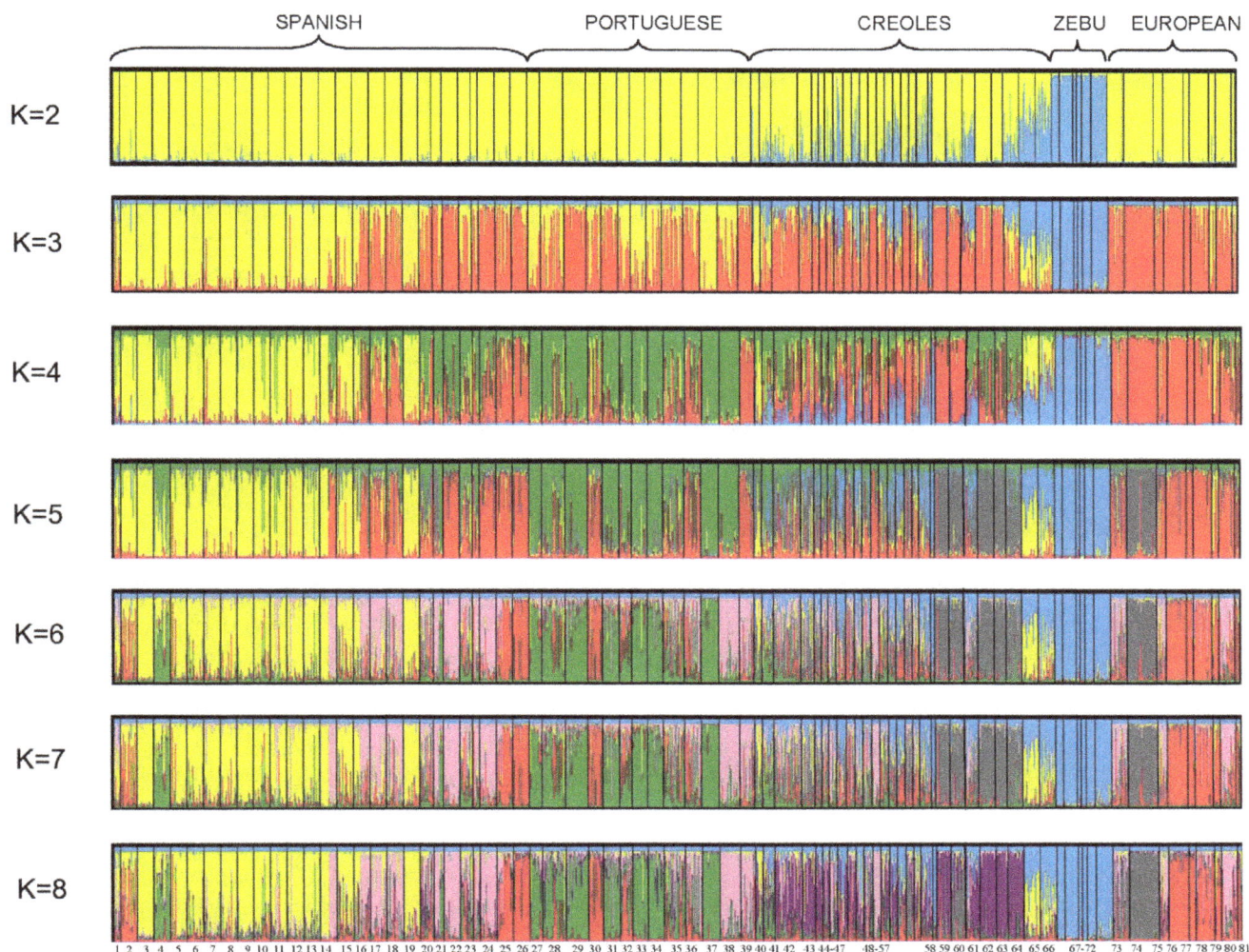

Figure 4. Population structure of 81 cattle breeds based on 19 microsatellite loci using STRUCTURE. Graphical representation of individual genotype membership coefficients (q) when K = 2 to K = 8. Each animal is represented by a single vertical line divided into K colours, where K is the number of clusters assumed and the coloured segment shows the individual's estimated membership proportions in that cluster. Black lines separate the populations. **SPANISH**. 1: Betizu (BET), 2: Toro de Lidia (TL), 3: Menorquina (MEN), 4: Alistana (ALS), 5: Sayaguesa (SAY), 6: Tudanca (TUD), 7: Asturiana de los Valles (ASV), 8: Asturiana de las Montañas (ASM), 9: Retinta (RET), 10: Morucha (MOR), 11: Avileña (AVI), 12: Pirenaica (PIRM), 13: Rubia Gallega (RGA), 14: Mallorquina (MALL), 15: Monchina (MON), 16: Serrana de Teruel (STE), 17: Parda de Montaña (PM), 18: Bruna de los Pirineos (BRP), 19: Pasiega (PAS), 20: Berrenda en Colorado (BC), 21: Berrenda en Negro (BN), 22: Marismeña (MAR), 23: Pajuna (PAJ), 24: Negra Andaluza (NAN), 25: Vaca Canaria (VCA), 26: Vaca Palmera (PAL); **PORTUGUESE**. 27: Alentejana (ALT), 28: Arouquesa (ARO), 29: Barrosã (BARR), 30: Brava de Lide (BRAV), 31: Cachena (CACH), 32: Garvonesa (GARV), 33: Marinhoa (MARI), 34: Maronesa (MARO), 35: Mertolenga (MERT), 36: Minhota (MINH), 37: Mirandesa (MIRA), 38: Preta (PRET), 39: Ramo Grande (RG); **CREOLE**. 40: Guabalá (GUA), 41: Guaymí (GY), 42: Texas Longhorn (TLH), 43: Criollo Poblano (CPO), 44: Criollo de Baja California (CBC), 45: Criollo de Chihuahua (CHU), 46: Criollo de Nayarit (CNY), 47: Criollo de Chiapas (CHI), 48: Blanco Orejinegro (BON), 49: Caqueteño (CAQ), 50: Sanmartinero (SM), 51: Romosinuano (RMS), 52: Costeño con Cuernos (CCC), 53: Chino Santandereano (CH), 54: Velasquez (VEL), 55: Lucerna (LUC), 56: Hartón del Valle (HV), 57: Criollo Casanareño (CC), 58: Criollo Ecuatoriano (EC), 59: Criollo Uruguayo (CUR), 60: Pampa Chaqueño (PA), 61: Criollo Pilcomayo (PIL), 62: Criollo Argentino (CARG), 63: Criollo Patagónico (PAT), 64: Caracú (CAR), 65: Cubano (CUB), 66: Siboney (SIB); **ZEBU**. 67: Gyr (GYR), 68: Brahman (BRH), 69: Sindi (SIN), 70: Guzerat (GUZ), 71 Nelore (NEL), 72: Zebu Cubano (CUZ); **BRITISH AND CONTINENTAL EUROPEAN**. 73: Friesian (FRI), 74: Hereford (HER), 75: Brown Swiss (BSW), 76: Aberdeen Angus (AA), 77: British White (BWC), 78: Charolais (CHAR), 79: Jersey (JER), 80: Limousin (LIM), 81: Shorthorn (SH).

The high genetic variability found in Creole cattle, even in populations considered as endangered, might reflect recent contributions of cattle from different origins, which are known to have been admixed with some Creole populations over the last century [40,41]. This result is in agreement with the analysis of mtDNA sequences and Y haplotypes, which have shown the genetic heterogeneity of Creole cattle, in which signatures of Iberian, European and Indian cattle are detected, and the direct influence of African cattle has also been claimed [35,37,42].

Recently, Gautier and Naves (2011) [43] used a high-density panel of SNPs to study genetic influences in Creole cattle from Guadeloupe and reported evidence of a direct African ancestry in this breed. In our study, no African samples were included, but some results may be interpreted as indicating a possible African influence on some Creole populations. For example, allele 123 in the BM2113 locus has previously been associated with West African taurine cattle [44], and is present at high frequencies in some Creole populations such as Caqueteño, Sanmartinero and

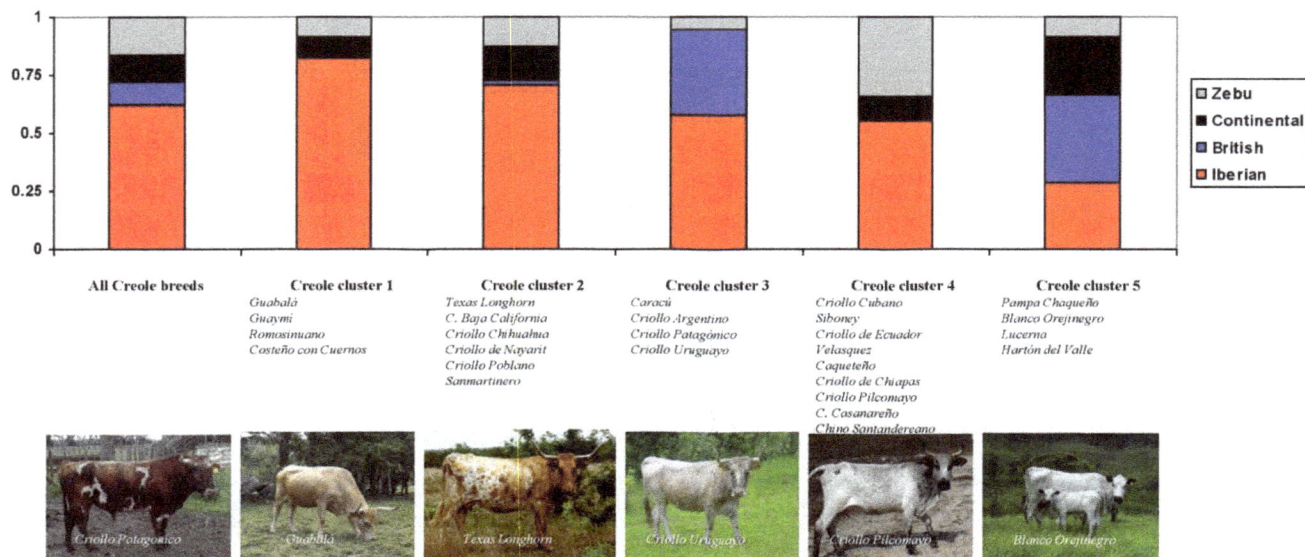

Figure 5. Genetic contributions from Iberian, British, Continental European and Zebu breeds to Creole cattle. Graphical representation of maximum-likelihood estimates of proportional genetic contributions form some groups of breeds to Creole cattle considered as a whole or grouped in five different clusters. The Creole breeds included in each cluster are listed below, and photos of animals representative of each cluster are also shown.

Pilcomayo. On the other hand, allele 143 at the same locus has been considered an indication of African Zebu influence [44], and it is present in Caracú, which could be regarded as further evidence of African influence on Creoles. It is not clear, however, if these African signatures correspond to a direct contribution of African cattle to Creoles, or rather to an indirect influence through Iberian cattle, given that some Iberian breeds in our study also have a high frequency of alleles considered to be African-specific [30]. Also, it has been suggested that the Zebu influence detected in Creoles may be in part due to *B. indicus* cattle imported from Africa during the colonial period [43]. Studies where mtDNA sequence variation was analyzed have confirmed the presence of African matrilines among Creoles [37,42,45], and Y-haplotypes have also revealed a possible West African signature in Creole cattle [35]. Overall, our results provide strong support to the conclusion of an Iberian influence on Creole cattle, but are not so clear in elucidating the possibility of a direct African influence on the different Creole groups. Further studies are necessary, covering a broad sample of African cattle breeds and a combination of different genetic markers, to clarify the African influence on Creoles.

The degree of genetic differentiation among all breeds studied indicates relatively low levels of gene flow and some level of reproductive isolation among most Creole breeds, probably as a result of geographic separation and differentiation. Taken together, Creole breeds differ more from *B. indicus* than from the remaining *B. taurus* breeds, with the lowest levels of differentiation for the Creole-Spanish and Creole-Portuguese pairs. Our results indicate that Creole cattle retain genetic signatures of their Iberian ancestry, in agreement with previous studies based on monoparental genetic markers [35].

The F-statistics, AMOVA and Factorial Correspondence Analysis results confirmed the closer proximity of Creoles to Iberian breeds, and their much larger differentiation from Zebu cattle. Nevertheless, Indian Zebus have had an important influence on some Creole breeds, as is clear from the Reynolds

Neighbor-net, the Bayesian approach adopted by STRUCTURE and the maximum-likelihood estimation of genetic contributions from parental populations carried-out with LEADMIX. However, the majority of Creoles seem to have been largely unaffected by the introduction of Zebus into South America in the 19th and 20th centuries. In general, Creole breeds from tropical areas (Siboney, Criollo Cubano, Criollo Ecuatorino, Criollo de Chiapas and some Colombian breeds), showed the highest degree of admixture with Zebu, but this influence extended as far south as the Criollo Pilcomayo from Paraguay.

The present study indicates that the influence of Iberian cattle was mostly due to the breeds from Portugal and Southern Spain, which had a closer relationship with Creoles, as detected in the Bayesian analysis with STRUCTURE. In the period when Central and South America were settled, cattle breeds from Portugal and Spain were probably not very distinct from each other, given that the major period of breed formation started in the late 18th century [46]. Also, Portugal and Spain were united in 1580 under Philip II, King of Spain, and kept together until 1640, and this corresponded to a period of major livestock shipments from the Iberian Peninsula to South America [48]. It is also known that the vast majority of the expeditions to South America departed from Lisbon (central Portugal) or from Cádiz and Seville (Southern Spain), which would explain the closer relationship of Creoles with cattle breeds from these regions. The shipment of cattle directly from Africa to the Americas could have occurred following slave routes and from intermediate ports in the Atlantic Islands.

The establishment of Iberian livestock in the "New World" followed two different migration routes, depending on the predominance of colonizers being Spanish or Portuguese. The first arrival of cattle was in 1493, when Columbus brought animals from Spain to the Caribbean Islands in his second trip, and the presence of hundreds of animals in Cuba was reported four decades later [3]. Cattle from the Caribbean Islands, e.g., Cuban Creole, could thus be considered a remnant of the first cattle brought from Spain, but it is widely recognized that Zebu cattle have been extensively used in crossbreeding with Caribbean cattle over the last century [4,40]; it is, therefore, difficult to identify a specific breed from this area as a good representative of the early Iberian stock.

From the Caribbean Islands, where animals were stocked and bred, cattle were brought to North America through the Mexican port of Veracruz, and from there they expanded throughout Mexico and towards the region corresponding to Texas. The Texas Longhorn and the different Mexican Creole populations could be regarded as representing this path of cattle dispersion.

An additional route of dispersion of animals from the Caribbean was into Central America through the ports of Panama or directly into the northern part of South America, through the port of Santa Marta in Colombia [47]. From these places, they were then distributed to the Northern Peruvian Vice Kingdom (today, Colombia, Ecuador and Peru). The Panamanian and some of the Colombian breeds could be considered as representatives of this path of cattle dispersion.

The Rio de la Plata was an important route of distribution of Iberian cattle in Southern South America, often disseminated by Jesuit missions which had a strong influence in this area [48]. From Rio de la Plata cattle dispersed through most of South America, from Patagonia up to the southern part of the Peruvian Vice Kingdom. In these areas, the Spanish path of dispersion was probably mixed with the Portuguese route, in which cattle were shipped from Portuguese ports into the Cape Verde Archipelago and from there to the Captaincies on the Brazilian coast, mainly to Pernambuco and Bahia in the North, and São Vicente (near Rio de Janeiro) in the South [48]. The southern part of Brazil, Uruguay and Argentina have milder climate, and the Zebu influence was probably less severe than in tropical Central America. Several breeds from Southern South America included in our studies confirm lower levels of Zebu admixture, including the Argentinean, Patagonian and Uruguayan Creoles as well as the Brazilian Caracú, even though the latter may have suffered some Zebu introgression in the past [35]. Nevertheless, it is known that British cattle, especially Hereford and Angus, were introduced and widely expanded in this region in the mid-19th century, and could have had some influence on Creole populations.

Our study provides evidence that Creole breeds still show important influences of Iberian cattle, which contributed with nearly two-thirds of the Creole genetic pool analyzed. Our results further indicate that large genetic differences exist among Creole sub-populations, reflecting the effects of genetic drift as well as the introduction of other breeds through the years after the initial arrival of Iberian cattle. The Neighbor-net representation of the pairwise Reynolds genetic distances generally supports the existence of five Creole clusters, which may correspond to different paths of cattle dispersion into the Americas or, in some cases, perhaps to a more recent admixture of germplasm from other breeds.

The first cluster corresponds to the Panamanian (Guabalá and Guaymi) and some of the Colombian breeds (Costeño con Cuernos and Romosinuano), and could represent the first route of cattle introduction into Central and South America from the Caribbean Islands, as the flow of ships between Panama and Colombia was very important in the 16th century [48]. This cluster presents the highest Iberian and possibly African influences, with minor contributions from Continental and Zebu breeds. Interestingly, a common origin was detected between the Costeño con Cuernos and Romosinuano from Colombia and the breeds from the Canary islands, lending support to the important role

played by this archipelago as a point of shipment and reload of animals taken to the Americas [3,48].

The second cluster is represented by most of the Mexican breeds, the Colombian Sanmartinero and the Texas Longhorn, which are near the centre of the radial net, and could correspond to one of the first paths of cattle dispersion into Central and North America. This cluster still shows a major signature of Iberian cattle, with minor influences from Continental and Zebu breeds. A very interesting result in our study was the detection of a common origin shared by this group of breeds and the Spanish Marismeña. It is generally believed that, since Columbian times, the Marismeña has been kept since Colombian times in semi-feral conditions in a natural park near the original point of departure of Spanish sailors [49]. Our results confirm that it could represent a remnant of the animals taken to the Americas in the early period of settlement.

The third cluster contains the Brazilian Caracú and the Argentinean, Patagonian and Uruguayan Creoles. This group is likely a representative of the Rio de la Plata and Brazilian routes of colonization and cattle flow into South America. Even though the Iberian contribution is very clear in this group, the later introduction of British cattle in the region probably resulted in some admixing, which is now detectable in the genetic pool of this cluster, particularly in the Uruguayan Creoles.

The fourth cluster includes breeds with a widely dispersed geographical distribution, such as the Creoles from Cuba and Ecuador, the Velasquez, Caqueteño, Casanareño and Chino Santandereño from Colombia, the Chiapas Creole from Southern Mexico and the Pilcomayo from Paraguay. All these breeds still show a major Iberian contribution but also a strong Zebu influence, confirming the impact of B. indicus on the genetic make-up of Creole cattle in many tropical areas during the last century.

The fifth cluster includes three Colombian breeds (Blanco Orejinegro, Lucerna and Harton del Valle) and the Paraguayan Pampa Chaqueño. This cluster shows some proximity with British and Continental breeds, and a minor, but still detectable, representation of the Iberian contribution.

Our results generally support the historical descriptions of cattle introduction and dispersion throughout the Americas [3,48]. Nearly 500 years after their arrival, strong genetic signatures of Iberian cattle are still present in Creoles. The major ancestral contributions are from breeds of Southern Spain and Portugal, in agreement with the historical ports of departure of ships sailing towards the Western Hemisphere. Furthermore, the role of the Canary Islands in the flow of cattle to the Americas was confirmed.

Even though the term "Creole" has been used since early colonial times in Latin America in reference to both people and animals born in the newly-discovered land from parents of Iberian origin [7], it is clear that there is more genetic variability among Creole cattle in comparison to breeds from other geographic regions. This diversity results from differential genetic contributions from several parental populations, genetic drift and some admixing with other breeds over time. Based on the information derived from our study, it is possible to summarize the gene flow that gave origin to the different Creole populations, confirming the influence of Iberian, British, Continental and Zebu breeds.

The major feature that should be retained is the predominant influence of Iberian cattle on Creoles still present today. Signatures of African cattle are also represented in many Creole breeds, and which can result from either direct contributions or indirect influences through Iberian cattle. Evidence of admixing with British breeds is visible in some Colombian and Paraguayan breeds and, to a lesser extent, in Creoles from southern South America. Creoles from tropical areas, especially those from the

Caribbean, show clear admixture with Zebu, which contribute high tolerance to hot and humid climates, and resistance to parasites. Some Creole populations still show a close proximity to their distant Iberian ancestors, and efforts should be made to avoid their extinction or further genetic erosion through admixture with other breeds, which would compromise five centuries of selective adaptation to environmental conditions which range from the deserts of Texas and Mexico, to the mountains of Patagonia.

Overall, our study indicates that: 1) several centuries after the introduction of Iberian cattle into America, Creole breeds still show strong and predominant signatures of Iberian influence; 2) Creole breeds differ widely from each other, both in their genetic structure and in the genetic influences received from other breeds; 3) in some Creole breeds, especially those from tropical regions, the impact of *B. indicus* is very clear, even though the Iberian influence is still prevalent; 4) a few Creole breeds from Colombia and Paraguay have a major influence from British and Continental breeds.

This study provides significant genetic information about cattle populations in the Americas that are remnants of historical colonization. Our findings reveal the evolutionary trajectories of cattle in close association with human dispersal and confirm Creoles as legitimate representatives of cattle from the discoveries. Furthermore, our results provide the means to identify the Creole breeds with different genetic signatures, which will be useful for the development of global and local conservation of cattle genetic diversity.

Materials and Methods

Samples

The study included biological samples of 3,333 animals representing 81 cattle breeds from 12 different countries (Table S1). The origin of the breeds studied (Figure 1) was either Creole (a comprehensive sample of 27 breeds, representing a wide range of Creole cattle, from North America to Patagonia), Iberian (39 native breeds from Portugal and Spain, including 3 breeds from the Atlantic Islands), European breeds (9 *B. taurus* breeds from the British Isles and Continental Europe which have been widely used throughout the world) and Zebu breeds (6 breeds representing the *B. indicus* group).

Semen samples were obtained from germplasm banks. Blood and hair root samples were collected by qualified veterinarians through their routine practice, in the framework of oficial programs aimed at the identification, health control and parentage confirmation of the breeds and populations included in our study. Therefore, the legal restrictions defined in "Spanish Law 32/2007 of November 7, on the care of animals in their husbandry, transportation, testing and sacrifice" do not apply, as they are waved in the case of non-experimental procedures and routine veterinary practices with livestock species, in Article 3d of the above-mentioned Law.

Molecular Markers

Six laboratories were involved in this study (Universidad de Córdoba, Universidad Complutense de Madrid and Universidad de Zaragoza from Spain, Instituto Nacional dos Recursos Biológicos from Portugal, University of California in Davis from the United States of America, and Universidad Nacional de Colombia in Palmira from Colombia).

A common set of 19 microsatellites were selected from a panel of 30 markers recommended for genetic diversity studies by the International Society for Animal Genetics (ISAG) / Food and Agriculture Organization of the United Nations (FAO) working group [50]: *BM1818, BM1824, BM2113, CSRM60, CSSM66, ETH3, ETH10, ETH185, ETH225, HAUT27, HEL9, ILSTS006, INRA032, INRA063, MM12, SPS115, TGLA53, TGLA122* and *TGLA227*.

DNA Amplification, and Genotyping

Genomic DNA was extracted using procedures previously described [13,29,51] The 19 microsatellite markers were amplified in multiplex polymerase chain reactions (PCRs) using fluorescence-labelled primers [13]. PCR products were separated by electrophoresis on ABI instruments (3730, 3130 and 377XL, Applied Biosystems, Foster City, CA) according to manufacturer recommendations and allele sizing was accomplished by using the internal size standards GeneScan^{TM}-500 LIZ^{TM} and GeneScan-400HD ROX (Applied Biosystems, Warrington, UK).

Allele nomenclature was standardized following a former European research project on cattle genetic diversity (EU RESGEN CT 98–118, for further details on the project outcome Dr J. A. Lenstra has to be contacted: J.A.Lenstra@uu.nl). To assure compatibility of results from different equipments and laboratories, a total of 30 samples representing the entire allele range for this set of markers was exchanged and genotyped in all laboratories. Allele sizing was standardized across laboratories based on these reference samples. Moreover, reference samples (2) were included in each assay to control for variation between electrophoresis.

Statistical Analysis

Data used in this paper have been archived at Dryad (www.datadryad.org): doi:10.5061/dryad.17 gk0.

Mean number of alleles (Am), observed (Ho) and unbiased expected (He) estimates of gene diversity [52] and their standard deviations were obtained with the MICROSATELLITE TOOLKIT software [53]. Distribution of genetic variability within and between breeds was studied by analysing F-statistics [54] as implemented in GENETIX v4.04 [55]. The within-breed inbreeding coefficient (F_{IS}) was calculated with a 95% confidence interval obtained by 10000 bootstraps across loci. The effective number of alleles (Ae) and allelic richness (Ar) over all loci per breed were calculated with POPGENE [56] and FSTAT v. 2.9.3 [57], respectively. Deviations from Hardy–Weinberg equilibrium (HWE) were assessed with GENEPOP v. 3.4 software [58]. Both global tests across populations and loci as well as tests per locus per breed were carried-out using the method of Guo & Thompson (1992) [59] and the p-values were obtained using a Markov chain of 10000 dememorization steps, 100 batches, and 5000 iterations.

After defining groups of breeds by geographic origin and ancestry (i.e., Creole, Spanish, Portuguese, British, Continental European and Zebu), a hierarchical analysis of variance was performed to partition the total genetic variance into components due to inter-individual and inter-breed differences. Variance components were used to compute fixation indices and their significance was tested using a non-parametric permutation approach [60]. Computations were carried out using the AMOVA (Analysis of Molecular Variance) module of ARLEQUIN 3.01 [61].

Genetic divergence among breeds was estimated by calculating the Reynolds distances [62] with the POPULATIONS software [63]. A Neighbor-net was constructed with the Reynolds distances using SPLITSTREE 4 [64] to graphically represent the relationships between breeds and to depict evidence of admixture.

Factorial Correspondence Analysis [65] was performed using the function "AFC 3D sur populations" of GENETIX v4.04.

The STRUCTURE v.2.1 software [20] was used to investigate the genetic structure of the 81 cattle populations, in order to identify population substructure and admixture, and to assign individuals to populations. Runs of 10^6 iterations after a burn-in period of 300000 iterations were performed for each K to determine the most probable number of clusters, as inferred from the observed genotypic data. Ten independent simulations for K equal to 2 to 81 were performed, and the method of Evanno *et al.* (2005) [66] was used to identify the most probable K, by determining the modal distribution of ΔK. The DISTRUCT v.1.1 software [67] was used to obtain a graphical display of individual membership coefficients in each ancestral population, considering the run with the highest posterior probability of the data at each K value.

In order to assess the relative genetic contributions of breeds from different regions (Iberian, British, Continental European and Zebu) in the development of Creoles, a maximum likelihood estimation of admixture proportions was carried out with the LEADMIX software, following the principles described by Wang (2003) [21]. These analyses were conducted for the full group of Creoles, and for five different Creole clusters, as revealed by the Reynolds genetic distances and the corresponding dendrogram.

Supporting Information

Figure S1 Graphical representation of ΔK values for K = 2 to K = 81. Representation for 81 Cattle breeds based on STRUCTURE results following Evanno criterion.
(TIF)

Figure S2 Population structure of 81 cattle breeds using STRUCTURE when K = 71. Graphical representation of individual genotype membership coefficients (q) when K = 71. Each animal is represented by a single vertical line divided into 71 coloured segments using only 6 colours showing the individual's estimated membership proportions in that cluster. 1: Betizu (BET), 2: Toro de Lidia (TL), 3: Menorquina (MEN), 4: Alistana (ALS), 5: Sayaguesa (SAY), 6: Tudanca (TUD), 7: Asturiana de los Valles (ASV), 8: Asturiana de las Montañas (ASM), 9: Retinta (RET), 10: Morucha (MOR), 11: Avileña (AVI), 12: Pirenaica (PIRM), 13: Rubia Gallega (RGA), 14: Mallorquina (MALL), 15: Monchina (MON), 16: Serrana de Teruel (STE), 17: Parda de Montaña (PM), 18: Bruna de los Pirineos (BRP), 19: Pasiega (PAS), 20: Berrenda en Colorado (BC), 21: Berrenda en Negro (BN), 22: Marismeña (MAR), 23: Pajuna (PAJ), 24: Negra Andaluza (NAN), 25: Vaca Canaria (VCA), 26: Vaca Palmera (PAL), 27: Alentejana (ALT), 28: Arouquesa (ARO), 29: Barrosã (BARR), 30: Brava de Lide (BRAV), 31: Cachena (CACH), 32: Garvonesa (GARV), 33: Marinhoa (MARI), 34: Maronesa (MARO), 35: Mertolenga (MERT), 36: Minhota (MINH), 37: Mirandesa (MIRA), 38: Preta (PRET), 39: Ramo Grande (RG); ***CREOLE***. 40: Guabalá (GUA), 41: Guaymí (GY), 42: Texas Longhorn (TLH), 43: Criollo Poblano (CPO), 44: Criollo de Baja California (CBC), 45: Criollo de Chihuahua (CHU), 46: Criollo de Nayarit (CNY), 47: Criollo de Chiapas (CHI), 48: Blanco Orejinegro (BON), 49: Caqueteño (CAQ), 50: Sanmartinero (SM), 51: Romosinuano (RMS), 52: Costeño con Cuernos (CCC), 53: Chino Santandereano (CH), 54: Velasquez (VEL), 55: Lucerna (LUC), 56: Hartón del Valle (HV), 57: Criollo Casanareño (CC), 58: Criollo Ecuatoriano (EC), 59: Criollo Uruguayo (CUR), 60: Pampa Chaqueño (PA), 61: Criollo Pilcomayo (PIL), 62: Criollo Argentino (CARG), 63: Criollo

Patagónico (PAT), 64: Caracú (CAR), 65: Cubano (CUB), 66: Siboney (SIB); ***ZEBU***: 67: Gyr (GYR), 68: Brahman (BRH), 69: Sindi (SIN), 70: Guzerat (GUZ), 71 Nelore (NEL), 72: Zebu Cubano (CUZ), 73: Friesian (FRI), 74: Hereford (HER), 75: Brown Swiss (BSW), 76: Aberdeen Angus (AA), 77: British White (BWC), 78: Charolais (CHAR), 79: Jersey (JER), 80: Limousin (LIM), 81: Shorthorn (SH).
(TIF)

Table S1 Breeds, samples and origins. Breed names, acronyms (Acron.), sample sizes (N), sample type, breed type, genetic group (GG), country of sampling and region of origin (Reg) of the 81 breeds included in this study.
(PDF)

Table S2 Genetic diversity for 81 Cattle breeds. Number of individuals per breed (N), mean number of alleles/locus (Am), mean effective number of alleles/locus (Ae), mean allelic richness per locus corrected for sample size (Ar), mean observed heterozygosity (Ho) and mean expected heterozygosity (He) and their standard deviations, within-breed inbreeding coefficient (FIS) and corresponding confidence interval.
(PDF)

Table S3 Table S3. Estimated membership coefficients in each cluster (q), as inferred by STRUCTURE for K = 71. Contribution of the more important cluster per breed is represented in bold.
(PDF)

Table S4 Genetic contributions from Iberian, British, Continental European and Zebu breeds to Creole cattle. Maximum-likelihood estimates of proportional genetic contributions from Iberian, British, Continental European and Zebu breeds to Creole cattle, considered as a whole or grouped in five different clusters. The SD was obtained from 1000 bootstrapping samples (over loci). **Creole cluster 1**: Guabalá, Guaymí, Romosinuano, Costeño con Cuernos; **Creole cluster 2**: Texas Longhorn, Criollo Baja California, Criollo Chihuahua, Criollo de Nayarit, Criollo Poblano, Sanmartinero; **Creole cluster 3**: Caracú, Criollo Argentino, Criollo Patagónico, Criollo Uruguayo; **Creole cluster 4:** Criollo Cubano, Siboney, Criollo de Ecuador, Velasquez, Caqueteño, Criollo de Chiapas, Criollo Pilcomayo, Criollo Casanareño, Chino Santandereano; **Creole cluster 5:** Pampa Chaqueño, Blanco Orejinegro, Lucerna, Hartón del Valle.
(DOC)

Acknowledgments

The authors gratefully thank the different breeders' associations and research groups who kindly provided the biological samples used in this study and the members of the CYTED XII-H and CONBIAND networks for valuable cooperation over the years.

Author Contributions

Conceived and designed the experiments: AMM CG SD VL IM-B MCTP CR. Performed the experiments: AMM CG SD VL IM-B CR LAA. Analyzed the data: AMM LTG VL. Contributed reagents/materials/analysis tools: AMM JC JVD MCTP CR JLV-P AA LAA EC OC JRM ORM RDM LM GM-V JEM AP JQ PS OU AV DZ PZ. Wrote the paper: AMM LTG JVD. Reviewed and edited the manuscript: JC CG IM-B MCTP JLV-P LAA OC JQ PS PZ.

References

1. Stahl PW (2008) Animal Domestication in South America. In: Handbook of South American archaeology. Springer. 121–130.

2. Crosby AW (1973) The Columbian Exchange: Biological and Cultural Consequences of 1492. Greenwood. 268 p.

3. Rodero E, Rodero A, Delgado JV (1992) Primitive andalusian livestock an their implications in the discovery of America. Arch Zootec 41: 383–400.

4. Rouse JE (1977) The Criollo: Spanish cattle in the Americas. University of Oklahoma Press. 303 p.

5. Willham RL (.982) Genetic improvement of beef cattle in the United States: cattle, people and their interaction. J Anim Sci 54: 659–666.

6. Santiago AA (1978) Evolution of Zebu cattle in Brazil. The Zebu Journal 1: 6.

7. De Alba J (1987) Criollo Cattle of Latinamerica. In: Animal genetic resources. Strategies for improved use and conservation. Available:http://www.fao.org/docrep/010/ah806e/AH806E06.htm. Accessed 5 November 2011.

8. Rischkowsky B, Pilling D, Commission on Genetic Resources for Food and Agriculture (2007) The state of the world's animal genetic resources for food and agriculture. Roma: Food & Agriculture Organization of the United Nations (FAO). 554 p.

9. Delgado JV, Martínez AM, Acosta A, Álvarez LA, Armstrong E, et al. (2012) Genetic characterization of Latin-American Creole cattle using microsatellite markers. Anim Genet 43: 2–10. doi:10.1111/j.1365–2052.2011.02207.x.

10. Cañón J, Alexandrino P, Bessa I, Carleos C, Carretero Y, et al. (2001) Genetic diversity measures of local European beef cattle breeds for conservation purposes. Genet Sel Evol 33: 311–332. doi:10.1051/gse:2001121.

11. García D, Martínez A, Dunner S, Vega-Pla JL, Fernández C, et al. (2006) Estimation of the genetic admixture composition of Iberian dry-cured ham samples using DNA multilocus genotypes. Meat Sci 72: 560–566. doi:10.1016/j.meatsci.2005.09.005.

12. Tapio I, Värv S, Bennewitz J, Maleviciute J, Fimland E, et al. (2006) Prioritization for conservation of northern European cattle breeds based on analysis of microsatellite data. Conserv Biol 20: 1768–1779. doi:10.1111/j.1523-1739.2006.00488.x.

13. Ginja C, Telo Da Gama L, Penedo MCT (2010) Analysis of STR markers reveals high genetic structure in Portuguese native cattle. J Hered 101: 201–210. doi:10.1093/jhered/esp104.

14. Li M-H, Kantanen J (2010) Genetic structure of Eurasian cattle (Bos taurus) based on microsatellites: clarification for their breed classification. Anim Genet 41: 150–158. doi:10.1111/j.1365–2052.2009.01980.x.

15. Beja-Pereira A, Caramelli D, Lalueza-Fox C, Vernesi C, Ferrand N, et al. (2006) The origin of European cattle: Evidence from modern and ancient DNA. Proc Natl Acad Sci USA 103: 8113–8118. doi:10.1073/pnas.0509210103.

16. Edwards CJ, Ginja C, Kantanen J, Pérez-Pardal L, Tresset A, et al. (2011) Dual Origins of Dairy Cattle Farming – Evidence from a Comprehensive Survey of European Y-Chromosomal Variation. PLoS ONE 6: e15922. doi:10.1371/journal.pone.0015922.

17. Gibbs RA, Taylor JF, Van Tassell CP, Barendse W, Eversole KA, et al. (2009) Genome-wide survey of SNP variation uncovers the genetic structure of cattle breeds. Science 324: 528–532. doi:10.1126/science.1167936.

18. Gautier M, Lalcë K, Moazami-Goudarzi K (2010) Insights into the genetic history of French cattle from dense SNP data on 47 worldwide breeds. PLoS ONE 5. Available:http://www.ncbi.nlm.nih.gov/pubmed/20927341. Accessed 11 November 2011.

19. Lewis J, Abas Z, Dadousis C, Lykidis D, Paschou P, et al. (2011) Tracing cattle breeds with principal components analysis ancestry informative SNPs. PLoS ONE 6: e18007. doi:10.1371/journal.pone.0018007.

20. Pritchard JK, Stephens M, Donnelly P (2000) Inference of population structure using multilocus genotype data. Genetics 155: 945–959.

21. Wang J (2003) Maximum-likelihood estimation of admixture proportions from genetic data. Genetics 164: 747–765.

22. Egito AA, Paiva SR, Albuquerque M do SM, Mariante AS, Almeida LD, et al. (2007) Microsatellite based genetic diversity and relationships among ten Creole and commercial cattle breeds raised in Brazil. BMC Genet 8: 83. doi:10.1186/1471-2156-8-83.

23. Ulloa-Arvizu R, Gayosso-Vázquez A, Ramos-Kuri M, Estrada FJ, Montaño M, et al. (2008) Genetic analysis of Mexican Criollo cattle populations. J Anim Breed Genet 125: 351–359. doi:10.1111/j.1439-0388.2008.00735.x.

24. Martínez-Correa G, Alvarez LA, Martínez GC (2009) Conservación, Caracterización y Utilización de los bovinos Criollos en Colombia. In: X Simposio Iberoamericano sobre Conservación y Utilización de Recursos Zoogenéticos. Palmira, Colombia.

25. Villalobos Cortés A, Martinez AM, Vega-Pla JL, Delgado JV (2009) Genetic characterization of the Guabala bovine population with microsatellites. Arch Zootec 58: 485–488.

26. Villalobos Cortés AI, Martínez AM, Escobar C, Vega-Pla JL, Delgado JV (2010) Study of genetic diversity of the Guaymi and Guabala bovine populations by means of microsatellites. Livest Sci 131: 45–51. doi:10.1016/j.livsci.2010.02.024.

27. Martín-Burriel I, García-Muro E, Zaragoza P (1999) Genetic diversity analysis of six Spanish native cattle breeds using microsatellites. Anim Genet 30: 177–182.

28. Mateus JC, Penedo MCT, Alves VC, Ramos M, Rangel-Figueiredo T (2004) Genetic diversity and differentiation in Portuguese cattle breeds using microsatellites. Anim Genet 35: 106–113. doi:10.1111/j.1365–2052.2004.01089.x.

29. Martín-Burriel I, Rodellar C, Lenstra JA, Sanz A, Cons C, et al. (2007) Genetic diversity and relationships of endangered Spanish cattle breeds. J Hered 98: 687–691. doi:10.1093/jhered/esm096.

30. Martín-Burriel I, Rodellar C, Cañón J, Cortés O, Dunner S, et al. (2011) Genetic diversity, structure, and breed relationships in Iberian cattle. J Anim Sci 89: 893–906. doi:10.2527/jas.2010-3338.

31. Lara MAC, Contel EPB, Sereno JRB (2005) Genetic characterization of Zebu populations using molecular markers. Arch Zootec 206/207: 295–303.

32. Dani MAC, Heinneman MB, Dani SU (2008) Brazilian Nelore cattle: a melting pot unfolded by molecular genetics. Genet Mol Res 7: 1127–1137.

33. European Cattle Genetic Diversity Consortium (2006) Marker-assisted conservation of European cattle breeds: An evaluation. Anim Genet 37: 475–481. doi:10.1111/j.1365–2052.2006.01511.x.

34. Giovambattista G, Ripoli MV, De Luca JC, Mirol PM, Lirón JP, et al. (2000) Male-mediated introgression of Bos indicus genes into Argentine and Bolivian Creole cattle breeds. Anim Genet 31: 302–305.

35. Ginja C, Penedo MCT, Melucci L, Quiroz J, Martínez López OR, et al. (2010) Origins and genetic diversity of New World Creole cattle: inferences from mitochondrial and Y chromosome polymorphisms. Anim Genet 41: 128–141. doi:10.1111/j.1365–2052.2009.01976.x.

36. Magee DA, Meghen C, Harrison S, Troy CS, Cymbron T, et al. (2002) A partial african ancestry for the creole cattle populations of the Caribbean. J Hered 93: 429–432.

37. Carvajal-Carmona LG, Bermudez N, Olivera-Angel M, Estrada L, Ossa J, et al. (2003) Abundant mtDNA diversity and ancestral admixture in Colombian criollo cattle (Bos taurus). Genetics 165: 1457–1463.

38. Lirón JP, Bravi CM, Mirol PM, Peral-García P, Giovambattista G (2006) African matrilineages in an American Creole cattle: evidence of two independent continental sources. Anim Genet 37: 379–382. doi:10.1111/j.1365–2052.2006.01452.x.

39. Achilli A, Bonfiglio S, Olivieri A, Malusà A, Pala M, et al. (2009) The Multifaceted Origin of Taurine Cattle Reflected by the Mitochondrial Genome. PLoS ONE 4: e5753. doi:10.1371/journal.pone.0005753.

40. Uffo O, Martín-Burriel I, Martinez S, Ronda R, Osta R, et al. (2006) Caracterización genética de seis proteínas lácteas en tras razas bovinas cubanas. AGRI 39: 15–24.

41. Martinez AM, Llorente RV, Quiroz J, Martínez RD, Amstrong E, et al. (2007) Estudio de la influencia de la raza bovina Marismeña en la formación de los bovinos Criollos. VIII Simposio Iberoamericano sobre Conservación y utilización de Recursos Zoogenéticos. Quevedo, Ecuador.

42. Miretti MM, Dunner S, Naves M, Contel EP, Ferro JA (2004) Predominant African-derived mtDNA in Caribbean and Brazilian Creole cattle is also found in Spanish cattle (Bos taurus). J Hered 95: 450–453. doi:10.1093/jhered/esh070.

43. Gautier M, Naves M (2011) Footprints of selection in the ancestral admixture of a New World Creole cattle breed. Molecular Ecology 20: 3128–3143. doi:10.1111/j.1365-294X.2011.05163.x.

44. MacHugh DE, Shriver MD, Loftus RT, Cunningham P, Bradley DG (1997) Microsatellite DNA Variation and the Evolution, Domestication and Phylogeography of Taurine and Zebu Cattle (Bos Taurus and Bos Indicus). Genetics 146: 1071–1086.

45. Mirol PM, Giovambattista G, Lir[oacute]n JP, Dulout FN (2003) African and European mitochondrial haplotypes in South American Creole cattle. Heredity 91: 248–254. doi:10.1038/sj.hdy.6800312.

46. Lush JL (1943) Animal Breeding Plans. Ames, Iowa, USA: The Iowa State College Press. 457 p.

47. Villalobos Cortés A, Martinez AM, Vega-Pla JL, Delgado JV (2009) History of Panama bovines and their relationships with other Iberoamerican popultions. Arch Zootec 58: 121–129.

48. Primo AT (2004) América: conquista e colonização: a fantástica história dos conquistadores ibéricos e seus animais na era dos descobrimentos. Porto Alegre, Brazil: Movimento. 192 p.

49. Martínez AM, Calderón J, Camacho E, Rico C, Vega-Pla JL, et al. (2005) Genetic characterisation of the Mostrenca cattle with microsatellites. Arch Zootec 206: 357–361.

50. FAO (2004) Secondary Guidelines for Development of National Farm Animal Genetic Resources Management Plans: Management of Small Populations at Risk. Roma: Food & Agriculture Organization of the United Nations (FAO). 225 p.

51. Martínez AM, Delgado JV, Rodero A, Vega-Pla JL (2000) Genetic structure of the Iberian pig breed using microsatellites. Anim Genet 31: 295–301.

52. Nei M (1973) Analysis of gene diversity in subdivided populations. Proc Natl Acad Sci USA 70: 3321–3323.

53. Park SDE (2001) The Excel Microsatellite Toolkit. Trypanotolerance in West African Cattle and the Population Genetic Effects of Selection [Ph.D. thesis]. University of Dublin. Available:http://animalgenomics.ucd.ie/sdepark/ms-toolkit/. Accessed 5 November 2011.

54. Weir BS, Cockerham CC (1984) Estimating F-Statistics for the Analysis of Population Structure. Evolution 38: 1358–1370. doi:10.2307/2408641.

55. Belkhir K, Borsa P, Chikhi L, Raufaste N, Bonhomme F (2004) GENETIX 4.05, logiciel sous Windows TM pour la génétique des populations. Laboratoire Génome, Populations, Interactions, CNRS UMR 5000, Université de Montpellier II, Montpellier (France). GENETIX INTRODUCTION. Available:http://www.genetix.univ-montp2.fr/genetix/intro.htm. Accessed 5 November 2011.

56. Yeh FC, Boyle TJB (1997) Population genetic analysis of co-dominant and dominant markers and quantitative traits. Belg J Bot 129: 157.

57. Goudet J (1995) FSTAT, a program to estimate and test gene diversities and fixation indices. Department of Ecology & Evolution, Biology Building, UNIL, CH-1015 LAUSANNE, Switzerland. Available:http://www2.unil.ch/popgen/softwares/fstat.htm. Accessed 5 November 2011.

58. Raymond M, Rousset F (1995) GENEPOP (Version 1.2): Population Genetics Software for Exact Tests and Ecumenicism. J Hered 86: 248–249.

59. Guo SW, Thompson EA (1992) Performing the exact test of Hardy-Weinberg proportion for multiple alleles. Biometrics 48: 361–372.

60. Excoffier L, Smouse PE, Quattro JM (1992) Analysis of molecular variance inferred from metric distances among DNA haplotypes: application to human mitochondrial DNA restriction data. Genetics 131: 479–491.

61. Excoffier L, Laval G, Schneider S (2005) Arlequin (version 3.0): an integrated software package for population genetics data analysis. Evol Bioinform Online 1: 47–50.

62. Reynolds J, Weir BS, Cockerham CC (1983) Estimation of the coancestry coefficient: basis for a short-term genetic distance. Genetics 105: 767–779.

63. Langella O (1999) Populations 1.2.31 CNRS UPR9034. Available:http://www.bioinformatics.org/~tryphon/populations/. Accessed 5 November 2011.

64. Huson DH, Bryant D (2006) Application of phylogenetic networks in evolutionary studies. Mol Biol Evol 23: 254–267. doi:10.1093/molbev/msj030.

65. Lebart L, Morineau A, Warwick KM (1984) Multivariate descriptive statistical analysis: correspondence analysis and related techniques for large matrices. Wiley. 266 p.

66. Evanno G, Regnaut S, Goudet J (2005) Detecting the number of clusters of individuals using the software STRUCTURE: a simulation study. Mol Ecol 14: 2611–2620. doi:10.1111/j.1365-294X.2005.02553.x.

67. Rosenberg NA (2004) DISTRUCT: a program for the graphical display of population structure. Molecular Ecology Notes 4: 137–138. doi:10.1046/j.1471-8286.2003.00566.x.

Expression of VP7, a Bluetongue Virus Group Specific Antigen by Viral Vectors: Analysis of the Induced Immune Responses and Evaluation of Protective Potential in Sheep

Coraline Bouet-Cararo[1], Vanessa Contreras[2], Agathe Caruso[3], Sokunthea Top[3], Marion Szelechowski[4], Corinne Bergeron[1], Cyril Viarouge[1], Alexandra Desprat[1], Anthony Relmy[1], Jean-Michel Guibert[5], Eric Dubois[5], Richard Thiery[5], Emmanuel Bréard[1], Stephane Bertagnoli[3], Jennifer Richardson[1], Gilles Foucras[3], Gilles Meyer[3], Isabelle Schwartz-Cornil[2], Stephan Zientara[1], Bernard Klonjkowski[1]*

1 UPE, ANSES, INRA, ENVA, UMR 1161 ANSES/INRA/ENVA, Maisons-Alfort, France, 2 Virologie et Immunologie Moléculaires, UR 892 INRA, Jouy-en-Josas, France, 3 INRA, UMR1225, IHAP, Université de Toulouse, INP, ENVT, Toulouse, France, 4 Centre de Physiopathologie de Toulouse Purpan, INSERM U1043, CNRS U5282, Université Paul-Sabatier, Toulouse, France, 5 Unité de pathologie des petits ruminants, ANSES, Sophia-Antipolis, France

Abstract

Bluetongue virus (BTV) is an economically important *Orbivirus* transmitted by biting midges to domestic and wild ruminants. The need for new vaccines has been highlighted by the occurrence of repeated outbreaks caused by different BTV serotypes since 1998. The major group-reactive antigen of BTV, VP7, is conserved in the 26 serotypes described so far, and its role in the induction of protective immunity has been proposed. Viral-based vectors as antigen delivery systems display considerable promise as veterinary vaccine candidates. In this paper we have evaluated the capacity of the BTV-2 serotype VP7 core protein expressed by either a non-replicative canine adenovirus type 2 (Cav-VP7 R^0) or a leporipoxvirus (SG33-VP7), to induce immune responses in sheep. Humoral responses were elicited against VP7 in almost all animals that received the recombinant vectors. Both Cav-VP7 R^0 and SG33-VP7 stimulated an antigen-specific CD4$^+$ response and Cav-VP7 R^0 stimulated substantial proliferation of antigen-specific CD8$^+$ lymphocytes. Encouraged by the results obtained with the Cav-VP7 R^0 vaccine vector, immunized animals were challenged with either the homologous BTV-2 or the heterologous BTV-8 serotype and viral burden in plasma was followed by real-time RT-PCR. The immune responses triggered by Cav-VP7 R^0 were insufficient to afford protective immunity against BTV infection, despite partial protection obtained against homologous challenge. This work underscores the need to further characterize the role of BTV proteins in cross-protective immunity.

Editor: Ding Xiang Liu, Nanyang Technological University, Singapore

Funding: This work was supported by the FP7 project ORBIVAC (contract N °. 245266) and the UE Network of Excellence, EPIZONE (Contract N °. FOOD-CT-2006-016236). The funders had no role in study design, data collection and analysis, decision to publish, or preparation of the manuscript.

Competing Interests: The authors have declared that no competing interests exist.

* Email: bklonjkowski@vet-alfort.fr

Introduction

Bluetongue virus (BTV) is an arthropod-borne virus transmitted to ruminants by blood-sucking *Diptera* of the genus *Culicoides*. BTV is the prototype species of the genus *Orbivirus* which belongs to the family *Reoviridae* with twenty-six serotypes described so far [1]. The non-enveloped Bluetongue virion has a complex icosahedral capsid structure that packages a genome composed of 10 double-stranded (ds) RNA segments [2]. The BTV core comprises two major (VP3 and VP7) and three minor (VP1, VP4 and VP6) proteins. The outer capsid is exclusively composed of VP2 and VP5. Four additional non-structural proteins (NS1, 2, 3/3A and 4) are produced during the viral cycle. The antigenic variability of the 26 serotypes of the virus is determined by VP2. VP7, whose sequence is relatively well-conserved between isolates, is the most abundant structural protein and the major immunogenic serogroup-reactive protein [3].

To control BTV infection in domestic animals, vaccination strategies mainly rely on inactivated or live-attenuated vaccines. The inactivated vaccines have proven to be effective and economically feasible for the control of Bluetongue during the last decades [4]. Nevertheless, live attenuated vaccines are subject to criticism, since residual virulence has been reported in experimental animals as well as in the field [5,6]. Abortions and teratological effects have also been reported [7,8], prohibiting their use for pregnant females. Reassortment of genome segments has also been reported between two different vaccine strains in Italy [9]. For these reasons, live-attenuated vaccines tend to be replaced by safer, inactivated vaccines, which can elicit complete protection for one year in sheep after a single injection [10]. However, both inactivated and live-attenuated vaccines allow neither differentiation between infected and vaccinated animals nor broad cross-protection across serotypes. In areas where several serotypes occur, multivalent vaccines are needed to achieve efficient protection. Alternative approaches such as the use of viral-based vectors as antigen delivery systems are needed to face the unmet requirement for a broad-spectrum, effective, safe and economical vaccine strategy [11].

Poxviruses encoding VP2 and VP5 have been shown to induce a neutralizing antibody response that generally protects animals against homologous virus [12,13]. However, in addition to the B cell response, cytotoxic T lymphocytes (CTL) play a role in protection against BTV [14]. Among BTV proteins, VP7, NS1 and VP2 contain major T-cell epitopes [15,16] but only the VP7 protein contains CD8$^+$ T-cell epitopes that are conserved amongst BTV serotypes [17]. A recombinant capripoxvirus expressing VP7 of BTV-1 has proved to confer partial protection against a heterologous BTV-3 challenge [18]. No neutralizing antibody could be detected, suggesting that cellular immunity was involved in protection against BTV. Furthermore, IFNAR ($-/-$) mice immunized with a cocktail of modified vaccinia Ankara vectors expressing VP2 and VP7 associated with VP5 or NS1 antigens were protected against lethal challenge with different BTV serotypes [19–21]. Even though the potential of such multivalent vaccines has not as yet been evaluated in target species, evidence of cross-reactive immunity in ruminants has been noted. In particular, sheep experimentally immunized with an inactivated BTV-1 vaccine were partially protected against BTV-23 challenge through a cross-reactive CTL response [22]. That provides a rationale for the development of multiserotypic vaccines against BTV.

The present study addressed the ability of the major core protein (VP7) of BTV to elicit CD4$^+$ and CD8$^+$ T-cell responses in sheep and to protect animals against challenge with both homologous and heterologous BTV serotypes. We generated two recombinant vectors expressing the VP7 of BTV-2, Cav-VP7 R^0 and SG33-VP7, derived from the canine adenovirus type 2 (CAV2) and the other from *Myxomavirus* (MYXV), respectively. CAV2 is a promising antigen delivery system based on its vaccine efficacy in various animal species and lack of pre-existing immunity in non-host species [23]. Canine adenovirus vectors expressing the rabies G protein have been shown to induce high levels of protective neutralizing antibodies against rabies in all immunized sheep [24]. MYXV-derived vectors were previously shown to express recombinant antigen in several ruminant immune cell types [13,25] and to induce a humoral immune response in sheep against a biologically relevant antigen [25]. Immunization of sheep with a MYXV-based vector expressing the VP2 protein of BTV-8 afforded substantial protection against a highly virulent homologous BTV challenge [13]. In this work, we demonstrate that both vectors were able to induce a VP7-specific CD4$^+$ response whereas only Cav-VP7R^0 stimulated a CD8$^+$ T-cell response in sheep. Thereafter, sheep were immunized using Cav-VP7R^0 and challenged with homologous and heterologous BTV serotypes. We show that the immune responses elicited against VP7 in sheep were not sufficient to significantly protect animals against BTV infection.

Materials and Methods

Cells

Dog kidney cells expressing the CAV2 E1 region (DK-E1) [23] were grown as monolayers in Dulbecco's modified Eagle's medium supplemented with 10% fetal calf serum, 100 IU of penicillin per ml, and 100 µg of streptomycin per ml. Baby Hamster Kidney cells (BHK, ATCC CRL-6281) were grown as monolayers in minimum essential medium, (MEM) supplemented with 5% fetal

calf serum, 100 IU of penicillin per ml, 100 µg of streptomycin per ml, and 1% L-glutamine. Rabbit kidney cells (RK13, ATCC CCL-37), CHO cells expressing the human coxsackievirus/ adenovirus receptor (CHO-CAR) [26] and Vero cells (ATCC CCL-81) were grown as described for DK-E1 cells.

Vectors

Generation of the Cav-VP7 R0 vector. Following isolation of RNA from BHK cells infected with a Corsican strain of BTV serotype 2 [27], the VP7 gene (Genbank accession number AF 346302) was amplified by RT-PCR. The open reading frame (ORF) of VP7 was ligated into the cloning plasmid pCR2.1 by using commercially available reagents (OneStep RT-PCR kit, Qiagen and TA cloning Kit, Invitrogen, respectively). The VP7 ORF of the BTV serotype 2 was subcloned into a CAV2-shuttle vector, within an expression cassette under the control of the CMV promoter; a detailed description of the final plasmid is available upon request. The VP7 gene was entirely sequenced in both directions. This construct was used to generate the pCav-VP7 R^0 genome by homologous recombination in BJ 5183 bacteria [28], as previously described [29]. An E1-deleted Cav-VP7 R^0 vector expressing the VP7 of BTV-2 was then rescued by transfection of DK-E1 cells with the recombinant genome pCav-VP7 R^0. Cav-G R^0 has been described [24] and is isogenic to Cav-VP7 R^0 except that the transgene encodes the rabies glycoprotein. Vectors were amplified in DK-E1 cells, purified by double banding on CsCl gradient and titrated by end-point dilution as previously described [30]. Infectious titers were expressed as TCID$_{50}$ ml^{-1}.

Generation of the SG33-VP7 vector. The recombinant myxomavirus SG33-VP7, expressing the VP7 ORF under the control of the strong early/late vaccinia virus P7.5 promoter, was derived from SG33, the attenuated vaccine strain of MYXV. The donor plasmid pAT-VP7 was derived from pAT-GPT, containing the M009L and M012L sequences of SG33 required for homologous recombination and the Ecogpt selection cassette [25]. Briefly, the VP7 gene was cloned into the pAT-GPT plasmid between the M009L and M012L sequences, yielding the donor plasmid pAT-VP7. The SG33-VP7 vector was then generated by homologous recombination within the SG33 M009L and M012L genes between the donor plasmid and SG33. Recombinant viruses were selected using mycophenolic acid [31]. Five sequential rounds of plaque purification followed by end-point dilution were performed to purify recombinant from parental viruses. For each round, the purity was confirmed by PCR for the absence of parental SG33 M011L and M010L genes and for the presence of the recombinant *L2* or *M5* genes (data not shown). SG33-VP7 was propagated in RK13 cells grown in OptiMEM supplemented with 2% FCS and antibiotics as above. Titration of recombinant SG33 viruses was performed as previously described [31].

Transgene expression

To confirm expression of VP7 protein, CHO-CAR cells cultivated in 96-well plates were transduced with 1 TCID$_{50}$ per cell of Cav-VP7 R^0. Two days after transduction, cells were fixed with cold acetone:ethanol (4:1; v/v). Staining was performed by using anti-VP7 monoclonal antibody (1/500; kindly provided by ID Vet, Montpellier, France), followed by Alexa Fluor 546-coupled anti-mouse IgG (1/1000; Invitrogen). Labeling was observed with a fluorescent microscope (Nikon, eclipse TE300). Non-transduced CHO-CAR cells were used as a negative control. The expression of VP7 by SG33-VP7 was tested in RK13 cells using multichamber slide flasks. Cells were infected with recombinant virus at an m.o.i. of 1, then fixed 24 hours later with 4%

(w/v) paraformaldehyde in PBS and permeabilized with 0.1% (v/v) Triton X-100 in PBS. Primary staining was performed using a rabbit anti-VP7 hyperimmune serum and secondary staining was performed with FITC-coupled swine anti-rabbit IgG (1/200; Dako). Samples were observed by using a confocal LSM Olympus microscope.

Vaccination trials

Animal experiments were performed following the recommendations of the "Charte nationale portant sur l'éthique en expérimentation animale" established by the "Comité national de réflexion éthique sur l'expérimentation animale, Ministère de l'Enseignement Supérieur et de la Recherche et Ministère de l'Agriculture et de la Pêche". The protocol was approved by the "Comité d'éthique sur l'expérimentation animale" of the" Ecole Nationale Vétérinaire d'Alfort" (Permit number: 2009/09). All sheep used for these experiments were purchased from herds free of BTV infection, and this status was assessed by ELISA performed on sera prepared at Day 0, prior to immunization.

In a first experiment, two-year-old female Préalpes sheep were purchased from the experimental breeding of Brouessy (INRA), and randomly assigned to two groups of 3 animals to be inoculated with either Cav-VP7 R^0 or, as a negative control, Cav-G R^0. Sheep received two injections of 4×10^8 TCID$_{50}$ of CAV2 vectors in 1 ml of PBS at a 4-week-interval: half of the viral suspension was injected by the intramuscular route, and half by the subcutaneous route. Sheep were housed indoors until euthanasia and lymph node collection at day 60.

In a second experiment, 16 twelve- to twenty-four-month old Lacaune sheep were randomly assigned to two groups. Twelve sheep were immunized twice at a four-week interval, by the intradermal route with 2×10^7 PFU of SG33-VP7. The last four control animals were not immunized. Sera were collected at D0, D28 and D60 for evaluation of the antibody response against MYXV and BTV.

In a third experiment, four- to ten-month-old male and female sheep were housed in the experimental station of Sophia-Antipolis (Anses) and randomly assigned to five groups. Two groups of 5 animals were inoculated with Cav-VP7 R^0 and challenged with either BTV-2 or BTV-8. Two groups of 5 and 3 animals were inoculated with Cav-G R^0 (as negative control) and challenged with BTV-2 and BTV-8, respectively. The last group of 3 sheep received only PBS and was challenged with BTV-8. Each group contained at least one male. Sheep received two injections of 4×10^8 TCID$_{50}$ of CAV2 vectors in 1 ml of PBS, or only PBS, at a 4-week-interval, using intramuscular and subcutaneous routes. Viral challenge was performed on day 56 by the intradermal route with wild-type strains of serotypes 2 and 8 of BTV (MERIAL SAS, Drs Hudelet and Hamers). Before viral challenge, blood samples were taken by jugular puncture every week and sera were prepared and stored at $-20°C$. After viral challenge, on day 56, blood samples were taken every two days until day 64 and then once a week. Total blood samples were stored at $-80°C$. Sheep were housed indoors in separate conventional pens and were daily monitored for health status (rectal temperature and clinical signs) for three months.

Lymph node collection and evaluation of cell-mediated immune responses

For assessing the VP7-specific response elicited by the Cav-VP7R^0 vector, the left lymph nodes draining the inoculation sites were excised at D60 post-priming from each immunized sheep. Lymph node cells (LNC) were prepared from the collected lymph nodes and labeled with CFSE (500 nM, Invitrogen, Cergy-Pontoise, France) as previously described [24]. LNC were cultured in X-Vivo 15 medium (Biowhittaker) supplemented with 2% FCS and 100 IU/ml penicillin and 100 μg/ml streptomycin (complete medium), either alone (basal proliferation) or with 3 μg/ml recombinant VP7 generously provided by P. Pourquier (ID-Vet). Stimulation with concanavalin A (conA, 1 μg/ml) was performed for each LNC preparation as a positive control. After 5 days of culture, cells were labeled for detection of CD4 (GC50A1 mAb, [IgM]) and CD8 (7C2 mAb, [IgG2a]) markers using appropriate isotype controls as previously described [32]. The 7-amino-actinomycin D negative (7-AAD$^-$) CD4$^+$ and CD8$^+$ T cells that had divided were analyzed by flow cytometry on a FACSCalibur with the CELLQuest Software (200,000 events per sample, Becton Dickinson). The percent (%) of VP7-induced specific proliferation was evaluated by the [% divided T cells cultured with VP7 protein] – [% divided T cells cultured in complete medium alone]. Horizontal bars represent the means of proliferation for each group. Differences between groups were determined using an unpaired Student T test using the Graphpad software (La Jolla, USA). Results were considered significant if $p < 0.05$.

For assessing the VP7-specific CD4$^+$ T cell response elicited by SG33 vectors, blood samples were collected at D0 and D60. PBMCs were isolated and labeled with CFSE dye as described [33]. They were incubated with a complete medium containing 10% FCS and seeded into 24-well plates at a density of 3×10^6 cells/well in the presence of either medium or recombinant virus SG33-VP7 or SG33 empty vector (m.o.i. of 1). Five days later, the cells were labeled with PE- anti-CD4- and A647-anti-CD2 antibodies (Abd Serotec, Düsseldorf, Germany) for 20 minutes at 4°C, and viability was determined using propidium iodide (BD Biosciences Pharmingen, Le Pont de Claix, France) at 1 μg/ml. Acquisition was performed using a FACScalibur flow cytometer (Becton Dickinson), and the data were analyzed with FlowJo software. A Mann-Whitney test was used for statistical analyses using Graphpad software (La Jolla, USA).

Serological assays

For the Cav-VP7 R^0 trial. The antibody response against VP7 elicited by administration of CAV2 vectors and by BTV challenge was monitored by an indirect ELISA. The anti-VP7 response was assessed against immobilized recombinant VP7 from 96-well microplates contained in the ELISA ID Screen Bluetongue Competition kit (ID VET, Montpellier, France). Plates were blocked by incubation with 100 μl PBS- 0.05% Tween containing 2% BSA and 3% milk for 45 min at 37°C. After washing, 50 μl of a 1/20 dilution of ovine sera diluted in PBS-Tween were added and incubated for 45 min at 37°C. After washing, 50 μl of a 1/8000 dilution of HRP-conjugated rabbit anti-sheep IgG (Dako-Cytomation) were added and incubated for 45 min at 37°C. After washing with PBS–Tween, 100 μl of DAB substrate (SIGMA-FAST 3,3'-Diaminobenzidine tablets, Sigma Aldrich) were added and color development was measured by spectrophotometry at 492 nm after addition of 100 μl of 2N H$_2$SO$_4$. Assays were performed in duplicate, and results were expressed as means of optical density (OD). Seroconversion against VP7 is considered significant when ODs are higher than the mean value + two standard deviations observed on day 0 (OD>m+2SD).

For the SG33-VP7 trial. Detection of MYXV and BTV-specific antibodies in serum was carried out using indirect ELISAs. Briefly, 96-well plates were coated with SG33-VP7 (1 μg/well) or VP7 protein (1 μg/well) in PBS- 0.1% Tween-3% BSA and incubated overnight. After washing with PBS-Tween, serial dilutions of test sera were incubated for 1 hour at 37°C. Plates were washed and HRP-conjugated donkey anti-sheep IgG (1/

5000, Serotec) was added for 1 hour at 37°C. After washing, substrate solution (tetramethylbenzidine at 2.4 mg/ml in citrate buffer, Sigma, France) was added for 15 minutes and color development was measured by spectrophotometry at 492 nm after addition of sulphuric acid (1N). Titers were determined by comparison with a hyperimmune serum as reference and expressed in arbitrary units. The higher point range was arbitrarily defined as 1000 units.

Viral burden

Viral burden in plasma was determined by real-time quantitative reverse transcription PCR (RT-PCR) according to the protocol developed by Toussaint et al. [34]. RNA was extracted from 100 µl of plasma by using commercially available reagents (QIAamp Viral RNA Mini Kit, Qiagen) with an automated sample preparation system (QIAcube, Qiagen). Seven µl of extracted RNA were mixed with 0.7 µl of dimethylsulfoxide (DMSO) and heated at 95°C for 3 minutes. Each 25 µl RT-PCR mixture contained 20 µl of TaqVet Blue-Tongue Virus RT-PCR mix (LSI, Lyon, France) and 5 µl of sample or RNA standard. Reverse transcription and amplification were performed by using a 7300 Real Time PCR System (Applied Biosystems). The BTV copy number per RT-PCR mixture was evaluated by the cycle threshold (Ct)value.

Clinical observations

Clinical examination of animals was performed daily from D3 before to D7 after each vaccination. In the third experiment, body temperature and clinical signs were recorded daily from D56 to D70 and then on D77, D84 and D90. General behavior of the animals was scored as follows: good (0); apathic (1); prostrated (2) and decubitus (3) and the most frequent clinical signs observed during Bluetongue infection were also recorded as previously described [4].

Clinical score variable was calculated by the sum of the individual daily clinical scores for each animal, divided by the number of days of clinical examination. Mean individual scores were then averaged for each group. Mortality was not taken into account in the clinical score but was indicated.

Statistical analyses

Statistical analyses were carried out using GraphPad PRISM version 5.00 for WINDOWS to calculate mean, SD and p values (GraphPad Software, San Diego, CA).

Results

Expression of VP7 from recombinant vectors

Recombinant vaccine vectors encoding the core VP7protein from a Corsican strain of BTV serotype 2, Cav-VP7 R^0 and SG33-VP7, were derived from the canine adenovirus serotype 2 (CAV2) or the attenuated vaccine strain of myxomavirus, SG33, respectively. Expression of VP7 was sought by immunostaining following transduction of the CHO-CAR and RK13 cells with Cav-VP7 R^0 or SG33-VP7, respectively. Labeling was observed in transduced cells by using antibodies directed against VP7 (Figure 1, A and C) without any expression in the control groups (Figure 1, B and D).

Following transduction of DK-E1 cells with Cav-VP7 R^0, expression of VP7 was detected by Western blotting using serum of a sheep infected with serotype 2 of BTV (data not shown). These data confirm the expression of VP7 from both CAV2- and SG33 based vectors.

Antibody responses in vaccinated sheep

Sheep were immunized twice by the intramuscular and subcutaneous routes or the intradermal route with Cav-VP7 R^0 and SG33-VP7, respectively. Cav-G R^0 was used as a control in CAV2 trials. Ovine sera were assayed for anti-VP7 immunoglobulins by ELISA. In sheep that received recombinant vectors, antibodies directed against VP7 were detectable on day 60 after initial vaccination (average OD: 0.351 and 0.578 for Cav-G R^0 and SG33-VP7, respectively), in contrast to control vaccinated sheep (average OD: 0.1813 and 0.205 for Cav-G R^0 and SG33-VP7, respectively) (Figure 2). Seroconversion against VP7 was considered significant when ODs are higher than the mean value + two standard deviations observed with the control virus (OD > 0.305 and OD > 0.343 for Cav-G R^0 and SG33-VP7, respectively).

Cell-mediated responses in vaccinated sheep

To establish whether Cav-VP7 R^0 and SG33-VP7 were able to induce T-cell responses against VP7, sheep were immunized twice, 4 weeks apart, with Cav-VP7 R^0, Cav-G R^0 or SG33-VP7. One month after the second injection, prescapular node cells (Cav experiment) or peripheral blood mononuclear cells (SG33 experiment) were harvested and the proliferative T-cell response elicited against VP7 was evaluated (Figure 3). The VP7-specific response observed in lymph node CD8+ T cells was significantly higher in the group vaccinated with Cav-VP7 R^0 (average: 21.21%±3.532) than in the control group that received Cav-G R^0 (average: 3.630%±0.8869, p = 0.0085, Figure 3A). Moreover, the VP7-specific CD4+ T-lymphocyte response in the draining lymph nodes tended to be higher in sheep inoculated with Cav-VP7 R^0 (average: 8.182%±4.899) than sheep inoculated with Cav-G R^0 (average: 0.7800%±0.3922, p = 0.2065, Figure 3B). Similarly, the CD4+ T-cell proliferation of PBMCs from sheep immunized with SG33-VP7 was higher in the presence of SG33-VP7 (average: 6.817%±3.701) than in the presence of the control vector SG33 (average: 1.233%±0.4865). These results suggested that immunization with SG33-VP7 and Cav-VP7 R^0 primed a CD4+ T-cell response against VP7. Furthermore, data show that Cav-VP7 R^0 vector induces CD8+ T-cell to proliferate when restimulated in vitro by the VP7 protein.

Homologous challenge in vaccinated sheep with Cav-VP7 R^0

Since Cav-VP7 R^0 had been shown to induce both CD4+ and CD8+ T-cell responses, its capacity to afford protection against virulent challenge was assessed in a vaccination trial. Two groups of sheep were inoculated with Cav-VP7 R^0 according to the vaccination protocol followed for the first experiment. While control sheep remained seronegative until challenge, all vaccinated sheep seroconverted against VP7 (Figure 4). By twenty-one days after challenge with BTV-2 (Figure 4A) or BTV-8 (Figure 4B), all controls had developed a high antibody response against VP7. The level of VP7-specific antibody was also boosted in the vaccinated groups after challenge.

The ability of the Cav-VP7 R^0 vector to inhibit BTV replication in sheep was assessed by using a BTV-specific quantitative real-time RT-PCR assay. Blood samples were used to detect BTV at days D56, D59, D62, D64, D77 and D90. Relative viral burden was expressed as the threshold cycle number (Ct) necessary for the detection of BTV genome in blood samples.

For the group challenged with BTV-2 (Figure 5A), the vaccinated animals exhibited a slightly higher Ct value than those of the controls. Given that the amplification product is approximately doubled upon each PCR cycle, we noted that, from D6 to

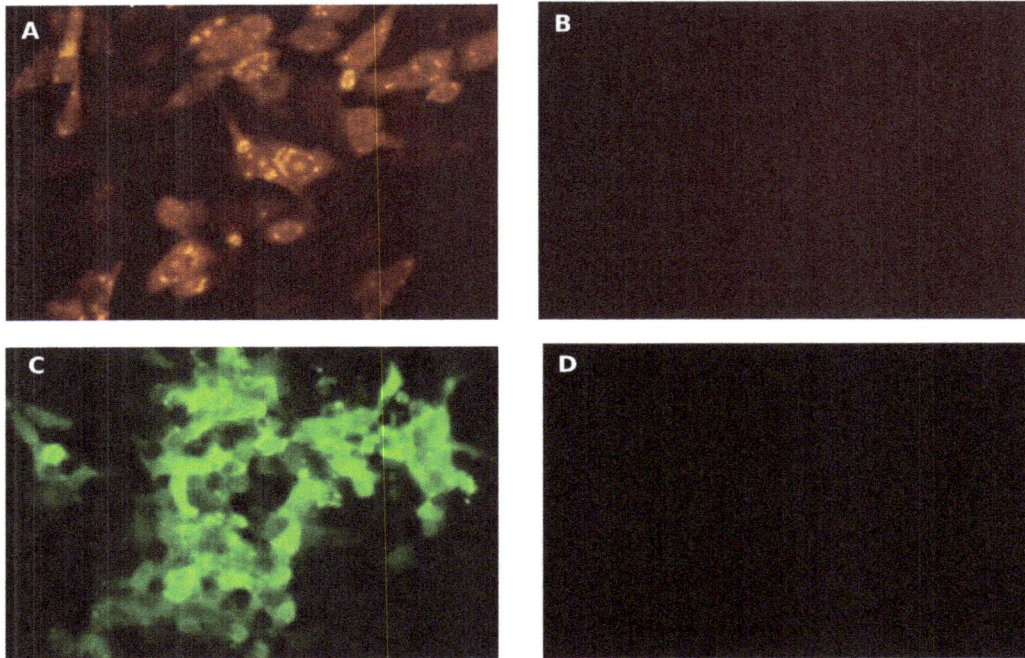

Figure 1. Expression of VP7 antigen. Following transduction of CHO-CAR cells with Cav-VP7 R^0, expression of bluetongue antigen was sought and detected by immunostaining using a MAb directed against the VP7 protein (A). Following transduction of RK13 cells with SG33-VP7, VP7 antigen was labeled using an anti-VP7 hyperimmune rabbit serum (C). Non-transduced cells were used as negative control (B and D).

D34 after challenge, viraemia was reduced by a factor of 10 in vaccinated sheep compared to the control group. Viral loads for these sheep remained significantly lower than those observed in the control group until the end of the experiment ($p<0.03$), suggesting a partial inhibition of the BTV replication.

For the group challenged with BTV-8 (Figure 5B), no difference could be observed between viral loads of animals inoculated with Cav-VP7 R^0 and sheep of the control group.

Heterologous challenge in vaccinated sheep with Cav-VP7 R^0

Clinical follow-up allowed us to establish scores for each animal from D56 (day of viral challenge) until the day of death to evaluate the protection afforded by inoculation of Cav2-VP7 R^0. In addition, rectal temperatures were regularly recorded throughout the experiment.

During the experiment, after the virulent challenge with BTV-2, 2 of 5 sheep in the group inoculated with Cav2-VP7 R^0 had to be euthanized due to a marked deterioration of their condition on D70. In the control group, 1 of 4 sheep was also euthanized on D70. Clinical signs typically observed during BTV infection were

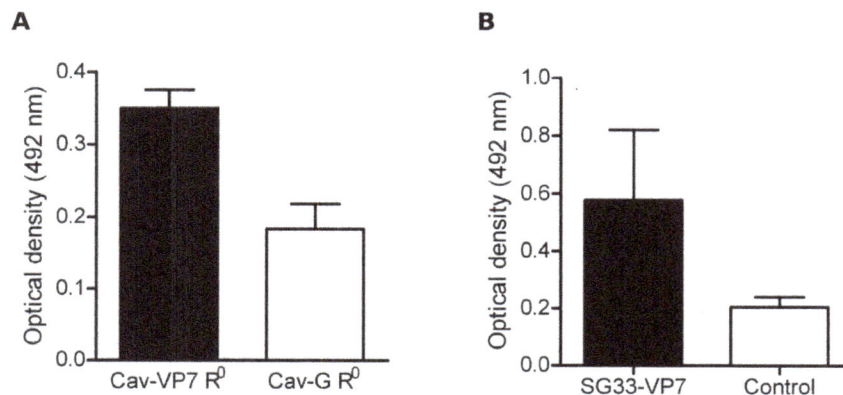

Figure 2. Serum antibodies induced by Cav-VP7 R^0 and SG33-VP7 in sheep. Sheep were immunized twice with recombinant vectors. Antibody responses elicited against the VP7 protein were assessed by ELISA at day 60 following the first inoculation of Cav-VP7 R^0 (A) or SG33-VP7 (B). Data are presented as means (with standard deviation) of individual ODs obtained for sheep inoculated with recombinant vector (black bar) or for control animals (white bar).

A

B

C

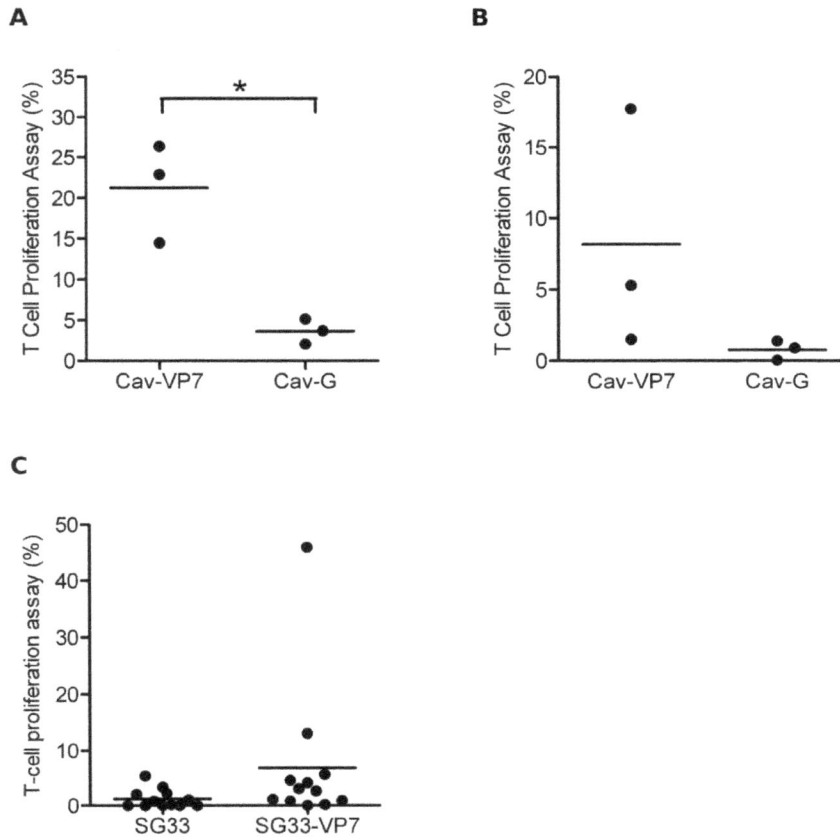

Figure 3. Inoculation of sheep with Cav-VP7 R⁰ and SG33-VP7 elicits T-cell responses. Sheep were immunized twice at a four-week interval with recombinant vectors and cellular immune responses were analyzed at D60. Prescapular lymph node cells were isolated and cultured in complete medium alone or with the VP7 recombinant protein. The specific proliferation was evaluated as [the % divided T cells cultured with VP7 protein] − [the % divided T cells cultured in medium alone]. (A) The percentage of CD8$^+$ T cells that had divided in the absence or presence of VP7 is shown (* statistical difference between immunized groups $p = 0.0085$). (B) The percentage of CD4$^+$ T cells that had divided in the absence or presence of VP7 is shown. (C) PBMCs were labeled with CFSE and cultured in complete medium alone -or with either SG33 or SG33-VP7. VP7-specific CD4$^+$ T cell proliferation was evaluated as [the % divided CD4$^+$ T cells cultured with recombinant vector] − [the % divided CD4$^+$ T cells cultured in medium alone]. Horizontal bars represent the means of proliferation for each group.

recorded in all groups tested with BTV-2. These included congestion of mucous membranes, generalized congestion, salivation, facial swelling, mucopurulent nasal discharge with crusting snout, anorexia, apathy, dyspnea and, in severe cases, ulcers or death.

Individual clinical scores were calculated by adding the scores obtained daily during the time of observation, from D56 to D70 (Figure 5C). For sheep #2 and #3, from the group inoculated with Cav-VP7 R⁰, clinical scores were very low (≤10). Notably, these two animals presented the lowest viral load (data not shown).

For the group challenged with BTV-8, no differences were observed between sheep inoculated with Cav-VP7 R⁰ and control groups.

Discussion

Bluetongue virus shows substantial antigenic variability, with 26 different serotypes described to date. Vaccination has proven very effective in BTV control and conventional vaccines based on either inactivated or attenuated virus are available for many serotypes. Cross-protective immunity between BTV serotypes is very limited, so that a specific vaccine must be manufactured for each serotype. Conventional vaccines confer protection mainly

through induction of neutralizing antibodies. Immunization of sheep with inactivated vaccine is not consistently associated with T-cell immunity, depending on adjuvant and/or booster dose [22,35]. Both CD4$^+$ and CD8$^+$ T-cell epitopes, recognized by T-cells from infected sheep, have been identified within the VP7 antigen [16]. VP7 is a major BTV group antigen and sheep vaccinated with VP7 encoded by a capripox vector were partially protected against heterologous challenge [18]. Development of vaccines eliciting a cross-reactive T-cell response and capable of affording protection against multiple serotypes of BTV would represent a major advance in management of BTV disease.

The aim of the present study was to characterize the immune responses induced in sheep by two vaccine vectors, CAV2 and MYXV, expressing VP7. We have shown that a CAV2 vector encoding the Rabies virus glycoprotein was able to elicit a protective level of rabies neutralizing antibodies in a short period of time as well as CD4$^+$ and CD8$^+$ T-cell responses against the glycoprotein in all immunized sheep [24]. It is also well-established that poxviruses elicit a strong cellular immune response, and indeed an MYXV vector expressing the VP2 of BTV-8 was shown to elicit neutralizing antibody and CD4$^+$ T-cell responses and afford protection in sheep [13]. Moreover, these vectors fulfil criteria for a DIVA vaccine.

A

B

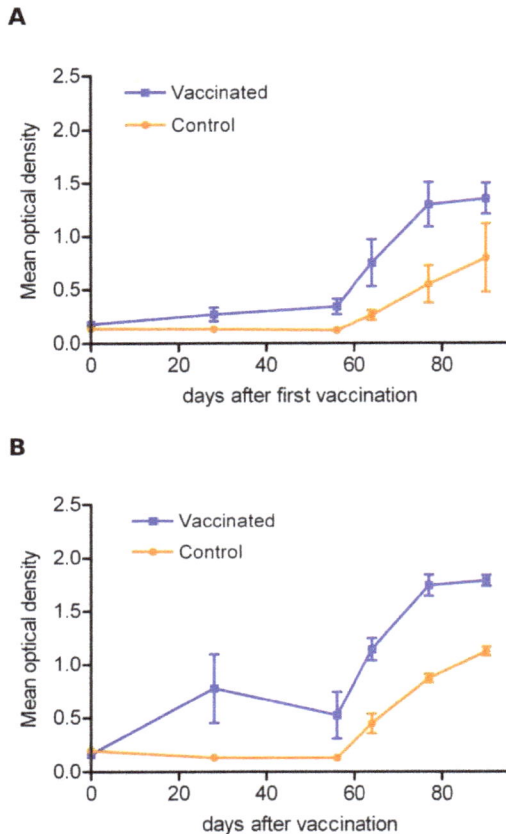

Figure 4. Serum antibodies induced by Cav-VP7 R⁰ or Cav-G R⁰ in sheep. Sheep received two doses of Cav-VP7 R⁰ or Cav-G R⁰ or PBS (control groups) on D0 and D28; the sheep were challenged with two wild type strains of BTV on day D56. Sera were collected at days 0, 28, 56 and every two days until day 64. Antibody responses elicited against the VP7 protein were assessed by ELISA performed on plates coated with recombinant VP7 for 1/20 dilutions of individual sera are shown, for the group challenged with BTV2 (A) and with BTV8 (B).

Here we have shown that both non-replicative Cav-VP7 R⁰ and SG33-VP7 vectors were able to produce the conserved BTV VP7 antigen in an efficient manner and to induce both humoral and T-

cell responses against VP7. These results confirm the role of the BTV VP7 protein in priming both CD4⁺ and CD8⁺ T-cell responses [16]. In agreement with a previous study [18], the vectors elicited detectable levels of VP7-binding but not neutralizing antibodies after immunization of sheep. If Cav-VP7 R⁰ vectors expressing the rabies antigen or VP7 are compared, the immune response elicited by VP7 was moderate immune, relatively delayed, and somewhat variable. Individual variability was also encountered after immunization with SG33-VP7, and only a fraction of sheep (10 of 12) developed a VP7-specific antibody response. Nevertheless, all sheep developed a strong antibody response against MYXV antigens (data not shown), attesting to an efficient vaccine delivery.

Cav-VP7 R⁰ also induced both VP7-specific CD8⁺ and CD4⁺ T-cell responses. Of note, all immunized sheep developed a CD8⁺ T-cell response, whereas the CD4⁺ T-cell proliferation was comparatively weaker and detected only in 2 of the 3 vaccinated sheep. This suggests that CAV2 vectors have a propensity for triggering potent CD8⁺ T-cell responses, as has been observed for human adenovirus-based vaccines [36]. The SG33-VP7 primed a CD4⁺ T-cell response against VP7 antigen in immunized sheep. Despite some evidence of CD8⁺ T-cell proliferation in SG33-VP7 group, no significant differences were observed between the vaccinated and control sheep, suggesting a non-specific proliferation (data not shown).

Once the immunogenicity of Cav-VP7 R⁰ had been established, its ability to afford protection against either BTV-2 or BTV-8 was assessed in sheep. Before challenge on D56, seroconversion against VP7 was observed in the majority of animals immunized with Cav-VP7 R⁰, and on D64, a booster effect due to the viral challenge was observed in all animals, demonstrating that immunity against VP7 was primed. The virological analysis performed by real-time RT-PCR showed partial albeit limited protection after homologous challenge (BTV-2). The decrease in Ct was equivalent to a 10-fold decrease in viral load compared with the average observed in the control group. Clinical examinations supported conclusions drawn from virological analysis, namely, a weak protection. For animals challenged with the heterologous BTV-8 serotype, no virological or clinical differences were observed between the group inoculated with Cav-VP7 R⁰ and the control groups, indicating that immunized sheep. We compared the VP7 amino acid sequence from BTV serotypes 2 and 8. The proteins are highly similar (97.99%), suggesting that antigenic polymorphism cannot explain the

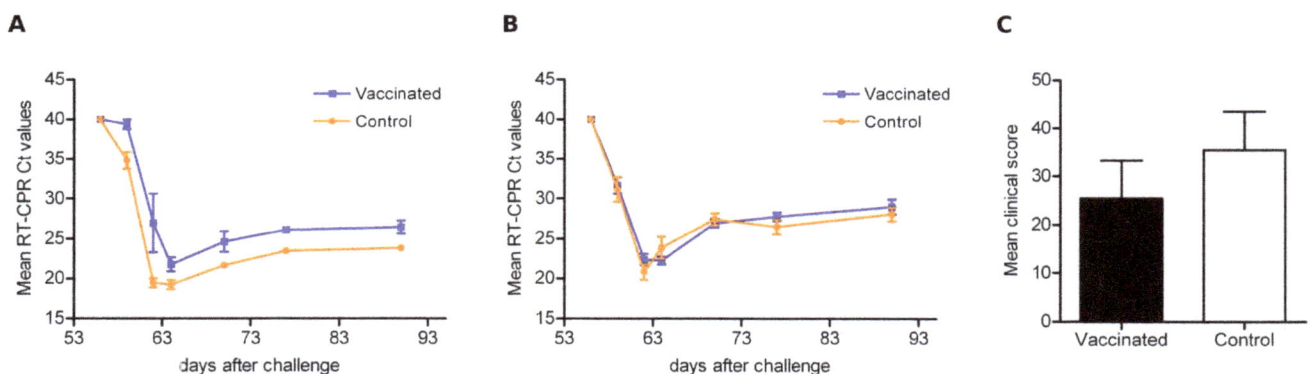

A

B

C

Figure 5. Evaluation of protection following vaccination with Cav-VP7 R⁰ and virulent challenge. (A) Evolution of mean threshold cycle (Ct) values (with standard deviation) of BTV-2 specific real-time RT-PCR (a) or BTV-8 specific real-time RT-PCR (b). (B) Mean cumulative scores (with standard deviation) after BTV-2 challenge in vaccinated (black bar) and control (white bar) groups.

observed difference after homologous and heterologous challenge as regards viral load. In conclusion, the immune response elicited against VP7 by the Cav-VP7 R^0 vector does not afford complete protection against homologous or heterologous challenge, despite detection of CD8[+] and CD4[+] T-cell responses. These results are similar to those observed with capripoxvirus vectors [37]. Absence of protection was also noted after three administrations of SG33-VP7 in sheep and subsequent challenge with BTV-8 (data not shown). These results contrast with previously published results suggesting that VP7 can elicit partial cross-protection against BTV challenge [18]. In this study, protection was only demonstrated for lethality, since all of the sheep immunized with a capripoxvirus expressing the VP7 protein from BTV-1 showed clinical signs after a heterologous challenge (BTV-3), but recovered fully from the disease in contrast to control animals. Differences between the experiments could be attributable to the BTV challenge strain or the gene transfer vectors used; of note, the capripoxvirus was replicative in ruminants and potentially expressed a higher level of VP7 protein than the non-replicative CAV2 and SG33 vectors. It is possible that the level of VP7 antigen expressed by our recombinant vectors is insufficient to prime sheep to elicit a protective immune response, as it has previously been suggested for other BTV antigens expressed in MYXV [13] or in capripoxvirus [37].

Wade Evans *et al.* were unable to detect a neutralizing antibody response after immunization of sheep with VP7. Since neither antibody dependant cell-mediated cytotoxicity (ADCC) nor complement-mediated ADCC was detected in BTV-infected sheep [38], the authors suggested that the cellular immune response was responsible for heterologous protection, although this possibility was not addressed in the study. Despite the many studies demonstrating the immunodominance of VP7 [39], we also have found that VP7-binding antibodies detected in ELISA were not associated with a protective immune response. The role of the Th1 response in protection against bluetongue disease is controversial. Adoptive transfer of lymphocytes partially protected monozygotic sheep from subsequent BTV challenge [14] and the CTL response after vaccination was suggested to protect animals against a broad spectrum of BTV serotypes [15,22,40]. While the protective efficacy of virus-like particles (VLPs) of BTV produced by baculoviruses has been thoroughly addressed, the protective efficacy of core-like particles (CLPs) formed by VP3 and VP7, which lack the outer membrane proteins, has not been extensively investigated [41]. A recent report described the protection afforded by VLPs and CLPs derived from a BTV-1 strain (RSA) of the western lineage against challenge with an virulent BTV-1 strain of the eastern lineage. The results indicated

that an intact outer capsid consisting of VP2 and VP5 is crucial for preventing BT disease and viraemia in sheep. In contrast, CLPs containing VP7 and VP3 were unable to elicit complete protection from disease. The CLPs did, however, mitigate disease severity, suggesting that VP7 with VP3 can contribute to protection. In the present study, the VP7 protein was expressed without VP3 and its conformation might have diverged from that found in native virions. In the BTV particle, VP7 assembles as a trimer and interacts with other viral proteins, and notably with the other core protein, VP3. VP7 has been shown to modify the cellular distribution pattern of VP3, from the proteasome to perinuclear zone [42]. We do not know whether interactions between VP7 and VP3 of BTV have a major impact on antibody or cellular immune responses elicited against VP7. VP7 appeared to improve the protection afforded by a vaccine consisting of a DNA prime and a vaccinia vector boost expressing VP2 and VP5 against homologous challenge in a mouse model [19]. Recently, co-expression of VP2, VP7 and NS1 has been shown to induce neutralizing activity against the homologous challenge strain (BTV4) and CD8[+] T-cell responses against these 3 BTV antigens and to afford cross-protection (BTV1 and BTV-8), which reinforced the presumption of a role for NS1 in multiserotype protection [20].

The humoral and cellular immunity induced in sheep by recombinant vectors expressing the VP7 encourages continued development of vaccine candidates derived from CAV2 or SG33 to protect ruminants against bluetongue. Regarding the choice of antigens to be expressed, and in view of achieving multi-serotype protection, it will be of interest to improve vaccine-elicited immunity against BTV by combining the Cav-VP7 R^0 vector with a vector expressing other viral proteins, such as NS1 and VP2.

Acknowledgments

We want to thank Annie Fournier for technical expertise. We are grateful to Merial France and ID-Vet for providing important materials. We also thank the personnel of the animal facility at the Ecole Nationale Vétérinaire d'Alfort, Thomas Lilin and Benoît Lecuelle, for care of animals.

Author Contributions

Conceived and designed the experiments: CB-C GF GM IS-C SZ BK. Performed the experiments: CB-C VC AC ST CB CV AD AR J-MG. Analyzed the data: CB-C VC ST ED RT EB GF GM IS-C BK. Contributed reagents/materials/analysis tools: MS SB. Wrote the paper: CB-C GF GM JR IS-C SZ BK.

References

1. Maan NS, Maan S, Belaganahalli MN, Ostlund EN, Johnson DJ, et al. (2012) Identification and differentiation of the twenty six bluetongue virus serotypes by RT-PCR amplification of the serotype-specific genome segment 2. PLoS One 7: e32601.

2. Schwartz-Cornil I, Mertens PP, Contreras V, Hemati B, Pascale F, et al. (2008) Bluetongue virus: virology, pathogenesis and immunity. Vet Res 39: 46.

3. Wilson WC, Ma HC, Venter EH, van Djik AA, Seal BS, et al. (2000) Phylogenetic relationships of bluetongue viruses based on gene S7. Virus Res 67: 141–151.

4. Breard E, Belbis G, Hamers C, Moulin V, Lilin T, et al. (2011) Evaluation of humoral response and protective efficacy of two inactivated vaccines against bluetongue virus after vaccination of goats. Vaccine 29: 2495–2502.

5. Savini G, MacLachlan NJ, Sanchez-Vizcaino JM, Zientara S (2008) Vaccines against bluetongue in Europe. Comp Immunol Microbiol Infect Dis 31: 101–120.

6. Veronesi E, Darpel KE, Hamblin C, Carpenter S, Takamatsu HH, et al. (2010) Viraemia and clinical disease in Dorset Poll sheep following vaccination with live attenuated bluetongue virus vaccines serotypes 16 and 4. Vaccine 28: 1397–1403.

7. MacLachlan NJ, Osburn BI, Stott JL, Ghalib HW (1985) Orbivirus infection of the bovine fetus. Prog Clin Biol Res 178: 79–84.

8. Murray PK, Eaton BT (1996) Vaccines for bluetongue. Aust Vet J 73: 207–210.

9. Batten CA, Maan S, Shaw AE, Maan NS, Mertens PP (2008) A European field strain of bluetongue virus derived from two parental vaccine strains by genome segment reassortment. Virus Res 137: 56–63.

10. Hamers C, Rehbein S, Hudelet P, Blanchet M, Lapostolle B, et al. (2009) Protective duration of immunity of an inactivated bluetongue (BTV) serotype 2 vaccine against a virulent BTV serotype 2 challenge in sheep. Vaccine 27: 2789–2793.

11. Brun A, Albina E, Barret T, Chapman DA, Czub M, et al. (2008) Antigen delivery systems for veterinary vaccine development. Viral-vector based delivery systems. Vaccine 26: 6508–6528.

12. Boone JD, Balasuriya UB, Karaca K, Audonnet JC, Yao J, et al. (2007) Recombinant canarypox virus vaccine co-expressing genes encoding the VP2 and VP5 outer capsid proteins of bluetongue virus induces high level protection in sheep. Vaccine 25: 672–678.

13. Top S, Foucras G, Deplanche M, Rives G, Calvalido J, et al. (2012) Myxomavirus as a vector for the immunisation of sheep: protection study against challenge with bluetongue virus. Vaccine 30: 1609–1616.

14. Jeggo MH, Wardley RC, Brownlie J (1984) A study of the role of cell-mediated immunity in bluetongue virus infection in sheep, using cellular adoptive transfer techniques. Immunology 52: 403–410.

15. Andrew M, Whiteley P, Janardhana V, Lobato Z, Gould A, et al. (1995) Antigen specificity of the ovine cytotoxic T lymphocyte response to bluetongue virus. Vet Immunol Immunopathol 47: 311–322.

16. Rojas JM, Rodriguez-Calvo T, Pena L, Sevilla N (2011) T cell responses to bluetongue virus are directed against multiple and identical CD4+ and CD8+ T cell epitopes from the VP7 core protein in mouse and sheep. Vaccine 29: 6848–6857.

17. Takamatsu H, Jeggo MH (1989) Cultivation of bluetongue virus-specific ovine T cells and their cross-reactivity with different serotype viruses. Immunology 66: 258–263.

18. Wade-Evans AM, Romero CH, Mellor P, Takamatsu H, Anderson J, et al. (1996) Expression of the major core structural protein (VP7) of bluetongue virus, by a recombinant capripox virus, provides partial protection of sheep against a virulent heterotypic bluetongue virus challenge. Virology 220: 227–231.

19. Calvo-Pinilla E, Rodriguez-Calvo T, Sevilla N, Ortego J (2009) Heterologous prime boost vaccination with DNA and recombinant modified vaccinia virus Ankara protects IFNAR(−/−) mice against lethal bluetongue infection. Vaccine 28: 437–445.

20. Calvo-Pinilla E, Navasa N, Anguita J, Ortego J (2012) Multiserotype protection elicited by a combinatorial prime-boost vaccination strategy against bluetongue virus. PLoS One 7: e34735.

21. Jabbar TK, Calvo-Pinilla E, Mateos F, Gubbins S, Bin-Tarif A, et al. (2013) Protection of IFNAR (−/−) mice against bluetongue virus serotype 8, by heterologous (DNA/rMVA) and homologous (rMVA/rMVA) vaccination, expressing outer-capsid protein VP2. PLoS One 8: e60574.

22. Umeshappa CS, Singh KP, Pandey AB, Singh RP, Nanjundappa RH (2010) Cell-mediated immune response and cross-protective efficacy of binary ethylenimine-inactivated bluetongue virus serotype-1 vaccine in sheep. Vaccine 28: 2522–2531.

23. Tordo N, Foumier A, Jallet C, Szelechowski M, Klonjkowski B, et al. (2008) Canine adenovirus based rabies vaccines. Dev Biol (Basel) 131: 467–476.

24. Bouet-Cararo C, Contreras V, Fournier A, Jallet C, Guibert JM, et al. (2011) Canine adenoviruses elicit both humoral and cell-mediated immune responses against rabies following immunisation of sheep. Vaccine 29: 1304–1310.

25. Pignolet B, Boullier S, Gelfi J, Bozzetti M, Russo P, et al. (2008) Safety and immunogenicity of myxoma virus as a new viral vector for small ruminants. J Gen Virol 89: 1371–1379.

26. Spiller OB, Goodfellow IG, Evans DJ, Hinchliffe SJ, Morgan BP (2002) Coxsackie B viruses that use human DAF as a receptor infect pig cells via pig CAR and do not use pig DAF. J Gen Virol 83: 45–52.

27. Zientara S, Sailleau C, Dauphin G, Roquier C, Remond EM, et al. (2002) Identification of bluetongue virus serotype 2 (Corsican strain) by reverse-transcriptase PCR reaction analysis of segment 2 of the genome. Vet Rec 150: 598–601.

28. Chartier C, Degryse E, Gantzer M, Dieterle A, Pavirani A, et al. (1996) Efficient generation of recombinant adenovirus vectors by homologous recombination in Escherichia coli. J Virol 70: 4805–4810.

29. Klonjkowski B, Klein D, Galea S, Gavard F, Monteil M, et al. (2009) Gag-specific immune enhancement of lentiviral infection after vaccination with an adenoviral vector in an animal model of AIDS. Vaccine 27: 928–939.

30. Szelechowski M, Fournier A, Richardson J, Eloit M, Klonjkowski B (2009) Functional organization of the major late transcriptional unit of canine adenovirus type 2. J Gen Virol 90: 1215–1223.

31. Blanie S, Mortier J, Delverdier M, Bertagnoli S, Camus-Bouclainville C (2009) M148R and M149R are two virulence factors for myxoma virus pathogenesis in the European rabbit. Vet Res 40: 11.

32. Hemati B, Contreras V, Urien C, Bonneau M, Takamatsu HH, et al. (2009) Bluetongue virus targets conventional dendritic cells in skin lymph. J Virol 83: 8789–8799.

33. Lyons AB, Parish CR (1994) Determination of lymphocyte division by flow cytometry. J Immunol Methods 171: 131–137.

34. Toussaint JF, Sailleau C, Breard E, Zientara S, De Clercq K (2007) Bluetongue virus detection by two real-time RT-qPCRs targeting two different genomic segments. J Virol Methods 140: 115–123.

35. Perez de Diego AC, Sanchez-Cordon PJ, Las Heras AI, Sanchez-Vizcaino JM (2012) Characterization of the immune response induced by a commercially available inactivated bluetongue virus serotype 1 vaccine in sheep. Scientific-WorldJournal 2012: 147158.

36. Millar J, Dissanayake D, Yang TC, Grinshtein N, Evelegh C, et al. (2007) The magnitude of the CD8+ T cell response produced by recombinant virus vectors is a function of both the antigen and the vector. Cell Immunol 250: 55–67.

37. Perrin A, Albina E, Breard E, Sailleau C, Prome S, et al. (2007) Recombinant capripoxviruses expressing proteins of bluetongue virus: evaluation of immune responses and protection in small ruminants. Vaccine 25: 6774–6783.

38. Jeggo MH, Wardley RC, Taylor WP (1984) Role of neutralising antibody in passive immunity to bluetongue infection. Res Vet Sci 36: 81–86.

39. Gumm ID, Newman JF (1982) The preparation of purified bluetongue virus group antigen for use as a diagnostic reagent. Arch Virol 72: 83–93.

40. Janardhana V, Andrew ME, Lobato ZI, Coupar BE (1999) The ovine cytotoxic T lymphocyte responses to bluetongue virus. Res Vet Sci 67: 213–221.

41. Stewart M, Dovas CI, Chatzinasiou E, Athmaram TN, Papanastassopoulou M, et al. (2012) Protective efficacy of Bluetongue virus-like and subvirus-like particles in sheep: presence of the serotype-specific VP2, independent of its geographic lineage, is essential for protection. Vaccine 30: 2131–2139.

42. Kar AK, Iwatani N, Roy P (2005) Assembly and intracellular localization of the bluetongue virus core protein VP3. J Virol 79: 11487–11495.

Trace Element Distribution in Selected Edible Tissues of Zebu (*Bos indicus*) Cattle Slaughtered at Jimma, SW Ethiopia

Veronique Dermauw[1]*, Marta Lopéz Alonso[2], Luc Duchateau[3], Gijs Du Laing[4], Tadele Tolosa[5,6], Ellen Dierenfeld[7], Marcus Clauss[8], Geert Paul Jules Janssens[1]

1 Laboratory of Animal Nutrition, Faculty of Veterinary Medicine, Ghent University, Merelbeke, Belgium, **2** Departemento de Patoloxía Animal, Universidade de Santiago de Compostela, Lugo, Spain, **3** Department of Comparative Physiology and Biometrics, Faculty of Veterinary Medicine, Ghent University, Merelbeke, Belgium, **4** Laboratory of Analytical Chemistry and Applied Ecochemistry Ghent University, Faculty of Bioscience Engineering, Ghent, Belgium, **5** School of Veterinary Medicine, College of Agriculture and Veterinary Medicine of Jimma University, Jimma, Ethiopia, **6** M-team and Mastitis Quality Research Unit, Department of Reproduction, Obstetrics, and Herd Health, Faculty of Veterinary Medicine, Ghent University, Merelbeke, Belgium, **7** Adjunct Faculty, Division of Animal Sciences, University of Missouri-Columbia, Columbia, Missouri, United States of America, **8** Clinic for Zoo Animals, Exotic Pets and Wildlife, Vetsuisse Faculty, University of Zurich, Zurich, Switzerland

Abstract

The amount of trace elements present in edible bovine tissues is of importance for both animal health and human nutrition. This study presents data on trace element concentrations in semitendinosus and cardiac muscles, livers and kidneys of 60 zebu (*Bos indicus*) bulls, sampled at Jimma, Ethiopia. From 28 of these bulls, blood samples were also obtained. Deficient levels of copper were found in plasma, livers, kidneys and semitendinosus muscles. Suboptimal selenium concentrations were found in plasma and semitendinosus muscles. Semitendinosus muscles contained high iron concentrations. Trace elements were mainly stored in the liver, except for iron and selenium. Cardiac muscles generally contained higher concentrations of trace elements than semitendinous muscles except for zinc. A strong association was found between liver and kidney concentrations of copper, iron, cobalt and molybdenum. Liver storage was well correlated with storage in semitendinosus muscle for selenium and with cardiac muscle for cobalt and selenium. Plasma concentrations of copper, selenium, cobalt were well related with their respective liver concentrations and for cobalt and selenium, also with cardiac muscle concentrations. The data suggest multiple trace element deficiencies in zebu cattle in South-West Ethiopia, with lowered tissue concentrations as a consequence. Based on the comparison of our data with other literature, trace element concentrations in selected edible tissues of *Bos indicus* seem quite similar to those in *Bos taurus*. However, tissue threshold values for deficiency in *Bos taurus* cattle need to be refined and their applicability for *Bos indicus* cattle needs to be evaluated.

Editor: Carlos Eduardo Ambrosio, Faculty of Animal Sciences and Food Engineering, University of São Paulo, Pirassununga, SP, Brazil, Brazil

Funding: This research was performed within the doctoral research project of Veronique Dermauw, funded by the Agency for the Promotion of Innovation through Science in Flanders (IWT-Vlaanderen), grant no. 093348. The funders had no role in study design, data collection and analysis, decision to publish, or preparation of the manuscript.

Competing Interests: The authors have declared that no competing interests exist.

* E-mail: veronique.dermauw@ugent.be

Introduction

Deficiencies in trace elements, such as selenium (Se) and zinc (Zn), are frequently observed in humans in tropical regions such as Ethiopia [1], with severe health consequences (e.g. stunted growth, lowered antioxidant status) [2]. Meat and organ consumption form an important contribution to human nutrition, as these tissues have the capacity to store high amounts of trace elements [3]. However, in Ethiopia, the world's fifth largest cattle holder (FAOSTAT 2013 data, http://faostat3.fao.org), zebu (*Bos indicus*) (*B. indicus*) cattle are typically free ranging on poor pastures and bovine trace element shortages (e.g. copper (Cu) deficiencies) are also very common [4]. Unfortunately, data on trace element concentrations in edible tissues (such as meat, liver, kidney, heart) of *B. indicus* cattle, the most commonly used cattle type in the tropics [5,6], and more specifically in Ethiopia, are absent.

In *Bos taurus* (*B. taurus*) cattle, the liver is considered the main indicator organ for status evaluation of several essential trace elements, assuming that it forms the main storage depot and is the most responsive tissue to dietary trace element supply [7]. On the contrary, in *B. taurus* cattle, the distribution of trace elements over other tissues, such as muscle is still not well understood. Essential trace elements seem to distribute differently over different types of muscles [8], possibly related to muscle activity and fat content [9]. Furthermore, the relation between liver trace element status and the distribution of trace elements in other tissues is not fully unravelled. A good relationship between liver and muscle cobalt (Co) and Zn concentrations was found in earlier research with cattle with an adequate status [9], whereas such a relationship was not noticed for other trace elements.

For at least some elements (e.g. Cu), especially at lower concentrations, a reasonable link of liver with plasma concentrations is present [10]. Consequently, plasma trace element concentrations are often used as proxy for liver concentrations and thus, trace element status [11]. The relationship between trace element concentrations in plasma and edible tissues, other than liver, however, was not studied before. The latter could be very important for human nutrition, as plasma concentrations might form a more practical tool for early evaluation of trace element concentrations in meat, essential for optimal human health.

Consequently, the objectives of the present study were to: i) present data on trace element concentrations in selected edible tissues and plasma of Ethiopian *B. indicus* cattle and ii) evaluate the association of liver and plasma trace element concentrations, as indicators of trace element status with other tissue concentrations.

Materials and Methods

Study area, animals and samples

The study was conducted in Jimma, the largest town in the Gilgel Gibe catchment area, Ethiopia, where bovine trace element deficiencies were previously recognized [4,12]. The local abattoir receives animals from the urban Jimma zone as well as from surrounding areas and distributes meat and organs within the city of Jimma, for consumption and sale in small restaurants and butcheries. During ten different days, adult zebu (*B. indicus*) bulls ($n = 60$) were randomly selected here, using the lottery method. Their origin was noted and is presented in Figure 1. Thereafter, 28 out of 60 bulls were randomly selected for immediate post-mortem blood sampling, using two sodium heparin tubes (VT-100SH, Venoject®). Subsequently, from all 60 bulls, the cranial part of the left kidney, caudal lobe of the liver, semitendinosus and cardiac muscle (apex of the heart) were sampled. We greatly acknowledge Keraa abattoir and Jimma municipality for their kind permission to sample carcasses. Samples were immediately cooled and transported to the laboratory. Plasma was obtained through centrifugation at $1500 \times g$ for 10 minutes and excessive fat was removed from tissue samples, where necessary. Samples were stored at $-20°C$ until further analysis.

Mineral analyses

Muscle, kidney and liver samples were oven dried at $65°C$ until constant weight and ground through a 2-mm screen. Afterwards, samples were ashed through microwave destruction with 10 ml HNO_3 (Ultrapure analytical grade for trace element analysis) in open vessels followed by filtration. Finally, all samples were analysed for Zn, Cu, iron (Fe), Se, molybdenum (Mo), Co and manganese (Mn) concentrations through inductively coupled plasma optical emission spectrometry (ICP-OES) (Vista MPX radial, Varian, Palo Alto, USA) and inductively coupled plasma mass spectrometry (ICP-MS) (Elan DRC-e, PerkinElmer, Sunnyvale, CA, USA). All glassware and microwave vessels were pre-rinsed with diluted HNO_3. A quality control program was employed throughout trace element analyses. Trace element recovery rates from the sampled matrices (plasma, liver, kidney, semitendinosus and cardiac muscle) spiked with two different concentrations of the studied trace elements (in the range of the determined concentrations), were measured. Average recovery was 98%, with a range between 82% (Zn in plasma) and 109% (Mo in kidney). Detection limits in acid digest, determined by the method of Hubaux and Vos [13], were: Mn 0.35 µg/l, Cu 0.25 µg/l, Mo 0.33 µg/l, Se 0.13 µg/l, Fe 21.4 µg/l, Zn 16.4 µg/l and Co 0.14 µg/l. Standards were run frequently alongside samples and all analytical results were blank-corrected.

Statistical analysis and reference value calculations

To detect whether differences were present for trace element concentrations between liver, kidney, semitendinosus and cardiac muscle, tissue concentrations were compared using a signed rank test at the 5% significance level with Bonferroni's adjustment technique for pairwise comparisons. Median and the first (Q1) and third (Q3) quartiles are reported. Spearman rank correlation coefficients (r) were used to determine the association between liver and plasma and other tissue concentrations of trace elements.

Diagnostic threshold concentrations for *B. taurus* cattle stated in literature [14] and [7] are expressed on wet weight (WW) basis, and in order to compare with the current data, were recalculated to dry weight (DW) basis by multiplying with the conversion factors stated by the authors: 3.5 for liver, 4.5 for other tissues [14]; 3.3 for liver respectively [7].

Literature search

In order to compare our data with earlier work, we searched for studies mentioning trace element concentrations in the selected tissues in zebu cattle. Based on the zebu distribution maps of Caramelli [5] and Bradley et al. [6], we considered cattle to be zebu-typed if originating from India, Chad, Sudan, Eritrea, Djibouti, Ethiopia, Somalia, Kenya, Tanzania, Mozambique, Malawi or Central African Republic. Other African countries were only partially inhabited with indicine cattle in addition to crossbreds or even taurine breeds [5,6], we, therefore, excluded these from our literature search. Furthermore, data from Yemen, Oman and the Emirates were added. Studies performed in Brazil were also included based on the presence of the indicine Nellore and Guzerat breeds [15]. Any other studies specifically stating the use of zebu cattle were also taken into account. Afterwards, studies investigating the same topic in taurine cattle were also searched.

Results

Summary statistics on trace element concentrations in selected edible tissues and plasma are presented in Table 1 and 2 respectively. Table 3, 4, 5 and 6 demonstrate a comparison of our data with results from literature. Upon comparison of liver concentrations with diagnostic criteria for deficiency in *B. taurus* cattle, 42% of animals ($n = 60$) were considered severely Cu deficient (<19 mg/kg DW; [7]) (Table 1). Plasma Cu concentrations in 29% of animals ($n = 28$) reflected this deficiency (Table 2). Furthermore, semitendinosus muscle and kidney Cu concentrations were below concentrations considered adequate in *B. taurus* cattle in 97% and 100% of animals ($n = 60$). Liver samples did not reveal deficiencies for Fe, Se ($n = 60$; all but one >150, all >0.07 mg/kg DW respectively; [7]) or Co, Zn and Mn (all >0.018, >70 and >3.5 mg/kg DW respectively, [14]). On the contrary, plasma concentrations did indicate a severe Mn and Fe deficiency in 29% and 11% of animals ($n = 28$) respectively. For Se, although none of the animals ($n = 28$) had plasma concentrations below diagnostic thresholds for deficiency according to Suttle [7], 82% of animals had plasma Se concentrations considered at least marginally deficient (<0.06 mg/l), according to Puls [14]. When comparing kidney and muscle concentrations of the latter trace elements with adequate ranges for *B. taurus* cattle, semitendinosus muscle Fe concentrations registered above this range in 73% of animals ($n = 60$), whereas 60% of animals ($n = 60$) had semitendinosus muscle Se concentrations below the adequate range.

Liver contained the highest concentrations of trace elements compared to kidney, cardiac and semitendinosus muscle (Table 1), except for Se, of which concentrations were highest in kidney (all

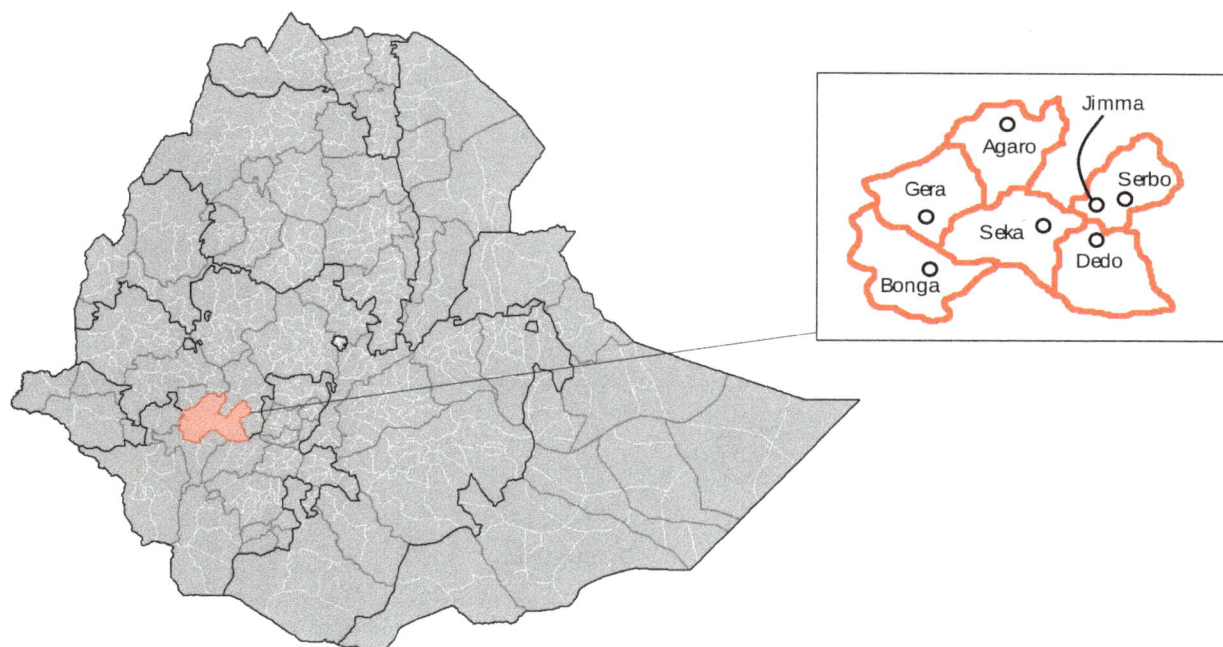

Figure 1. Map presenting the cities of origin of zebu (*Bos indicus*) bulls (*n* = 60) sampled at Jimma, South-West Ethiopia. The red lines depict the woredas, an administrative unit in Ethiopia, to which the cities belong = Bonga (Gimbo), Gera (Gera), Seka (Seka Chekorsa), Agaro (Goma), Jimma, Serbo (Kersa) and Dedo (Dedo).

$p<0.010$), and for Fe, for which we found no difference between liver and kidney concentrations ($p = 0.035$). Cardiac muscles systematically contained higher concentrations of trace elements than semitendinosus muscles, except for Zn of which concentrations were lower in the cardiac muscle samples (all $p<0.001$, for Mo: $p = 0.001$). Dry matter concentrations in our study averaged 29% (range: 23–35%) for liver samples, 24% (17–42%) for kidney, 23% (15–31%) for semitendinosus and 22% (18–27%) for cardiac muscle samples.

A strong association was found between liver and kidney concentrations of Cu, Fe, and Co ($r = 0.53$, $r = 0.65$, $r = 0.80$ respectively; all $p<0.001$) (Table 7), whereas liver and kidney concentrations of Mn, Zn and Mo were weakly correlated ($r = 0.28$, $p = 0.03$; $r = 0.38$, $p = 0.003$; $r = 0.39$, $p = 0.002$ respectively). There was a strong relation between liver and semitendinosus muscle concentrations of Se ($r = 0.57$, $p<0.001$) and a weak correlation for Mn ($r = 0.36$, $p = 0.005$) whereas for other elements, no significant association were found between these two tissues. Cardiac muscle concentrations of Co and Se were

Table 1. Trace element concentrations (mg/kg DW) in zebu (*Bos indicus*) bull (*n* = 60) tissues sampled at Jimma, Ethiopia with median, first quartile (Q1) and third quartile (Q3) as summary statistics.

Element	Liver								Kidney								Muscle										
																	Semitendinosus				Cardiac						
	Median	Q1	-	Q3	Adequate‡				Median	Q1	-	Q3	Adequate‡				Median	Q1	-	Q3	Median	Q1	-	Q3	Adequate‡		
Cu	28.3[a]	7.6	-	65.6	88	-	350		13[b]	12	-	16	18	-	27		2.9[d]	2.0	-	3.6	16[c]	15	-	17	5.4	-	6.8
Fe	306[a]	245	-	415	158	-	1050		369[a]	286	-	503	135	-	675		80[c]	53	-	105	210[b]	194	-	239	45	-	54
Mn	13[a]	11	-	14	8.8	-	21		5.2[b]	4.6	-	5.6	5.4	-	9.0		1.1[d]	0.8	-	1.8	2.4[c]	1.9	-	3.0	2.0	-	3.8
Zn	152[a]	121	-	201	88	-	350		107[b]	94.5	-	126	81	-	113		103[b]	83	-	148	81[c]	75	-	85		-	
Co	0.47[a]	0.38	-	0.65	0.07	-	0.30		0.39[b]	0.25	-	0.59		-			ND[d]	ND	-	ND	0.23[c]	0.14	-	0.37		-	
Mo	3.8[a]	3.3	-	4.2	0.49	-	4.9		1.9[b]	1.8	-	2.2	1.0	-	2.6		ND[d]	ND	-	0.32	0.33[c]	0.26	-	0.42		-	
Se	0.76[b]	0.65	-	0.88	0.88	-	1.8		4.8[a]	4.3	-	5.3	4.5	-	6.8		0.37[d]	0.23	-	0.49	0.74[c]	0.59	-	0.86	0.32	-	0.68

DW = dry weight.
‡Adequate range for cattle [14].
[a,b]Medians sharing a same letter do not differ significantly from each other ($p<0.050$).

Table 2. Trace element concentrations in zebu (*Bos indicus*) bull (*n* = 28) plasma sampled at Jimma, Ethiopia with median, first quartile (Q1) and third quartile (Q3) as summary statistics.

Mineral	Median	Q1	-	Q3	Adequate‡	-		Threshold value[1]
Cu, mg/l	0.7	0.5	-	0.8	0.8	-	1.5	0.6
Fe, mg/l	1.7	1.2	-	2.0	1.3	-	2.5	1.0
Mn, µg/l	45	18	-	60	6	-	70	20
Zn, mg/l	1.2	1.1	-	1.3	0.8	-	1.4	0.6
Co, µg/l	3.9	2.8	-	5.2		-		0.9[a]
Mo, µg/l	26	19	-	34	10	-	50	100[b]
Se, µg/l	45	36	-	54	80	-	300	20

‡Adequate range for cattle [14].
[1]upper threshold value indicating a deficiency risk in *Bos taurus* cattle [7].
[a]Co: lower boundary of normal Co concentrations in *Bos taurus* cattle [14].
[b]Mo: lower boundary of Mo concentrations in *Bos taurus* cattle considered elevated [14].

strongly correlated with liver concentrations of the same elements (r = 0.71, r = 0.75, respectively; both p<0.001). Additionally, there was a weak positive association between liver and cardiac concentrations of Fe and Mo (r = 0.28, p = 0.03; r = 0.37, p = 0.003) and a weak negative association for Zn (r = −0.34; p = 0.007).

Plasma concentrations of Se, Cu, Co were strongly associated with liver concentrations of the same elements (r = 0.74, r = 0.68, r = 0.61; all p<0.001) whereas for Mn and Zn, only a weak relation was present, which was even negative for Zn (r = 0.42, r = −0.39; both p<0.050). For kidney, only Co concentrations were significantly associated with plasma concentrations (r = 0.82, p<0.001). Furthermore, only Se semitendinosus muscle concentrations were associated with plasma concentrations (r = 0.71; p<0.001). Finally, Se and Co cardiac muscle concentrations were strongly related with plasma Se and Co concentrations (r = 0.83, r = 0.69; both p<0.001).

Discussion

To the best of our knowledge, this is the first study presenting data on trace element concentrations in all selected edible tissues in zebu (*B. indicus*) cattle. Bovine Cu deficiency in this region, as found in previous research [4,12] was confirmed by both plasma and liver concentrations. On the contrary, it was not clear whether or not a Mn, Fe or Se deficiency was present in the area due to conflicting interpretations based upon plasma and liver concentrations [7,14], as well as a wide range in threshold values found in literature, mentioned earlier [4]. In this respect, especially threshold values for Se both in liver and plasma vary largely among authors [7,14,16]. Thus, our data suggest a multiple trace element deficiency in zebu (*B. indicus*) bulls sampled at Jimma, Ethiopia. Yet, they also point to the clear and urgent need for refinement of bovine plasma and liver thresholds of deficiency, as a practical reference to evaluate the need for supplementation, considering discrepancies were present upon evaluation of plasma and liver concentrations based on diagnostic criteria for deficiency

Table 3. Literature review of average trace element concentrations in bovine liver (mg/kg WW).

Reference	Country	Cu	Mo	Fe	Zn	Mn	Se	Co
Bos taurus								
[20]	Morocco	32	-	-	37	-	-	-
[21][1]	Spain	40	1.1	70	49	2.4	0.2	0.10
[23]	Spain	90	1.4	44	54	3.5	0.2	0.07
[24]	Canada	28	-	-	45	-	0.3	-
[25]	Spain	60	-	-	60	-	-	-
[26]	Jamaica	20	-	-	30	-	0.4	-
[27]	Czech Republic	-	-	-	-	-	0.1	-
[28]	Belgium	80	-	-	40	-	-	-
Bos indicus[2]								
[29,30]	Kenya	21	-	-	37	-	0.1	-
[31]	Ethiopia	4	-	293	42	4.1	-	-
[32]	Sudan	67	-	-	-	-	-	-
This study	**Ethiopia**	**18**	**1.1**	**118**	**47**	**3.8**	**0.2**	**0.15**

[1]Geometric mean,
[2]Presumably *Bos indicus* cattle based upon location or mentioned as such.

Table 4. Literature review of average trace element concentrations in bovine kidney (mg/kg WW).

Reference	Country	Cu	Mo	Fe	Zn	Mn	Se	Co
Bos taurus								
[20]	Morocco	7.3	-	-	20	-	-	-
[21][1]	Spain	3.1	0.3	51	15	0.7	1.0	0.04
[23]	Spain	4.6	0.5	59	26	1.2	1.4	0.03
[24]	Canada	5.4	-	-	22	-	0.8	-
[25]	Spain	3.7	-	-	22	-	-	-
[26]	Jamaica	3.9	-	-	20	-	1.0	-
[28]	Belgium	5.0	-	-	18	-	-	-
Bos indicus×Bos taurus								
[33][2]	Burundi	3.4	-	-	23	-	1.4	-
Bos indicus								
This study	**Ethiopia**	**3.3**	**0.5**	**97**	**27**	**1.3**	**1.1**	**0.10**

[1]Geometric mean,
[2]Presumably *Bos indicus* crossbred cattle based upon location or mentioned as such.

in *B. taurus* cattle [7,14,16]. Further, in literature (Table 3, 4, 5 and 6), reported tissue concentrations of Cu concentrations seem to be generally low or at the lower border of adequacy reported by Puls [11], especially in *B. indicus* cattle. On the contrary, in literature, liver, kidney and muscle Fe seem to be at the higher end of or above adequacy ranges stated by the same author [14]. Adequate concentrations are generally observed for Zn, whereas Mn concentrations are rather low and Se often too low in comparison with ranges for adequacy of Puls [14]. This all might point to a generalized trace element imbalance in the cattle sampled in

Table 5. Literature review of average trace element concentrations in bovine muscle (mg/kg WW).

Source	Country	Cu	Mo	Fe	Zn	Mn	Se	Co
Bos taurus								
[8][1]	Spain	0.8	0.13	19	35	0.1	0.10	0.004
[20][4]	Morocco	1.0	-	-	27	-	-	-
[21][b,3]	Spain	1.7	0.09	39	50	0.2	ND	0.016
[25][3]	Spain	1.3	-	-	53	-	-	-
[27][3]	Czech Republic	-	-	-	-	-	0.04	-
[28][4]	Belgium	1.6	-	-	43	-	-	-
[34][1]	Uruguay	0.4	-	42	25	0.2	0.62	-
[35][a,2]	Venezuela	0.8	-	19	41	0.3	-	-
[36][a,4]	USA	0.7	-	20	41	0.1	0.18	-
[37][a,2]	USA	-	-	17	41	-	-	-
[38][2]	Brazil	-	-	13	34	-	-	-
Bos indicus×Bos taurus								
[33][c, 4]	Burundi	1.1	-	-	54	-	0.20	-
[34][1]	Uruguay	0.6	-	38	24	0.5	0.55	-
[38][2]	Brazil	-	-	13	35	-	-	-
[39][2]	Venezuela	0.9	-	18	38	0.1	-	-
Bos indicus								
This study[1]	**Ethiopia**	**0.7**	**0.07**	**29**	**27**	**0.8**	**0.10**	**0.020**

[a]Unsure whether *Bos taurus* or crossbred,
[b]Geometric mean,
[c]Presumably *Bos indicus×Bos taurus* crossbred cattle based on location,
[1]Semitendinosus muscle,
[2]Longissimus dorsi thoracis muscle,
[3]Diafragm muscle,
[4]Not-specified.

Table 6. Literature review of trace element concentrations in bovine heart (mg/kg WW).

Reference	Country	Cu	Mo	Fe	Zn	Mn	Se	Co
Bos taurus								
[8]	Spain	4.4	0.02	45	17	0.3	0.3	0.01
Bos indicus×*Bos taurus*								
[33][1]	Burundi	4.0	-	-	20	-	0.3	-
Bos indicus								
This study	**Ethiopia**	**3.5**	**0.09**	**49**	**18**	**0.6**	**0.2**	**0.06**

[1]Presumably *Bos indicus*×*Bos taurus* crossbred cattle based on location.

literature. However, it might also indicate that the ranges mentioned by Puls [14] need to be re-evaluated, as they might not reflect bovine trace element status well. The need for clarification is urgent as such reported ranges have a large impact on how we evaluate tissue concentrations in cattle.

An additional hurdle when comparing tissue concentrations with threshold values or other comparative data is the wide variability in dry matter content, both between and within tissues. Because of this, if stated concentrations are expressed in a different unit, e.g. on fresh matter basis, conversion to this unit using a single conversion factor, might create bias. We therefore recommend authors stating data and diagnostic threshold concentrations, to at least mention average dry matter concentrations per tissue.

It also remains unclear whether or not the mentioned thresholds values and adequate ranges are to be extrapolated from *B. taurus* to *B. indicus* cattle as mentioned by Dermauw et al. [12], seeing that even within *B. taurus* cattle, differences in breed sensitivity to deficiency are present [17,18]. When comparing limited literature data available in *B. taurus* and *B. indicus* cattle, the trace element concentrations in the selected tissues seem similar, although liver and kidney Fe concentrations as well as cardiac Mo concentrations seem higher in *B. indicus* than in *B. taurus* cattle. Moreover, seemingly, tissue Co concentrations are often higher in *B. indicus* than in *B. taurus* cattle (Table 3, 4, 5 and 6). However, to fully unravel potential similarities or differences in trace element distribution between *B. taurus* and *B. indicus* cattle types, further

research needs to investigate their trace element storage in a comparable environment during depletion and repletion phases.

Liver contained the highest concentrations of trace elements compared to kidney, cardiac and semitendinosus muscle. Despite significantly higher Se concentrations in the kidney, the liver may still function as a main storage tissue for Se due its larger weight, in agreement with previous research [11]. Fe seemed evenly distributed over liver and kidney, but the tissue weight may again result in the liver being the main storage entity. Liver contained the highest Zn concentrations, but Zn concentrations did not seem to vary widely among liver, kidney and muscle. as observed earlier [19]. Further, the systematically higher trace element concentrations in cardiac than in semitendinosus muscle demonstrate a profound difference in micromineral profile between muscle types. Comparative data in *B. indicus* cattle are absent, and data in *B. taurus* cattle are rare since sampled muscles are often not specified. However, our findings are generally in agreement with earlier research in *B. taurus* cattle [8]. Molybdenum formed an exception, considering the higher concentrations in cardiac muscle as compared to semitendinosus muscle in the current study, which contradicts the earlier data [8]. This may be explained by higher Mo concentrations in the environment and forages in the region [12] leading to a higher accumulation in the *B. indicus* cattle in general and in cardiac muscle more specifically, although none of the sampled tissues contained Mo concentrations beyond the normal ranges stated for *B. taurus* cattle [14]. Overall, the unequal distribution of trace elements amongst the tissues sampled,

Table 7. Spearman rank correlation coefficient between liver and other tissue concentrations of trace elements in zebu (*Bos indicus*) bulls (n = 60) at Jimma, Ethiopia.

Element	Plasma vs. liver	Liver vs.			Plasma vs.		
		Kidney	Muscle		Kidney	Muscle	
			Semitendinosus	Cardiac		Semitendinosus	Cardiac
Cu	0.68***	0.53***	0.01	0.08	0.25	−0.14	0.16
Fe	−0.06	0.65***	0.19	0.28*	0.29	0.11	0.19
Mn	0.42*	0.28*	0.36**	0.19	0.22	0.37	0.16
Zn	−0.39*	0.38**	−0.24	−0.34**	−0.32	0.08	0.18
Co	0.61***	0.80***	0.14	0.71***	0.82***	0.24	0.69***
Mo	0.25	0.39**	0.05	0.37**	−0.05	−0.31	−0.21
Se	0.74***	0.17	0.58**	0.75***	0.15	0.71***	0.83***

*p<0.050,
**p<0.010,
***p<0.001.

generally in line with earlier data, relates to the specific function and metabolism of these elements, and i.e. the organ-specific abundance of ligands, such as metallothionein (Cu, Zn uptake) and ceruloplasmin (Cu), binding the trace elements and regulating their uptake [7,20].

In general, trace element concentrations in the liver, the main storage organ and indicator of trace element status, seemed to correlate reasonably well with storage in other tissues, especially kidney and cardiac muscle. For most trace elements, there was a reasonable association between liver and kidney concentrations, which contradicts previous research [21]. Considering the relation between liver and muscle storage, Blanco Penedo et al. [9] earlier found a strong association for Co and Zn, but not for Se concentrations, the latter being in contrast to our study. Plasma is often presented as a practical sample to assess trace element status [11]. Overall, plasma Cu, Co and Se seemed to associate well with their respective liver concentrations and reasonably well for Mn. This was not found for Fe and Mo or even negative for Zn concentrations. Although based on a small sample size, our data suggest that plasma Co and Se are probably very suitable for evaluation of liver status whereas Cu concentrations were well associated but should probably only be used in the case of expected low liver Cu concentrations [10]. In this regard, the current results discourage the use of plasma Mo, Zn, and especially Fe concentrations, although mentioned previously [11]. Finally, plasma concentrations of Co and Se were well related with cardiac muscle concentrations, which was not reported before.

Overall, our results confirm the differential distribution patterns of Se, Co vs. the other elements, explained by different sites of homeostatic control mechanisms. Homeostatic control of Se and Co is mediated by renal excretion, causing a continuing rise in transport (plasma) and storage (kidney, liver, muscle) pool concentrations, related with rising dietary intake. For the other elements, the intestine is the site of homeostatic control, and consequently, transport and storage pool concentrations might plateau when requirements are met [22].

The lack of information on trace element concentrations in the consumed diets of the sampled B. indicus bulls could be considered a limitation of our study. However, in this trial, we focussed on the link between a certain trace element status and the distribution of trace elements in tissues. For liver, it is known that micromineral supply directly affects storage in the organ [7]. Both in literature and in practice, plasma concentrations are often employed as proxy for liver status, whenever sampling liver is impossible due to practical issues [11]. Indeed, our statements concerning the deficient bovine status in Cu and possibly Se and Mn based on concentrations both in liver and plasma are in line with earlier dietary data from the region, mentioning levels of Cu and Se considered inadequate for B. taurus cattle and excessive Fe levels [12]. Considering this multiple trace element deficiency present in the sampled B. indicus cattle in the area, we recommend for a future risk analysis study to be performed in order to define both animal and public health risks involved with this condition.

Conclusion

A multiple trace element deficiency was present in B. indicus cattle sampled in South-West Ethiopia. Based on the comparison of our data with literature, zebu cattle seemed to have similar trace element concentrations in edible tissues as B. taurus cattle. Overall, the liver was the main storage organ for trace elements and correlated well with concentrations of the same elements in the other tissues. Cardiac muscles generally contained higher amounts of trace elements than semitendinosus muscles. Further, different distribution patterns in edible tissues of Cu, Zn, Mn, Mo and Fe versus Co and Se were observed. Within the ranges observed in our study, plasma values were well related with liver status of Cu, Se and Co and even with muscle Se and Co concentrations.

Acknowledgments

We would like to thank the IUC-JU programme of VLIR-UOS for the logistic support, Bashahun Gebremichael Dirar and Eva Gossieaux for assistance during sampling and Ria Van Hulle and Joachim Neri for performing mineral analyses.

Author Contributions

Conceived and designed the experiments: VD GPJJ ED MC. Performed the experiments: VD TT. Analyzed the data: VD LD MLA. Contributed reagents/materials/analysis tools: GDL. Wrote the paper: VD.

References

1. Amare B, Moges B, Fantahun B, Tafess K, Woldeyohannes D, et al. (2012) Micronutrient levels and nutritional status of school children living in Northwest Ethiopia. Nutr J 11: 108.

2. World Health Organization (1996) Trace elements in human nutrition and health. Geneva: World Health Organization.

3. Berger MM (2005) Can oxidative damage be treated nutritionally? Clin Nutr 24: 172–183.

4. Dermauw V, Yisehak K, Belay D, Van Hecke T, Du Laing G, et al. (2013) Mineral deficiency status of ranging zebu (Bos indicus) cattle around the Gilgel Gibe catchment, Ethiopia. Trop Anim Health Prod 45: 1139–1147.

5. Caramelli D (2006) The origins of domesticated cattle. Hum Evol 21: 107–122.

6. Bradley DG, Loftus RT, Cunningham P, MacHugh DE (1998) Genetics and domestic cattle origins. Evol Anthropol Issues, News, Rev 6: 79–86.

7. Suttle NF (2010) Mineral Nutrition of Livestock. 4th Edition. Wallingford, Oxfordshire: CABI Publishing.

8. García-Vaquero M, Miranda M, Benedito JL, Blanco-Penedo I, López-Alonso M (2011) Effect of type of muscle and Cu supplementation on trace element concentrations in cattle meat. Food Chem Toxicol 49: 1443–1449.

9. Blanco-Penedo I, López-Alonso M, Miranda M, Hernández J, Prieto F, et al. (2010) Non-essential and essential trace element concentrations in meat from cattle reared under organic, intensive or conventional production systems. Food Addit Contam Part A Chem Anal Control Expo Risk Assess 27: 36–42.

10. Claypool DW, Adams FW, Pendell HW, Hartmann Jr NA, Bone JF (1975) Relationship between the level of copper in the blood plasma and liver of cattle. J Anim Sci 41: 911–914.

11. Herdt TH, Hoff B (2011) The use of blood analysis to evaluate trace mineral status in ruminant livestock. Vet Clin North Am Food Anim Pract 27: 255–283i.

12. Dermauw V, Yisehak K, Dierenfeld ES, Du Laing G, Buyse J, et al. (2013) Effects of trace element supplementation on apparent nutrient digestibility and utilisation in grass-fed zebu (Bos indicus) cattle. Livest Sci 155: 255–261.

13. Hubaux A, Vos G (1970) Decision and detection limits for calibration curves. Anal Chem 42: 849–855.

14. Puls R (1988) Mineral levels in animal health. Diagnostic data. Clearbrook, BC: Sherpa International.

15. McTavish EJ, Decker JE, Schnabel RD, Taylor JF, Hillis DM (2013) New World cattle show ancestry from multiple independent domestication events. Proc Natl Acad Sci U S A 110: E1398–406.

16. Kincaid RL (2000) Assessment of trace mineral status of ruminants: A review. J Anim Sci 77: 1–10.

17. Ward JD, Spears JW, Gengelbach GP (1995) Differences in copper status and copper metabolism among Angus, Simmental, and Charolais cattle. J Anim Sci 73: 571–577.

18. Mullis L, Spears J, McCraw R (2003) Effects of breed (Angus vs Simmental) and copper and zinc source on mineral status of steers fed high dietary iron. J Anim Sci 81: 318–322.

19. López Alonso M, Benedito JL, Miranda M, Castillo C, Hernández J, et al. (2000) Arsenic, cadmium, lead, copper and zinc in cattle from Galicia, NW Spain. Sci Total Environ 246: 237–248.

20. Sedki A, Lekouch N, Gamon S, Pineau A (2003) Toxic and essential trace metals in muscle, liver and kidney of bovines from a polluted area of Morocco. Sci Total Environ 317: 201–205.

21. López Alonso M, Prieto Montaña F, Miranda M, Castillo C, Hernández J, et al. (2004) Interactions between toxic (As, Cd, Hg and Pb) and nutritional essential (Ca, Co, Cr, Cu, Fe, Mn, Mo, Ni, Se, Zn) elements in the tissues of cattle from NW Spain. Biometals 17: 389–397.

22. Windisch W, E=tfie T (2008) Limitations and possibilities for progress in defining trace mineral requirements of livestock. In: Schleger P, Durosoy S, Jongbloed AW, editors. Trace elements in animal production systems. Wageningen: Wageningen Academic Publishers. pp. 187–201.

23. Blanco-Penedo I, Cruz JM, López-Alonso M, Miranda M, Castillo C, et al. (2006) Influence of copper status on the accumulation of toxic and essential metals in cattle. Environ Int 32: 901–906.

24. Korsrud GO, Meldrum JB, Salisbury CD, Houlahan BJ, Saschenbrecker PW, et al. (1985) Trace element levels in liver and kidney from cattle, swine and poultry slaughtered in Canada. Can J Comp Med 49: 159–163.

25. López Alonso M, Benedito J, Miranda M, Castillo C, Hernandez J, et al. (2000) Toxic and trace elements in liver, kidney and meat from cattle slaughtered in Galicia (NW Spain). Food Addit Contam 17: 447–457.

26. Nriagu J, Boughanen M, Linder A, Howe A, Grant C, et al. (2009) Levels of As, Cd, Pb, Cu, Se and Zn in bovine kidneys and livers in Jamaica. Ecotoxicol Environ Saf 72: 564–571.

27. Pavlata L, Pechová A, Bečvář O, Illek J (2001) Selenium status in cattle at slaughter: analyses of blood, skeletal muscle, and liver. Acta Vet Brno 70: 277–284.

28. Waegeneers N, Pizzolon J-C, Hoenig M, De Temmerman L (2009) Accumulation of trace elements in cattle from rural and industrial areas in Belgium. Food Addit Contam Part A Chem Anal Control Expo Risk Assess 26: 326–332.

29. Frøslie A, Ulvund M, Maina J, Norheim G (1983) Trace elements in grass, cattle liver and sheep liver from districts surrounding Karatina, Kenya. I. Selenium. Nord Vet Med 35: 209–212.

30. Frøslie A, Ulvund M, Maina J, Norheim G (1983) Trace elements in grass, cattle liver and sheep from districts surrounding Karatina, Kenya II. Copper, molybdenum, zinc and sulphur. Nord Vet Med 35: 213–218.

31. Khalili M, Lindgren E, Varvikko T (1993) A survey of mineral status of soil, feeds and cattle in the Selale Ethiopian highlands. II. Trace elements. Trop Anim Health Prod 25: 193–201.

32. Tartour G (1975) Copper status in livestock, pasture and soil in Western Sudan. Trop Anim Health Prod 7: 87–94.

33. Benemariya H, Robberecht H, Deelstra H (1993) Zinc, copper, and selenium in milk and organs of cow and goat from Burundi, Africa. Sci Total Environ 128: 83–98.

34. Cabrera MC, Ramos a, Saadoun a, Brito G (2010) Selenium, copper, zinc, iron and manganese content of seven meat cuts from Hereford and Braford steers fed pasture in Uruguay. Meat Sci 84: 518–528.

35. Huerta-Leidenz N, Arenas de Moreno L, Moron-Fuenmayor O, Uzcátegui-Bracho S (2003) Mineral composition of raw longissimus muscle derived from beef carcasses produced and graded in Venezuela. Arch Latinoam Nutr 53: 96–101.

36. Leheska JM, Thompson LD, Howe JC, Hentges E, Boyce J, et al. (2008) Effects of conventional and grass-feeding systems on the nutrient composition of beef. J Anim Sci 86: 3575–3585.

37. Duckett SK, Neel JPS, Fontenot JP, Clapham WM (2009) Effects of winter stocker growth rate and finishing system on: III. Tissue proximate, fatty acid, vitamin, and cholesterol content. J Anim Sci 87: 2961–2970.

38. de Freitas AK, Lobato JFP, Cardoso LL, Tarouco JU, Vieira RM, et al. (2014) Nutritional composition of the meat of Hereford and Braford steers finished on pastures or in a feedlot in southern Brazil. Meat Sci 96: 353–360.

39. Giuffrida-Mendoza M, Arenas de Moreno L, Uzcátegui-Bracho S, Rincón-Villalobos G, Huerta-Leidenz N (2007) Mineral content of longissimus dorsi thoracis from water buffalo and Zebu-influenced cattle at four comparative ages. Meat Sci 75: 487–493.

Toll-Like Receptor Responses to *Peste des petits ruminants* Virus in Goats and Water Buffalo

Sakthivel Dhanasekaran[1], Moanaro Biswas[2], Ambothi R. Vignesh[1], R. Ramya[1], Gopal Dhinakar Raj[1], Krishnaswamy G. Tirumurugaan[1], Angamuthu Raja[1], Ranjit S. Kataria[3], Satya Parida[4]*, Elankumaran Subbiah[2]*

1 Department of Animal Biotechnology, Madras Veterinary College, Tamil Nadu Veterinary and Animal Sciences University, Chennai, Tamil Nadu, India, **2** Department of Biomedical Sciences and Pathobiology, Center for Molecular Medicine and Infectious Diseases, Virginia-Maryland Regional College of Veterinary Medicine, Virginia Polytechnic Institute and State University, Blacksburg, Virginia, United States of America, **3** Animal Genetics Division, National Bureau of Animal Genetic Resources, Karnal (Haryana), India, **4** Head of FMD Vaccine Differentiation Group, The Pirbright Institute, Surrey, United Kingdom

Abstract

Ovine rinderpest or goat plague is an economically important and contagious viral disease of sheep and goats, caused by the *Peste des petits ruminants* virus (PPRV). Differences in susceptibility to goat plague among different breeds and water buffalo exist. The host innate immune system discriminates between pathogen associated molecular patterns and self antigens through surveillance receptors known as Toll like receptors (TLR). We investigated the role of TLR and cytokines in differential susceptibility of goat breeds and water buffalo to PPRV. We examined the replication of PPRV in peripheral blood mononuclear cells (PBMC) of Indian domestic goats and water buffalo and demonstrated that the levels of TLR3 and TLR7 and downstream signalling molecules correlation with susceptibility vs resistance. Naturally susceptible goat breeds, Barbari and Tellichery, had dampened innate immune responses to PPRV and increased viral loads with lower basal expression levels of TLR 3/7. Upon stimulation of PBMC with synthetic TLR3 and TLR7 agonists or PPRV, the levels of proinflammatory cytokines were found to be significantly higher while immunosuppressive interleukin (IL) 10 levels were lower in PPRV resistant Kanni and Salem Black breeds and water buffalo at transcriptional level, correlating with reduced viralloads in infected PBMC. Water buffalo produced higher levels of interferon (IFN) α in comparison with goats at transcriptional and translational levels. Pre-treatment of Vero cells with human IFNα resulted in reduction of PPRV replication, confirming the role of IFNα in limiting PPRV replication. Treatment with IRS66, a TLR7 antagonist, resulted in the reduction of IFNα levels, with increased PPRV replication confirming the role of TLR7. Single nucleotide polymorphism analysis of TLR7 of these goat breeds did not show any marked nucleotide differences that might account for susceptibility vs resistance to PPRV. Analyzing other host genetic factors might provide further insights on susceptibility to PPRV and genetic polymorphisms in the host.

Editor: Nagendra Hegde, Ella Foundation, India

Funding: Support was provided by TLR.NAIP (http://www.naiptlr.com/NAIP-TLR.pdf). The sponsors have no role in the study design, data collection and analysis, decision to publish or preparation of the manuscript.

* Email: satya.parida@pirbright.ac.uk (SP); kumarans@vt.edu (ES)

Introduction

Peste des petits ruminants (PPR), also known as ovine rinderpest or goat plague, is an acute, highly contagious viral disease of goats and sheep, caused by the *Peste des petits ruminants* virus (PPRV), a *morbillivirus* in the family *Paramyxoviridae*. The disease is characterized by high fever, nasal and ocular discharges, pneumonia, necrotic and ulcerative lesions of the mucus membranes and inflammation of the gastro-intestinal tract [1]. PPRV infection results in great economic losses and affects productivity of sheep and goats subsequent to the global eradication of Rinderpest [2]. For example, in 2004, the economic cost of PPRV in India was estimated to be 1800 million Indian rupees (US$ 39 million) per year [2,3]. PPRV replication and seroconversion has been demonstrated in large ruminants [4]. There is a solitary report on

clinical PPRV occurring in water buffalo [5], although it has not been confirmed in later studies. In September 2004, outbreaks of PPR in Sudan affected both sheep and camels [6].

PPR is generally considered a more serious disease in goats than sheep, however, increased susceptibility of sheep, goat and outbreaks involving both sheep and goats have been equally reported [2,3,7,8,9]. Goats appear not to be affected in some outbreaks, while sheep suffer with high rates of mortality and morbidity [10]. Strain specific virulence of PPRV has been reported when the same breed of goats were experimentally infected [11], and different breeds of goat have been shown to respond differently to infection with the same virus [12]. Species-specific disease occurrence has been observed with foot and mouth disease, where cattle were highly affected while sheep had less severe infection with the virus [13]. Epizootic haemorrhagic disease virus affects cattle but sheep do not suffer from this disease [14]. It is well recognised that ducks were generally resistant to avian influenza virus (AIV) whereas chickens suffer from severe disease with rapid death following infection with highly pathogenic AIV [15]. The reason for this species specificity is unclear at present.

The natural susceptibility to PPRV in goats could be attributed to several host-derived or virus-derived factors. One such host-derived factor could be the differential presence or distribution of specific viral receptors in these species, such as the signalling lymphocyte activation molecule (SLAM) that has previously been observed to be associated with PPRV and other morbilli viruses such as measles virus and canine distemper virus [16,17,18]. Host immune mechanisms could also account for this differential susceptibility, although this has not been explored in detail in ruminant species or breeds.

Toll like receptors (TLR) are type 1 transmembrane proteins expressed in almost all cell types and activate the innate immune system upon sensing pathogen associated molecular patterns (PAMPs). Intracellular TLR that sense viral nucleic acids include TLR3 (double stranded RNA), TLR7 and TLR8 (single stranded RNA) and TLR9 (CpG motifs in DNA) [19]. Imiquimod and poly I:C are standard agonists used to induce TLR7 and TLR3 respectively leading to the production of inflammatory cytokines including type 1 interferons (IFN) and immune cell maturation [20,21]. TLR are differentially expressed in various tissues and immune cells of water buffalo and goats, and have been shown to induce differential immune responses [22,23]. The cell specific location and basal expression levels of TLR mRNA could indicate the natural PAMP load of that tissue as well the innate host resistance to pathogens [24].

In addition to the differential expression profiles of TLR, ligand induced downstream cytokine profiles and/or levels could also play a role in the innate disease resistance of a species or breed. For example, mastitis is an economically important inflammatory disease of the udder that has been shown to be more prevalent in the Holstein breed of cows than in Jersey cows [25]. This has been linked to temporal differences in the onset and duration of immune responses, including cytokines such as tumor necrosis factor alpha (TNFα) and IFNγ, following intramammary inoculation of pathogenic E. coli.

India has 23 genetically well characterized indigenous breeds of goats (http://www.icar.org.in/en/node/4688). Barbari is a common breed of goat reared for meat and milk production in northern India. Tellicherry, Kanni and Salem black are indigenous breeds of goats prevalent in southern India [26]. Outbreaks of PPR have been reported in newly introduced Barbari goats to southern India with mortality rates of 16.67% to 65.0% [27]. Recently, a severe outbreak of PPR was reported in Tellicherry breed of goats with 100% mortality in kids and 87.5% mortality in adults [28]. PPRV infection appears to be subclinical in Kanni and Salem black breeds of goats. Similarly, native water buffalo in India are resistant to PPRV. Although such anecdotal evidence and field observations on differential resistance within goat breeds and water buffalo are available, experimental evidence is lacking for the observed differences in susceptibility.

We hypothesized that the differential susceptibility of various Indian goat breeds and native water buffalo to PPRV could be related to innate immune resistance mechanisms. Infection of peripheral blood mononuclear cells (PBMC) from four breeds of goats and water buffalo resulted in differential viral replication kinetics and inflammatory cytokine profile including IFNα, IFNγ and TNFα with differential activation of TLR3 and TLR7.

Analysis of single nucleotide polymorphisms (SNPs) in the complete gene sequences of TLR7 between goat breeds did not show any differences that could account for this.

Results

TLR3 and TLR7 mRNA expression and PPRV replication in goats

We examined the relative basal expression levels of TLR3 and TLR7 mRNA in naive PBMC by quantitative real-time RT-PCR (qRT-PCR) in nine individual animals per breed (n = 9). The Kanni and Salem Black breeds revealed significantly higher basal TLR3 ($p < 0.001$) and TLR7 ($p < 0.001$) transcripts than Barbari goats while there were no significant difference ($p > 0.05$) between Tellicherry and Barbari breeds (Figure 1). To understand their contribution to virus replication, PBMC from each of these four goat breeds (n = 5) were infected with $1 \times 10^{3.0}$ mean tissue culture infective dose (TCID$_{50}$) of PPRV and the virus load analyzed at 24 h post infection (PI) by qRT-PCR, using primers specific to the PPRV-H gene and TCID$_{50}$. PBMC from Barbari and Tellicherry goats supported significantly ($p < 0.01$) higher PPRV replication than those from Kanni and Salem Black (Figure 2A) with the yields being similar in these two breeds. There was a significant reduction in virus yield by one log$_{10}$ in Kanni/Salem Black PBMC ($p < 0.05$) compared to Barbari/Tellicherry goats (Figure 2A).

We then examined if pre-stimulation of PBMC with the TLR agonists poly I:C and imiquimod would result in reduction in PPRV replication. A difference of about 6–8 Ct values in PPRV-H gene mRNA levels was observed between unstimulated and poly I:C treated PBMC (Figure 2B). In the case of pre-stimulation with imiquimod, this difference was about 2–3 Ct values for Barbari and Tellicherry breeds and about 4–5 Ct values for Kanni and Salem black breeds (Figure 2C). Comparison of viral load (PPRV-H gene mRNA levels and infective viral titres) in all the breeds following imiquimod treatment indicates a significant ($p < 0.05$) reduction of viral load in Kanni and Salem black breeds than Barbari and Tellicherry breeds. TCID$_{50}$ was not determined for poly I:C treated and PPRV infected PBMC since the 40-Ct values of PPRV H mRNA expression levels across these breeds were not significantly different (Figure 2B).

Water buffalo PBMC are less permissive to PPRV replication

In order to investigate the species specific disease outcome, we compared the permissiveness of goat or buffalo PBMC's to PPRV replication in vitro. Water buffalo supported significantly lower (about 4.89 fold) replication of PPRV than goat PBMC ($p < 0.001$). Approximately 2 log$_{10}$ difference in the virus yield was evident between buffalo and goat PBMC upon PPRV infection (Figure 3A & B).

Cytokine mRNA expression

To understand the mechanistic basis for this differential permissiveness for virus replication in different goat breeds and goat vs water buffalo, we analyzed the downstream effector molecules of TLR3 and TLR7 engagement. PBMC stimulated with poly I:C, or imiquimod and further infected with PPRV were analyzed for the expression of IL1β, IL6, IL8, IL10, IL12p40, TNFα, IFNγ and IFNα mRNA (Figure 4A–D). Though upregulation of both pro- and anti-inflammatory cytokines were observed in all goat breeds, levels of TNFα was seen to be consistently higher in Kanni and Salem black breeds in all treatment groups (poly I:C, imiquimod and PPRV). This effect was prominent in the case of PPRV infected PBMC, where levels of TNFα and more

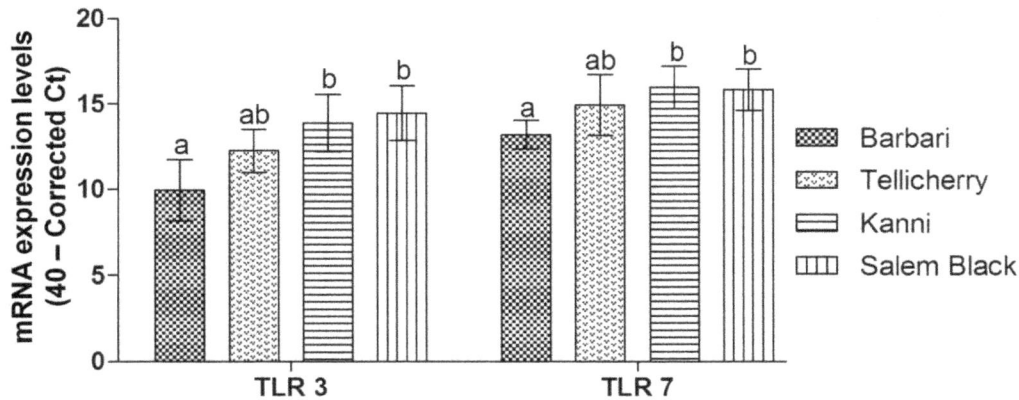

Figure 1. Basal expression levels of TLR3 and 7 mRNA in Barbari, Tellicherry, Kanni and Salem Black goat breeds. Significantly higher basal TLR3 ($p < 0.001$) and TLR7 ($p < 0.001$) mRNA expression levels observed in Kanni and Salem Black breeds, compared to Barbari. Bars with the same superscript do not differ significantly. Values represent mean $\pm$ SD of 40-corrected CT of TLR3 and 7 in nine individual animals per breed (n = 9).

importantly, IFNα and IFNγ, were significantly higher in Kanni and Salem black breeds as compared to Barbari (Figure 4C). Significant differences were not observed between these breeds and Tellicherry in the levels of pro-inflammatory cytokine mRNA. Consistent with this, levels of the immunosuppressive cytokine IL10 were significantly lower in these breeds than in Barbari (Figure 3c). Lower mRNA expression levels of IL10 in Kanni and Salem black breeds were also observed on stimulating TLR3 and TLR7 with poly I:C and imiquimod (Figure 4A, B).

We further validated these results by cytokine ELISA for TNFα (Figure 5A), IFNα (Figure 5B) and IFNγ (Figure 5C). Cytokine levels were similar to the mRNA expression profiles observed by qRT-PCR (Figure 5A–C). PPRV infected PBMC from Kanni and Salem black breeds had higher production of TNFα, IFNα and IFNγ than Barbari breed. Poly I:C and imiquimod treated PBMC, similarly, showed higher production of TNFα and IFNα in Kanni and Salem black breeds than in Barbari suggesting TLR3 and TLR7 engagement by PPRV.

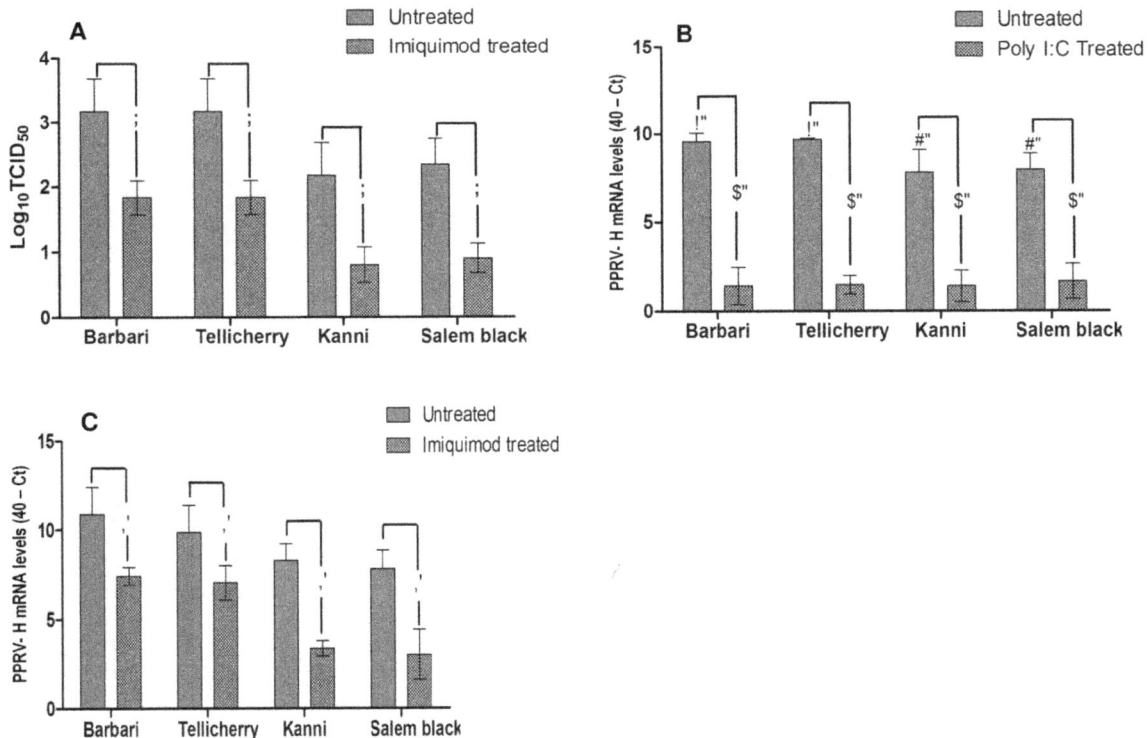

Figure 2. Virus replication in PBMCs stimulated with TLR3 and TLR7 agonists. a) Reduction in $TCID_{50}$ values of PPRV in goat PBMC on imiquimod treatment. Reduction in PPRV H gene expression levels in goat PBMC on treatment with b) poly I:C and c) imiquimod. Bars with the same superscript do not differ significantly. Significance is indicated when $p < 0.05$. Values represent mean $\pm$ SD of 40-corrected CT in 5 individual animals per goat breed. Significantly higher PPRV viral loads observed in PBMC of Barbari and Tellicherry breeds as compared to Kanni and Salem Black. Significant reduction in PPRV levels on poly I:C and imiquimod treatment in all breeds.

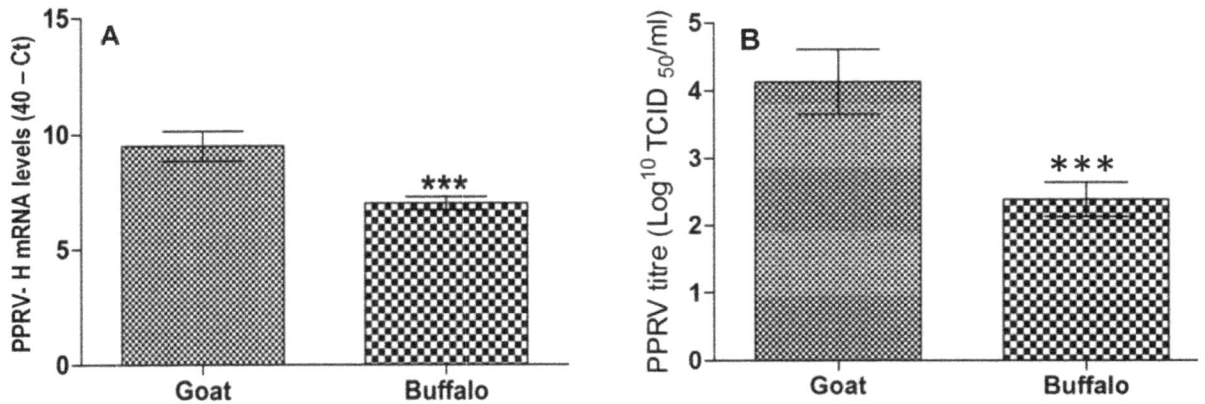

Figure 3. PPRV replication in goat and buffalo PBMC. Buffalo PBMC were less permissive to PPRV replication as observed by the significantly higher PPRV viral loads (P<0.001) in PBMC of goat as compared to buffalo PBMC. A) Significantly higher PPRV H gene mRNA levels and B) Significantly higher viral load in goat as compared to buffalo PBMC.

To determine whether there are differences in the induction of pro-inflammatory cytokines between water buffalo and goats, we examined the levels of these after infecting respective PBMC with PPRV. Increased mRNA levels of IL1β and IFNα in buffalo

PBMC and IL10, IL12p40 and IFNγ in goat PBMCs were observed. IFNα activation was greater in buffalo than in goats, 2360 vs 1498 fold, respectively (Figure 4D). To further confirm this observation, IFNα protein levels were assayed from culture

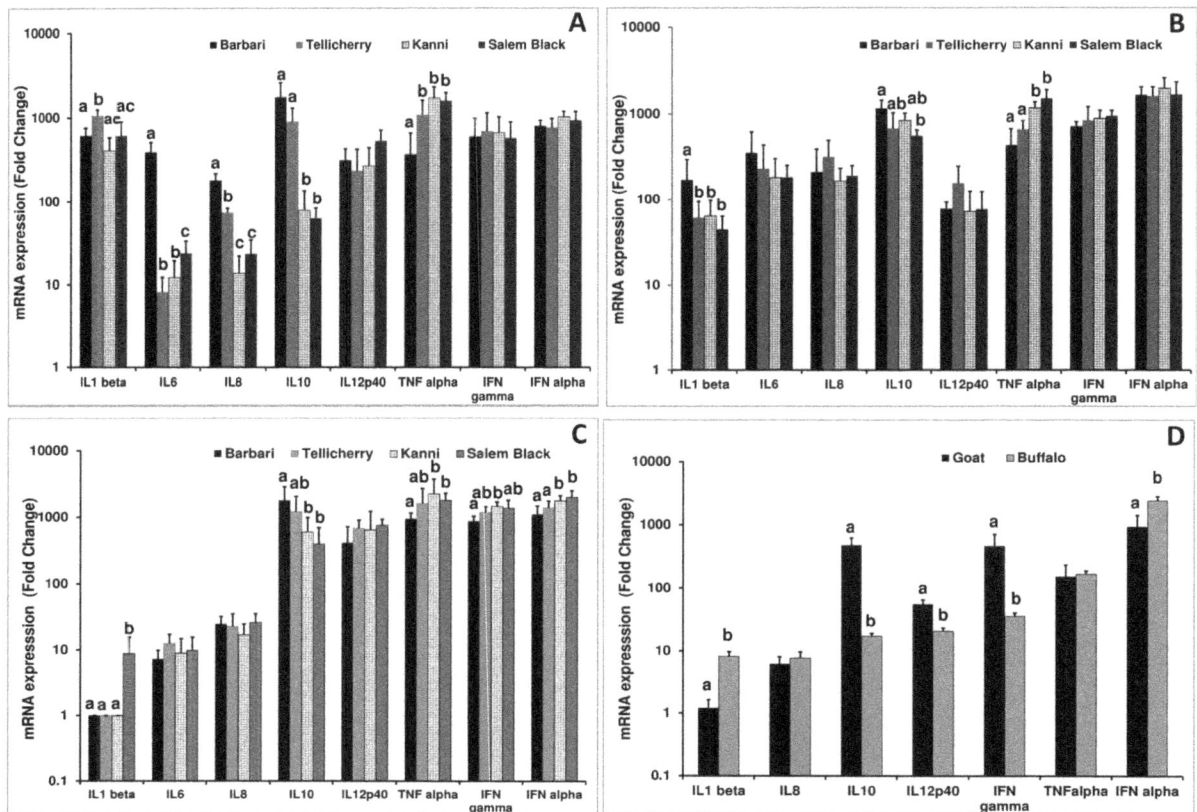

Figure 4. Induction of cytokine genes of goat PBMCs with TLR3 and TLR7 agonists and PPRV. A) Fold changes in mRNA expression levels of IL1β, IL6, IL8, IL10, IL12p40, TNFα, IFNγ and IFNα in goat PBMC stimulated with a) poly I:C, b) imiquimod or c) infected with PPRV d) Goat and buffalo PBMCs infected with PPRV. Fold change was determined by the $2^{-\Delta\Delta Ct}$ formula. An upregulation in TNFα expression levels, consistent with a downregulation in IL10 levels, is observed in Kanni and Salem Black breeds of all treatment groups. In addition, upregulation of IFNα and IFNγ was observed in the PBMC of Kanni and Salem Black breeds after PPRV infection. PPRV stimulation resulted in an upregulation of IL1β and IFNα in buffalo PBMC and IL10, IL12p40 and IFNγ in goat PBMCs. Bars with the same superscript do not differ significantly. Significance is indicated when p<0.05. Values represent mean ± SD of 5 individual animals per goat breed.

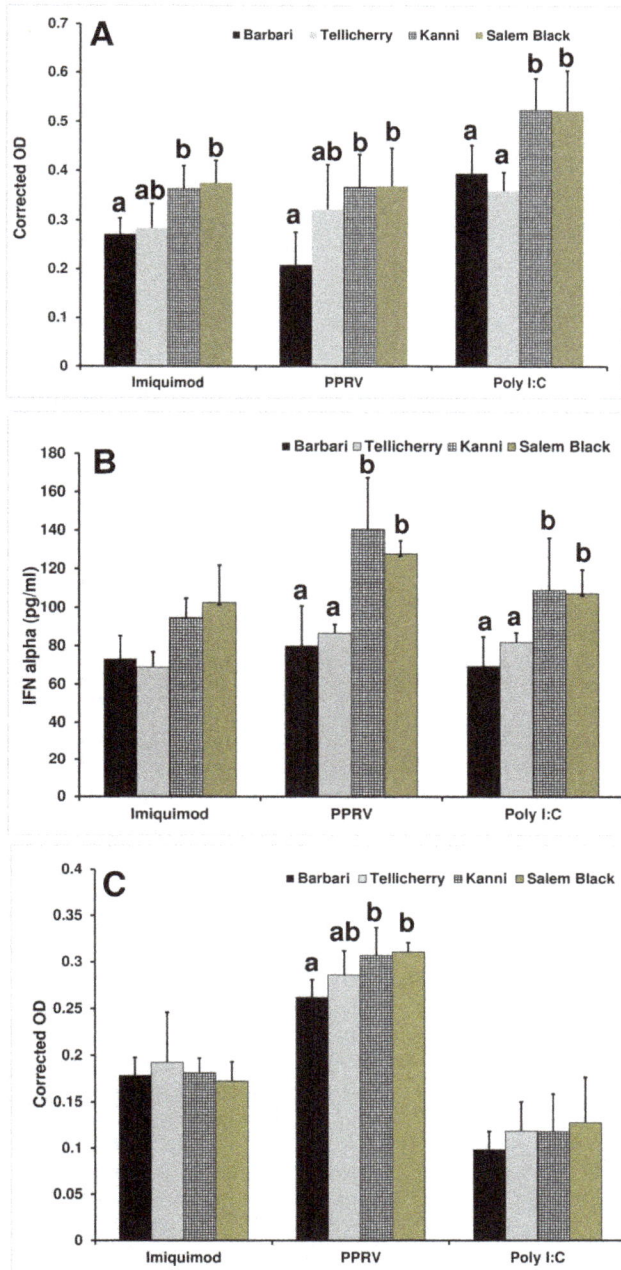

Figure 5. Cytokine levels in imiquimod, poly I:C or PPRV treated PBMCs. a) TNFα, b) IFNα, and c) IFNγ from supernatants of PPRV infected PBMC of different goat breeds. Bars with the same superscript do not differ significantly. Significance is indicated when p< 0.05. PPRV infected PBMC from Kanni and Salem black breeds show higher production of TNFα, IFNα and IFNγ than Barbari. Poly I:C and imiquimod treated PBMC, similarly, show higher production of TNFα and IFNα in Kanni and Salem black breeds than in Barbari. TNFα and IFNγ levels are expressed as the corrected mean ± SD of optical density [OD] of treatment groups from which the OD of mock infected supernatants is subtracted. IFNα concentrations in the experimental samples are expressed as pg/ml.

supernatants of buffalo or goat PBMC infected with PPRV. IFNα levels in infected buffalo or goat PBMC was 86.20±4.45 and 58.13±7.48 units (p<0.01), respectively.

To further define the association between differential PPRV replication and IFNα expression, the ability of IFNα in limiting the viral replication was confirmed in cells treated with human recombinant IFNα. The mean 40 - corrected Ct value of PPRV H gene expression in Vero cells without IFNα treatment was 12.10±0.42. Upon pre-treatment of Vero cells with increasing concentrations of IFNα, PPRV H gene expression decreased. Even the lowest dose of IFNα tested (100 ng/ml) could reduce the replication by 451 fold (Figure 6).

To confirm the role of TLR7 signalling in IFNα induction by PPRV unequivocally, goat and buffalo PBMCs were treated with a predetermined optimum concentration (10 µg/ml) of a TLR7 antagonist IRS 661 [] 24 h prior to imiquimod treatment or PPRV infection. IFNα mRNA levels were significantly reduced by IRS 661 treatment, when PBMC were stimulated either with Imiquimod or PPRV (Figure 7A, B). To further confirm the role of TLR7 induced IFNα in limiting virus replication, conditioned medium (CM) from PPRV infected cells were used to pre-treat PBMC before virus infection (Figure 7C). The percent inhibition in virus replication was significantly higher (p<0.01) in CM pre-treated buffalo PBMC than goat PBMC (Figure 7C).

Genetic polymorphisms in TLR7 across goat breeds

To determine whether the differential IFN and pro-inflammatory cytokine production between Kanni/Salem vs Barbari/Tellicherry breeds of goats are dependent on single nucleotide polymorphisms (SNPs) in TLR genes, we examined the complete gene sequence of TLR7. The TLR7 gene was amplified and seven overlapping PCR amplicons were sequenced to obtain the full length TLR7 gene using the goat TLR7 sequence from GenBank as a template (Accession # GU289401). The sequences obtained for Barbari, Tellicherry and Kanni goat breeds were submitted to GenBank (Accession # KC127658, KC127659, KC127660, KC127661, KC127662, KC127663, KC127657). Nucleotide variations were detected across goat breeds. Sequence analysis revealed five nucleotide changes in the TLR7 coding region and two nucleotide changes in the 3′ untranslated region (3′UTR). Changes were observed at nucleotide positions 777 (C/T in Salem while C/C in other breeds), 2082 (A/A in Salem while A/G in other breeds), 2730 (C/T in Barbari while C/C in other breeds), 3006 (A/A in Tellicherry while A/G in other breeds) and 3069 (G/A in Barbari while G/G in other breeds) that corresponded to amino acid positions 259, 694, 910, 1002 and 1023. However, all of these changes were synonymous and no amino acid changes were observed between different goat breeds. In addition, two changes in the 3′UTR, at nucleotide positions 151 (G/A in Barbari and Salem while G/G in other two breeds) and 161 (G/A in Barbari while G/G in other breeds) were also observed.

Discussion

Following eradication of rinderpest, focus has shifted to PPRV as a potential virus for eradication efforts. The single serotype of PPRV, availability of an efficacious live attenuated vaccine, sensitive diagnostic tests and the economic impact of this disease, coupled with its ability to spread into new geographical regions, render it an attractive target [30]. PPR was first reported in India in sheep with a lineage IV virus in 1989 [31], with no further reports until 1996, when a massive outbreak occurred in goats throughout northern India with a lineage IV virus [32]. Three live attenuated vaccines, specific to lineage IV PPRV strains have so far been tested in India [33].

Goats have been reported to be more susceptible to PPRV than sheep, while cattle and buffalo do not contract clinical disease

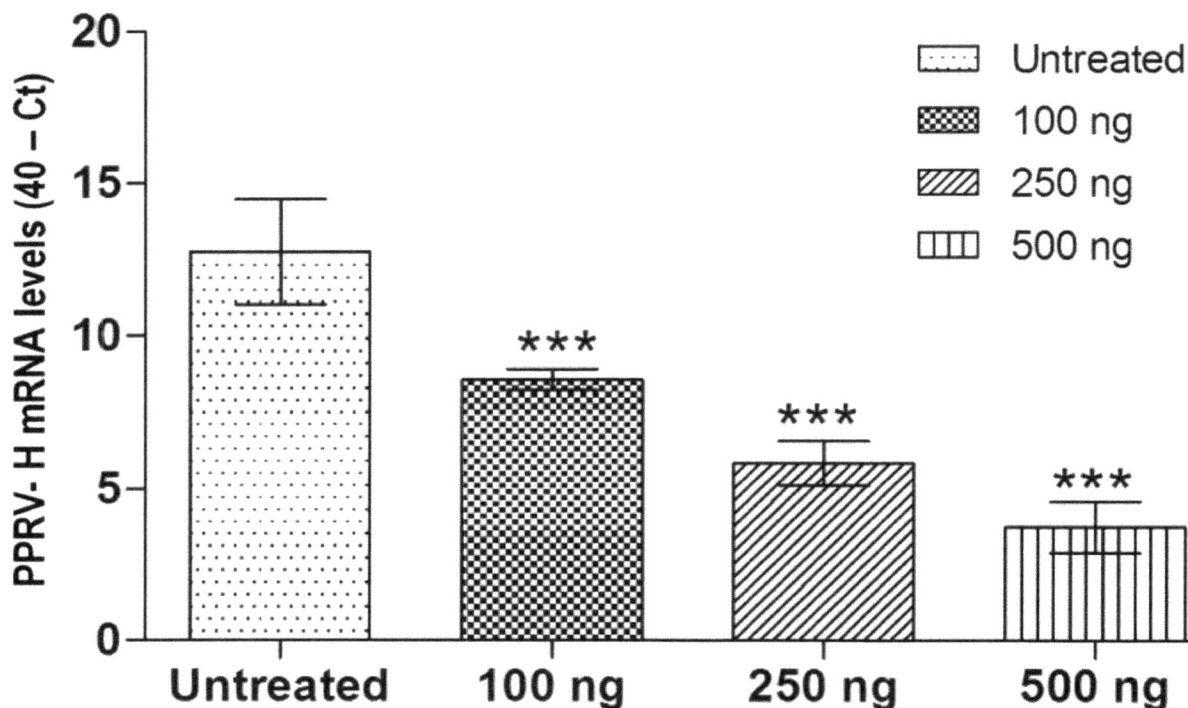

Figure 6. Antiviral activity of human IFNα against PPRV. Reduction in PPRV viral load was observed in Vero cells pretreated with different concentrations of human IFNα. All the doses tested significantly reduced PPRV viral load (mean ± SD of 40-Ct). Statistical significance was defined as follows: ***$P<0.001$.

[5,32,34]. Increased mortality in lambs/kids, and increased susceptibility of West African goats, especially dwarf goats, compared to their European counterparts have been documented [11]. Differential susceptibility of goat breeds within India have also been reported [35]. Host genetic factors, in particular the major histocompatibility complex (MHC) genes, may influence susceptibility to disease. Virus recognition can be influenced by genetic mutations in the interaction domains between virus and host receptors. In particular, non-MHC genetic variations in host TLR may cause reduced pathogen recognition and hamper innate immune activation [36]. Studies on Maedi-Visna infection in sheep indicate that breed dependent susceptibility to the disease as well as individual susceptibility within the breed may be defined by specific polymorphisms in TLR7 and TLR8 genes [37]. In a related morbillivirus, SNPs in TLR 3, 4, 5, 6 and associated signalling molecules like Myeloid differentiation primary response gene 88 (MyD88) and MD2 affected immune responses to the measles vaccine in human subjects [38].

Single and double stranded RNA are recognized by TLR7/8 and TLR3, respectively. TLR3 is a key sensor of viral infection, as most viruses will produce dsRNA at some stage of its life cycle. TLR7 is highly expressed in immune cells like plasmacytoid dendritic cells (pDC), which produce substantial amounts of type I IFNs in response to viral RNA. In our study, the basal levels of TLR3 and TLR7 were significantly higher in the PBMCs of PPRV resistant goat breeds, Kanni and Salem Black. Engagement of both TLR3 and TLR7 with the synthetic ligands poly I:C and imiquimod respectively, led to the suppression of PPRV RNA and infectious virus yield in PBMC of goats. This indicates that TLR3 and 7 play a role in the recognition of PPRV RNA by goat PBMC, though the role of cytosolic RNA sensors like Retinoic acid-inducible gene 1 (RIGI) and Melanoma differentiation-associated

protein 5 (MDA5) have not been analyzed in this study and cannot be ruled out. Almost complete abrogation of viral gene expression was observed after stimulation by poly I:C. This may be because poly I:C can also be recognized by other sensors, including RIGI, MDA5 and Protein kinase R (PKR) [39].

If factors other than the receptor expression for PPRV determine clinical disease, this should be at the level of virus replication and clearance by innate and adaptive responses. The cell surface receptor for PPRV, the SLAM is expressed at lower levels in buffalo than goats [16]. However, we did see virus replication in infected PBMC although at considerably reduced levels suggesting that the cells of water buffalo are permissive to PPRV infection. Therefore, we questioned whether these differences are reflected at the level of multi-cycle virus replication. PPRV replication in water buffalo PBMC was significantly lower than in goats, possibly because of enhanced type I IFN production in these species upon virus infection. PPRV replication in human IFNα pre-treated Vero cells or in PBMC pre-treated with CM from virus infected cells was significantly lower even at very low doses, confirming the role of type I IFN in limiting virus replication. Human IFNα has been shown to effectively suppress the replication of bovine viral diarrhea virus and bovine parainfluenza virus [40,41].

Cytokines play a pivotal role in the induction and modulation of immunological responses. TLR signaling events lead to the activation of nuclear factor kapp-light-chain-enhancer of activated B cells (NFκB) and interferon regulatory factor (IRF), which switch on expression of a specific panel of pro-inflammatory cytokines and chemokines such as TNFα, IL6, IL8 and Regulated on activation, Normal T cells (RANTES) [19]. Activation of TLR by viruses also results in the production and release of type I IFNs [39]. TLR3 and TLR7 engagement by synthetic ligands lead to

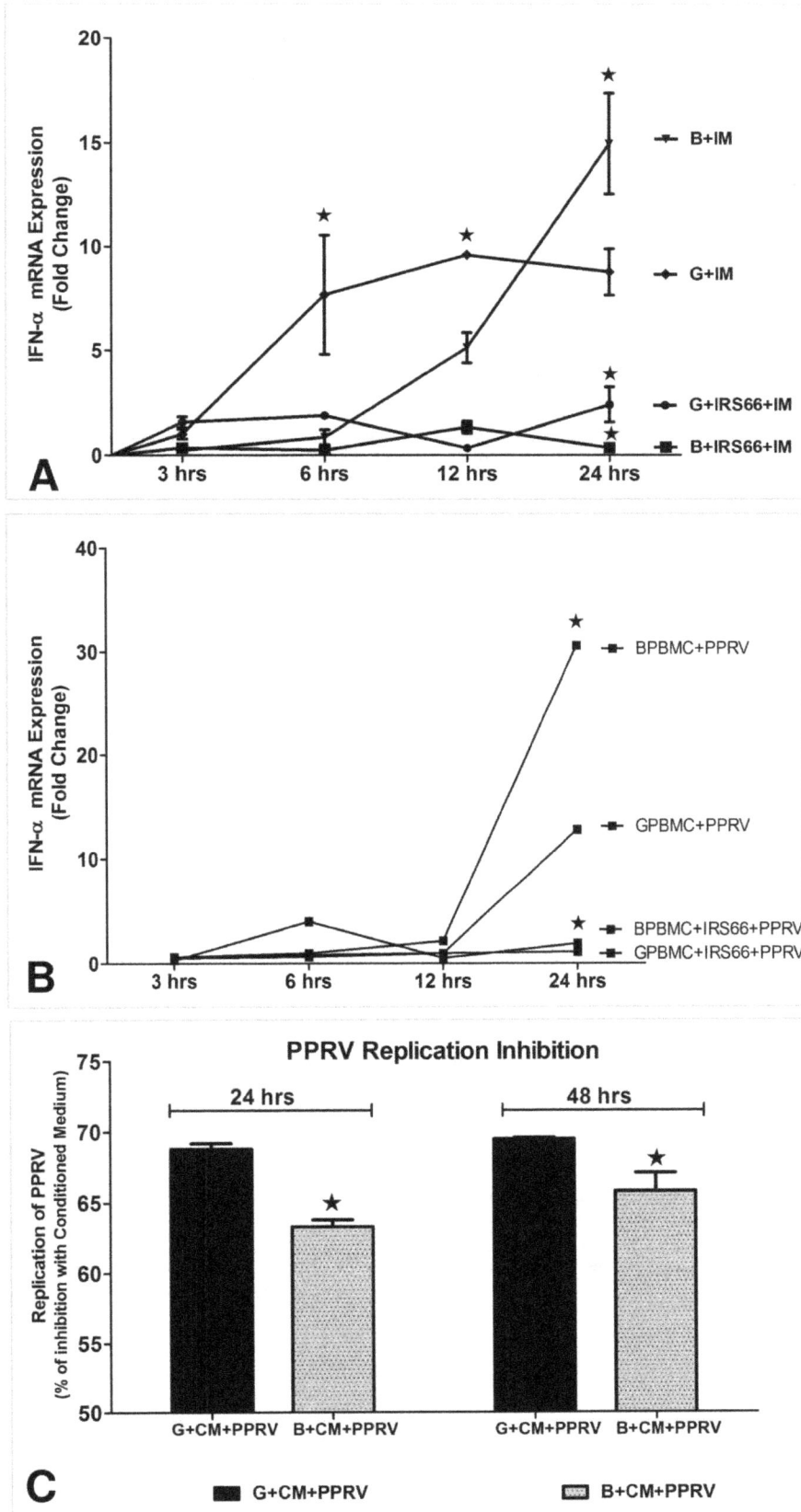

Figure 7. Role of type I IFN in limiting PPRV replication. a) Imiquimod or b) PPRV induced IFN-alpha mRNA expression in buffalo and goat PBMCs in the presence and absence of the TLR7 antagonist (IRS 661). Cytokine mRNA expression was quantified at 3, 6, 12 and 24 h post stimulation

by qRT-PCR assays using SYBR Green chemistry. Fold change in mRNA expression induced by Imiquimod or PPRV stimulation was calculated using mock induced cytokine mRNA expression levels as a calibrator. c) PPRV replication in goat and buffalo PBMC (at 24 h and 48 h) in the presence of conditioned medium (CM) from PPRV infected cells. The expression levels of viral hemagglutinin (H) mRNA levels expressed as a percentage inhibition in viral replication in the presence of CM when compared with the control (PBMC+PPRV). Statistical significance at $P<0.01$. Values are mean ± SD of fold change/percent inhibition. B: water buffalo, G: goat, BPBMC: buffalo PBMC, GPBMC: goat PBMC.

cytokine expression profiles similar to PPRV infection except for a weak IL1β, IL6 and IL8 production in goat PBMC. A predominantly inflammatory cytokine repertoire, with expression of TNFα, IFNα and IFNγ was observed at both mRNA and protein levels. Thus, it could be inferred that TLR engagement upon PPRV infection results in inflammatory cytokine production via the canonical NFκB pathway and type I IFN production via the activation of IRFs [42]. Stimulation of TLR7 with synthetic RNA oligonucleotides has earlier been shown to induce production of IL-12, TNFα and IFNγ in PBMC of cattle [43]. Interestingly, in our study, IFNγ levels were higher in PPRV infected PBMC, compared to the engagement of TLR3/7 by their respective agonists. IFNγ production by NK cells can be induced by IL12 secreted by TLR stimulated DCs [44]. In buffalos, approximately 1.5 fold higher levels of IFNα at mRNA and protein levels were induced in PBMC compared to goats after infection with PPRV suggesting that type I IFN may play a role in limiting virus replication in buffalo. Further, we found that TLR7 mediated IFNα production is critical because TLR7 antagonist inhibited IFNα production both in Imiquimod or PPRV-treated goat and buffalo PBMC. This effect was more prominent in buffalo PBMC suggesting that TLR7 mediated IFNα production determines PPRV replication efficiency in this species.

Consistent with the inflammatory cytokine environment induced by PPRV infection, expression of the immunomodulatory cytokine IL10 was also observed, but its levels were high in the PPRV susceptible goat breeds, Barbari and Tellicherry. IL10 is a key regulatory cytokine with immunosuppressive properties that helps to regulate an uncontrolled inflammatory response [45]. In addition to preventing the maturation of antigen presenting cells, IL10 can also regulate the proliferation and differentiation of Th1 cells, which induces T cell-dependent suppression of antiviral responses [46,47]. Dexamethasone, a well-known immunosuppressive drug, induces immunosuppression by altering the expression levels of IL10 and TNFα [48]. Experimental immunosuppression of goat with dexamethasone and challenge with virulent PPRV indicated that, immunosuppressed goats had a shorter viremia, more extensive and severe disease advancement, significant decrease in the number of leucocytes, high antigen load

in various organs and higher mortality rate than the non-immunosuppressed goats [49,50]. Taken together, it appears that a higher basal expression of TLR3 and TLR7 in Kanni and Salem breeds may correlate with increased inflammatory cytokine expression with lower levels of immunomodulatory cytokines leading to an enhanced antiviral state thus affording reduced susceptibility to PPRV infection. Similarly, in buffalo, the TLR-7 mediated type I IFN response upon infection may afford resistance to PPRV.

The goat TLR7 gene is 3.4 Kb long, with a 3141 nucleotide open reading frame (ORF), coding for 1046 amino acids. Nucleotide sequence homology studies have shown a close relationship with other ruminant species, particularly sheep TLR7 [51]. Earlier studies in 12 Indian goat breeds have shown a total of 22 polymorphic sites, out of which 19 were present within the coding region and three in the 3'UTR [41]. The Toll/interleukin-1 receptor (TIR) domain sequence is highly conserved between species, as it plays a crucial role in TLR downstream signaling [52]. In our study, sequence analysis revealed five nucleotide changes in the TLR7 coding region and two nucleotide changes in the 3'UTR. All the changes were synonymous and it is difficult to establish a correlation with specific SNP and altered susceptibility to PPRV in the goat breeds examined. There were no differences in the leucine repeat regions of TLR7 between different breeds of goats. Though, we were unable to demonstrate a positive association between SNP and differential susceptibility to PPRV in the goat TLR7 gene, analyses of other immune genes including TLR3 and TLR4 may indicate the relationship between susceptibility to PPRV infection and genetic polymorphisms in the host. Earlier studies in buffalo TLR7 gene have also reported four different polymorphic positions (A/G-1400, A/G-1480 (D234N), C/T-2029 (L417F), A/G-2640), of which two were non synonymous SNP's in leucine rich repeats (LRR) [53]. Given the presence of SNP's in LRR of buffalo TLR7 gene, which is responsible for sensing the PAMP's, these SNP's could be associated with differential sensing of PPRV. Further studies are required to correlate the reported SNP's in buffalo TLR7 with observed differential response of buffalo PBMC to PPRV.

Table 1. TaqMan primer/Probe sequences for assessing basal expression levels of TLR3 and TLR7 mRNA.

Target gene	Primer/Probe sequence 5'-3' (Forward/Reverse)	Efficiency	Slope	Accession number
TLR7	GCTCCAAATGCCCATGTGATT	0.953	−3.43	HQ263216
	AGGAATACCTCCAGGAATTTCTGTCA			
	FAM 5'CTGCACAGACAAACTT 3' NFQ			
TLR3	GTCCTTGACCTCGGCCTTAA	0.976	−3.38	H263210
	CCCCATTCTTGGCCTGTGA			
	FAM 5' TTCTTGCCCAATTTCA 3' NFQ			
GAPDH	GGCGCCAAGAGGGTCAT	0.942	−3.46	AJ431207
	GTTCACGCCCATCACAAACAT			
	FAM 5'CTTCTGCTGATGCCCC3' NFQ			

We also found a role for basal and induced TLR3 expression levels in PPRV infection in our study. In addition to TLR, variations in downstream intracellular signaling molecules such as MyD88 and MD2 may also play a role. In conclusion, our studies suggest that higher basal levels of TLR3/7 and augmented innate antiviral immune responses upon infection may afford resistance to PPRV infection in Kanni and Salem breeds of goats compared to Barbari and Tellicherry breeds. Compared to goats, elevated type I IFN levels after PPRV infection in water buffaloes may afford reduced virus replication and possibly early virus clearance. LRR prediction revealed goat TLR7 has 21 LRRs and buffalo TLR7 has 15 LRRs. The difference in LRR numbers could be a critical factor in determining the signaling responses of goat and buffalo TLR7. Future studies may provide insights into understanding the immunogenetic mechanisms underlying variations in the immune response to PPRV.

Materials and Methods

Animals, virus and reagents

All animal studies have been conducted as approved by the Ethics Committee of the Tamil Nadu Veterinary and Animal Sciences University, Chennai-600 051, India. Apparently healthy, 12–18 month old, Barbari, Tellicherry, Kanni and Salem Black breeds of goats of either sex were maintained under similar conditions at the Livestock Research Station, Tamil Nadu Veterinary and Animal Sciences University, India. These animals were not vaccinated against PPRV and had no record of any other disease during the course of the study. PBMC from nine animals of each breed were used for TLR mRNA expression analysis, while PPRV infection and TLR stimulationstudies were carried out on PBMC from five animals of each breed. Live attenuated PPRV (strain AR-87) was obtained from the department of veterinary microbiology, Madras Veterinary College, India and has been described elsewhere [31]. Imiquimod R837 (2.5 μg/ml) and polyionosinic-polycytidylic acid, a synthetic analog of dsRNA (poly I:C) (25 μg/ml) (InVivoGen, San Diego, CA) were diluted in endotoxin-free water. Aliquots were tested by the E-toxate kit (*Limulus*am ebocyte lysate assay, Sigma Aldrich, St. Louis, MO) and found to be free of endotoxins.

TLR stimulation and PPRV infection

Blood was collected aseptically from the jugular vein into sterile ethylene diamine tetra acetic acid (EDTA) coated vacutainer tubes (Becton Dickenson, Cambridge, UK) and processed for PBMC isolation. Briefly, 5 ml of anti-coagulated blood was diluted in equal volumes of RPMI-1640 (Invitrogen, Paisley, UK) medium containing antibiotic and antimycotic solution, overlaid on 2.5 ml of Histopaque, specific gravity 1.077 (Sigma Aldrich, St. Louis, MO) and centrifuged at 1500×g for 25 min. Mononuclear cells were collected from the interface and washed three times in RPMI-1640 by centrifugation at 200×g for 10 min. Viability was determined by trypan blue dye exclusion method. PBMC were stimulated with predetermined doses of TLR3 and TLR7 agonists, poly I:C and imiquimod (R837), respectively. Cells were harvested at 24 h for cytokine transcript analysis.

In a separate experiment, TLR ligand stimulated PBMC were infected with $10^{3.0}$ $TCID_{50}$ of PPRV. Virus yields from TLR 3/7 stimulated and un-stimulated PBMC were assessed at 24 h PI by SyBr Green quantitative real time reverse transcription polymerase chain reaction (qRT-PCR), using primers specific for the PPRV-H gene [35]. Infective virus in the supernatants of PPRV infected PBMC cultures were determined on Vero cells and expressed as $TCID_{50}$/mL [54].

Analysis of basal TLR transcript levels

Total RNA from PBMCs was extracted using TRI reagent solution (Sigma-Aldrich, St. Louis, MO) as permanufacturer's instructions. RNA concentration and purity was determined using the BioPhotometer plus (Eppendorf, Hamburg, Germany). Two μg of total RNA was reverse transcribed with Oligo $(dT)_{18}$ primers using the High Capacity cDNA Archive kit (Applied Biosystems, Carlsbad, CA). Basal expression levels of TLR3 and TLR7 mRNA in PBMC were determined using gene-specific primers and TaqMan probes (FAM-NFQ) (Applied Biosystems, Carlsbad, CA) (Table 1). qRT-PCR was performed in triplicate under the following cycle conditions, 2 min at 50°C, 10 min at 95°C and 40 cycles of 95°C for 15 sec and 60°C for 1 min (Applied Biosystems 7500 Real time PCR System, Carlsbad, CA).

Cytokine transcripts after stimulation with TLR agonists or infection with PPRV

Cytokine gene expression levels were compared by SyBr Green qRT-PCR using gene specific primers (Applied Biosystems, Carlsbad, CA) (primer sequences available upon request). Unstimulated PBMC were used as control. Corrected Ct was calculated as:

Ct + (Nt − Ct')×S/S', where Ct is the mean sample Ct, Nt is the mean of the house keeping genes GAPDH/actin from the control group, Ct' is the mean of the GAPDH/actin from treatment, S is the target gene slope, and S' is the GAPDH/actin slope. The slope values were calculated using serial dilutions of cDNA and the respective Ct values for each dilution and PCR efficiency $(E = 10^{-1}/slope)$ was determined. Results were expressed as 40-Ct values [54,55]. Changes in cytokine expression were expressed as fold change $(2^{-\Delta\Delta Ct})$ over the respective basal levels of mock-induced PBMCs after normalizing for the endogenous control gene and using the corrected Ct.

Differential Enzyme linked immunosorbent assay (ELISA) for TNFα, IFNα and IFNγ

Antigen capture ELISA kits for TNFα (AbDSerotec, Kidlington, UK), IFNα (Mabtech, Sweden) and IFNγ (CUSABIO Biotech, China) were used to determine cytokine concentrations in the culture supernatants of TLR ligand stimulated and/or PPRV infected PBMC. ELISA was performed according to the manufacturer's instructions and values were obtained spectrophotometrically on an ELISA reader (Epoch Micro-Volume Spectrophotometer System, Biotek, Winooski, VT) at 492 nm. Mock infected cell culture supernatant served as a control. TNFα and IFNγ levels were expressed as the corrected optical density [OD] of TLR-ligand stimulated or PPRV infected culture supernatants from which the OD of mock infected supernatants is subtracted. IFNα concentrations in the experimental samples were extrapolated from the values generated from standards.

Detection of single nucleotide polymorphisms in TLR7 gene

Blood samples were collected from Barbari, Tellicherry, Kanni and Salem black breeds of goat and genomic DNA was isolated using the Blood DNA isolation kit (Biobasic, USA). The concentration of extracted DNA was determined using biophotometer plus (Eppendorf, Germany). Seven pairs of overlapping primers were designed to amplify the full-length TLR7 gene (primers available upon request) and the PCR fragments were directly sequenced in both directions using the Big Dye Terminator v3.1 cycle sequencing kit (Applied Biosystems, Carlsbad, CA). Sequences were assembled into a 3.4 kb contig, which contained a 3141 bp open reading frame. Sequence contigs of TLR7 from each animal were further subjected to multiple alignments to identify nucleotide variations, using the Lasergene software (DNASTAR, Madison, WI). Heterozygous nucleotides were scored manually across samples from different breeds by visualizing the individual chromatogram in Chromas Lite 2.01 (Technelysium, Queensland, Australia). Each polymorphic nucleotide was further analyzedfor its amino acid position and change.

Statistical analysis

Statistical analysis was performed with the GraphPad Prism software. The 40-corrected Ct values of TLR3 and TLR7 mRNA, fold changes in the ligand induced cytokine mRNA expression, PPRV H gene levels and virus yield estimated by $TCID_{50}$ determination was compared by two-way ANOVA with Bonferroni test for multiple comparisons. ELISA values for each cytokine and PPRV-H gene levels upon $IFN\alpha$ treatment were compared by one-way ANOVA with Bonferroni test for multiple comparisons. Means were considered significantly different when the p value was <0.05.

Acknowledgments

We gratefully acknowledge the funding received through VirginiaTech's Open Access subvention fund to defray the cost of open access publication of this manuscript. This work was partially supported through BBSRC-DBT fund # BB/L004801/1 ---awarded to SP and GDR.]

Author Contributions

Conceived and designed the experiments: ES GDR SP SD. Performed the experiments: SD AV RR KGT AR RK. Analyzed the data: SD MB AV RR GDR KGT AR RK SP ES. Contributed reagents/materials/analysis tools: GDR SP ES. Contributed to the writing of the manuscript: SD MB GDR KG SP ES. Provided reagents and materials: GDR. Provided analysis tools: SP ES.

References

1. Lefevre PC, Diallo A (1990) Peste des petits ruminants. Rev Sci Tech 9: 935–981.
2. Singh RP, Saravanan P, Sreenivasa BP, Singh RK, Bandyopadhyay SK (2004) Prevalence and distribution of peste des petits ruminants virus infection in small ruminants in India. Rev Sci Tech 23: 807–819.
3. Banyard AC, Parida S, Batten C, Oura C, Kwiatek O, et al. (2010) Global distribution of peste des petits ruminants virus and prospects for improved diagnosis and control. J Gen Virol 91: 2885–2897.
4. Khan HA, Siddique M, Sajjad ur R, Abubakar M, Ashraf M (2008) The detection of antibody against peste des petits ruminants virus in sheep, goats, cattle and buffaloes. Trop Anim Health Prod 40: 521–527.
5. Govindarajan R, Koteeswaran A, Venugopalan AT, Shyam G, Shaouna S, et al. (1997) Isolation of pestes des petits ruminants virus from an outbreak in Indian buffalo (Bubalus bubalis). Vet Rec 141: 573–574.
6. Saeed IK, Ali YH, Khalafalla AI, Rahman-Mahasin EA (2010) Current situation of Peste des petits ruminants (PPR) in the Sudan. Trop anim Health Prod 42: 89–93.
7. Roeder PL, Abraham G, Kenfe G, Barrett T (1994) Peste des petits ruminants in Ethiopian goats. Trop Anim Health Prod 26: 69–73.
8. Taylor WP, Abegunde A (1979) The isolation of peste des petits ruminants virus from Nigerian sheep and goats. Res Vet Sci 26: 94–96.
9. Taylor WP, Diallo A, Gopalakrishna S, SreeramaluP, Wilsmore AJ, et al. (2002) Peste des petits ruminants has been widely present in southern India since, if not before, the late 1980s. Prev Vet Med 52: 305–312.
10. Yesilbag K, Yilmaz Z, Golcu E, Ozkul A (2005) Peste des petits ruminants outbreak in western Turkey. Vet Rec 157: 260–261.
11. Couacy-Hymann E, Bodjo C, Danho T, Libeau G, Diallo A (2007) Evaluation of the virulence of some strains of peste-des-petits-ruminants virus (PPRV) in experimentally infected West African dwarf goats. Vet J 173: 178–183.
12. Diop M, Sarr J, Libeau G (2005) Evaluation of novel diagnostic tools for peste des petits ruminants virus in naturally infected goat herds. Epidemiol Infect 133: 711–717.
13. Anderson EC, Anderson J, Doughty WJ, Drevmo S (1975) The pathogenicity of bovine strains of foot-and-mouth disease virus for impala and wildebeest. J Wildl Dis 11: 248–255.
14. Kedmi M, Levi S, Galon N, Bomborov V, Yadin H, et al (2011) No evidence for involvement of sheep in the epidemiology of cattle virulent epizootic hemorrhagic disease virus. Vet Microbiol 148: 408–12.
15. Kuchipudi SV, Dunham SP, Nelli R, White GA, Coward VJ, et al. (2012) Rapid death of duck cells infected with influenza: a potential mechanism for host resistance to H5N1. Immunol Cell Biol 90: 116–123.
16. Pawar RM, Dhinakar Raj G, Balachandran C (2008a) Relationship between the level of signaling lymphocyte activation molecule mRNA and replication of Peste-des-petits-ruminants virus in peripheral blood mononuclear cells of host animals. Acta virol 52: 231–236.
17. Sidorenko SP, Clark EA (2003) The dual-function CD150 receptor subfamily: the viral attraction. Nature Immunol 4: 19–24.
18. Tatsuo H, Ono N, Yanagi Y (2001) Morbilliviruses use signaling lymphocyte activation molecules (CD150) as cellular receptors. J Virol 75: 5842–5850.
19. Kawai T, Akira S (2007) TLR signaling. Sem Immunol 19: 24–32.
20. Booth JS, Buza JJ, Potter A, Babiuk LA, Mutwiri GK (2010) Co-stimulation with TLR7/8 and TLR9 agonists induce down-regulation of innate immune responses in sheep blood mononuclear and B cells. Dev Comp Immunol 34: 572–578.
21. Hemmi H, Kaisho T, Takeuchi O, Sato S, Sanjo H, et al. (2002) Small anti-viral compounds activate immune cells via the TLR7 MyD88-dependent signaling pathway. Nature Immunol 3: 196–200.
22. Vignesh AR, Dhanasekaran S, Raj GD, Balachandran C, Pazhanivel N, et al. (2012) Transcript profiling of pattern recognition receptors in a semi domesticated breed of buffalo, Toda, of India. Vet Immunol Immunopathol 147: 51–9.
23. Tirumurugaan KG, Dhanasekaran S, Raj GD, Raja A, Kumanan K, et al (2010) Differential expression of toll-like receptor mRNA in selected tissues of goat (Capra hircus). Vet Immunol Immunopathol 133: 296–301.
24. Menzies M, Ingham A (2006) Identification and expression of Toll-like receptors 1–10 in selected bovine and ovine tissues. Vet Immunol Immunopathol 109: 23–30.
25. Bannerman DD, Kauf AC, Paape MJ, Springer HR, Goff JP (2008) Comparison of Holstein and Jersey innate immune responses to Escherichia coli intramammary infection. J Dairy Sci 91: 2225–2235.
26. Joshi MB, Rout PK, Mandal AK, Tyler-Smith C, Singh L, et al. (2004) Phylogeography and origin of Indian domestic goats. Mol Biol Evol 21: 454–462.
27. Rita NP, Gopu S, Baegan S, Barathidasan (2008) Clinical management in an outbreak of peste des petits ruminants in Barbari goats. Vet World 1: 81–82.
28. Roy P, Vairamuthu S, Thangavelu A, Chitradevi S, Purushothaman V, et al. (2010) An outbreak of peste des petits ruminants among Thelichery breed of goats. Int J Appl Res Vet Med 8: 155–160.
29. Barrat FJ, Meeker T, Chan JH, Guiducci C, Coffman RL (2007) Treatment of lupus-prone mice with a dual inhibitor of TLR7 and TLR9 leads to reduction of autoantibody production and amelioration of disease symptoms. Eur J Immunol 37: 3582–6.
30. Baron MD, Parida S, Oura CA (2011) Peste des petits ruminants: a suitable candidate for eradication? Vet Rec 169: 16–21.
31. Shaila MS, Purushothaman V, Bhavasar D, Venugopal K, Venkatesan RA (1989) Peste des petits ruminants of sheep in India. Vet Rec 125: 602.
32. Nanda YP, Chatterjee A, Purohit AK, Diallo A, Innui K, et al. (1996) The isolation of peste des petits ruminants virus from northern India. Vet Microbiol 51: 207–216.
33. Saravanan P, Sen A, Balamurugan V, Rajak KK, Bhanuprakash V, et al. (2010) Comparative efficacy of peste des petits ruminants (PPR) vaccines. Biologicals 38: 479–485.
34. Mornet P, Orue J, Gilbert Y, Thiery G, Mamadou S (1956) La peste des petite ruminants en Afrique occidentale française ses rapports avec la peste bovine. Revue d'élevage et de médecine vétérinaire des pays tropicaux 9: 313–342.
35. Pawar RM, Raj GD, Kumar TM, Raja A, Balachandran C (2008b) Effect of siRNA mediated suppression of signaling lymphocyte activation molecule on replication of peste des petits ruminants virus in vitro. Virus Res 136: 118–123.
36. Schroder NW, Schumann RR (2005) Single nucleotide polymorphisms of Toll-like receptors and susceptibility to infectious disease. Lancet Infect Dis 5: 156–164.
37. Mikula I, Bhide M, Pastorekova S, Mikula I (2010) Characterization of ovine TLR7 and TLR8 protein coding regions, detection of mutations and Maedi Visna virus infection. Vet Immunol Immunopathol 138: 51–59.
38. Dhiman N, Ovsyannikova IG, Vierkant RA, Ryan JE, Pankratz VS, et al. (2008) Associations between SNPs in toll-like receptors and related intracellular signaling molecules and immune responses to measles vaccine: preliminary results. Vaccine 26: 1731–1736.
39. Honda K, Taniguchi T (2006) IRFs: master regulators of signalling by Toll-like receptors and cytosolic pattern-recognition receptors. Nature Rev Immunol 6: 644–658.
40. Sentsui H, Takami R, Nishimori T, Murakami K, Yokoyama T, et al (1998) Anti-viral effect of interferon-alpha on bovine viral diarrhea virus. J Vet Med Sci 60: 1329–33.

41. Panigrahi P, Mohanty SB, Maheshwari RK, Friedman RM (1988) Effect of cloned human interferon-alpha 2a on bovine parainfluenza-3 virus. Brief report. Arch Virol 98: 107–15.

42. Uematsu S, Akira S (2007) Toll-like receptors and Type I interferons. J Biol Chem 282: 15319–15323.

43. Buza J, Benjamin P, Zhu J, Wilson HL, Lipford G, et al. (2008) CD14+ cells are required for IL-12 response in bovine blood mononuclear cells activated with Toll-like receptor (TLR) 7 and TLR8 ligands. Vet Immunol Immunopathol 126: 273–282.

44. Hart OM, Athie-Morales V, O'Connor GM, Gardiner CM (2005) TLR7/8-mediated activation of human NK cells results in accessory cell-dependent IFN-gamma production. J Immunol 175: 1636–1642.

45. Filippi CM, von Herrath MG (2008) IL-10 and the resolution of infections. J Pathol 214: 224–230.

46. Brooks DG, Trifilo MJ, Edelmann KH, Teyton L, McGavern DB, et al. (2006) Interleukin-10 determines viral clearance or persistence in vivo. Nature Med 12: 1301–1309.

47. Fiorentino DF, Zlotnik A, Vieira P, Mosmann TR, Howard M, et al. (1991) IL-10 acts on the antigen-presenting cell to inhibit cytokine production by Th1 cells. J Immunol 146: 3444–3451.

48. Harrison SM, Tarpey I, Rothwell L, Kaiser P, Hiscox JA (2007) Lithium chloride inhibits the coronavirus infectious bronchitis virus in cell culture. Avian Pathol 36: 109–114.

49. Franchimont D, Martens H, Hagelstein MT, Louis E, Dewe W, et al. (1999) Tumor necrosis factor alpha decreases, and interleukin-10 increases, the sensitivity of human monocytes to dexamethasone: potential regulation of the glucocorticoid receptor. J Clin Endocrinol Metab 84: 2834–2839.

50. Jagtap SP, Rajak KK, Garg UK, Sen A, Bhanuprakash V, et al. (2012) Effect of immunosuppression on pathogenesis of peste des petits ruminants (PPR) virus infection in goats. Microb. Pathogenesis. 52, 217–226.

51. Goyal S, Dubey PK, Tripathy K, Mahajan R, Pan S, et al. (2012) Detection of polymorphism and sequence characterization of Toll-like receptor 7 gene of Indian goat revealing close relationship between ruminant species. Anim Biotechnol 23: 194–203.

52. Xu Y, Tao X, Shen B, Horng T, Medzhitov R, et al. (2000) Structural basis for signal transduction by the Toll/interleukin-1 receptor domains. Nature 408: 111–115.

53. Banerjee P, Gahlawat SK, Joshi J, Sharma U, Tantia MS, et al. (2012) Sequencing, Characterization and Phylogenetic analysis of TLR genes of *Bubalus bubalis*. DHR-IJBLS 3: 137–158.

54. Dhinakar Raj G, Nachimuthu K, Mahalinga Nainar A (2000) A simplified objective method for quantification of peste des petits ruminants virus or neutralizing antibody. J Virol Methods 89: 89–95.

55. Gopinath VP, Raj GD, Raja A, Kumanan K, Elankumaran S (2011) Rapid detection of Newcastle disease virus replication in embryonated chicken eggs using quantitative real time polymerase chain reaction. J Virol Methods 171: 98–101.

A Dynamic Spatio-Temporal Model to Investigate the Effect of Cattle Movements on the Spread of Bluetongue BTV-8 in Belgium

Chellafe Ensoy[1]*, **Marc Aerts**[1], **Sarah Welby**[2], **Yves Van der Stede**[2,3], **Christel Faes**[1]

1 Interuniversity Institute for Biostatistics and statistical Bioinformatics, Universiteit Hasselt, Hasselt, Belgium, **2** Veterinary and Agrochemical Research Centre, Brussels, Belgium, **3** University of Ghent, Laboratory of Vet. Immunology, Merelbeke, Belgium

Abstract

When Bluetongue Virus Serotype 8 (BTV-8) was first detected in Northern Europe in 2006, several guidelines were immediately put into place with the goal to protect farms and stop the spreading of the disease. This however did not prevent further rapid spread of BTV-8 across Northern Europe. Using information on the 2006 Bluetongue outbreak in cattle farms in Belgium, a spatio-temporal transmission model was formulated. The model quantifies the local transmission of the disease between farms within a municipality, the short-distance transmission between farms across neighbouring municipalities and the transmission as a result of cattle movement. Different municipality-level covariates such as farm density, land composition variables, temperature and precipitation, were assessed as possibly influencing each component of the transmission process. Results showed a significant influence of the different covariates in each model component, particularly the significant effect of temperature and precipitation values in the number of infected farms. The model which allowed us to predict the dynamic spreading of BTV for different movement restriction scenarios, also affirmed the significant impact of cattle movement in the 2006 BTV outbreak pattern. Simulation results further showed the importance of considering the size of restriction zones in the formulation of guidelines for animal infectious diseases.

Editor: Alex R. Cook, National University of Singapore, Singapore

Funding: Support from the IAP Research Network P7/06 of the Belgian State (Belgian Science Policy) and the Flemish Research Council (G064710). The funders had no role in study design, data collection and analysis, decision to publish, or preparation of the manuscript.

Competing Interests: The authors have declared that no competing interests exist.

* E-mail: chellafe.ensoy@uhasselt.be

Introduction

For decades, the livestock industry has been battling the emergence and recurrence of various infectious animal diseases. The negative social, economic and environmental impact brought on by these diseases is a major concern not only for this industry, but also for the countries involved and the international community [1]. Prevention and control measures are then usually prepared at the national and international level. One of the important control measures is the restriction of animal movement between farms and/or countries during an outbreak [2]. This control measure comes from the knowledge that movement provides an important route of transmission of infectious diseases [3], [4], [5]. As in the case of the foot-and-mouth disease (FMD) epidemic in 2001, Févre et al. [3] reported that the spread of FMD from the north of England to France and the Netherlands was due to animal movement. Other animal diseases also have the potential to be spread through animal movements, such as rabies [3], bovine tuberculosis [3], Coxiellosis [5] and bluetongue [6].

Bluetongue (BT) is a non-contagious, infectious, vector-borne disease of ruminants caused by the bluetongue virus (BTV) and is transmitted between hosts by bites of *Culicoides* midges. Over the past decade BT has become one of the most important diseases of livestock following a series of incursions in Europe [7]. In particular, the first cases of BTV serotype 8 (BTV-8) in northern Europe were reported near Maastricht in the Netherlands in July/

August 2006, with subsequent cases reported in Belgium, Germany, France and Luxembourg. In May 2007, BTV-8 re-emerged and caused major outbreaks across the previously-affected countries and spread into new areas [8].

Since the time it was detected up to the present, several studies have been conducted which explained why and how the BTV outbreak occurred. Some of these studies looked into the risk factors associated with BTV such as climatic conditions, land composition [9],[10],[11], while others looked in greater detail on the effect of animal movement on the spread of the virus [6], [12], [13], [14]. In this study, a spatio-temporal transmission model was formulated using data of the 2006 BTV outbreak in cattle farms in Belgium. The proposed model quantifies the local transmission of the disease between farms within a municipality, the transmission between farms across neighbouring municipalities and transmission as a result of the movement/transport of animals. Municipality-level factors influencing the transmission process were also investigated.

In the subsequent section, an overview of the BTV-8 outbreak, risk-factor and cattle movement data are given. This is then followed by a detailed description of the proposed spatio-temporal transmission model and the procedure for model selection. Results of the model fitting are then presented along with simulation results followed by a brief discussion.

Materials and Methods

Data Sources

Outbreak Data. The 2006 Bluetongue outbreak information for cattle in Belgium used in this study was obtained from the Veterinary and Agrochemical Research Centre (CODA-CERVA), Belgium. Figures 1 and 2 detail the observed spatial and temporal trend of the BTV-8 outbreak in Belgium for 2006 and 2007, respectively. Farms having at least 1 observed infected animal were considered as infected. The onset of infection that was used in this study was the date that the infection was thought to have occurred similar to the dates used by Faes et al. [10]. We have used the date that the disease symptoms were first observed as the date of infection as we have no knowledge on when the infection actually occurred.

Risk Factor Data. Different covariates deemed influential to the spread of BTV [15], [12], [10] were investigated. These risk factors (Figure 3) include:

- Farm and animal (cattle and sheep) density per municipality,
- Proportion of forest, crop, urban, and pasture area per municipality,
- Temperature, and
- Precipitation.

Land use variables (proportion of forest, crop, urban, and pasture area per municipality) are highly correlated owing to the fact that each of the variables convey relative information to the whole. To deal with this issue, these covariates were transformed using the compositional data technique based on the additive log-ratio [16]. The additive log-ratio transformation works by taking the log of the ratio of a covariate and another reference covariate. Applying this transformation to the compositional data resulted in less correlated variables with values which can vary over the entire real number range. In this study, the crop variable was taken as the reference and the other variables were transformed into: $forestT = \log(forest/crop)$, $urbanT = \log(urban/crop)$ and $pastureT = \log(pasture/crop)$. The problem of zero value for the crop proportion was handled by adding a small constant

(0.0001) to all variables. Only the transformed forest, urban, and pasture variables were entered into the model, and all of them were interpreted in terms of the proportion of crop land.

For the temperature and precipitation, it has been shown that seasonal variations in weather affect the spread of *Culicoides* and therefore also affect the spread of BTV [17], [9], [11]. Temperature and precipitation data (daily mean temperature (°C) and precipitation (cm)) from all weather station of Belgium were obtained and then summarized to average weekly readings (black circles in Figure 3D and 3E with black solid dots signifying values during the outbreak period). Since Belgium is a very small country (total area of 30,528 sq. km.), the temperature and precipitation reading were observed not to differ much from one weather station to the next and hence assumed to be also the same in all non-station locations. Readings from all weather stations were aggregated to average weekly readings and both variables were assumed constant throughout the whole country. Based on results from various studies, [11], [15], [18], a moving average of these values at time lag of 1 to 4 weeks (black solid line in the figure) was used to ensure a smooth trend and to consider the time needed for the vector population to develop a competent population, and to account for the uncertainty of the date of the infectious bite.

Transport of Animals. Purchase of an animal results in movement of the animal from one farm to the other. If the farm of origin has cases of BT infection, there is a certain probability that the animal was also infected thereby increasing the chance that the animals in the destination farm will also, via vectors, be infected (see [6] and [14]). To account for this source of infection, the number of animal movements across different municipalities was explored. Animal movement in this paper refer to cattle movement only. Although sheep movement might also be an important source of infection, it was not included in this paper as there is no available information for sheep movement in Belgium. Furthermore, in Belgium, sheep are raised for meat and breeding as a hobby [19], hence movement concern only ovines of high genetic performance, between large or high producer ovine herds, which constitutes only a minority of the sheep herds in Belgium.

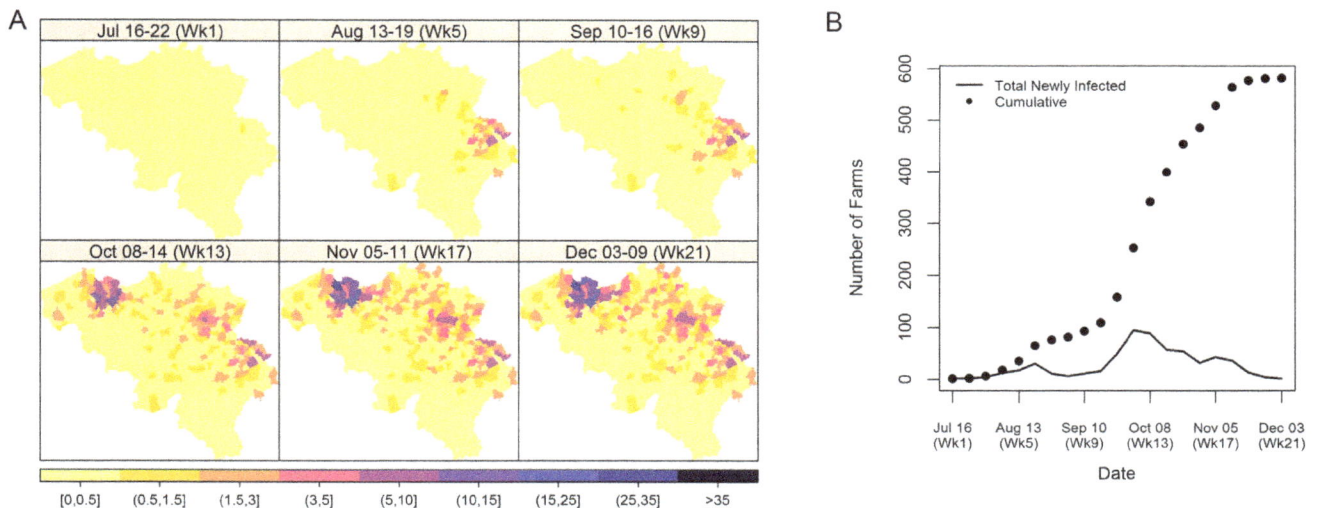

Figure 1. Spatial and temporal trend of the BTV-8 outbreak in Belgium for 2006. Figures are based on the weekly data with (A) giving the spatial trend of the cumulative number of infected farms and (B) giving the temporal trend of the weekly new infections and cumulative number of infected farms. The onset of infection was the date that disease symptoms were first observed, assumed 3 to 4 weeks before confirmation of report.

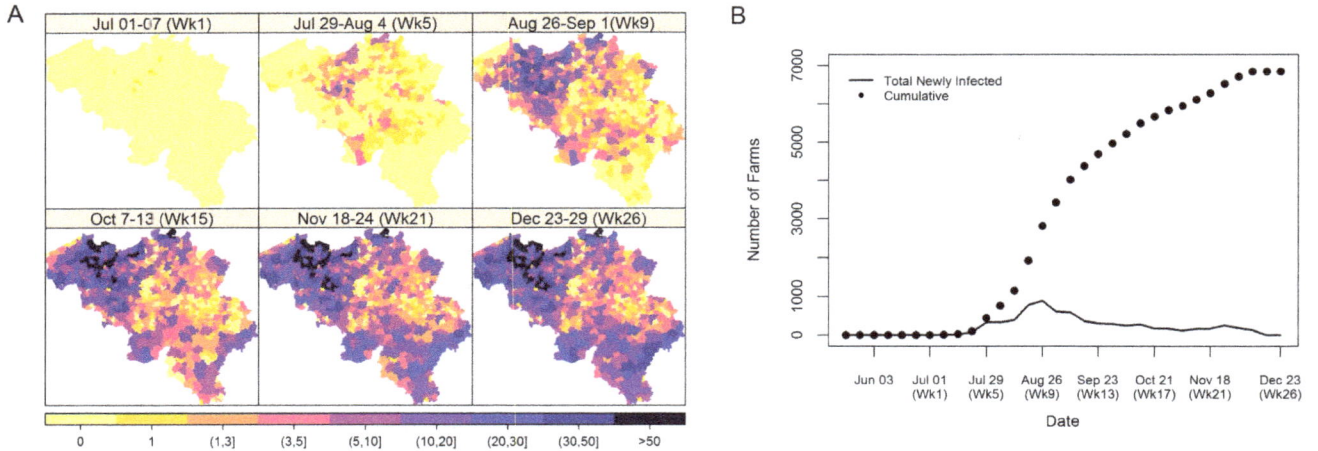

Figure 2. Spatial and temporal trend of the BTV-8 outbreak in Belgium for 2007. Figures are based on the weekly data with (A) giving the spatial trend of the cumulative number of infected farms and (B) giving the temporal trend of the weekly new infections and cumulative number of infected farms.

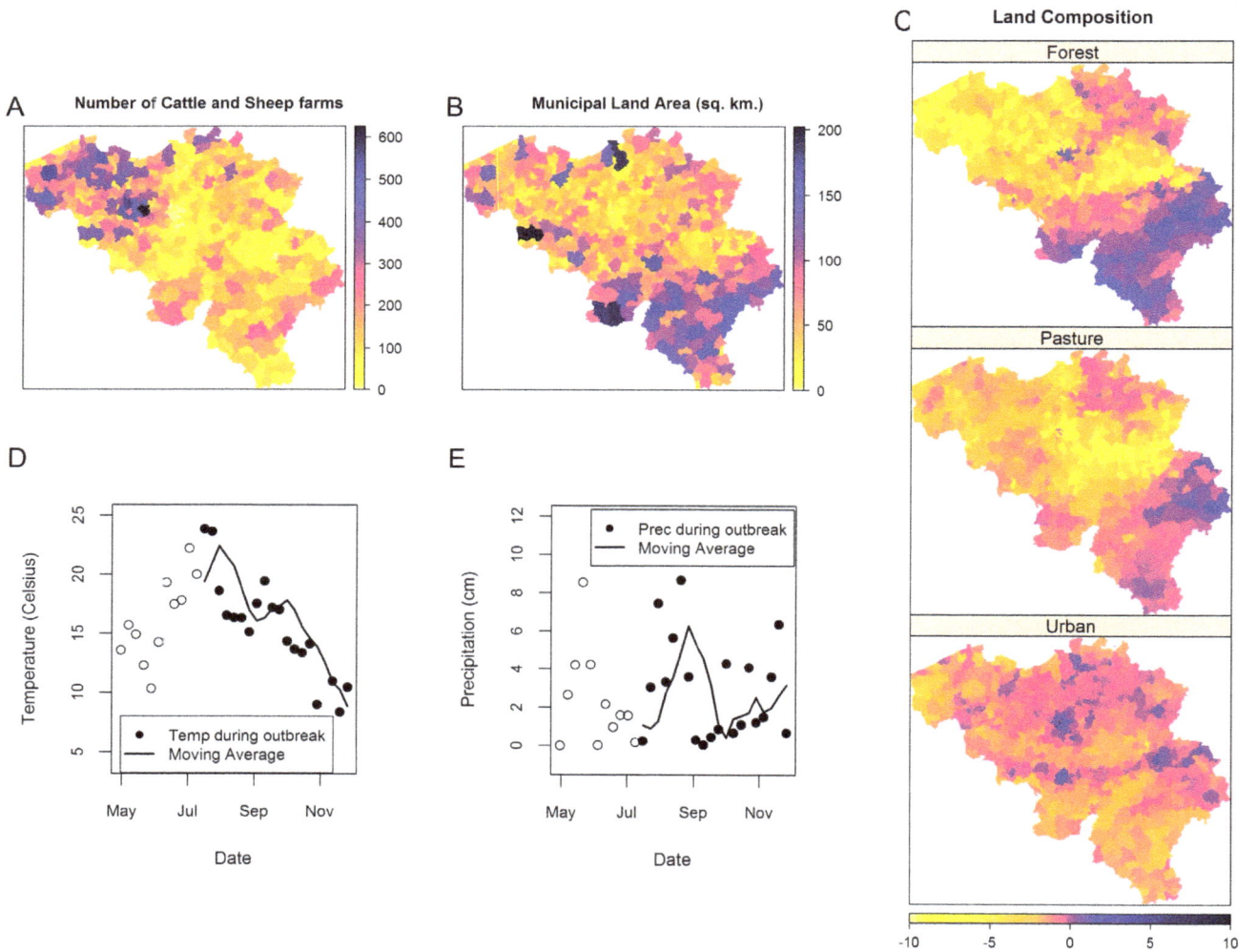

Figure 3. Risk factor data. (A) Number of cattle and sheep farms; (B) Land area per municipality (in square km); (C) Land composition variables with respect to proportion of crop land; (D) and (E): Average weekly temperature and precipitation in Belgium during the 2006 outbreak. With the moving average computed as the average of the lag 1 to 4 weeks values.

Movement information was extracted from the cattle birth and purchase information in Belgium. Birth data were however only available from 2005, thereby limiting the number of cattle movement which could be traced. It was thus decided to use a constant general pattern of movement based on the 2005 to 2009 data given the fact that a rather similar trend of purchasing over the years was observed (Figure S1).

Defining cattle movement as the farm-to-farm transfer of cattle through purchasing, the number of animal movements for two different municipalities was counted. Thus, movement of cattle is defined if cattle were transferred/transported/purchased from a farm in municipality j to a farm in municipality i. Due to the restriction on the resolution of the data, transport between farms in the same municipality was not considered as movement in this case. Two ways of quantifying the movement were explored:

- Presence of movement (Binary indicator of transport, taking the value of 1 if at least one animal is transported and 0 otherwise),
- Relative movement which show the abundance of movement (Proportion of animals transported).

Figure 4 shows the general pattern of contacts derived from the binary definition of transport and shows that most movements originate from the Walloon area and end up in the Flemish region, and in particular, in the provinces of Antwerp, East and West Flanders.

Infection Model

Mathematical modeling of an infectious disease is a tool used to describe the dynamics of the spread of that infectious disease and can be used to evaluate different control strategies (e.g. movement restriction). One of the most common models used is the SIR (Susceptible → Infected → Recovered) model. The SIR model postulates that an individual or unit starts in the susceptible class, can become infected by the disease, thus moving to the infected class, and after a while, recover from it. Different variations of this mathematical model are available (e.g. SIS, SIRS and SEIS (Susceptible → Exposed → Infected → Susceptible)), depending on the disease in question [20].

In the case of BTV, a modified SEI and SEIR model can be found in the literature [7], [9], [14]. In this paper, using the farm as an individual unit and assuming that once infected, a farm is infectious until the end of the outbreak, a SI model was applied.

The assumption of no recovery of infected farms within the outbreak period was based on the long recovery time of an infected cattle, with BTV-8 virus still detected in cattle 1–2 months after infection [21], [22], [23].

Each farm was classified into either susceptible **S** (no cattle infected with BTV), or infected and infectious **I** (at least one reported case of infected cattle). An infectious farm can then infect another farm through the vector (*Culicoides* midges). Although it would have been preferable to build an individual cattle-based model rather than a farm-based model, the reporting procedure (owners report only the first observed infection) constrained the analysis to farm level. Furthermore, the unavailability of the vector data constrained the analysis to a basic SI model.

The SI model for BT is a closed population model, where $S_{i,t} + I_{i,t} = N_i$, and $S_{i,t}$ and $I_{i,t}$ are the number of susceptible and infectious farms respectively, in week t at municipality i and N_i is the total number of farms for each municipality i. The number of susceptible farms can then be written as the difference of the total number of farms and the number of infectious farms, while the number of infectious farms at week t is just the sum of the total number of newly infected farms (Y) until week t and is given by:

$$S_{i,t} = N_i - I_{i,t}, \qquad (1)$$

$$I_{i,t} = \sum_{k=1}^{t} Y_{i,k}. \qquad (2)$$

Several authors have proposed various ways of modelling the number of infected farms. Held et al.[24] proposed a Poisson branching process model, Knorr-Held and Richardson [25] used a hierarchical hidden Markov model, while Schrödle et al. [5] used parameter-driven and observation-driven models to link the movement and spreading of diseases. In this study, the number of newly infected farms (Y) was modelled as a binomial random variable which depends on the number of susceptible farms at the previous week ($S_{i,t-1}$) and a parameter $\theta_{i,t}$, $Y_{i,t} \sim \text{Bin}(S_{i,t-1}, \theta_{i,t})$. The parameter $\theta_{i,t}$ was formulated as a function of the previous infectious population via the following equation similar to the method by Hooten et al. [26]:

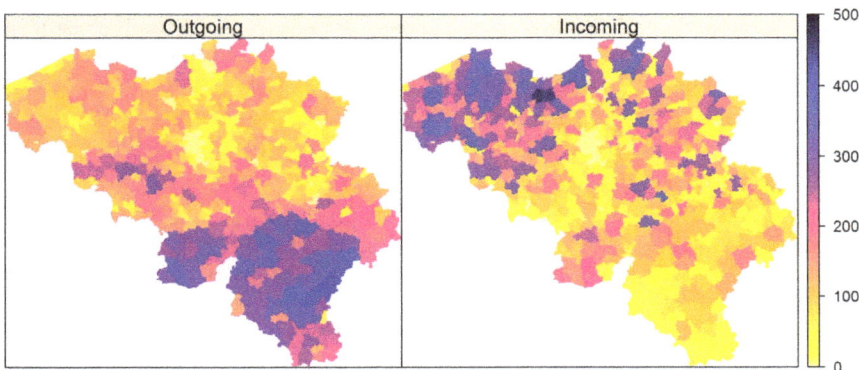

Figure 4. Spatial pattern of cattle movement in Belgium. The total number of outgoing and incoming cattle movement between municipalities obtained from the 2005–2009 cattle birth and purchase information in Belgium. A municipality (*i*) is defined to have an outgoing movement if there is a transport of cattle from that municipality (*i*) to municipality (*j*). Municipality (*j*) is then defined to have an incoming movement.

$$logit(\theta_{i,t}) = [\beta_1 \mathbf{Ind}(I_{i,t-1}=0) + \beta_2 \mathbf{Ind}(I_{i,t-1}>0) + \mathbf{X}\beta_{int}]$$

$$+ (\mathbf{X}_W\beta_W)I_{i,t-1} + (\mathbf{X}_B\beta_B)\sum_{j=1}^{N} b_{i,j}(I_{j,t-1}) \qquad (3)$$

$$+ \mathbf{X}_A\beta_A \sum_{j=1}^{N} a_{i,j}(I_{j,t-1}).$$

This dynamic infection model contains at most four additive terms representing the different transmission scenarios. The first term, $(\beta_1\mathbf{Ind}(I_{i,t-1}=0) + \beta_2\mathbf{Ind}(I_{i,t-1}>0) + \mathbf{X}\beta_{int})$ represents the background risk of the municipality which depends on municipality-specific covariates and where the overall risk is increased or decreased depending on whether or not there was an infectious farm in the municipality at the previous time. Similarly, the second term $(\mathbf{X}_W\beta_W)I_{i,t-1}$ is only present when an infection was observed previously, and corresponds to the within-municipality transmission or the local spread. This term expresses the belief that the number of infected farms in municipality i at the current week is a function of the number of infectious farms at the previous week and some covariates $(\mathbf{X}_W)$. The coefficient β_W represents the contribution of each covariate to the local transmission.

The third term, $(\mathbf{X}_B\beta_B)\sum_{j=1}^{N} b_{i,j}(I_{j,t-1})$, deals with the neighbourhood or between-municipality transmission, representing the effect of the infectious state of neighbouring municipalities the previous week, together with some municipality-level covariates. The spatial weight $b_{i,j}$ was derived based on contiguity between municipalities i and j. Municipalities considered as neighbours take the value of 1, otherwise they take the value of 0. This results in a symmetric weight matrix with a 0 diagonal. The binary weighting then ensures that a municipality with more infected neighbours is given more weight in the transmission model [27].

The final term $(\mathbf{X}_A\beta_A)\sum_{j=1}^{N} a_{i,j}(I_{j,t-1})$ corresponds to long-distance transmission through animal movements. The $a_{i,j}$ in this equation quantifies the movement of animals from municipality j to municipality i and is defined as explained above. Unlike the neighbourhood matrix, the movement weight matrix is asymmetric. This is based on the fact that the number of transports from municipality j to i can be different from municipality i to j. To ensure that only the long-distance transmission is reflected in this part of the model, local (movement between farms within the same municipality) and neighbourhood movements were taken out of the weight matrix (since the local and neighbourhood effect are already accounted for by the second and third term of the model). Similar to the second and third term, different environmental factors were included in this component. The argument behind this is that movement of infected cattle alone does not ensure transmission, it is the combination of movement and presence of vectors in the area.

Model Selection and Exploration

Using the average weekly temperature, precipitation and their interaction, transformed pasture, forest and urban areas, farm density and total land area as covariates, equation (3) was fitted using Proc NLMIXED in SAS, version 9.2 (SAS Institute, Cary NC). Various representations (binary indicator, actual count, log-transformed count) of the infection $(I_{j,t-1})$ and movement $(a_{i,j})$ status in the model were explored. Model selection was done using the Akaike information criterion (AIC) with the model having the

Table 1. Comparison of the AIC for different model component choices.

Infection status ($I_{j,t-1}$) in Component 3 and 4[a]	Movement ($a_{i,j}$)		
	Binary	Count	Log-Count
Binary	3435.1	3444.2	3443.9
Count	3576.3	3594.6	3594.5

The spatio-temporal model is fitted to the 2006 BTV-8 outbreak data.
[a] Infection status in neighbourhood and movement components.

smallest AIC value selected as the best model. Based on the AIC values given in Table 1, the best fitting model was the model with a binary indicator for infection status $(I_{j,t-1}^b)$ in the neighbourhood and movement components along with a binary movement weight $(a_{i,j})$.

Some of the parameters in the full model were not significantly different from zero and hence the model could be reduced. Model reduction using the AIC criterion resulted in the retention of 24 parameters from the 37 parameters in the full model, and the AIC value was decreased by 15.5 (from 3435.1 to 3419.6). Parameter estimates for this reduced model are given in Table 2.

Results

Figure 1 shows that, in 2006 BTV-8 in Belgium first appeared around the area of Liège during the 3rd week of July (considered as week 1 of the outbreak). The infection then spread within Liège and around Limburg and neighbouring provinces and reached its peak during the 6th week of the outbreak (August 20–26). It was also during this week that the whole country of Belgium was declared as BTV-8 infected and thus movement restriction were lifted [13]. After this peak, a dying-out phase was observed with the total outbreak size of 82 farms during the 8th week. However, in the week of September 10–16 (week 9), a jump to the East Flanders area was observed, with the first case appearing in the municipality of Destelbergen, a neighbouring municipality of Ghent. It then quickly spread to other municipalities in the province (i.e. Ghent, Nevele and Deinze) during the succeeding weeks. By the end of 2006, out of the 40 141 cattle farms in Belgium (partitioned across 576 municipalities), a total of 582 cases of infected farms from 205 different municipalities was observed.

During the winter period, many hoped that BTV-8 had disappeared [28]. However, BTV-8 re-emerged in the first week of July 2007 and by the end of that year, 6 840 farms (9.5% of the total) across 90% of the municipalities, had notified an infection. This second episode was much larger than the one in 2006. It also involved areas which were previously not affected by the disease, notably municipalities in the southern part of Belgium.

Fitting the infection model to the data, results show that for any given farm in a municipality, the background odds of contracting BTV is increased by 2.50 (95% CI [1.89, 3.11]) if an infection had already occurred in that same municipality during the previous week. If infection was not observed in any municipalities at the previous time point, Figure 5 shows the inherent susceptibility of the different municipalities in acquiring BTV. The map for background transmission shows that depending on covariate values, some municipalities have higher odds of acquiring BT than others. Most areas with increased odds are found in Liège and in the provinces of Antwerp and Limburg, where infection was mostly observed.

Table 2. Parameter Estimates for the reduced spatio-temporal model fitted to the 2006 BTV-8 outbreak data.

Component		Covariate	Estimate	95% CI
Background	β_1	Intercept 1	-0.20	$[-2.82, 2.42]$
	β_2	Intercept 2	0.71	$[-1.90, 3.33]$
	β_{int}	Temperature (°C)	-0.38	$[-0.53, -0.24]$
		Precipitation (cm)	-2.92	$[-3.97, -1.86]$
		Temp x Prec	0.16	$[0.10, 0.22]$
		PastureT^a	0.12	$[0.04, 0.21]$
		UrbanT^a	0.46	$[0.33, 0.59]$
		Farm Density^b	-0.11	$[-0.16, -0.06]$
		Land Area (sq. km.)	-1.09	$[-1.46, -0.71]$
Within-municipality	β_W	Intercept	0.73	$[0.26, 1.20]$
		Temperature (°C)	-0.06	$[-0.09, -0.03]$
		Precipitation (cm)	-0.27	$[-0.44, -0.09]$
		Temp x Prec	0.02	$[5.78 x 10^{-3}, 0.03]$
		UrbanT^a	-0.05	$[-0.09, -0.01]$
		Land Area (sq. km.)	0.20	$[0.10, 0.29]$
Between-municipality	β_B	Intercept	-1.59	$[-1.97, -1.21]$
		Temperature (°C)	0.10	$[0.08, 0.12]$
		Precipitation (cm)	0.06	$[0.01, 0.10]$
		PastureT^a	-0.06	$[-0.06, -0.01]$
		UrbanT^a	-0.17	$[-0.23, -0.11]$
Movement	β_A	Intercept	-0.06	$[-0.12, -0.01]$
		Temperature (°C)	$5.31 x 10^{-3}$	$[1.94 x 10^{-3}, 8.68 x 10^{-3}]$
		Precipitation (cm)	0.02	$[-8.20 x 10^{-4}, 0.04]$
		Temp x Prec	$-1.68 x 10^{-3}$	$[-3.04 x 10^{-3}, -3.30 x 10^{-4}]$

^{a} Log-ratio transformed ^{b} Number of cattle and sheep farms per municipality.

Conversely, if infection was already observed in at least 1 municipality at the previous time point, the within-municipality, between-municipality and movement transmission comes into effect. Figure 6 (A, B and C) shows the temporal trend of within-municipality, between-municipality, and movement transmission contributions depending on the infection status at previous time points. The plots show a non-monotone change in transmission values, starting from 0 at week 1 and increasing or decreasing depending on the temperature and precipitation values at 1 to 4 weeks prior to the investigated time point. A smooth temporal pattern per municipality was furthermore observed for the movement transmission, coming from the fact that in the final model, no municipality-specific covariate for the movement transmission was retained, unlike in the within-municipality and between-municipality transmission. Table 2 shows that the within-municipality transmission was found to be significantly influenced not only by temperature, precipitation and their interaction, but also by the land area and proportion of urban area relative to crop area. The neighbourhood or between-municipality transmission on the other hand was significantly influenced by the proportion of pasture and urban areas relative to crop area, aside from temperature and precipitation.

Figure 6 (D, E and F) shows that during the second peak of the outbreak (12th week of outbreak in 2006), municipalities around Namur, Luxembourg and West Flanders have high odds of within-municipality transmission (Figure 6, D), while areas around Liège, Limburg and East Flanders have high odds of between-municipality

transmission as compared to other areas (Figure 6, E). The maps clearly show that areas with low values for local transmission have high between-municipality transmission and vice versa. However, there are areas with high local and between-municipality transmission, although for these areas the odds of local transmission is slightly below or equal to 1. With regards to transmission through movements, Figure 6 (F) shows that areas with more incoming movements (Figure 4) have increased risk of BTV transmission. Hot spots were found in the provinces of Antwerp, East Flanders and Limburg. The pattern seen on the maps implies that during the peak of the 2006 outbreak, the spread of BTV was more due to the between-municipality and movement transmission rather than the within-municipality transmission.

One-step-ahead and Long-term Prediction

A one-step-ahead deterministic prediction and long-term stochastic prediction based on the parameters from the reduced model (Table 2) are depicted in Figure 7. The deterministic prediction (A) gives the current week predicted values based on the parameters of the reduced model (Table 2) and observed values of the previous week. The stochastic prediction (B and C), on the other hand, starts with an initial condition (e.g. introduction of one case in an area) and predicts future events by generating observations from a binomial distribution based on the predicted probabilities from the model. In this study, parameter estimates from the fitted model (Table 2) and data up until the 7th week of

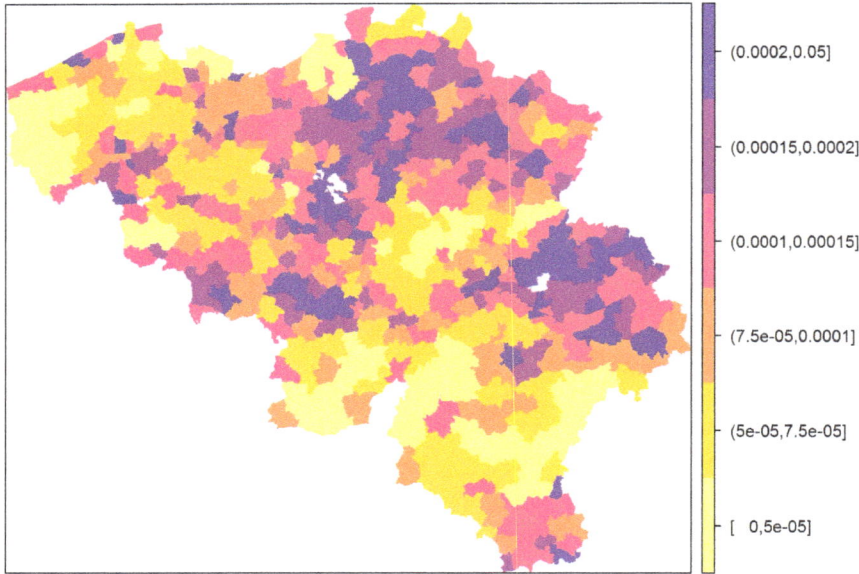

Figure 5. Spatial structure of the background odds in acquiring BTV. The map gives the odds of transmission during the start of the outbreak which is computed using $\exp(\beta_1 + X\beta_{int})$.

the outbreak (79 observed cases) were used as an initial condition and the model was then allowed to predict the rest of the outbreak period. The choice of the 7 weeks data coincided with the time (a week) after the lifting of the movement restriction. A total of 1000 simulations was done (gray lines in Figure 7) with the median stochastic prediction given by the black line. The model managed

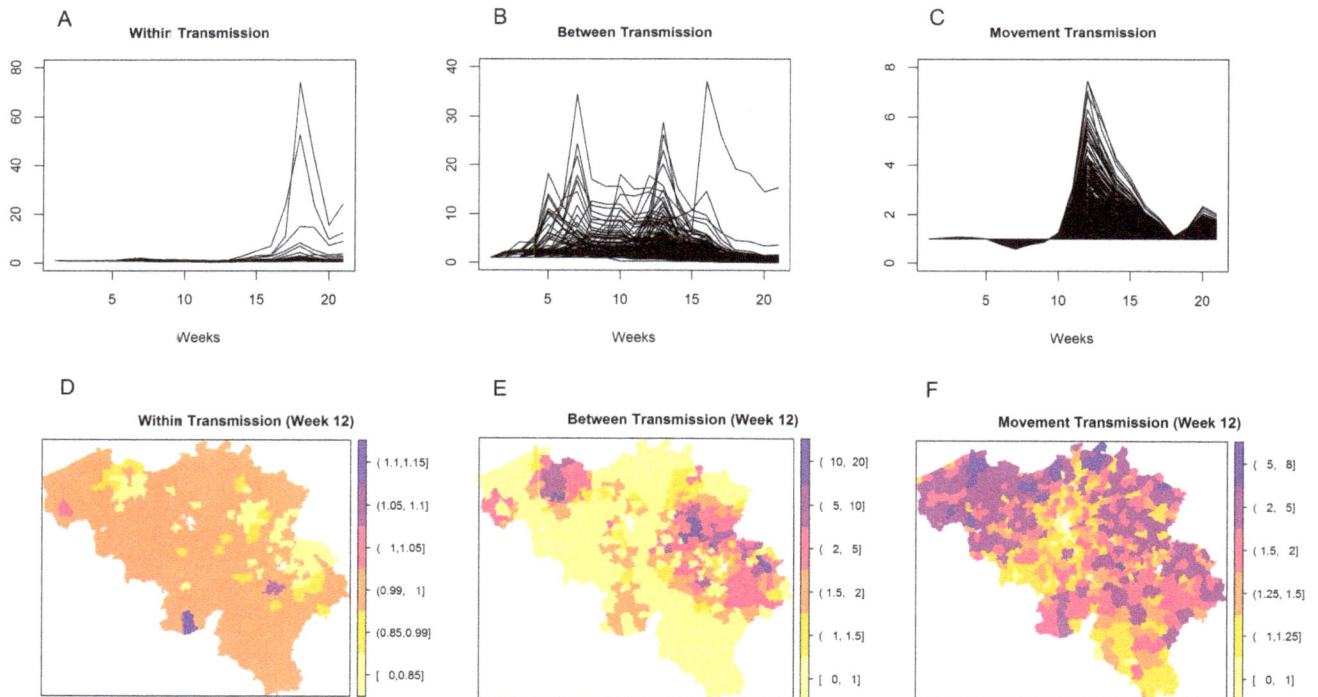

Figure 6. Odds of BT transmission within and between municipalities and through cattle movement. The top 3 figures (A, B and C) gives the temporal trend cf the contributions of each model component to the BT transmission while the maps at the bottom (D, E and F) gives the spatial structure of the contributions of each component during the peak of the outbreak (week 12 of 2006 outbreak, where week 1 is on 20–26 July). The odds were computed based on the within-municipality transmission, $\exp((X_W\boldsymbol{\beta}_W)I_{i,t-1})$; Between-municipality transmission, $\exp\left((X_B\boldsymbol{\beta}_B)\sum_{j=1}^{N} b_{i,j}I_{j,t-1}^b\right)$; Movement transmission, $\exp\left((X_A\boldsymbol{\beta}_A)\sum_{j=1}^{N} a_{i,j}I_{j,t-1}^b\right)$.

A

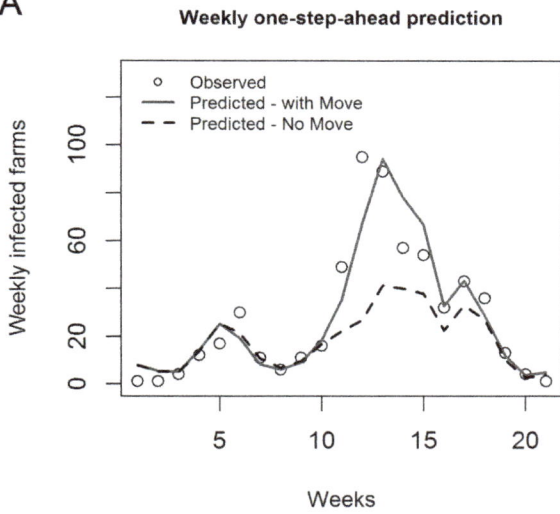

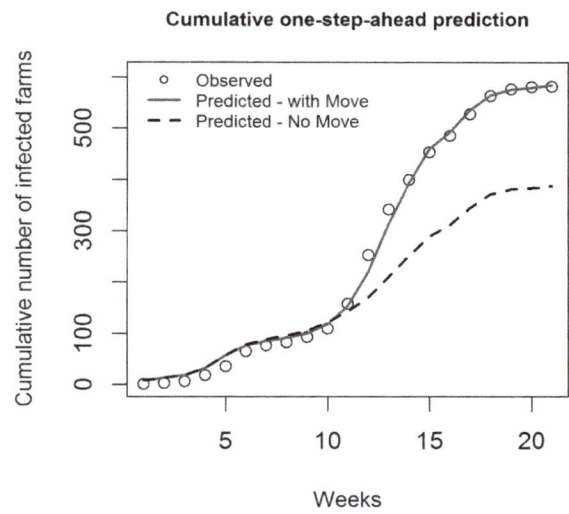

B

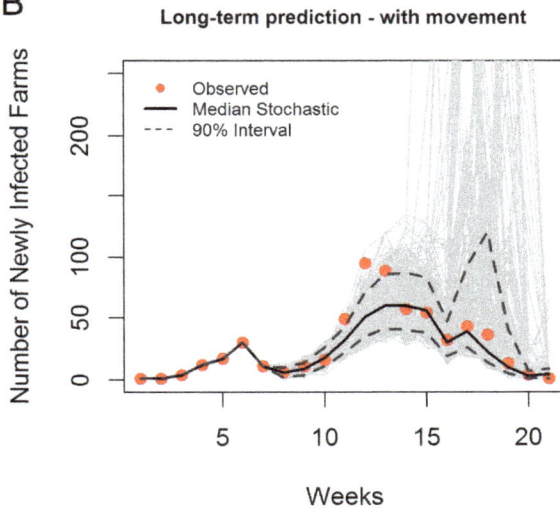

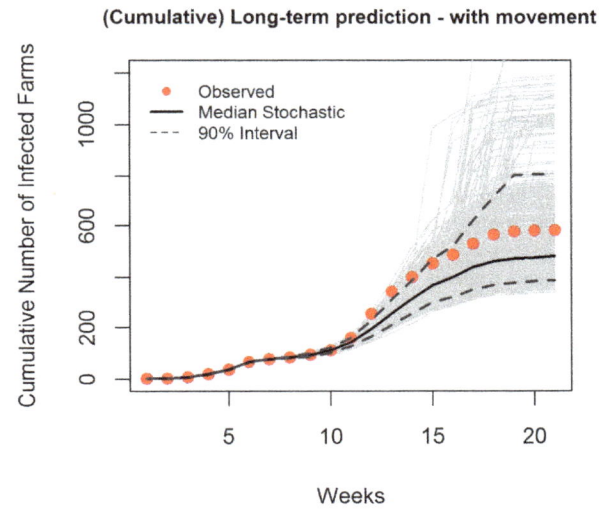

C

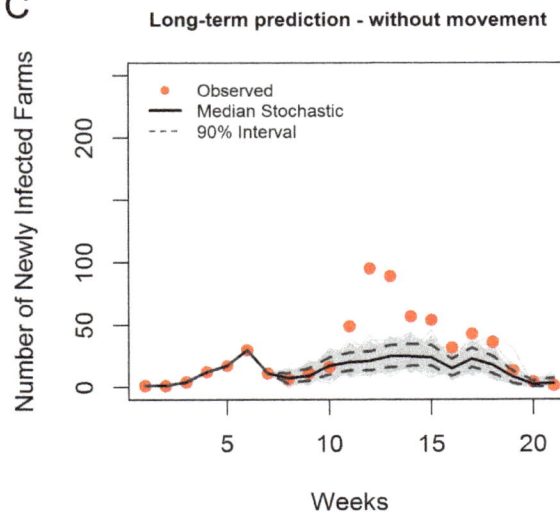

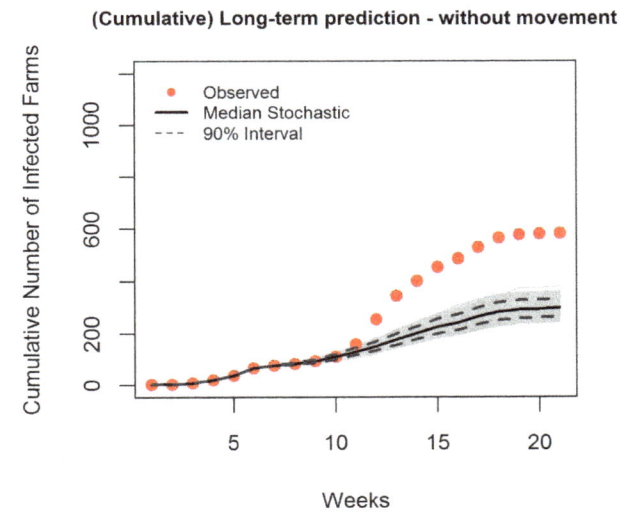

Figure 7. Predicted temporal trend of the 2006 outbreak. The weekly number of predicted infections and the cumulative number of infections are based on one-step-ahead (deterministic) predictions (A) and long-term (stochastic) predictions (B) with and (C) without cattle movements (complete ban). The gray lines are the predictions from 1000 simulations.

to capture fairly well both the temporal (Figure 7B) and spatial trend (Figure 8A) of the infection, with the true trend falling within the 90% interval.

To investigate the impact of movement of animals in the transmission of BTV, Figure 7 shows the one-step-ahead and long-term prediction from the model with and without movement restrictions. In the deterministic plot (A), complete absence of cattle movement throughout the outbreak duration was assumed, while for the stochastic prediction (C), complete movement restriction in the whole of Belgium was assumed to start from week 8 of the outbreak. We can see in the temporal plots the reduction of the predicted number of newly infected cases, and hence reduction in the number of cumulative infections per municipality when there is complete movement restriction (see Figure 8).

The median stochastic final size without animal movement was estimated to be 295 farms with a 90% confidence interval of 262–330 farms. This was significantly lower than the true outbreak final size of 582 farms. If there were no restriction, on the other hand, the median stochastic final size was estimated to be 480 farms with a 90% confidence interval of 387–810 farms. This reduction in the predicted number of cases suggests that animal movements have a significant impact on the spread of BTV. It can also be observed from the figures that with complete movement restriction, the noticeable jump in the Ghent region was not predicted by the model. This implies that the outbreak becomes limited only to the surrounding municipalities and provinces and the long-distance transmission of the virus does not occur with movement restrictions.

To further investigate the effect of different restriction scenarios on the spread of BT, a stochastic prediction was simulated for 2 different types of restriction established within a certain radius (e.g. 20 km) around an infected farm. Restriction 1 denotes movement restriction within the zone, while in restriction 2, movement within the zone is allowed and only movement outside the zone is prohibited. In other words, when movement restriction 1 is in place, no movement of cattle is allowed within the restriction zone

and from the restriction zone to outside the restriction zone, but for the rest of the country, cattle movement is allowed. Figure 9 shows that restricting the movement resulted in a significant reduction in the predicted final size of the outbreak, although it is apparent in the plot that this depends on both the type of restriction and the size of the restriction zone. For restriction 1, a 15 km restriction zone is already as effective as a total ban of movement and increasing the radius of the zone no longer leads to significant decrease in the outbreak size. Based on the bootstrap confidence interval however, a 10 km zone seems to be already sufficient. Restriction 2 on the other hand, is only effective up to around 10 km, increasing the restriction zone further also increases the predicted outbreak size. It should be noted that the restriction zone which was set-up during the start of the 2006 epidemic and lifted during August was similar to the second type of restriction investigated here and covers a radius of 20 km.

Application to the 2007 Outbreak

To validate the performance of the model, the 2007 BTV outbreak was simulated using the model fitted to the 2006 data. Stochastic prediction results for simulations initialized using the observed 6 weeks data (761 cases of observed infection) are presented in Figure S2. The median stochastic final size of the outbreak was estimated to be 10 117 farms with 90% interval of 8 971–11 705, which was much higher than the observed 2007 outbreak size of 6 840 farms. In effect, the stochastic prediction estimated that around 25.2% of farms in Belgium would be infected by BTV at the end of 2007. However, this prediction assumes that the parameters underlying the two outbreaks are the same, hence it is not surprising that the model did not predict well the 2007 outbreak. Another way of performing model validation is to update the model with new observations and predict the values k-week (s) ahead. Hence, a one-week-ahead, two-weeks-ahead and final size prediction was preformed based on the model fitted not only to the 2006 data but also to the 2007 data at different weeks. Figures 10 A, B, and C show the performance of the model, where the one-week-ahead prediction is generally not far-off from the

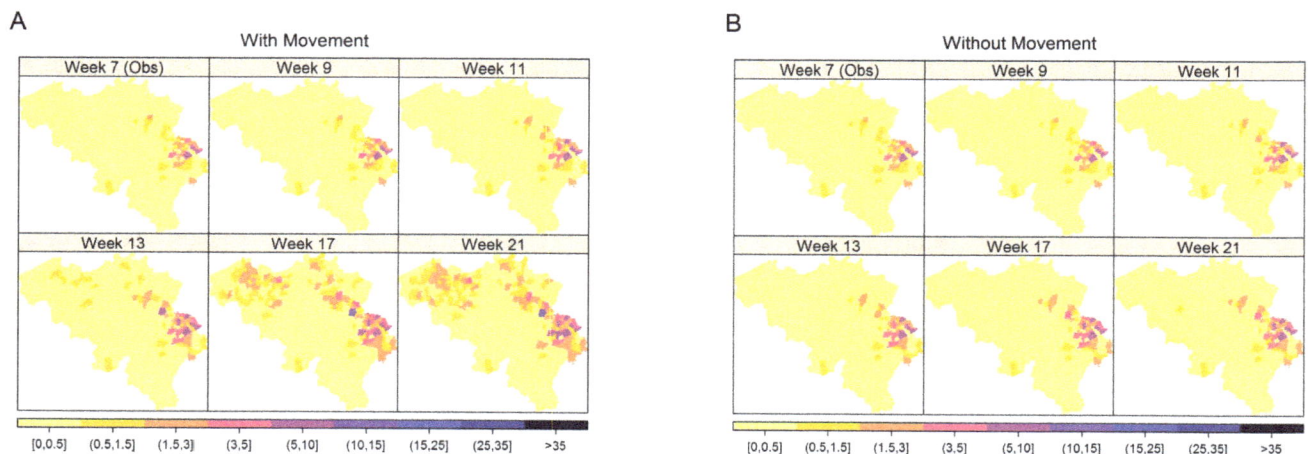

Figure 8. Predicted spatial trend of the 2006 outbreak. The maps show the median cumulative number of infected farms with (A) and without (B) animal movements based on 1000 stochastic simulations.

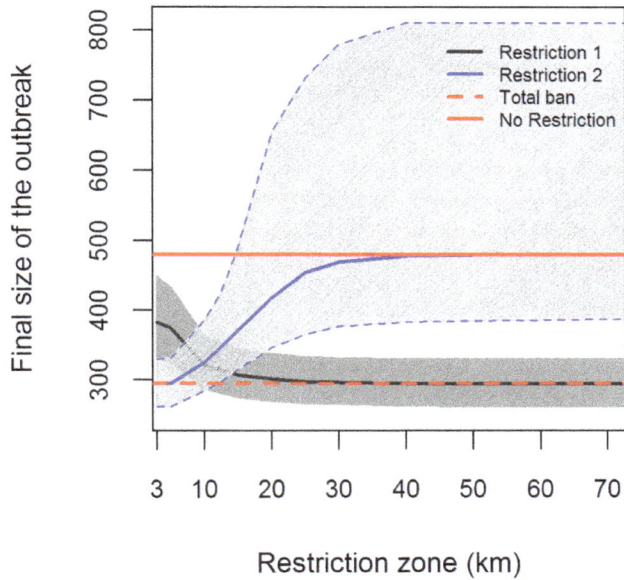

Figure 9. Final outbreak size as a function of movement restriction radius. Restriction 1 denotes movement restriction within the zone, while in restriction 2, movement within the zone is allowed. Values were based on 1000 stochastic simulations from the reduced spatio-temporal model. Data until week 6 of the 2006 outbreak was used and the model was allowed to predict the rest of the outbreak period.

observed values, while the prediction two-weeks ahead of time is far from the observed value in some weeks. The final outbreak size prediction shows that the model prediction starts to stabilize from week 15 of the 2007 outbreak (using data from 2006 until week 15 of 2007 outbreak).

Discussion

This paper presents a modelling framework for the transmission of BTV-8 across different farms in Belgium for 2006. Results from the model fit suggests that temperature, precipitation, farm density, land area, proportion of pasture and urban areas relative

to crop area are important for describing the BT dynamics. Municipality-specific covariates explain the varying level of susceptibility of the different municipalities while the increase and decrease in transmission is explained by the temperature and precipitation values in the preceding 4 weeks. These results are not surprising given the fact that BTV is transmitted via a vector which thrives on certain climatic conditions, specifically, high temperature values [13] and high precipitation level [29], [30]. A risk factor which would have been interesting to include is the number of wind events during the outbreak. Hendrickx et al. [31] and Faes et al. [10] have found that wind was a significant contributor to the spread of the infection. But due to the unavailability of the data, it was left out of the model.

Fitting the spatio-temporal model also allowed the BTV-8 transmission process to be divided into different components: background, within-municipality, between-municipality, and movement transmission. The background transmission quantifies the inherent susceptibility of a municipality to bluetongue infection where the different covariate effects (except the intercept) does not depend on the previous infection status of the municipality. This measures the susceptibility of the municipalities at the beginning of the outbreak, and as such it is an important component of the model. The within-municipality transmission quantifies the susceptibility of municipalities to local spread of BTV given that infection has already been detected at the previous week in the municipality. The two possible routes for a municipality to contract BTV as stipulated in the model is through between-municipality transmission and movement of an infected animal. The between-municipality transmission quantifies the influence of the infection status of neighbouring municipalities (those with shared borders). This happens even though BTV-8 is non-contagious since presence of infection in neighbouring municipalities implies that a vector with the virus might be present, travel to the neighbouring municipality and may bite the animal in that municipality, causing the transmission of BTV. The neighbourhood assumption based on contiguous regions was deemed appropriate, given the fact that although *Culicoides* midges can be dispersed by the wind to great distances, dispersal over land follow a hopping pattern, i.e. with intermediary stops [31], and with the midges being able to fly only a maximum distance of 2 km [21]. The transmission through movements, on the other hand, quantifies the effect of animal transports in the transmission of

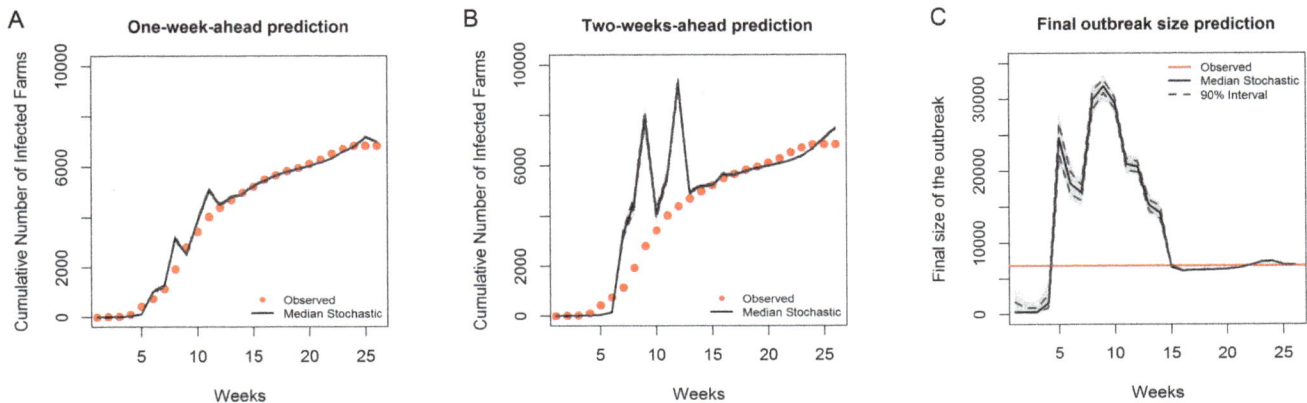

Figure 10. One-week and two-weeks-ahead prediction and the predicted final size of the 2007 BTV-8 outbreak in Belgium based on the model fitted to various time points. The weekly cumulative number of infected cases in (A) is the prediction at $t+1$ while (B) is the predicted cases at $t+2$, (C) on the other hand, is the predicted final outbreak size. Predictions are based on the model fitted to the 2006 outbreak data until week t of 2007. It was assumed that at the beginning of the 2007 outbreak, all farms are susceptible again.

BTV which allows us to study the effect of movement restrictions. Animal transport was quantified as a binary matrix weight signifying presence or absence of transport between municipalities. A constant general weight matrix was used. A weekly movement matrix was considered (data not shown) but offered no improvement over the constant assumption. The spatio-temporal model could be simplified by combining the within- and between-municipality components as one component representing the overall transmission by the vector thereby reducing the model components from four to three. However, as it is expected that *Culicoides* midges would cause more transmission in shorter distances (within-municipality) as compared to longer distances (between-municipality), we opted to keep the two components separate.

To investigate the effect of movement restriction on the spread of BTV, a one-step-ahead deterministic prediction and long-term stochastic prediction based on the model was performed with and without movement. For the one-step-ahead deterministic prediction, predicted weekly values were based on the previous week observed data and the parameter estimates from the model fitted to the whole data. The disadvantage of this procedure is the absence of uncertainty estimates especially for the weekly outbreak size. Stochastic prediction obtained by generating observations from a binomial distribution based on the predicted probabilities allowed the quantification of uncertainty for the predicted values. For this paper, parameter estimates from the model and data until the 7th week of outbreak were used as an initial condition and the model was allowed to predict the rest of the outbreak period. This coincides around the time period that the whole country of Belgium was declared as BTV-8 infected which resulted to the lifting of livestock transport restrictions [13].

Deterministic predictions showed a significant contribution of movement to the BTV outbreak at the end of 2006. Movement restriction would result to 200 fewer farms infected with BTV-8. Stochastic predictions also showed that movement restriction resulted only in local spreading and no infection in the West and East Flanders. This is an important result since it implies that movement of cattle caused the introduction of BTV-8 to these Flemish areas. De Koeijer et al. [13] in fact pointed out that the lifting of movement restrictions in Belgium resulted in long distance transmission and spatial pattern of transmission that was different from that of the Netherlands and Germany. Furthermore, simulating the effect of targeted restriction, specifically, restricting movement within a certain radius of the observed infection showed that it is effective in reducing the outbreak size. In fact, results have suggested that up to a certain radius around the infected farms, 10–15 km in this case, the movement restriction is as effective as the total ban all over Belgium in reducing the outbreak size. This was also observed by Turner et al. [14] for the BTV in England. This finding is important especially in guidelines formulation since a small restriction zone (e.g. 15 km) would lead to less adverse economic impact to the cattle industry than a larger zone (e.g. 70 km) or a total movement ban [32].

The model was validated with the 2007 BTV outbreak in Belgium. However, using the estimates from the model fitted to the 2006 outbreak gives a completely different predicted pattern. This might be due to the different nature of the outbreak in the two years. In 2007, BTV-8 was already present the year before and the model does not take this into account since it was built on the 2006 data before which infection had not previously occurred. Furthermore, reporting biases might have had an effect on the results and we did not take this into account. An ideal approach would have been to validate the model on data from other countries like the Netherlands and Germany which also experienced the BTV outbreak for the first time in 2006. Since these data from other countries are not available, a different validation approach was done, which was based on refitting the model to the new data each week and predicting one-week and two-weeks-ahead and the final size of the outbreak. Simulations show that the model performs well in terms of short-term prediction, but does not perform well in the long-term. However, as more data become available, the model was able to adapt to the new outbreak.

This study establishes the importance of movement restriction in reducing the outbreak size and preventing the long-distance transmission of BTV. This study also showed how to estimate different effects (local, neighbourhood and long-distance effects) in order to understand more what is happening during the outbreak. It also showed the importance of proper guidelines, especially in terms of the size of the restriction zones, in the reduction of outbreak size. It would be interesting to see an application of this model to other livestock diseases such as the recently discovered Schmallenberg virus [33] and to see the interplay of each component in the spreading of the virus.

Supporting Information

Figure S1 Spatial structure of the yearly total cattle purchases per municipality in Belgium for 2005–2009. (TIF)

Figure S2 Prediction of the 2007 BTV-8 outbreak in Belgium based on the model fitted to the 2006 data. The weekly number of predicted cases (A) and the cumulative number of cases (B and C) is based on 1000 stochastic predictions done using data until week 6 of the 2007 outbreak (July 01– August 05, 2007) and the model was then allowed to predict the outbreak until the end of 2007. The gray lines are the predictions from 1000 simulations. (TIF)

Acknowledgments

We thank the the the referees and the associate editor for the comments raised, which greatly improved the manuscript.

Author Contributions

Analyzed the data: CE MA CF. Wrote the paper: CE MA SW YVdS CF.

References

1. FAO (2001) The state of food and agriculture. Rome: Food and Agriculture Organization of the United Nations. Available: http://www.fao.org/docrep/003/x9800e/x9800e00.htm. Accessed 24 Sep 2013.

2. (2007) Commission Regulation (EC) No 1266/2007 of 26 October 2007 on implementing rules for Council Directive 2000/75/EC as regards the control, monitoring, surveillance and restrictions on movements of certain animals of susceptible species in relation to bluetongue; OJ L 283/37.

3. Févre E, de C Bronsvoorta B, Hamilton K, Cleaveland S (2006) Animal movements and the spread of infectious diseases. Trends in Microbiology 14: 125–131.

4. Volkova V, Howey R, Savill N,Woolhouse M (2010) Sheep movement networks and the transmission of infectious diseases. PLoS ONE 5(6): e11185.

5. Schrodle B, Held L, Rue H (2012) Assessing the impact of a movement network on the spatiotemporal spread of infectious diseases. Biometrics 68(3): 736–744.

6. Mintiens K, Méroc E, Mellor P, Staubach C, Gerbier G, et al. (2008) Possible routes of introduction of bluetongue serotype 8 virus into the epicentre of the 2006 epidemic in north-western europe. Prev Vet Med 87: 131–144.

7. Szmaragd C, Wilson A, Carpente S, Wood J, Mellor P, et al. (2009) A modeling framework to describe the transmission of bluetongue virus within and between farms in great britain. PLoS ONE 4(11): e7741.

8. Saegerman C, Berkvens D, Mellor P (2008) Bluetongue epidemiology in the European Union. Emerg Infect Dis 14: 539–544.

9. Gubbins S, Carpenter S, Baylis M, Wood J, Mellor P (2008) Assessing the risk of bluetongue to UK livestock: uncertainty and sensitivity analyses of a temperature-dependent model for the basic reproduction number. J R Soc Interface 5(20): 363–371.

10. Faes C, van der Stede Y, Guis H, Staubach C, Ducheyne E, et al. (2013) Factors affecting bluetongue serotype 8 spread in northern europe in 2006: The geographical epidemiology. Prev Vet Med 110(2): 149–158.

11. Pioz M, Guis H, Crespin L, Gay E, Calavas D, et al. (2012) Why did bluetongue spread the way it did? environmental factors influencing the velocity of bluetongue virus serotype 8 epizootic wave in france. PLoS ONE 7(8): e43360.

12. Purse B, Brown H, Harrup L, Mertens P, Rogers D (2008) Invasion of bluetongue and other orbivirus infections into europe: the role of biological and climatic processes. Rev Sci Tech 27(2): 427–442.

13. de Koeijer A, Boender G, Nodelijk G, Staubach C, Meroc E, et al. (2011) Quantitative analysis of transmission parameters for bluetongue virus serotype 8 in Western Europe in 2006. Veterinary Research 42: 53.

14. Turner J, Bowers R, Baylis M (2012) Modelling bluetongue virus transmission between farms using animal and vector movements. Sci Rep 2: 319.

15. Purse B, Baylis M, Tatem A, Rogers D, Mellor P, et al. (2004) Predicting the risk of bluetongue through time: climate models of temporal patterns of outbreaks in Israel. Rev sci tech Off int Epiz 23(3): 761–775.

16. Aitchison J (1986) The Statistical Analysis of Compositional Data. London, UK: Chapman & Hall.

17. Wilson A, Mellor P (2008) Bluetongue in europe: vectors, epidemiology and climate change. Parasitology Research 103(1): 69–77.

18. Ward M, Thurmond M (1995) Climatic factors associated with risk of seroconversion of cattle to bluetongue viruses in queensland. Prev Vet Med 24(2): 129–136.

19. Peeters A (2010). Country pasture/forage resource profile for belgium. Available: http://www.fao.org/ag/agp/AGPC/doc/Counprof/Belgium/belgium.htm. Accessed 24 Sep 2013.

20. Diekmann O, Heesterbeek J (2000) Mathematical Methodology of Infectious Diseases: Model Building, Analysis and Interpretation. Chichester, UK: John Wiley & Sons.

21. Singer R, MacLachlan N, Carpenter T (2001) Maximal predicted duration of viremia in bluetongue virus-infected cattle. J Vet Diag Invest 13: 43–49.

22. Vandenbussche F, Vanbinst T, Verheyden B, Van Dessel W, Demeestere L, et al. (2012) Evaluation of antibody-ELISA and real-time RT-PCR for the diagnosis and profiling of bluetongue virus serotype 8 during the epidemic in Belgium in 2006. Vet Microbiol 129(1–2): 15–27.

23. Di Gialleonardo L, Migliaccio P, Teodori L, Savini G (2011) The length of btv-8 viraemia in cattle according to infection doses and diagnostic techniques. Research in Veterinary Science 91(2): 316–320.

24. Held L, Hohle M, Hofmann M (2005) A statistical framework for the analysis of multivariate infectious disease surveillance counts. Stat Model 5: 187–199.

25. Knorr-Held L, Richardson S (2003) A hierarchical model for spacetime surveillance data on meningococcal disease incidence. Appl Stat 52: 169–183.

26. Hooten M, Anderson J, Waller L (2010) Assessing north american influenza dynamincs with a statistical sirs model. Spatial and Spatio-temporal Epidemiology 1: 177–185.

27. Bivand R, Pebesma E, Gmez-Rubio V (2008) Applied Spatial Data Analysis with R. New York, USA: Springer.

28. Wilson A, Mellor P (2009) Bluetongue in europe: past, present and future. Phil Trans R Soc B 364: 2669–2681.

29. Blackwell A (1997) Diel flight periodicity of the biting midge *Culicoides impunctatus* and the effects of meteorological conditions. Medical and Veterinary Entomology 11(4): 361–367.

30. Brugger K, Rubel F (2013) Bluetongue disease risk assessment based on observed and projected *Culicoides obsoletus* spp. vector densities. PLoS ONE 8(4): e60330.

31. Hendrickx G, Gilbert M, Staubach C, Elbers A, Mintiens K, et al. (2008) A wind density model to quantify the airborne spread of culicoides species during north-western europe bluetongue epidemic, 2006. Prev Vet Med 87: 162–181.

32. Schley D, Gubbins S, Paton D (2009) Quantifying the risk of localised animal movement bans for foot-and-mouth disease. PLoS ONE 4(5): e5481.

33. Kupferschmidt K (2012). New animal virus takes northern europe by surprise. Available: http://news.sciencemag.org/sciencenow/2012/01/new-animal-virus-takes-northern-.html. Accessed 24 Sep 2013.

The Effect of Deltamethrin-treated Net Fencing around Cattle Enclosures on Outdoor-biting Mosquitoes in Kumasi, Ghana

Marta Ferreira Maia[1,2]*, **Ayimbire Abonuusum[3]**, **Lena Maria Lorenz[1]**, **Peter-Henning Clausen[5]**, **Burkhard Bauer[5]**, **Rolf Garms[4]**, **Thomas Kruppa[4]**

1 London School of Hygiene and Tropical Medicine, Department of Disease Control, London, United Kingdom, 2 Ifakara Health Institute, Bagamoyo, Pwani Region, United Republic of Tanzania, 3 Kumasi Centre for Collaborative Research in Tropical Medicine, Kumasi, Ghana, 4 Bernhard-Nocht Institute for Tropical Medicine, Hamburg, Germany, 5 Free University of Berlin, Faculty of Veterinary Medicine Institute for Parasitology and Tropical Veterinary Medicine, Berlin, Germany

Abstract

Classic vector control strategies target mosquitoes indoors as the main transmitters of malaria are indoor-biting and – resting mosquitoes. However, the intensive use of insecticide-treated bed-nets (ITNs) and indoor residual spraying have put selective pressure on mosquitoes to adapt in order to obtain human blood meals. Thus, early-evening and outdoor vector activity is becoming an increasing concern. This study assessed the effect of a deltamethrin-treated net (100 mg/m^2) attached to a one-meter high fence around outdoor cattle enclosures on the number of mosquitoes landing on humans. Mosquitoes were collected from four cattle enclosures: Pen A – with cattle and no net; B – with cattle and protected by an untreated net; C – with cattle and protected by a deltamethrin-treated net; D – no cattle and no net. A total of 3217 culicines and 1017 anophelines were collected, of which 388 were *Anopheles gambiae* and 629 *An. ziemanni*. In the absence of cattle nearly 3 times more *An. gambiae* (p<0.0001) landed on humans. The deltamethrin-treated net significantly reduced (nearly three-fold, p<0.0001) culicine landings inside enclosures. The sporozoite rate of the zoophilic *An. ziemanni*, known to be a secondary malaria vector, was as high as that of the most competent vector *An. gambiae*; raising the potential of zoophilic species as secondary malaria vectors. After deployment of the ITNs a deltamethrin persistence of 9 months was observed despite exposure to African weather conditions. The outdoor use of ITNs resulted in a significant reduction of host-seeking culicines inside enclosures. Further studies investigating the effectiveness and spatial repellence of ITNs around other outdoor sites, such as bars and cooking areas, as well as their direct effect on vector-borne disease transmission are needed to evaluate its potential as an appropriate outdoor vector control tool for rural Africa.

Editor: Clive Shiff, Johns Hopkins University, United States of America

Funding: The study was funded by Vestergaard Frandsen Co. (www.vestergaard-frandsen.com). The funders had no role in study design, data collection and analysis, decision to publish, or preparation of the manuscript.

Competing Interests: Flight expenses to Kumasi, construction of the animal enclosures and provision of the experimental nets was sponsored by the Vestergaard Frandsen.

* E-mail: Marta.Maia@lshtm.ac.uk

Introduction

Within the last decade, great advances have been made in the fight against vector borne diseases with malaria decreasing considerably in sub-Saharan Africa [1,2,3,4]. However, despite well-planned vector control programs, diagnostics and artemisinin-combination therapies (ACTs), malaria transmission continues to persist in some settings. A large-scale integrated malaria control intervention was implemented on Bioko Island, a small, contained island situated off the coast of Equatorial Guinea, to eliminate the disease [5]. However, contrary to the classical behaviour pattern of *Anopheles gambiae* sensu stricto (s.s.), a large proportion of malaria vectors were found biting outdoors, and malaria elimination failed [6]. This has also been observed in other settings suggesting that the application of vector control tools that strictly and aggressively target mosquitoes biting and resting indoors may increasingly shift malaria transmission loci from in-to outdoors [7,8]. This may occur due to genotype bottlenecking of individuals within a species under selection from insecticidal pressure from Insecticide-Treated Nets (ITNs) and Indoor Residual Spraying (IRS), favouring those vectors that adapt to outdoor or early biting, or through species replacement. Recently, a cryptic subgroup of wholly exophilic *An. gambiae* was found in West Africa, reminding the scientific community on how little is known, or how much is assumed, on the diversity and behaviour of this species complex [9]. In addition, several regions have reported an increase in malaria transmission by secondary malaria vectors [10,11]. Most secondary malaria vectors are known to feed outdoors and preferably on animals, this way eluding indoor vector control interventions. Their zoophagy allows them to find refuge in cattle and maintain vectorial fitness despite interventions. There is only limited knowledge of the dynamics of vectorial systems and their contribution to malaria transmission after the implementation of large-scale interventions targeting indoor feeding vectors. Local and up-to-date information on vector systems is essential to understand malaria epidemiology and design appropriate vector control programs.

In order to tackle malaria elimination and the elimination of other neglected tropical diseases such as filariasis and dengue, new innovative outdoor vector control tools must be developed. One of the only methods of protection against outdoor-biting mosquitoes are topical repellents [12]. These require regular compliance by the user and offer only individual protection. In contrast, spatial or area repellents create a protective area by volatilizing repellent or low-dose insecticides into the air, thus enabling the protection of several individuals within an area. Mosquito coils are the classical example of a spatial repellent; pyrethroids are dispersed into the air through slow volatilization when the coil is lit. Similarly to topical repellents, this intervention requires nightly compliance as well as regular purchasing. Alternatively, pyrethroid-treated nets, besides killing mosquitoes, also repel them [13,14,15]. One of the most successful vector control interventions in history was IRS using DDT, which also works as a spatial repellent by deterring mosquitoes from sprayed households. The use of spatial repellency combined with insecticidal toxicity could reduce insect pressure outdoors if a pyrethroid-treated net is placed strategically at locations where people gather in the evening, for example a public venue or an outdoor domestic area.

In the present study, cattle enclosures were used to simulate a highly attractive outdoor area to measure whether the outdoor application of a pyrethroid-treated net 1) reduces mosquito densities within the enclosures, and 2) reduces mosquitoes in its vicinity through spatial repellence. In addition, cattle can be used to divert mosquitoes from humans to cows (zooprophylaxis). This has been applied with variable success in the past, with some cases resulting in increased mosquito densities (zoopotentiation) rather than zooprophylaxis, as the availability of blood meals increases the probability of mosquito survival and fecundity [16,17,18,19]. However, the combination of zooprophylaxis and chemical control may present a valuable strategy for controlling outdoor biting mosquitoes [20,21]. Studies have shown that by regularly treating cattle with insecticidal pour-on, mosquito densities and disease incidence can be reduced [22]. Additionally, usage of pour-on solutions increases the cost-effectiveness of animal production for farmers through control of ticks and other harmful insects [23,24,25]. The use of ITNs around cattle stables reduces nuisance and biting flies attacking enclosed animals by 80% compared to controls [26]. This new method of deploying ITNs may provide a low-cost alternative to insecticide-treated cattle (ITC) and become a further feasible vector-control tool in rural sub-Saharan Africa.

Methods

The study took place in Kumasi, Ghana at the Boadi Cattle Research Farm of the Kwame Nkrumah University for Science and Technology (KNUST). The study was conducted for six weeks during the months of October and November 2005. This period corresponded to the end of the rainy season; mosquito breeding sites were identified in the area prior to study initiation. Four similar sites located 500 m from each other were chosen, within which four identical cattle pens were built (Figure 1). The pen dimensions were 6×7 meters; the floor consisted of concrete and half of the pen was roofed with corrugated iron sheets. All pens were fenced with 1 m-high chicken wire. Pens were randomly assigned a treatment:

– Pen A – No netting protection and occupied by two zebu No netting and no animals.

– Pen B – Protected by an untreated net attached to the chicken wire fence (1 meter height) and occupied by two zebu.

Figure 1. Experimental animal enclosure at Boadi Cattle Research Farm, KNUST.

– Pen C – Protected by the same but deltamethrin-treated net – 100 mg/m^2 – attached to the chicken wire fence (1 meter height) and occupied by two zebu.

– Pen D –No netting protection and occupied by two zebu.

In all pens except pen D, two zebu bulls of comparable size and colour were introduced and kept in a zero-grazing system. These were weekly rotated to prevent biases due to differences in individual attraction. The netting material consisted of a black color, polyester fiber with a mesh width of 2×2 mm (Vestergaard-Frandsen Lausanne, Switzerland). The manufacturer incorporated an UV protection factor in the treated net to prevent the decay of deltamethrin by exposure to sunlight.

During 6 weeks, human landing catches (HLCs) were performed once a week within and 20 m away from each pen. Sixteen volunteers were divided into two shifts and performed HLCs from 18:00 to 06:00. Landing rates were recorded for each hour. An on-site weather station measured precipitation and minimum and maximum temperature during each collection night. Caught mosquitoes were taken to the Kumasi Centre for Collaborative Research (KCCR) where they were morphologically identified to culicines and anophelines. The latter were identified to species using morphological keys [27]. All the collected *Anopheles* females were examined for the presence of circumsporozoite *Plasmodium falciparum* antigen using head and thorax with enzyme-linked immunosorbent assay (ELISA)[28]. Sporozoite rates (SR) were calculated per location by dividing the number of infected mosquitoes by the total of mosquitoes caught each month. Daily entomological inoculation rates (EIR) were calculated for each study site by multiplying the sporozoite rate by the average human biting rate per night.

Samples from the deltamethrin-treated net in pen C were regularly taken for a period of nine months after project start and submitted to bioassays using lab-reared *Aedes aegypti*. Persistence of insecticidal activity in the treated net exposed to outdoor conditions was tested. Bioassays were conducted by lining the inside of a small cylindrical container with experimental net (5 cm diameter×10 cm height). The control container was lined with non-impregnated net. All net samples were tested twice; 30 mosquitoes were inserted into the container through a small hole and submitted to 10 seconds contact with the net. The mosquitoes were then released into a large cage and monitored for knockdown and mortality after 5 minutes, 10 minutes, 15 minutes, 6 hours and 24 hours.

Data analysis was performed using package 'lme4' [29] in R [30]. The total numbers of female mosquitoes caught by HLC were analysed separately for the three groups of mosquitoes (culicines, *An. gambiae* and *An. ziemanni*) with Poisson-lognormal mixed-effects models to account for over-dispersion in the count data. Model estimated means (MEM) were calculated using this model and used for comparison between pens. Treatment was set as the fixed effect and experimental day was set as the random effect. The reference pen was considered to be the unprotected pen with cattle (pen A). Comparison between pens A (cattle) and D (no cattle) showed whether zooprophylaxis was occurring. Comparisons between pens A (no net), B (untreated net), and C (treated net), respectively showed the effect of a deltamethrin-treated net on the density of host-seeking mosquitoes.. Separate analyses were performed for mosquitoes caught 1) indoors and 2) 20 m apart from the enclosures in order to be able to differentiate spatial repellence effects.

Ethical approval was obtained from Kumasi Centre of Collaborative Research Institutional Review Board (KCCR–IRB), certificate number: KCCR/IRB/063/11. Participants were enrolled on verbal informed consent as approved by the KCCR-IRB, this was chosen given the fact that most volunteers did not speak English and could not read and write. Following the IRB approved protocol a meeting was held with all participants in their local language explaining the project outline and participatory risk. The meeting was documented with the signature of all participants and of two witnesses. Diagnosis and treatment were offered during the entire period of the project and one month after conclusion, though none of the volunteers became sick during or at least four weeks after the catching period.

Results

Human landing catches collected 3217 culicines and 1017 anophelines of which 388 were *An. gambiae* and 629 *An. ziemanni*. Other studies conducted in the area attest that the only *An. gambiae* complex species present in Kumasi is *An. gambiae* s.s. [31,32]. In addition, PCR analysis of mosquitoes collected from the same study site a few months later confirmed the sole presence of *An. gambiae* s.s. (Abonuusum, unpublished). *Anopheles gambiae* showed a biting peak between 02:00 and 04:00 compared to *An. ziemanni*, which fed mostly between 00:00 and 02:00. The cumulative proportion of human landings between 18:00 and 22:00 was 9.3% and 18.3% for *An. gambiae* and *An. ziemanni* respectively. The presence of cattle reduced the number of human-host-seeking *An. gambiae* inside the enclosures by 66% (p<0.0001) (Table 1). The same was not observed for *An. ziemanni* or culicines. Cattle presence did not influence the number of mosquitoes caught 20 m from the enclosures (Table 2).

The HLCs performed inside the pen surrounded by the untreated net collected the highest number of *An. ziemanni* (N = 57;   MEM = 7.85;   p = 0.0257) and culicines (N = 514; MEM = 80.73; p = 0.0035; Table 1) compared to the reference pen that contained animals and was not protected by any type of netting. The model estimated mean of culicine human landings was 38% higher inside the pen (p = 0.0035) and 38% higher outside the pen (p = 0.0005; Table 1 and 2) surrounded by an untreated net than when the pen had no netting protection. On the other hand, the insecticide-treated net considerably reduced the density of host seeking culicines by 61% (N = 128; MEM = 19.84; p<0.0001) compared to the pen with cattle and no protection (Table 1). There was no effect of ITN use on

Table 1. Total number of mosquitoes, model estimated means (MEM) and 95% confidence intervals (95%CI) of HLC collections performed inside all experimental pens as well as sporozoite rates (SR) and entomological inoculation rates (EIR).

Species and location	N	MEM[1]	95% CI	p value	SR%	EIR[2]
Anopheles gambiae						
Pen A – Cattle, no net [3]	42	5.97	(3.24–10.97)	–	0.00	0.0
Pen B –Cattle, untreated net	51	6.84	(3.93–11.90)	0.629	1.96	0.1
Pen C – Cattle, treated net	28	3.76	(2.04–6.92)	0.138	3.57	0.1
Pen D – No Cattle, no net	121	17.54	(10.56–29.10)	<0.0001 ***	2.48	0.4
Anopheles ziemanni						
Pen A – Cattle, no net [3]	28	3.65	(1.89–7.04)	–	0.00	0.0
Pen B –Cattle, untreated net	57	7.85	(4.00–15.39)	0.0257 *	0.00	0.0
Pen C – Cattle, treated net	32	4.39	(2.15–8.97)	0.614	3.13	0.1
Pen D – No Cattle, no net	18	2.56	(1.18–5.56)	0.370	0.00	0.0
Culicines						
Pen A – Cattle, no net [3]	348	50.43	(35.64–71.36)	–	–	–
Pen B – Cattle, untreated net	514	80.73	(58.88–110.68)	0.0035 **	–	–
Pen C – Cattle, treated net	128	19.84	(13.96–28.19)	<0.0001 ***	–	–
Pen D – No Cattle, no net	379	60.66	(44.07–83.49)	0.257	–	–

[1]Model estimated mean.
[2]Daily EIR – number of infected bites per person per night.
[3]Reference Pen is A – no net and occupied by two zebu bulls.

mosquitoes collected 20 m from the pens, regardless of mosquito species (Table 2).

Sporozoite incrimination revealed infection in both *An. gambiae* and *An. ziemanni* with sporozoite rates (SR) ranging from 0 to 3.57% (mean = 2.02%) in *An. gambiae* and 0 to 3.13% in *An. ziemanni* (mean = 1.25%) (Table 1 and 2). The daily EIRs were calculated for both species. *Anopheles ziemanni* scored the highest EIR with 0.6 infective bites per person per night compared to 0.4 for *An. gambiae* (Table 1 and 2).

Bioassays using the ITN deployed for the study were conducted using lab-reared *Aedes aegypti*. More than 80% of the mosquitoes exposed to the treated nets were paralyzed after 15 minutes, with the exception of the net sampled 9 months later, where only 40% were paralyzed within 15 minutes (Table 3). However, even after 9 months of exposure, more than 80% of the mosquitoes were paralyzed after 6 hours. The percentage of recovering mosquitoes after 24 hours remained below 5% up to 9 months, after which it increased to 20%.

Table 2. Total number of mosquitoes, model estimated means (MEM) and 95% confidence intervals (95% CI) of HLC collections performed 20 m apart from all experimental pens as well as sporozoite rates (SR) and entomological inoculation rates (EIR).

Species and location	N	MEM[1]	95%CI	p value	SR%	EIR[2]
Anopheles gambiae						
Pen A – Cattle, no net [3]	45	5.83	(2.98–11.40)	–	2.20	0.1
Pen B –Cattle, untreated net	37	4.85	(2.53–9.30)	0.579	2.70	0.1
Pen C – Cattle, treated net	33	4.30	(2.22–8.33)	0.366	0.00	0.00
Pen D – No Cattle, no net	31	4.15	(2.13–8.09)	0.319	3.23	0.1
Anopheles ziemanni						
Pen A – Cattle, no net [3]	117	14.70	(8.15–26.51)	–	1.71	0.3
Pen B –Cattle, untreated net	125	17.00	(8.54–33.84)	0.680	1.60	0.3
Pen C – Cattle, treated net	77	10.01	(4.93–20.32)	0.287	1.30	0.1
Pen D – No Cattle, no net	175	25.08	(12.73–49.42)	0.123	2.29	0.6
Culicines						
Pen A – Cattle, no net [3]	405	63.31	(47.59–84.24)	–	–	–
Pen B –Cattle, untreated net	656	102.16	(78.10–133.62)	0.0005 ***	–	–
Pen C – Cattle, treated net	321	52.09	(39.40–68.86)	0.171	–	–
Pen D – No Cattle, no net	520	82.24	(62.70–107.87)	0.059	–	–

[1]Model estimated mean.
[2]Daily EIR – number of infected bites per person per night.
[3]Reference Pen is A – no net and occupied by two zebu bulls.

Discussion

In the present study, the presence of cattle reduced the *An. gambiae* human biting rate inside the pens by 66% (p<0.0001), indicating that either mosquitoes were diverted from humans to cattle or *An. gambiae* were repelled by the cattle odours [33]. On the other hand, the animal presence attracted higher densities of zoophilic mosquitoes like *An. ziemanni* and diverse culicines, which then had the choice of a human or animal blood meal. The number of host-seeking culicines inside the pen protected by the untreated net was significantly higher than in the unprotected pen (p = 0.0035), the same was observed for *Anopheles ziemanni* (p = 0.0257). The untreated net might have acted as a physical barrier hindering mosquitoes that entered the pen from leaving. The ITN surrounding the cattle enclosure provided an added protection from nuisance mosquitoes, resulting in 61% fewer host seeking culicine mosquitoes (p<0.001) and 37% fewer *An. gambiae* (p = 0.138) inside the enclosures. Results for anophelines were not significant, probably because the number of *Anopheles* mosquitoes being caught was low and the study was therefore underpowered for malaria vectors. However, our results indicate that the introduction of ITN fences around animal enclosures could result in fewer host-seeking mosquitoes inside the enclosure. The insecticidal effect of deltamethrin remained effective for the study period of nine months following deployment despite the net material being exposed to weathering and sunlight. Twenty-four hours post-exposure to the treated net that had been in the field for nine months yielded 84% knockdown of tested *Aedes aegypti* mosquitoes. Further experiments should be devised to test the longevity of the insecticidal activity of this netting material against lab-reared malaria vectors. Though it is expected that results will be encouraging since *Aedes aegypti* are a very robust mosquito specimen, usually needing a stronger stimulus for knockdown than Anophelines.

Outdoor transmission of vector borne diseases is an increasingly important problem that needs to be addressed in order to reach disease elimination [6,7]. Traditional methods such as IRS and ITNs target mainly indoor-biting vectors and neglect outdoor-biting mosquitoes, for which insufficient tools are available. In Pakistan, the usefulness of zooprophylaxis as a malaria vector control tool was studied in a community-randomised clinical trial. Regular treatment of cattle with deltamethrin pour-on resulted in over 50% reduction of clinical malaria episodes [22]. The major disadvantages of ITC (insecticide-treated cattle) are the necessary regular re-treatments, reliance on compliance and the constant community-wise investment. Additionally, mistreatments, such as substituting pour-on solutions with old motor oil and household disinfectants, could jeopardize successful vector control [25]. In Kenya, ITNs were used to protect enclosed cattle from tsetse flies, which successfully reduced animal morbidity. As an additional benefit, villagers also reported fewer mosquitoes in their homes [34]. More recently tsetse flies were successfully controlled in the Eastern Region of Ghana [35] The use of ITNs surrounding cattle enclosures would provide a less costly alternative to ITC and, since the insecticidal activity lasted over nine months, a long-lasting tool for vector control. However, its application would require that cattle were either kept enclosed in zero-grazing units or summoned indoors in the evenings, potentially limiting the application of

Table 3. Percentage of active *Aedes aegypti* following 10 seconds exposure to the treated net samples collected from the field after 5 minutes, 10 minutes, 15 minutes, 6 hours and 24 hours.

Time after exposure	Control	2 months	5 months	7 months	8 months	9 months
5 min	98	97	94	97	94	99
10 min	98	61	49	17	47	83
15 min	98	6	0	2	19	59
6 h	97	0	0	0	2	16
24 h	93	0	4	0	2	16

insecticide-treated net fencing to locations where husbandry practices permit.

At the community level, a proportion of the mosquitoes seeking blood meals could be diverted from the cattle enclosures to humans due to a repellent effect of deltamethrin-treated nets. So far only one study has attempted to measure the diversion of mosquitoes from ITCs to humans [36]. However the methodology employed did not investigate the possibility of the treated-cattle acting as a source of spatial repellence since the distance between HLC and cattle was only two meters [33]. Spatial repellence of mosquito coils and ITCs works by exuding a chemical barrier that protects not only the cows from mosquitoes, but also everything else within a certain radius. Thus, it is impossible to evaluate mosquito diversion if measurements are made only at short distances from the repellent source. There is an urgent need to evaluate whether spatial repellents result in mosquito diversion to non-users. In this study, the number of mosquitoes caught 20 m away from the cattle enclosure did not differ between pens without a net or with ITNs. Therefore, no spatial repellence was measured in Kumasi. The 20 m distance chosen in this study may have been too far away from the cattle enclosures for this ITN brand. The protective radius of deltamethrin-treated nets (100 mg/m^2) is as yet unknown, so a range of distances from five to 20 meters for mosquito catches should be tested in further experiments. In addition, here, treatments were fixed to each location and not rotated. Future studies must consider this source of spatial variation and both replicate and rotate treatments between sites.

Many malaria elimination scenarios around Africa are confronted with the increase in secondary malaria vectors or species replacement due to selective pressure from the existing malaria control tools. In some areas, *An. arabiensis* have replaced the classical main vectors, *An. gambiae* s.s. and *An. funestus*, and are now considered the major malaria vectors [7,37]. *Anopheles arabiensis* presents highly variable feeding behaviour dependant on climate and location: it has been shown to feed on blood meals both from humans and animals, with early evening and late-evening habits [7,38,39,40]; these characteristics make it more difficult to control through indoor-targeted interventions. In southern Zambia, despite high bed-net coverage *An. arabiensis* remained highly anthropophilic with less than 6% of analysed blood-meals being taken from animal origin [41], proving that the behaviour of the local *An. arabiensis* precluded malaria control through ITN usage. In the present study, *An. ziemanni*, a zoophilic mosquito [42], was caught in very high numbers on human landings. Host seeking *An. ziemanni* were mostly caught outside the animal pens, which is consistent with the outdoor-biting behaviour of this species [42]. However, *An. ziemanni* has not been known to readily feed on humans as was observed in this study. A considerable number of

An. ziemanni were collected outside the enclosure that did not contain animals, indicating that *An. ziemanni* were probably attracted to the humans despite the absence of cattle odour cues. High numbers of *An. ziemanni* were also collected on humans sitting outside pens containing animals; in this case the mosquito had been given the choice between an animal or human blood meal and chose to bite the human host. In addition, sporozoite incrimination revealed surprisingly high sporozoite rates and EIR for this species (Table 2). The cumulative proportion of *An. ziemanni* bites from dusk until 22:00 was 18.3%; during this period people will not yet have retired to the protection of their bed net and might be exposed to the activity of outdoor-biting malaria vectors. In areas where people spend most of their evening time outdoors and cattle are present, *An. ziemanni* might be responsible for a much higher proportion of malaria transmission than previously expected through zoopotentiation. As such, these findings provide further evidence towards the notion that changes are occurring in malaria transmission in sub-Saharan Africa and call attention to new potential secondary vectors.

Ideally, ITN fences should be evaluated inside a semi-field system, where safe and controlled conditions allow a high-throughput method to measure spatial repellence, feeding inhibition, protective distance and delayed mortality. Other compounds or compound combinations aside from deltamethrin should also be evaluated, such as transfluthrin, metofluthrin or actellic. This experiment would grant a holistic perception on the impact of ITN fencing on host-seeking malaria vectors and other mosquito species. The results arising from such an evaluation would form the foundation to further studies investigating the efficacy of ITNs around outdoor sites, such as bars and cooking areas. Studies should be performed for longer periods of time to achieve sufficient power as well over different seasons of the year. If successful this intervention could present an innovative and appropriate tool against outdoor-biting and outdoor-resting disease vectors in rural Africa.

Acknowledgments

We kindly acknowledge all collaborators from the KNUST (Kwame Nkrumah University) as well as the KCCR (Kumasi Centre for Collaborative Research). In addition we wish kindly thank all the workers of the Boadi Cattle Research Farm who assisted us throughout the project.

Author Contributions

Conceived and designed the experiments: MFM AA BB PHC RG TK. Performed the experiments: MFM AA. Analyzed the data: MFM LL. Contributed reagents/materials/analysis tools: LL BB RG PHC TK. Wrote the paper: MFM LL.

References

1. O'Meara WP, Mangeni JN, Steketee RW, Greenwood B (2010) Changes in the burden of malaria in sub-Saharan Africa. Lancet Infect Dis 10: 545–555.

2. D'Acremont V, Lengeler C, Genton B (2010) Reduction in the proportion of fevers associated with *Plasmodium falciparum* parasitaemia in Africa: a systematic review. Mal J 9: 240.

3. Ceesay SJ, Casals-Pascual C, Nwakanma DC, Walther M, Gomez-Escobar N, et al. (2010) Continued decline of malaria in The Gambia with implications for elimination. PLoS ONE 5: e12242.

4. Steketee RW, Campbell CC (2010) Impact of national malaria control scale-up programmes in Africa: magnitude and attribution of effects. Mal J 9.

5. Kleinschmidt I, Torrez M, Schwabe C, Benavente L, Seocharan I, et al. (2007) Factors influencing the effectiveness of malaria control in Bioko Island, Equatorial Guinea. Am J Trop Med Hyg 76: 1027–1032.

6. Reddy M, Overgaard H, Abaga S, Reddy VP, Caccone A, et al. (2011) Outdoor host seeking behaviour of *Anopheles gambiae* mosquitoes following initiation of malaria vector control on Bioko Island, Equatorial Guinea. Mal J 10: 184.

7. Russell TL, Govella NJ, Azizi S, Drakeley CJ, Kachur SP, et al. (2011) Increased proportions of outdoor feeding among residual malaria vector populations

following increased use of insecticide-treated nets in rural Tanzania. Mal J 10: 80.

8. Geissbuhler Y, Chaki P, Emidi B, Govella NJ, Shirima R, et al. (2007) Interdependence of domestic malaria prevention measures and mosquito-human interactions in urban Dar es Salaam, Tanzania. Mal J 6: 126.

9. Riehle MM, Guelbeogo WM, Gneme A, Eiglmeier K, Holm I, et al. (2011) A cryptic subgroup of *Anopheles gambiae* is highly susceptible to human malaria parasites. Science 331: 596–598.

10. Antonio-Nkondjio C, Kerah CH, Simard F, Awono-Ambene P, Chouaibou M, et al. (2006) Complexity of the malaria vectorial system in Cameroon: contribution of secondary vectors to malaria transmission. J Med Entomol 43: 1215–1221.

11. Manh CD, Beebe NW, Van NT, Quang TL, Lein CT, et al. (2010) Vectors and malaria transmission in deforested rural communities in north-central Vietnam. Mal J 9.

12. Hill N, Lenglet A, Arnez AM, Carneiro I (2007) Randomised, double-blind control trial of p-menthane diol repellent against malaria in Bolivia. BMJ 55.

13. Arredondo-Jimenez JI, Rodriguez MH, Loyola EG, Bown DN (1997) Behaviour of *Anopheles albimanus* in relation to pyrethroid-treated bednets. Med Vet Entomol 11: 87–94.

14. Asidi AN, N'Guessan R, Hutchinson RA, Traore-Lamizana M, Carnevale P, et al. (2004) Experimental hut comparisons of nets treated with carbamate or pyrethroid insecticides, washed or unwashed, against pyrethroid-resistant mosquitoes. Med Vet Entomol 18: 134–140.

15. Darriet F, N'Guessan R, Koffi AA, Konan LY, Doannio J, et al. (2000) Impact of the resistance to pyrethroids on the efficacy of impregnated bednets used as a means of prevention against malaria: results of the evaluation carried out with deltamethrin SC in experimental huts.. Bull Soc Pathol Exot 93: 131–134.

16. Bogh C, Clarke SE, Pinder M, Sanyang F, Lindsay SW (2001) Effect of passive zooprophylaxis on malaria transmission in The Gambia. J Med Entomol 38: 822–828.

17. Bouma M, Rowland M (1995) Failure of passive zooprophylaxis: cattle ownership in Pakistan is associated with a higher prevalence of malaria. Trans R Soc Trop Med Hyg 89: 351–353.

18. Hewitt S, Kamal M, Muhammad N, Rowland M (1994) An entomological investigation of the likely impact of cattle ownership on malaria in an Afghan refugee camp in the North West Frontier Province of Pakistan. Med Vet Entomol 8: 160–164.

19. Bogh C, Clarke SE, Walraven GE, Lindsay SW (2002) Zooprophylaxis, artefact or reality? A paired-cohort study of the effect of passive zooprophylaxis on malaria in The Gambia. Trans R Soc Trop Med Hyg 96: 593–596.

20. Saul A (2003) Zooprophylaxis or zoopotentiation: the outcome of introducing animals on vector transmission is highly dependent on the mosquito mortality while searching. Mal J 2: 32.

21. Kawaguchi I, Sasaki A, Mogi M (2004) Combining zooprophylaxis and insecticide spraying: a malaria-control strategy limiting the development of insecticide resistance in vector mosquitoes. Proc Biol Sci 271: 301–309.

22. Rowland M, Durrani N, Kenward M, Mohammed N, Urahman H, et al. (2001) Control of malaria in Pakistan by applying deltamethrin insecticide to cattle: a community-randomised trial. Lancet 357: 1837–1841.

23. Bauer B, Amsler-Delafosse S, Clausen PH, Kabore I, Petrich-Bauer J (1995) Successful application of deltamethrin pour on to cattle in a campaign against tsetse flies (Glossina spp.) in the pastoral zone of Samorogouan, Burkina Faso. Trop Med Parasitol 46: 183–189.

24. Bauer B, Kabore I, Liebisch A, Meyer F, Petrich-Bauer J (1992) Simultaneous control of ticks and tsetse flies in Satiri, Burkina Faso, by the use of flumethrin pour-on for cattle. Trop Med Parasitol 43: 41–46.

25. Hlatshwayo M, Mbati PA (2005) A survey of tick control methods used by resource-poor farmers in the Qwa-Qwa area of the eastern Free State Province, South Africa. Onderstepoort J Vet Res 72: 245–249.

26. Maia MF, Clausen PH, Mehlitz D, Garms R, Bauer B (2010) Protection of confined cattle against biting and nuisance flies (Muscidae: Diptera) with insecticide-treated nets in the Ghanaian forest zone at Kumasi. Parasitol Res 106: 1307–1313.

27. Gillies MT, Coetzee M (1987) A Supplement to the Anophelinae of Africa South of the Sahara (Afrotropical region). Publications of the South African Institute for Medical Research 55: 143.

28. Wirtz R (1987) Comparative testing of monoclonal antibodies against *Plasmodium falciparum* sporozoites for ELISA development. Bull World Health Org: 39–45.

29. Bates D, Maechler M, Bolker B (2011) lme4: Linear mixed-effects models using S4 classes. R package version 0999375-39. R project website. Available at: http://CRAN.R-project.org/package = lme4. Accessed 2012 August 15.

30. R Development Core Team website (2011) R: A language and environment for statistical computing. R Foundation for Statistical Computing, Vienna. Available at: http://www.R-project.org/. Accessed 2012 August 15.

31. Afrane YA, Klinkenberg E, Drechsel P, Owusu-Daaku K, Garms R, et al. (2004) Does irrigated agriculture influence the transmission of malaria in the city of Kumasi, Ghana? Acta Trop 89: 125–134.

32. Abonuusum AG, Owusu-Daako K, Tannich E, May J, Garms R, et al. (2011) Malaria transmission in two rural communities in the forest zone of Ghana. Parasitol Res: 1465–1471.

33. Pates HV, Takken W, Stuke K, Curtis CF (2001) Differential behaviour of *Anopheles gambiae* sensu stricto (Diptera: Culicidae) to human and cow odours in the laboratory. Bull Entomol Res 91: 289–296.

34. Bauer B, Gitau D, Oloo FP, Karanja SM (2006) Evaluation of a preliminary title to protect zero-grazed dairy cattle with insecticide-treated mosquito netting in western Kenya. Trop Anim Health Prod 38: 29–34.

35. Bauer B, Holzgrefe B, Mahama CI, Baumann MP, Mehlitz D, et al. (2011) Managing tsetse transmitted trypanosomosis by insecticide treated nets–an affordable and sustainable method for resource poor pig farmers in Ghana. PLoS Negl Trop Dis 5: e1343.

36. Hewitt S, Rowland M (1999) Control of zoophilic malaria vectors by applying pyrethroid insecticides to cattle. Trop Med Int Health 4: 481–486.

37. Kitau J, Oxborough RM, Tungu PK, Matowo J, Malima RC, et al. (2012) Species Shifts in the *Anopheles gambiae* Complex: Do LLINs Successfully Control *Anopheles arabiensis*? PLoS ONE 7: e31481.

38. Mahande A, Mosha F, Mahande J, Kweka E (2007) Feeding and resting behaviour of malaria vector, *Anopheles arabiensis* with reference to zooprophylaxis. Mal J 6: 100.

39. Githeko AK, Adungo NI, Karanja DM, Hawley WA, Vulule JM, et al. (1996) Some observations on the biting behavior of *Anopheles gambiae* s.s., *Anopheles arabiensis*, and *Anopheles funestus* and their implications for malaria control. Exp Parasitol 82: 306–315.

40. Tirados I, Costantini C, Gibson G, Torr SJ (2006) Blood-feeding behaviour of the malarial mosquito *Anopheles arabiensis*: implications for vector control. Med Vet Entomol 20: 425–437.

41. Fornadel CM, Norris LC, Glass GE, Norris DE (2010) Analysis of *Anopheles arabiensis* blood feeding behaviour in southern Zambia during the two years after introduction of insecticide-treated bed nets. Am J Trop Med Hyg 83: 848–853.

42. Chandler JA, Boreham PF, Highton RB, Hill MN (1975) A study of the host selection patterns of the mosquitoes of the Kisumu area of Kenya. Trans R Soc Trop Med Hyg 69: 415–425.

Development of a Blocking ELISA Based on a Monoclonal Antibody against a Predominant Epitope in Non-Structural Protein 3B2 of Foot-and-Mouth Disease Virus for Differentiating Infected from Vaccinated Animals

Yuanfang Fu, Zengjun Lu*, Pinghua Li, Yimei Cao, Pu Sun, Meina Tian, Na Wang, Huifang Bao, Xingwen Bai, Dong Li, Yingli Chen, Zaixin Liu*

State Key Laboratory of Veterinary Etiological Biology, National Foot-and-Mouth Disease Reference Laboratory, Lanzhou Veterinary Research Institute, Chinese Academy of Agriculture Science, Lanzhou, Gansu, China

Abstract

A monoclonal antibody (McAb) against non-structural protein (NSP) 3B of foot-mouth-disease virus (FMDV) (3B4B1) was generated and shown to recognize a conserved epitope spanning amino acids 24–32 of 3B (GPYAGPMER) by peptide screening ELISA. This epitope was further shown to be a unique and predominant B cell epitope in 3B2, as sera from animals infected with different serotypes of FMDV blocked the ability of McAb 3B4B1 to bind to NSP 2C3AB. Also, a polyclonal antibody against NSP 2C was produced in a rabbit vaccinated with 2C epitope regions expressed in *E. coli*. Using McAb 3B4B1 and the 2C polyclonal antibody, a solid-phase blocking ELISA (SPB-ELISA) was developed for the detection of antibodies against NSP 2C3AB to distinguish FMDV-infected from vaccinated animals (DIVA test). The parameters for this SPB-ELISA were established by screening panels of sera of different origins. Serum samples with a percent inhibition (PI) greater than or equal to 46% were considered to be from infected animals, and a PI lower than 46% was considered to indicate a non-infected animal. This test showed a similar performance as the commercially available PrioCHECK NS ELISA. This is the first description of the conserved and predominant GPYAGPMER epitope of 3B and also the first report of a DIVA test for FMDV NSP 3B based on a McAb against this epitope.

Editor: Jagadeesh Bayry, Institut National de la Santé et de la Recherche Médicale U 872, France

Funding: This study was supported by a national "863" research project (grant no. 2012AA101304) and a research grant for public welfare from the General Bureau of Quality Supervision, Inspection, and Quarantine of the People's Republic of China (grant no. 201410061). The funders had no role in study design, data collection and analysis, decision to publish, or preparation of the manuscript.

Competing Interests: The authors have declared that no competing interests exist.

* Email: luzengjun@caas.cn (Z. Lu); liuzaixin@caas.cn (Z. Liu)

Introduction

Foot-and-mouth disease (FMD) is a highly contagious and economically devastating viral disease of cloven-hoofed animals. The causative agent, foot-and-mouth disease virus (FMDV), has a positive sense and single-stranded RNA genome of 8400 nucleotides that codes for 12 proteins. VP1–4 are structural proteins (SPs) that make up the capsid of the virus, which is the main component of the inactivated vaccine. L, 2A, 2B, 2C, 3A, 3B, 3C, and 3D are non-structural proteins (NSPs) that participate in viral replication and play additional roles within the host cell [1]. Differentiation of FMDV-infected animals from inactivated vaccine-inoculated animals (DIVA) mainly depends on the detection of antibodies against NSPs because most of these proteins are removed from the vaccine by some kind of purification method during production. Therefore, suboptimal vaccine purity may affect diagnostic specificity, as the presence of NSPs in some vaccine preparations may result in misclassification of animals that have been repeatedly vaccinated [2]. The FMDV antigen used to formulate the vaccine needs to be purified to reduce the NSP content so that the vaccine does not induce antibodies to NSPs. This effort is necessary in countries wishing to be recognized as FMD-free with vaccination. However, purification of the vaccine antigen increases the cost of production. Therefore, identification of conserved and immunodominant B cell epitopes in the NSPs of FMDV will provide a basis for the development of epitope-deleted marker vaccines and corresponding DIVA tests, which will significantly improve the efficiency of disease control programs.

Numerous FMDV-specific linear B cell epitopes have been identified in FMDV NSPs, including in 2B, 2C, 3A, 3B, and 3D [3]. Several DIVA tests have been developed based on detection of antibodies against these predominant B cell epitopes. One research group identified a linear B cell epitope (CELHEKVSSHPIFKQ) in 2C and developed a synthetic peptide ELISA based on this epitope [4]; they subsequently developed an epitope-blocking ELISA for DIVA diagnosis that uses a monoclonal antibody (McAb) targeting the core repeat epitope QKPLK in NSP 3B [5].

A recent study also described a competition ELISA using a McAb against an immunodominant linear epitope (MRKTKLAPT-VAHGVF) in NSP 3D [6]. The same group then mutated the above epitope in 3D and the core repeat epitope QKPLK in 3B to develop a negative marker vaccine with improved DIVA capabilities [7]. The FMDV NSP 3B exists as three similar but non-identical tandem repeats, 3B1 (VPg1), 3B2 (VPg2), and 3B3 (VPg3), that share a high amino acid sequence homology and a common core amino acid motif [QKPL (M) K] [3]. These proteins were found to have a high density of linear B cell epitopes, some of which, including GPYAGPMERQKPLK, PMERQKPLKVKAKA, and QKPLKVKAKAPVVK, are conserved and immunodominant and could potentially be used to develop DIVA tests [8].

In this study, we identified a McAb, 3B4B1, that recognizes a conserved and immunodominant epitope (^{24}GPYAGPMER^{32}) located in 3B2 of NSP 3B. Because 3B2 can be deleted without significantly affecting the replication of FMDV [9], we developed a solid-phase blocking ELISA (SPB-ELISA) using 3B4B1 as a competitive antibody and 2C3AB NSP produced in *E. coli* as the antigen [10]. This research provides a useful tool for DIVA diagnosis of FMD and a potential companion DIVA test for a negative marker vaccine with deleted 3B2.

Materials and Methods

Ethics statement

Before the beginning of the experiment, all animals were acclimatized for one week. Rabbits and mice for preparation of antibodies were bred in clean and spacious animal rooms. Swine, bovine and sheep for FMDV infection were raised in bio-safety level 3 (BSL-3) containment facility in Lanzhou Veterinary Research Institute (LVRI). All animals were handled humanely according to the rules described by the Animal Ethics Procedures and Guidelines of the People's Republic of China, and the study was approved by the Animal Ethics Committee of LVRI, Chinese Academy of Agricultural Sciences (Permit No. LVRIAEC2010-006). All animals used in the present study were humanely bred and bled. Swine, bovine and sheep were euthanized by exsanguination under deep anesthesia (intramuscular injection of chlorpromazine at 2–6 mg/kg) at the end of the experiment.

Preparation of Recombinant Proteins

The expression and purification of the NSPs 3A, 3B, 2C epitope region, 3D, 3ABC, and 2C3AB of FMDV were carried out according to previously described methods [10,11].

Production of Polyclonal Antibody against 2C Epitope Regions

The purified 2C epitope region was emulsified in Montanide ISA 206 adjuvant (Seppic, Paris, France) and used to immunize rabbits to produce specific polyclonal antibodies. The rabbit was immunized hypodermically with 200 μg/0.5 ml of 2C epitope protein vaccine three times at 2 week intervals. One week after the third injection, rabbits were bled to collect the sera. The 2C antibody titers were determined by an indirect ELISA using 2C3AB as the coating antigen. The polyclonal antibodies against 2C epitope region protein were purified from the sera of the immunized rabbits using an Affi-Gel protein G column (GE healthcare, OH, USA) and stored at −20°C for later use.

Production of Monoclonal Antibody (McAb) against 3B

McAbs were produced according to traditional protocols [12]. Briefly, female BALB/c mice were immunized three times with 100 μg of purified 3B protein emulsified in Montanide ISA 206 adjuvant (Seppic, Paris, France) at 2 week intervals. After 2 weeks, the mice were boosted intraperitoneally with 100 μg of purified protein without adjuvant. The mice were subsequently sacrificed 3 days after the final vaccination. The spleen cells of mice were fused with SP2/0 cells. After 2 weeks, supernatants from the hybridomas were screened using recombinant 3B in an indirect ELISA. Horseradish peroxidase (HRP)-conjugated McAb was prepared with the EZ-Link Plus Activated Peroxidase kit (Thermo, USA). The reactivity of McAbs to different recombinant NSPs was determined by indirect ELISA according to the procedure described previously [11].

Peptide ELISA Screening for Binding Epitopes of the McAbs

Six peptides were synthesized by Genscript Inc. (Nanjin, China) based on the complete amino acid sequence of 3B from FMDV O/CHA/99 (Table 1). The purity of these peptides was determined by HPLC to be ≥90%. The peptide ELISA was performed according to the method described by Hohlich et al. [3] to analyze the binding epitopes of different McAbs against 3B. Briefly, microtiter plates (Corning, Salt Lake City, USA) were coated with synthetic peptides. After blocking, hybridoma supernatants containing the McAbs were added to each plate. After washing, HRP-conjugated goat anti-mouse IgG (Sigma, USA) was added and TMB (3, 3', 5, 5'-tetramethyl-benzidine) substrate (SurModics, USA) was used for color development. The optical density (OD) was measured at 450 nm using an automated plate reader (Bio-Rad, USA).

Table 1. Synthetic peptides used to identify the FMDV-specific B cell epitope recognized by the McAbs.

Peptide no.	Location	Amino acid sequence
Pep 1	1–14	GPYTGPLERQKPLK
Pep 2	10–23	QKPLKVRAKLPQQE
Pep 3	24–37	GPYAGPMERQKPLK
Pep 4	29–42	PMERQKPLKVKVKA
Pep 5	43–56	PVVKEGPYEGPVKK
Pep 6	57–71	PVALKVKAKNLIVTE

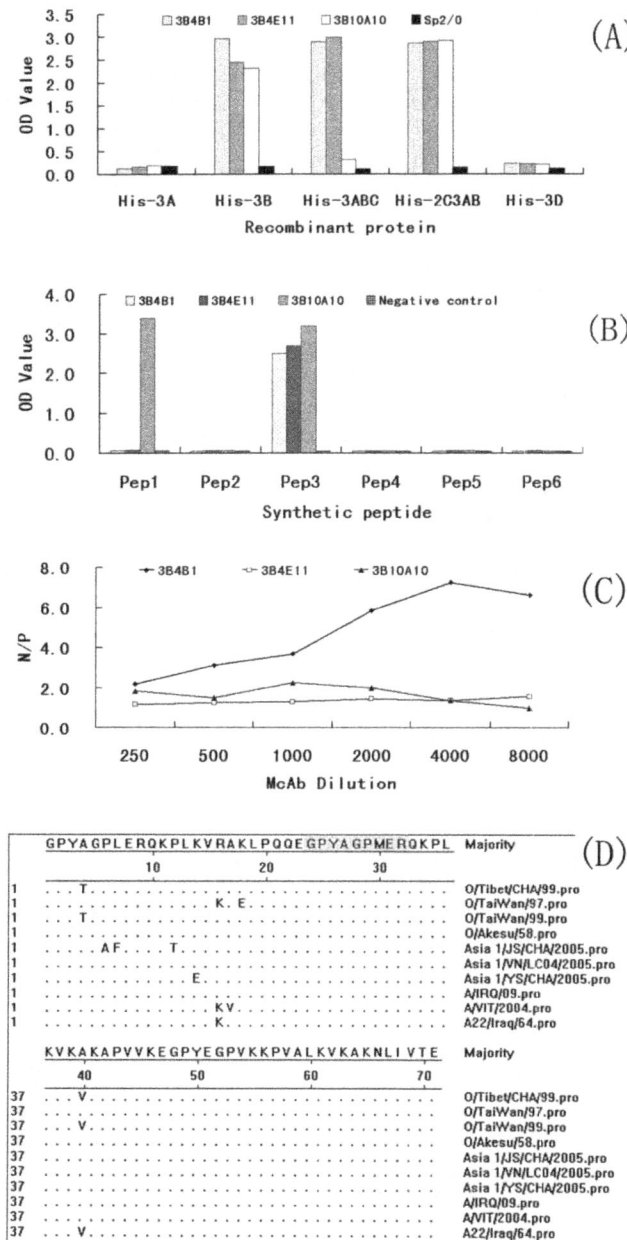

Figure 1. Characteristics of the monoclonal antibodies. (A) The reactivity of the McAbs with various recombinant non-structural proteins. (B) Identification of the epitopes recognized by the McAbs against NSP 3B by peptide ELISA. (C) The N/P values of the McAbs at different dilutions of serum. (D) Conservation of the motifs in FMDV NSP 3B.

Selection of the McAb Binding to a Native Epitope of NSP 3B

To determine whether or not the McAbs bound to a native epitope of NSP 3B, a blocking ELISA was performed by coating microtiter plates with recombinant NSP 2C3AB (3 μg/mL) in coating buffer overnight at 4°C. After three washes with PBST (0.01 M PBS and 0.05% Tween-20), test sera from FMDV-infected (positive control) or non-infected (negative control) cattle diluted 1:5 with serum dilution buffer (2.5 g/L casein and 10%

equine serum in PBS, pH 7.2) were added to the plate and incubated at 37°C for 1 h. After washing, serially diluted McAbs were added and incubated for 1 h at 37°C. After washing, an HRP-conjugated goat anti-mouse antibody was added to the wells and the plate was incubated at 37°C for 1 h. Color development and OD readings were done as described above. The ratios of OD values of negative to positive serum (N/P) were calculated. A greater N/P ratio indicates that sera from infected cattle have a greater ability to block the binding of the McAbs to recombinant FMDV NSP 2C3AB.

Conservation of B Cell Epitopes in 3B Protein Among Different FMDV Serotypes

The homology of 3B amino acid sequences was analyzed by alignment of sequences from different serotypes of FMDV. The selected reference virus strains and corresponding GenBank numbers were O/Tibet/CHA/99 (CAD62370.1), O/Akesu/58 (AAM44304.1), O/TAW/99 (CAD62208.1), O/TAW/99 (AAT01778.1), Asia 1/JS/CHA/2005 (ABM66095.1), Asia1/YS/CHA/2005 (ADU56664.1), Asia1/VN/LC04/2005 (ADC92544.1), A/IRQ/09 (AER28328.1), A/VIT/2004 (AEO16200.1), and A22/Iraq/64 (AAT01706.1). The amino acid sequences were aligned using DNASTAR Lasergene 7.1 software (DNASTAR, Inc., Madison, WI, USA).

Serum Samples of Different Origins

The following serum samples were used in this study: (1) sera from healthy, unvaccinated animals, including 152, 123, and 162 serum samples from cattle, sheep, and swine, respectively, which were found to have no antibodies (<1:4) against O or Asia 1 FMDV by liquid-phase blocking ELISA (LPBE); (2) sera from clinically healthy, vaccinated animals, including 24 serum samples from cattle and 48 serum samples from swine that had been inoculated twice with inactivated vaccine for type O/CHA/99 FMDV (PD$_{50}$>6.0); (3) sera from infected animals, including 121 serum samples from cattle infected with Asia 1/JS/CHA/05 or O/CHA/99 FMDV at 10–28 days postinfection (DPI), 40 serum samples from four cattle infected with A/WH/CHA/2009 FMDV at 0–229 DPI, 48 serum samples from two sheep infected with O/CHA/99 FMDV at 0–417 DPI, four serum samples from four sheep infected with Asia 1/JS/05 at 28 DPI, nine serum samples from nine pigs infected with Asia 1/JS/05 at 16–28 DPI, 38 serum samples from 38 pigs infected with O/CHA/99 at 11–60 DPI, and 16 serum samples from one pig infected with O/CHA/99 FMDV at 0–194 DPI; (4) field sera, including 200 serum samples collected from cattle in FMDV-endemic regions of China in 2009; and (5) control sera, including positive control serum samples derived from cattle, sheep, and swine infected with the O/CHA/99 strain of FMDV at 30 DPI, weakly positive serum samples derived from animals that developed a low-level antibody titer against both SP and NSP 3ABC 15 days after infection with Asia 1/JS/05 FMDV, and negative control serum samples derived from clinically healthy, unvaccinated animals.

Solid-Phase Blocking ELISA

The solid-phase blocking ELISA (SPB-ELISA) was performed according to the protocol of Sørensen et al. [13], with some modifications. Briefly, ELISA plates were coated with 2.0 μg/ml purified 2C polyclonal antibody diluted in carbonate buffer (pH 9.6) in a 100 μL volume and incubated overnight at 4°C. After three washes with PBST, the plates were sealed with 10 mg/ml gelatin and incubated for 45 min at 37°C. After three washes with PBST, purified 2C3AB protein was diluted to an optimal

Table 2. Specificity and sensitivity of the SPB-ELISA at different cutoff values.

Species	Cutoff value	Specificity Naive animals	Specificity Vaccinated animals	Sensitivity Infected animals
Cattle	40%	92.1% (140/152)	91.7% (22/24)	92.6% (112/121)
	46%	95.4% (145/152)	95.8% (23/24)	92.6% (112/121)
	50%	95.4% (145/152)	95.8% (23/24)	91.7% (111/121)
Swine	40%	97.5% (158/162)	97.9% (47/48)	96.8% (61/63)
	46%	98.8% (160/162)	100% (48/48)	95.2% (60/63)
	50%	98.8% (160/162)	100% (48/48)	93.7% (59/63)
Sheep	40%	100% (122/122)	–	94.2% (49/52)
	46%	100% (122/122)	–	94.2% (49/52)
	50%	100% (122/122)	–	94.2% (49/52)

concentration in PBST, 100 µl/well was added, and the plate was incubated for 60 min at 37°C. After five washes, serum samples were diluted 1:5 in dilution buffer (PBS containing 0.25% casein, 10% horse serum, and 3% *E. coli* lysate), 100 µl/well was added, and the plate was incubated on a plate rocker overnight at room temperature (20–25°C). Following washing, the optimal dilution of HRP-conjugated McAb in dilution buffer was added, and the plate was incubated at 37°C for 1 h. Color development and OD readings were done as described in section 2.4. The percent inhibition (PI) of the sample was derived according to the following formula: $PI = (1 - \text{test sample OD/negative control OD}) \times 100\%$.

Parameters of the SPB-ELISA

Parameters for the SPB-ELISA were established using panels of sera of different origins: sera from uninfected cattle (n = 152), sheep (n = 123), and swine (n = 162); and sera from infected cattle (n = 121), sheep (n = 52), and swine (n = 63). The PI for each sample was calculated. The cutoff PI value was selected based on the frequency distribution at different PIs to result in relatively high sensitivity and specificity. Diagnostic sensitivity was calculated as the proportion of the infected herd that tested positive, and diagnostic specificity was calculated as the proportion of vaccinated or non-vaccinated healthy animals that tested negative.

Validation of the SPB-ELISA

The performance of the SPB-ELISA was determined by comparison with a 3ABC ELISA [14] and the PrioCHECK NS ELISA (Prionics AG, Schlieren-Zurich, Switzerland) using a test panel of sera of different origins. Samples included 61 serum samples from cattle infected with O/CHA/99 or Asia 1/JS/05 FMDV at 10–28 DPI; 40 serum samples from swine infected with O/CHA/99 or Asia 1/JS/05 FMDV at 0–194 DPI; 41 serum samples from two sheep infected with O/CHA/99 FMDV at 0–417 DPI; 50 and 43 sera from healthy cattle and swine, respectively, that were determined to be antibody-negative against O and Asia 1 FMDV by LPBE; 24 serum samples from clinically healthy cattle aged 1–2 years; 48 serum samples from swine that had been vaccinated twice with inactivated vaccine, taken at 14 DPI; 200 serum samples from field cattle herds; 40 serum samples from four cattle infected experimentally with A/WH/CHA/2009 at 0–229 DPI; and 16 serum samples from one pig infected with O/CHA/99 at 0–194 DPI.

The 3ABC-ELISA antibody titer was expressed as the ratio of the experimental sample to the positive control. In this assay, a cutoff value of 0.2 has been reported to result in relatively high sensitivity and specificity [14]. The PrioCHECK NS ELISA (Prionics AG, Schlieren-Zurich, Switzerland) is a commercial blocking ELISA kit that can be applied to all susceptible animal species. In this test, the antibody titer was expressed as the PI compared with the negative control. A PI of 50% is the threshold for qualitative judgment of infection.

Results

Titer of the 2C Polyclonal Antibody

Polyclonal antibodies against the 2C protein were collected from the blood of immunized rabbits. The optimal concentration of the purified 2C polyclonal antibody for use as a coating antibody was determined to be 2.0 µg/ml in a capture ELISA using recombinant NSP 2C3AB as the antigen. This concentration was chosen for use in the SPB-ELISA.

Characteristics of McAbs

Three McAbs, 3B4B1, 3B4E11, and 3B10A10, were screened. McAbs 3B4B1 and 3B4E11 were classified as IgG1, and 3B10A10 was classified as IgG2b using a commercially available isotype classification kit. All three McAbs specifically bind to recombinant proteins His-3B, His-3ABC, and His-2C3AB, but not His-3A or His-3D, as determined by an indirect ELISA (Fig. 1A), indicating that the three McAbs recognize epitopes located within protein 3B. A peptide screening ELISA showed that 3B4B1 and 3B4E11 only react with peptide 3, not with other peptides containing the previously reported QKPLK core motif [7], indicating that 3B4B1 and 3B4E11 specifically bind to the [24]GPYAGPMER[32] epitope. McAb 3B10A10 reacted with peptide 3 and with peptide 1 (Fig. 1B), indicating that it might recognize a space epitope.

The McAb recognizing a native epitope in NSP 2C3AB was selected by using sera from FMDV-infected cattle to block the binding of the three McAbs to NSP 2C3AB, as described in the Materials and Methods. As shown in Figure 1C, only the N/P value of McAb 3B4B1 was greater than 2.1 at different serum dilutions, indicating that McAb 3B4B1 recognized the native epitope [24]GPYAGPMER[32] located in NSP 3B2.

Conservation of the Epitopes in FMDV 3B Protein

Sequence alignment showed that the epitope [1]GPYTGPLER[9] in 3B1 was not conserved among different serotypes of FMDV;

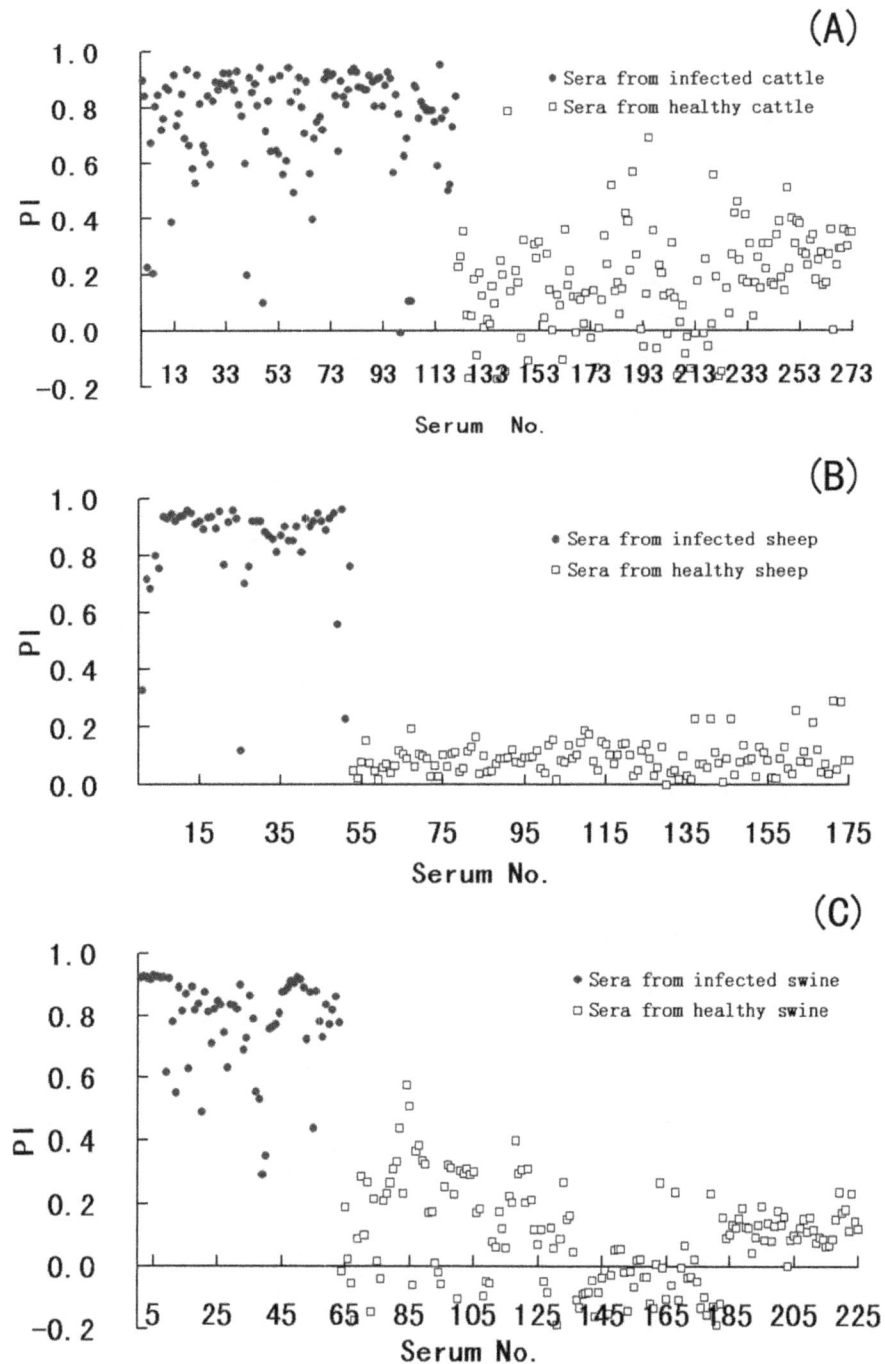

Figure 2. Frequency distribution of PI values obtained by SPB-ELISA. (A) Sera from cattle, (B) sera from sheep, and (C) sera from swine.

however, the epitope ^{24}GPYAGPMER^{32} in 3B2 was well-conserved in the O, A, and Asia 1 serotypes of FMDV (Fig. 1D).

The Cutoff Value, Sensitivity, and Specificity of the SPB-ELISA

The cutoff PI value for the SPB-ELISA was determined by testing 437 serum samples from healthy unvaccinated animals and 236 serum samples from infected animals. The frequency distributions of the resultant PI values are shown in Figure 2A–

C. Table 2 shows the specificity and sensitivity of the assay at three different cutoff values, 40%, 46%, and 50%. Using a cutoff value of 46%, the sensitivity and specificity of the assay were relatively high in the three animal species. Based on these data, serum samples with a PI equal to or greater than 46% were considered to be from infected animals, and samples with a PI lower than 46% were considered to be from non-infected animals. As a quality control measure taken in this assay, the PIs of the positive and

Table 3. Comparison of the concordance rates between the SPB-ELISA and the 3ABC-ELISA or the PrioCHECK FMDV-NS ELISA.

Origin of sera	Total no.	SPB-ELISA		3ABC-ELISA (no. of coincident)		PrioCHECK NSP (no. of coincident)	
		Positive	Negative	Positive	Negative	Positive	Negative
Infected cattle[a]	61	59	2	60 (56)	1 (0)	56 (55)	5 (0)
Infected swine[b]	40	36	4	36 (34)	4 (1)	37 (35)	3 (1)
Infected sheep[c]	41	39	2	35 (35)	6 (2)	38 (37)	3 (1)
Vaccinated cattle	24	0	24	0	24 (24)	/	/
Vaccinated swine	48	0	48	1 (0)	47 (47)	0	48 (48)
Non-infected cattle	50	0	50	0	50 (50)	0	50 (50)
Non-infected swine	43	0	43	0	43 (43)	0	43 (43)
Field cattle	200	22	178	19 (16)	181 (174)	17 (16)	183 (176)
Total no.	507	156	351	151 (141)	356 (341)	148 (143)	335 (319)

[a]Sera from cattle infected with O/CHA/99 or Asia 1/JS/05 FMDV at 10–28 DPI.
[b]Sera from swine infected with O/CHA/99 or Asia 1/JS/05 FMDV at 0–194 DPI.
[c]Sera from sheep infected with O/CHA/99 FMDV at 0–417 DPI.

weakly positive controls were required to be greater than 70% and 50%, respectively.

Using the cutoff value described above, specificities of 95.4% (145/152), 100% (123/123), and 98.8% (160/162) and sensitivities of 92.6% (112/121), 94.2% (49/52), and 95.2% (60/63) were obtained for non-infected and infected cattle, sheep, and swine sera, respectively. These results demonstrate that most animals infected with FMDV develop antibodies against the corresponding epitope of McAb 3B4B1. Thus, the McAb 3B4B1 binding epitope ^{24}GPYAGPMER^{32} is a predominant B cell epitope in NSP 3B2 of FMDV.

Comparison of the SPB-ELISA with two other ELISAs

As shown in Table 3, for 61 sera from infected cattle, 59, 60, and 56 samples were positive by SPB-ELISA, 3ABC-ELISA, and PrioCHECK NS ELISA, respectively. For 200 sera from field cattle, 22, 19, and 17 samples were positive by SPB-ELISA, 3ABC-ELISA, and PrioCHECK NS ELISA, respectively. For 40 sera from infected swine, 36, 36, and 37 samples were positive by SPB-ELISA, 3ABC-ELISA, and PrioCHECK FMDV-NS ELISA, respectively, and for 41 sera from infected sheep, 39, 35, and 38 samples were positive. For sera from uninfected animals, all three methods gave negative results, except the 3ABC-ELISA, which gave one positive result from a vaccinated swine. The coincident rates between the SPB-ELISA and the 3ABC-ELISA and between the SPB-ELISA and the PrioCHECK NS ELISA were 95.1% (482/507) and 95.7% (462/483), respectively.

Results from 97 sequential serum samples from four cattle (n = 40), two sheep (n = 41), and one swine (n = 16) infected with FMDV are shown in Figure 3A–G. Of the 40 serum samples from four infected cattle, 39 samples yielded identical results by the SPB-ELISA, 3ABC-ELISA, and PrioCHECK NS ELISA. Only one sample, taken from animal no. 4004 at 8 DPI, showed a positive result by SPB-ELISA (Fig. 3D). Of the 41 serum samples from two infected sheep, only one differential result, taken from animal no. 54 at 164 DPI, was observed between the SPB-ELISA and the PrioCHECK NS ELISA (Fig. 3E); however, differential results were observed between the 3ABC-ELISA and the other two ELISAs in four samples: the sample taken from sheep no. 40 at 164 DPI (Fig. 3F), and the samples taken from sheep no. 54 at 117, 132, and 194 DPI (Fig. 3E). Of the 16 sera from an infected swine, one negative result at 7 DPI by the SPB-ELISA, one negative result at 181 DPI by the PrioCHECK NS ELISA, and three negative results, at 22, 50, and 181 DPI, by the 3ABC-ELISA were observed (Fig. 3G).

Discussion

FMD is the most important viral infectious disease of livestock. Some ruminants can become virus carriers after infection even though a vaccination strategy has been adopted. Differentiation of naturally infected from vaccinated animals is necessary to evaluate the effects of disease control measures and to detect early signs of FMD incursion or transmission in epidemic regions. The development of a negative marker vaccine and its corresponding DIVA test will improve the efficacy of disease control strategies without the increased production costs of highly purified vaccines. Identification and screening of conserved and immunodominant epitopes are critically important for the development of negative marker vaccines and the corresponding DIVA tests. In this study, three McAbs against 3B NSP were prepared using prokaryotically expressed 3B protein. One McAb, 3B4B1, was determined to specifically bind to a conserved and immunodominant epitope of 3B2, ^{24}GPYAGPMER^{32}, which is distinct from the ^{1}GPYTG-

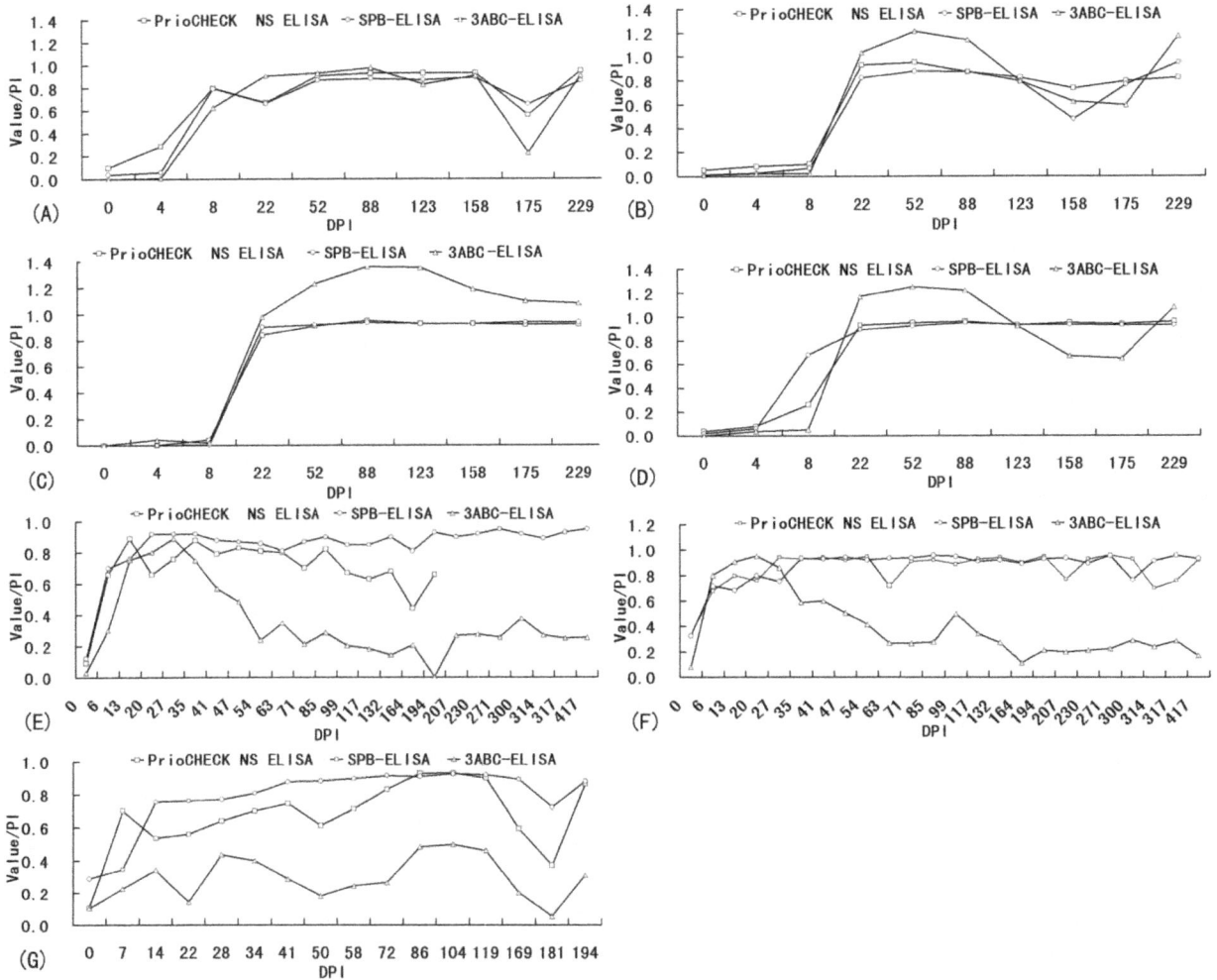

Figure 3. Detection of antibodies by SPB-ELISA and two commercial ELISA kits (3ABC-ELISA and PrioCHECK FMDV-NS ELISA) in sera collected sequentially from animals infected with FMDV. (A)–(D) show the results from cattle no. 4009, 0738, 4017, and 4004, respectively, (E) and (F) show the results from sheep no. 54 and 40, and (G) shows the result from pig no. 39.

PLER[14] epitope of 3B1 and the [48]GPYEGPVKK[56] epitope of 3B3. It seems that [27]A and [30]M are critical residues influencing the binding capacity of McAb 3B4B1 to the epitope [24]GPYAGP-MER[32]. Moreover, McAb 3B4B1 also bound to one native and immunodominant B cell epitope in NSP 2C3AB, and this binding capacity could be blocked by serum samples from FMDV-infected animals. A previous study indicated that 3B1, 3B2, and 3B3 tandem repeats share a common core motif, QKPL(M)K, which is the core motif of several B cell epitopes [3]. Animals infected with a genetically modified FMDV with a mutated core motif did not produce antibodies to the above motif [7], indicating that native B cell epitopes also exist within the core motif regions. Here, we show that [24]GPYAGPMER[32] in 3B2 is another unique and native B cell epitope that is well-conserved among different serotypes of FMDV and can be mutated to abolish binding by the corresponding McAb (unpublished data). These results provide more information on B cell epitopes in FMDV NSPs and provide a basis for the development of a negative marker vaccine against FMD.

To develop a blocking ELISA for the detection of an antibody response to the identified epitope of McAb 3B4B1, NSP 2C3AB protein, which contains the major B cell epitope regions of 2C and

the entire 3AB region of the NSP of FMDV [10,11], was used as the antigen. In addition, 2C polyclonal antibody produced in rabbits was used as a capture antibody, and McAb 3B4B1 was used as a detection antibody to develop a SPB-ELISA test for DIVA purposes. The cutoff PI value of this SPB-ELISA was determined to be 46% by screening panels of sera of different origins. Based on this threshold, the sensitivity of the assay was 92.6% in cattle, 95.2% in swine, and 94.2% in sheep, and the specificity for non-infected animals was 95.4% in cattle, 98.8% in swine, and 100% in sheep. When compared with two commercial ELISA kits, the 3ABC-ELISA and the PrioCHECK NS ELISA, the coincident rate of the SPB-ELISA with the PrioCHECK NS ELISA (95.7%) was higher than that of the SPB-ELISA with the 3ABC-ELISA (95.1%), indicating that the SPB-ELISA has a similar performance rate as the PrioCHECK NS ELISA and a higher specificity than the indirect 3ABC-ELISA.

Over the past 2 decades, detection of NSP antibodies has been accepted as a reliable method for evaluating the infectious status of animal herds with or without vaccination [13–17]. The accuracy of DIVA tests can be compromised by interference from NSPs remaining in the vaccine [16]. In South America, a combined system of an indirect ELISA (3ABC-ELISA) with an enzyme-linked immunoelectrotransfer blot assay was successfully used to support local FMD control programs under systematic vaccination [15]. However, development of a negative marker vaccine will provide an improved strategy to overcome the above difficulties.

In conclusion, a McAb target against the immunodominant epitope, [24]GPYAGPMER[32], located in 3B2, was identified and used to develop a DIVA test for FMDV surveillance. The ELISA described here is a potential companion DIVA method for an FMDV negative marker vaccine, which will greatly improve the efficacy of FMD control and eradication efforts in the future.

Author Contributions

Conceived and designed the experiments: YFF Z. Lu Z. Liu. Performed the experiments: YFF PHL Y. Cao MNT NW. Analyzed the data: YFF Z. Lu Z. Liu. Contributed reagents/materials/analysis tools: PS HFB XWB DL Y. Chen. Contributed to the writing of the manuscript: YFF Z. Lu.

References

1. Clavijo A, Wright P, Kitching P (2004) Developments in diagnostic techniques for differentiating infection from vaccination in foot-and-mouth disease. Vet J 167: 9–22.

2. Office International des Epizooties, OIE (2012) Foot and mouth disease. In: Standards Commission, O.I.E. (Ed), Manual of Standards for Diagnostic Tests and Vaccines. Office International des Epizooties, Paris, France (Chapter 2.1.5).

3. Höhlich BJ, Wiesmüller KH, Schlapp T, Haas B, Pfaff E, et al. (2003) Identification of foot-and-mouth disease virus-specific linear B-cell epitopes to differentiate between infected and vaccinated cattle. J Virol 77: 8633–9.

4. Oem JK, Kye SJ, Lee KN, Park JH, Kim YJ, et al. (2005) Development of synthetic peptide ELISA based on non-structural protein 2C of foot andmouth disease virus. J Vet Sci 6: 317–325.

5. Oem JK, Chang BS, Joo HD, Yang MY, Kim GJ, et al. (2007) Development of an epitope-blocking-enzyme-linked immunosorbent assay to differentiate between animals infected and vaccinated against foot-and-mouth disease virus. J Virol Methods 142: 174–81.

6. Yang M, Alfonso C, Li M, Kate H, Hilary H, et al. (2007) Identification of a major antibody binding epitope in the non-structural protein 3D of foot-and-mouth disease virus in cattle and the development of a monoclonal antibody with diagnostic applications. J Immunol Methods 321: 174–81.

7. Uddowla S, Hollister J, Pacheco JM, Rodriguez LL, Rieder E (2012) A safe foot-and-mouth disease vaccine platform with two negative markers for differentiating infected from vaccinated animals. J Virol 86: 11675–85.

8. Gao MC, Zhang RX, Li M, Li S, Cao YS, et al. (2012) An ELISA based on the repeated foot-and-mouth disease virus 3B epitope peptide can distinguish infected and vaccinated cattle. Appl Microbiol Biotechnol 93: 1271–9.

9. Arias A, Perales C, Escarmis C, Domingo E (2010) Deletion mutants of VPg reveal new cytopathology determinants in picornavirus. PloS ONE 5: e10735.

10. Lu ZJ, Zhang XL, Fu YF, Cao YM, Tian MN, et al. (2010) Expression of the major epitope regions of 2C integrated with the 3AB non-structural protein of foot-and-mouth disease virus and its potential for differentiating infected from vaccinated animals. J Virol Methods 170: 128–33.

11. Fu YF, Cao YM, Sun P, Bao HF, Bai XW, et al. (2011) Development of a dot immunoblot method for differentiation of animals infected with foot-and-mouth disease virus from vaccinated animals using non-structural proteins expressed prokaryotically. J Virol Methods 171: 234–40.

12. Earley EM, Rener JC (1985) Hybridoma technology I: murine monoclonal antibodies. J Tissue Cult Methods 9: 129–87.

13. Sørensen KJ, Stricker KD, Dyrting KC, Grazioli S, Haas B (2005) Differentiation of foot-and-mouth disease virus infected animals from vaccinated animals using a blocking ELISA based on baculovirus expressed FMDV 3ABC antigen and a 3ABC monoclonal antibody. Arch Virol 150: 805–14.

14. Lu ZJ, Cao YM, Guo JH, Qi SY, Li D, et al. (2007) Development and validation of 3ABC indirect ELISA for differentiation of foot-and-mouth disease virus infected from vaccinated animals. Vet Microbiol 125: 157–69.

15. Bergmann IE, Malirat V, Neitzert E, Panizzutti N, Sanchez C, et al. (2000) Improvement of a serodiagnostic strategy for foot-and-mouth disease virus surveillance in cattle under systematic vaccination: a combined system of an indirect ELISA-3ABC with an enzyme-linked immunoelectrotransfer blot assay. Arch Virol 145: 473–89.

16. Brocchi E, Bergmann IE, Dekker A, Paton DJ, Sammin DJ, et al. (2006). Comparative evaluation of six ELISAs for the detection of antibodies to the non-structural proteins of foot-and-mouth disease virus. Vaccine 24: 6966–79.

17. Paton DJ, Clercq KD, Greiner M, Dekker A, Brocchi E, et al. (2006) Application of non-structural protein antibody tests in substantiating freedom from foot-and-mouth disease virus infection after emergency vaccination of cattle. Vaccine 24: 6503–12.

13

Density of Wild Prey Modulates Lynx Kill Rates on Free-Ranging Domestic Sheep

John Odden*, Erlend B. Nilsen, John D. C. Linnell

Norwegian Institute for Nature Research, Trondheim, Norway

Abstract

Understanding the factors shaping the dynamics of carnivore–livestock conflicts is vital to facilitate large carnivore conservation in multi-use landscapes. We investigated how the density of their main wild prey, roe deer *Capreolus capreolus*, modulates individual Eurasian lynx *Lynx lynx* kill rates on free-ranging domestic sheep *Ovis aries* across a range of sheep and roe deer densities. Lynx kill rates on free-ranging domestic sheep were collected in south-eastern Norway from 1995 to 2011 along a gradient of different livestock and wild prey densities using VHF and GPS telemetry. We used zero-inflated negative binomial (ZINB) models including lynx sex, sheep density and an index of roe deer density as explanatory variables to model observed kill rates on sheep, and ranked the models based on their AICc values. The model including the effects of lynx sex and sheep density in the zero-inflation model and the effect of lynx sex and roe deer density in the negative binomial part received most support. Irrespective of sheep density and sex, we found the lowest sheep kill rates in areas with high densities of roe deer. As roe deer density decreased, males killed sheep at higher rates, and this pattern held for both high and low sheep densities. Similarly, females killed sheep at higher rates in areas with high densities of sheep and low densities of roe deer. However, when sheep densities were low females rarely killed sheep irrespective of roe deer density. Our quantification of depredation rates can be the first step towards establishing fairer compensation systems based on more accurate and area specific estimation of losses. This study demonstrates how we can use ecological theory to predict where losses of sheep will be greatest, and can be used to identify areas where mitigation measures are most likely to be needed.

Editor: Marco Festa-Bianchet, Université de Sherbrooke, Canada

Funding: At various stages of the study funding was provided by the Norwegian Ministry of Defence (www.regjeringen.no/nb/dep/fd.html), the Norwegian Directorate for Nature Management (www.dirnat.no), the Research Council of Norway (www.forskningsradet.no), the county governors' offices of Hedmark (www. fylkesmannen.no/Hedmark/), Akershus & Østfold (www.fylkesmannen.no/Oslo-og-Akershus/), and Buskerud (www.fylkesmannen.no/Buskerud/) counties as well as the Large Carnivore Management Boards for regions 2 and 3 and the Norwegian Institute for Nature Research (www.nina.no). All funding was provided based on detailed project descriptions submitted by the authors to the various funding mechanisms. Beyond this approval of the proposed study design and methodology, the funders had no input into the data collection, data analysis, interpretation of results, decision to publish, or preparation of this manuscript, with the exception of the occasional contribution to field work from game wardens of the State Nature Inspectorate (http://www.naturoppsyn.no/om-sno/), who are technically a unit of the Directorate for Nature Management.

Competing Interests: The authors have declared that no competing interests exist.

* E-mail: john.odden@nina.no

Introduction

Large carnivores live at low densities, and in many parts of the world their conservation is dependent on integrating carnivores into multi-use landscapes [1]. Predation on livestock is one of the major conflicts between humans and large carnivores, and depredation occurs where livestock overlap in distribution with large carnivores [2]. Livestock losses to carnivores are highly variable, both spatially and temporally. Among the factors contributing to this variation are the species and densities of predator and livestock, husbandry practises, size ratio between carnivores and livestock, landscape characteristics, and wild prey densities [2–5]. Understanding the dynamics of carnivore–livestock conflicts in such landscapes is vital to effectively achieve large carnivore conservation.

Among the factors that have received attention in previous studies is the effect that distribution and density of wild ungulate prey might have on livestock losses [6]. Previous studies have provided seemingly contradictory results. For example in Eurasian lynx *Lynx lynx* and wolf *Canis lupus* locally high densities of wild

prey can increase depredation [7–10], whereas other studies of wolves and coyotes have found the opposite [6,11–14]. We propose two models to explain these seemingly contradictory findings. The *attraction model* can explain elevated predation risk for livestock in areas with high densities of wild prey. The model proposes that locally high densities of wild prey will attract carnivores to these patches, and thus also induce elevated risk for livestock. This model assumes that carnivore habitat use is a function of wild prey densities, and that livestock are depredated if they are encountered by chance while the carnivore is searching for wild prey. To explain reduced predation on livestock in areas with high densities of wild prey, the *energetic model* predicts that regionally high densities of wild prey will reduce predation on livestock primarily because there is no need for predators to kill livestock to satisfy their energetic needs. An additional effect comes from the possibility that searching time is reduced at high wild prey density thus reducing the probability of chance encounters with domestic prey. Previous studies supporting the *energetic model* have been conducted by comparing losses in areas of different densities of wild prey [15–21] or through scat analysis [6,12–

14,22,23]. However, such results could also arise from differences in carnivore densities and are not an adequate test of the model.

The difference between the energetic and the attraction models may also simply be a matter of scale. In large areas (home range scales) with low densities of wild prey, carnivores may increase depredation on livestock to compensate for having few prey (thus supporting the *energetic model*), but in areas where wild prey is widely distributed and abundant, carnivores may spend most time in the most prey-rich patches, leading to high encounter rates, and therefore greater incidental depredation on livestock (thus supporting the *attraction model*). We have previously found support for the *attraction model* on a small scale [10]. To rigorously test the *energetic model*, we here use data on lynx kill rates on free-ranging domestic sheep from radio-collared lynx in south-eastern Norway, along a gradient of different livestock and wild prey densities.

Methods

Ethic Statement and Sampling

We captured 109 lynx between 1995 and 2011. Adult lynx and juveniles (>5 months) were captured in walk through box-traps, spring-loaded foot-snares, treed using trained dogs, or immobilized from cars and helicopters in a few cases. In addition, kittens were captured by hand in natal lairs. Lynx were immobilized with medetomidine-ketamine, following pre-established protocols [24–26]. All capture methods were constantly refined to minimise the risk of injury or death to the animals. In particular, the design and alarms of box traps and snares were modified to allow response time of less than 12 hours (average 5 hours), and 20 minutes, respectively, and a safety net was used when animals were treed with hounds. All capture and handling procedures were approved by the Norwegian Experimental Animal Ethics Committee and followed their ethical requirements for research on wild animals (permit numbers 2012/206992, 2010/161554, 2010/161563, 08/127430, 07/81885, 07/7883, 2004/48647, 201/01/641.5/FHB,

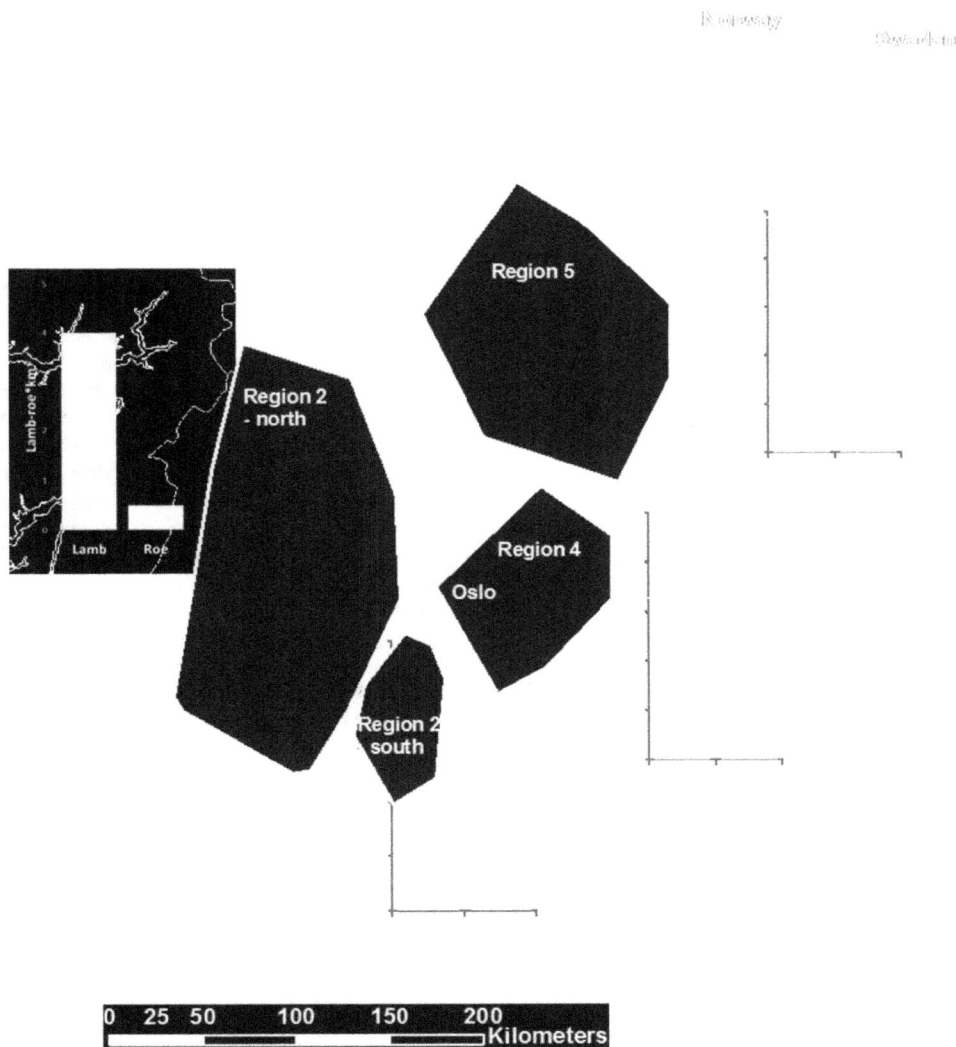

Figure 1. Map of the study area, showing the 4 areas of data collection of lynx (*Lynx lynx*) predation on sheep from 1995 to 2011.
The shaded areas indicate the four study areas, Region 5 (Hedmark County 1995–1999), Region 4 (Oslo, Akershus, Østfold Counties 2001–2006), Region 2 north (Buskerud, Telemark, and Oppland Counties 2006–2011), and Region 2 south (Buskerud, Telemark, and Vestfold Counties 2006–2011). For each of the 4 study areas the estimated roe deer and lamb densities are indicated as number of lambs or roe deer per square kilometre.

127/03/641.5/fhb, 1460/99/641.5/FBe, 1081/97/641.5/FBe, and NINA 1/95). In addition, permits to capture wild animals were provided by the Norwegian Directorate for Nature Management. Of the 109 lynx, 61 were equipped with VHF radio-collars (Telonics MOD-335 transmitter, Telonics Inc., Mesa, AZ, USA), 15 received free-floating intra-peritoneal implant transmitters (Telonics IMP/150/L and IMP/400/L implantable transmitter with mortality sensor), 7 received store-on-board GPS collars, (2 Lotek 3300SL, Lotek Wireless Inc., Ontario, Canada, and 5 Televilt Posrec 300, Followit AB., Lindesberg, Sweden), and 26 received GPS-GSM collars (4 Tellus 1C, Followit AB and 22 Vectronic GPS PLUS, Vectronic Aerospace GmbH, Berlin, Germany).

The Socio-economic System

The conflict between carnivores and livestock is relatively high in Norway because of a grazing system based on free-grazing unguarded sheep in carnivore habitat [27,28]. In Norway, there is a legal requirement that all losses to large carnivores should be fully compensated. An *ex post facto* compensation system [29] is based on an estimation of losses, and on a national level 334,159 sheep and lambs have been compensated as being killed by large carnivores and golden eagles (at a cost of EUR 82 579 801) during the last ten years [30]. Since 1996, an estimated lynx population of between 259 and 486 individuals has been held responsible for killing 6125 to 10 093 sheep annually, corresponding to an annual average of 22 sheep killed per lynx ($\pm$ SD 2.2) [31]. However, there is a large degree of uncertainty in the magnitude of depredation, because only a small fraction (4–9%) of the compensated sheep is subject to a formal post mortem. The remaining numbers are estimated based on various studies of mortality rates of sheep and lambs [32–35].

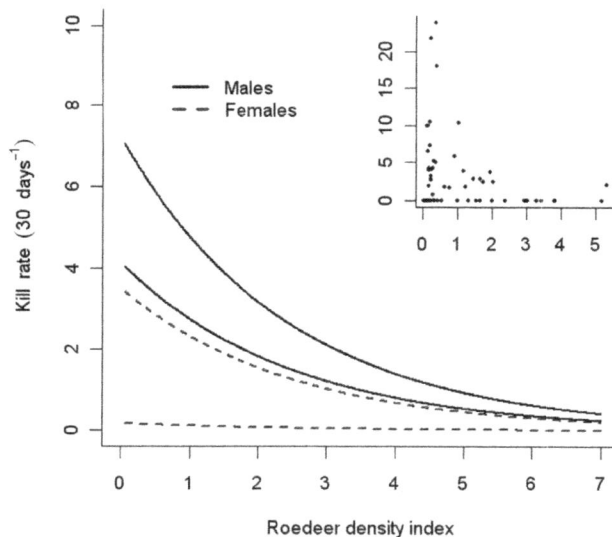

Figure 2. Predicted kill rates (i.e. number of sheep killed in 30 days) under different roe deer densities for male (solid black lines) and female (dashed red lines) lynx. Upper lines for each sex are predicted kill rates under high sheep density (95% percentile of observed lamb densities: 6.6 lamb km^{-2}) and lower lines are predictions for low sheep density (5% percentile of observed lamb densities: 0.1 lamb km^{-2}). Scatter plot inset in right corner represents the raw data.

Study Areas

Data on lynx kill rates on free-ranging domestic sheep were collected in a 57,000 km^2 study area in south-eastern Norway from 1995 to 2011. The area can roughly be divided into four parts (Figure 1) – whose boundaries refer to the present day lynx management regions in Norway [31]. The predation study started in the central parts of Hedmark County in 1995, hereafter called Region 5 [10,35,36]. In 2001 the study moved south to the area in and around Akershus and Østfold Counties (Region 4). The study in Region 4 ended in 2007. From 2006 to 2011 data were collected in two areas further west, in Buskerud, Telemark, Vestfold and Oppland Counties, hereafter called Region 2 north and Region 2 south [37].

The study area encompasses an environmental gradient as you move from south to north in the area (increased ruggedness and differences in elevation, increasing snow fall, decreasing proportion of farmland and decreasing human density). The northeastern portion of the study area (Region 5) is characterized by several river valleys at around 200–300 m, separated by hills reaching to 700–800 m. The forest is mainly composed of Norwegian spruce *Picea abies* and Scots pine *Pinus sylvestris* and most of it has been logged and regenerated throughout the last 100 years. The roe deer density in this portion of the study area is generally lower than further south. The south-eastern portion of the study area (Region 4) includes patches of both coniferous and deciduous forest, represented mainly by birch *Betula* spp. and the landscape is more human-modified, with the forest fragmented by cultivated land. The altitude is not higher than 300 m, and roe deer occur at higher densities than further north. The northwestern portion of the study area (Region 2 north) resembles region 5, but with deeper valleys, steeper terrain and higher mountains between the valleys. As we move to the coastal areas in the southwest (Region 2 south), the landscape is more human-modified, and the forest is fragmented by cultivated land. The roe deer density is lower in the northwest compared to further south. This density gradient reflects variation in habitat, snowfall and environmental productivity. Although a few red deer *Cervus elaphus* L. exist in small pockets in Region 5, red deer density is far higher in the two parts of Region 2, which reflects an on-going recolonisation process. Wild mountain reindeer *Rangifer tarandus* were seasonally available at higher altitudes (mainly above treeline) in the two northern parts (Region 2 north and Region 5) of the study area, and moose *Alces alces* occur in high numbers in all areas. Throughout the study area, a wide range of small prey species were available as prey. The most important are mountain hares *Lepus timidus*, red fox *Vulpes vulpes* and forest birds such as black grouse *Tetrao tetrix* and capercaillie *Tetrao urogallus*. We had no available data on small game density in the different areas.

All parts of the study area have free ranging sheep grazing in forest and alpine-tundra habitats from June until September, mainly without effective protection or constraints on their movements. However, the density and distribution of sheep vary considerably inside the area. Lynx depredation focuses almost exclusively on lambs so we expressed all densities as lamb density only. In this study 94% (156) of the sheep found killed by collared lynx where lambs, and found no difference in the proportion of lamb between the four areas ($\chi2 = 1.8$, d.f. = 3, P>0.05). Region 2 north has the widest distribution of grazing areas and the highest densities of sheep, with lamb densities inside each lynx summer home ranging from 0.4 to 6.5 lamb km^{-2}. In Region 5 lamb densities range from 0.1 to 10 lamb km^{-2}. As we move south (Region 2 south and Region 4), the density of sheep can still be high locally, but sheep grazing areas are smaller and more patchily

Table 1. Prey found at clusters from Eurasian lynx (*Lynx lynx*) in south-eastern Norway during summer, 1995–2011, grouped by study area (see text for explanation).

Prey species	Region 5 (1995–1999) Low roe / low lamb densities n (%)	Region 4 (2001–2006) High roe / low lamb densities n (%)	Region 2 - north (2007–2011) Low roe / high lamb densities n (%)	Region 2 - south (2008–2011) High roe / low lamb densities n (%)
Domestic sheep (*Ovis aries*)	32 (32)	10 (25)	117 (62)	16 (23)
Domestic goat (*Capra hircus*)	3 (3)			
Roe deer (*Capreolus capreolus*)	22 (22)	24 (60)	19 (10)	36 (52)
Red deer (*Cervus elaphus*)			1 (0.5)	5 (7)
Reindeer (*Rangifer tarandus*)	4 (4)		1 (0.5)	
Moose (*Alces alces*)[a]		1 (3)	1 (0.5)	
Mountain Hare (*Lepus timidus*)	16 (16)	1 (3)	28 (15)	5 (7)
Capercaillie (*Tetrao urogallus*)	12 (12)		9 (5)	4 (6)
Black grouse (*Tetrao tetrix*)	4 (4)	1 (3)	5 (3)	1 (2)
Red fox (*Vulpes vulpes*)		1 (3)	1 (0.5)	
Other mammals[b]	5 (5)			1 (2)
Other birds[c]	3 (3)		6 (3)	1 (2)
Scavenging events[d]		2 (5)	1 (0.5)	
Total prey	101	40	189	69
Lynx individuals	18	7	14	9
Monitoring days	483	446	854	378

Percentages are based on frequency of occurrence.
[a]Calves.
[b]Lemming (*Lemmus lemmus*), Brown rat (*Rattus norvegicus*), 3 red squirrels (*Sciurus vulgaris*), and 1 Eurasian beaver (*Castor fiber*).
[c]3 Wood Pigeon (*Columba palumbus*), 1 Meadow Pipit (*Anthus pratensis*), 6 unknown birds.
[d]2 moose carcasses, 1 roe deer carcass.

distributed. Lamb densities in these areas range from 0.2 to 6 sheep per km^{-2}.

Lynx Kill Rates

Lynx kill rates on sheep were sampled during the grazing season (June – September) following different sampling protocols as the telemetry technology developed over the years from 1995 to 2011.

Forty-eight individuals (24 males, 24 females) had access to free-ranging sheep inside their summer home ranges, and were included in this analysis. From 1995 to 2003 data on sheep predation were sampled using intensive VHF tracking. Lynx were located every 15 minutes during the night (8 to 12 hours), or during the entire 24 hour period using VHF tracking, and we searched all locations where the collared lynx stopped for at least

Table 2. Average number of sheep killed per 30 days, estimated lamb and roe deer densities, in four study areas in southern Norway during summer, 1995–2011.

Area	Sex	Lynx*season	Proportion (%) of lynx involved in depredation (n)	Average days	Average lamb per km^2	Average roe per km^2	Sheep killed per 30 days
Region 5	M	9	83% (6)	17 (±7)	1.3 (±2.6)	0.2 (±0.2)	7.9 (±8.6)
	F	14	8% (12)	24 (±14)	1.0 (±1.1)	1.1 (±1.3)	0.2 (±0.7)
Region 4	M	5	25% (4)	76 (±46)	1.8 (±2.5)	3.5 (±1.8)	0.4 (±1.8)
	F	3	33% (3)	23 (±12)	1.9 (±2.4)	2.2 (±1.1)	0.8 (±1.4)
Region 2 - north	M	8	100% (8)	63 (±34)	3.2 (±1.8)	0.6 (±0.4)	5.9 (±3.1)
	F	7	83% (6)	50 (±34)	5.2 (±3.0)	0.4 (±0.3)	2.4 (±1.8)
Region 2 - south	M	6	67% (6)	34 (±12)	1.1 (±0.7)	3.2 (±1.4)	1.9 (±1.6)
	F	3	33% (3)	59 (±19)	1.5 (±0.9)	2.7 (±2.1)	0.9 (±1.6)

Standard deviations in brackets.

Table 3. AICc values for all evaluated models, together with ΔAIC values.

	Model Specification	Df	AICc	dAICc	wi
Step I	Roe+Sex|Roe+Sex	7	214.0006	22.936	<0.001
	Roe+Sex|sex	6	216.4416	25.377	<0.001
	Roe+Sex|Roe	6	216.2099	25.145	<0.001
	Roe+Sex|1	5	217.1654	26.101	<0.001
Step II	Roe+Sex|Lam+Sex	7	191.0649	0.000	0.746
	Roe+Sex|Lam	6	202.6714	11.607	0.002
Step III	Sex|Lam+Sex	6	196.1979	5.133	0.057
	Roe|Lam+Sex	6	194.0824	3.018	0.165
	Lam+Sex|Lam+Sex	7	198.7967	7.732	0.016
	Lam|Lam+Sex	6	201.2709	10.206	0.005
	1|Lam+Sex	5	199.7381	8.673	0.010

The models are presented in the format of the R-language, so that $x_1+x_2|x_3+x_4$ is read so that the first part represents the negative binomial part of the model (x_1+x_2;"count process") whereas the latter part is the referring to the zero-inflated part of the model (x_3+x_4 ;excess zeros). [lam = lamb density, sex = lynx sex, roe = roe deer density index].

Table 4. Parameter estimates for the best model in the set of candidate models (presented in Table 3).

Model term	Parameter	Se	z-score	P-value
Negative Binomial model				
Intercept	−1.462	0.180	−8.119	<0.001
Sex [Females]	−0.705	0.282	−2.502	0.012
Roe	−0.356	0.130	−2.746	0.006
Zero-inflation model				
intercept	0.673	0.909	0.740	0.459
Sex [Females]	5.473	3.519	1.555	0.120
Lam	−2.962	2.076	−1.427	0.154

Parameters are presented on the link-scale (log-link in the negative binomial model; logit-link in the zero-inflation model).
*log($theta$) parameter of the negative binomial model estimated at 1.498 (se = 0.490).

one hour during its travels at night in order to locate potential kills [35,38]. From 2004 to 2008, data on sheep predation were sampled using GPS store-on board collars programmed to take 2 to 8 positions per day. Potential predation sites ("clusters") were identified from GPS-locations after the collar had dropped off the animal using GIS-software (ArcView 3.3, ESRI) and a web-based map-system for displaying telemetry data (http://www.dyreposisjoner.no). The number of locations required to define a cluster, later visited in the field, was at least 2 locations within 100 m. From 2008 to 2011, data on sheep predation were sampled using GPS-GSM collars programmed to take between 12 and 19 GPS-locations a day for 15 to 80 days per summer. The GPS-GSM technology allowed us to visit every cluster (at least 2 locations within 100 m) in the field directly after the lynx had left the area. In addition, all single locations around clusters were visited when logistically possible. For all 3 methods, dogs were often used to search for kills. Because of the store onboard collars (2004–2008), the time elapsed between the lynx had left a kill site and the field visits ranged from 1 to 369 days. However, >80% of the clusters were visited within 20 days after the lynx left the kill site. Lynx rarely scavenge [38–40], and 89 out of 185 sheep carcasses found at clusters were defined as probably lynx-killed when we found prey remains (e.g. hair, rumen, bones) that matched the date of the cluster and where there were no other signs of cause of death. The remaining 96 sheep carcasses had clean bite marks to the throat of a sheep when found, and were defined as confirmed lynx-kills.

Data on lynx kill rates were pooled per lynx per season, and expressed as number of sheep killed in 30 days. Number of days monitored per lynx per season varied from 10 to 122 (mean 39 days SD±31). To scale the kill rate to 30 days we used two procedures. To display raw data (e.g. in Figure 2) we divided number of kills detected by number of days with intensive monitoring, and multiplied by 30. In the zero-inflated models (see *Statistical analyses* under), we used the observed number of kills as the response variable, and the duration of the intensive monitoring as an offset variable (i.e. a factor for which there is an expected

slope equal to 1). The lines in Figure 2 are scaled so as to predict 30-day kill rats.

Roe Deer Density Index

We used 3 year average hunting bag statistics at the municipality level, available from Statistics Norway (www.ssb.no), as an index of local roe deer density. We assume that the spatial variation in our roe deer density index broadly reflects the variations in roe deer density. This assumption has previously been supported by [41] and used in several previous analyses [42,43]. For ease of interpretation, we rescaled our roe deer density index from number of roe deer shot per forested area to number of roe deer per forested area based on the assumption that the harvesting rate was constant across years and regions (~15%; based on mortality data from 299 radio-collared roe deer in the study area [44]). For the analysis, the roe deer density index was calculated as the average density across the municipalities in which the individual lynx travelled during summer. Although this density index is crude, we believe that it is functional across such a massive variation in roe deer density (>2 orders of magnitude) and across such a large study area.

Sheep Density

Spatial data on sheep grazing areas and on the associated area-specific numbers of sheep released in spring were obtained from the Norwegian Forest and Landscape Institute (http://www.skogoglandskap.no/) for 1995–2010. Because lynx depredation on sheep mainly affects lambs (94% of sheep killed in this study were lambs), we only included data on number of lambs available inside each lynx summer home range in our analyses. We used GIS-software (ArcView 3.3, ESRI) to calculate the number of lambs available inside each lynx's summer home range (June–September). When lynx only utilized parts of a grazing area we assumed that sheep were evenly distributed inside the grazing area, and calculated the number of lambs available to an individual lynx from the proportion of the grazing area that was overlapped by the lynx's home range.

Statistical Analyses

Our kill rate data contained many more zeros than expected from a Poisson distribution, thus zero-inflated negative binomial models (ZINB) [45] were used to model the effects of lynx sex, roe deer density and sheep density on observed kill rates on sheep. A ZINB model is a mixture model consisting of two parts; a binomial model (zero-inflation model), with or without covariates used to model the excess zeros, and a count process model including expected zeros modelled with a negative binomial error distribution. Both parts of the process can be modelled as a function of independent variables. ZINB models were appropriate because 1) the existence of excess zeros made a zero-inflated model necessary and 2) excessive variation also after accounting for the zero-inflation made a negative binomial distribution in the count part of the model more appropriate than a Poisson distribution. Although we had observations for 2 seasons for 7 lynx individuals, we did not deal with pseudoreplication [46] by treating individual as a random factor. Primarily this was because there is currently no generally accepted statistical method to deal with autocorrelated error structures in ZINB models [45] and secondly, other sources of spatial and temporal correlation that we could not account for would make the inclusion of individual identity as a random term arbitrary.

We included three independent variables (lynx sex, roe deer density and sheep density). To avoid overfitting (i.e. fitting models with too many parameters with respect to the data), we made two main restrictions: 1) We never fitted models with both roe deer density and sheep density in the same part of the model, and 2) we considered only main effects (although a "biological" interaction might become evident depending on the structure of the two parts of the ZINB model). Even with these restrictions, a total of 36 different models exist, and the most complex models have 7 parameters to be estimated. To ease model selection, we divided the selection process into discrete steps gradually improving the model fit to the data. In the first step, we compared models with lynx sex and roe deer density in both parts of the model and conducted model simplification in the zero-inflation part of the model only. In the second step, we substituted roe deer density with sheep density in zero-inflation model and continued to the procedure of selecting the best predictor variables for the zero-inflation part of the model. Then, in step three we used the "best model" (sensu [47]) from this set of models, but this time we were interested in finding the best predictor variables in the negative binomial part of the model. The performance of the models was investigated based on their AICc values [47], where the model with the lowest AICc value indicates the best model among the examined models, given the data. For each model we also computed Akaike weights (w$_i$) [47]. The Akaike weight is a measure of the relative support for each of the models in the subset, given the data and the model subset, and the sum of w$_i$ is equal to 1. All models were run in R 2.12.1 [48] with the add on library pscl [49] using the command zeroinfl. Likelihood ratio tests for comparing models were performed with the command lrtest in the add on library lmtest [50]. As the number of tracking days varied between the different study periods, we included log(number of tracking days) as an offset variable (i.e. a variable with a slope equal to one).

Results

We found 399 prey remains from 48 different lynx monitored during 2161 nights (Table 1). Free-ranging sheep were the most frequently killed prey in the two areas with low roe deer densities (Region 5 and Region 2-north), while roe deer was the most commonly killed prey species in the two areas with high roe deer density (Region 4 and Region 2-south). Twenty-six of the 48 monitored lynx (18 [75%] of the males and 8 [33%] of the females) killed sheep while being monitored. Fifteen (10%) of the 154 killing

events on domestic sheep involved multiple killing from 2 to 5 sheep, and all multiple killing were made by males. The highest kill rates on sheep were found in the two areas with the lowest roe deer density and the highest sheep density (Table 2), where male lynx on average killed 8 and 6 sheep per 30 days, respectively.

The ZINB model including the additive effects of lynx sex and sheep density in the zero-inflation model, and the additive effect of lynx sex and roe deer density in the negative binomial part of the model received most support (Table 3; Figure 2). The second best ranked model included the additive effects of lynx sex and sheep density in the zero-inflation model and the effect of roe deer density in the negative binomial part. This model was 4.5 times less likely to be the best model based on the Akaike's weights (w$_i$). In addition, a likelihood-ratio test clearly suggested that the effect of roe deer density was needed to adequately describe the data ($\chi^2 = 5.651$, df = 1, p = 0.017). Consequently, the best supported model suggested that the number of free-ranging sheep killed by an individual lynx in southern Norway is best understood as a combination of the individual effects of sheep density, roe deer density, and lynx sex (Table 4, Figure 2). Highest kill rates were found for males in areas with high sheep densities and low roe deer densities. Irrespective of sheep density and sex, the lowest kill rates occurred in areas with high density of roe deer. As roe deer density decreased, males killed sheep at higher rates, and this pattern holds for both high and low sheep densities. Similarly, females killed sheep at higher rates in areas with high densities of sheep and low densities of roe deer. In areas with low sheep densities, our model suggests that female lynx rarely kill free ranging sheep even when densities of their main wild prey (i.e. roe deer) are low.

Discussion

This study demonstrates how social and ecological factors interact to influence a predator's depredation rate on livestock. We found that variation in sheep density, roe deer density, and lynx sex are all needed to understand the variation in lynx depredation rates on free-ranging sheep in southern Norway. The direction of the results, that sheep depredation is inversely related to wild prey density, are in agreement with the predictions from the *energetic model*, with the highest kill rates on sheep in areas with low densities of their main prey, roe deer. This result does not necessarily contradict predictions from the *attraction model* which we have found support for in an early analysis [10]. The difference may simply be a matter of scale. As demonstrated here, and in several other studies, the depredation pressure on livestock can be higher when the overall wild prey densities are less abundant [6,15,16,18,22,51,52]. On a fine scale (within sections of a home range) depredation may be linked to patches of high ungulate densities, because carnivores may just spend more time in the most prey rich patches, leading to higher encounter rates with livestock and therefore more incidental depredation [8–10,53].

When we first started to study lynx depredation on sheep in Region 5 (1995–1999) we concluded that lynx depredation on sheep seemed to be more due to chance encounters between lynx and sheep, rather than the results of lynx actively seeking sheep as prey [10,35,36]. In this area sheep densities were greater than roe deer densities within most of the monitored lynx home ranges. Still sheep only constituted about 26% of the digestible biomass in lynx diet during summer estimated from the frequency of lynx kills and the relative consumption of each kill [10]. Here, the avoidance of abundant free-ranging sheep seemed to reflect some intrinsic aversion to sheep [35] indicating that depredation on livestock and predation on wild prey are based on different processes. However, when adding another 10 years of data from an expanded study area, the data suggest that lynx can switch from roe deer, to sheep depending on their relative abundance. In a multiple prey system, an opportunistic predator should respond to variations in prey densities by changing its relative use of prey [54]. Decreasing densities of prey are likely to result in decreased encounter rates and thus increased searching time. The diet shift observed in our study might be explained by changes in lynx-sheep encounter rates with changes in the relative abundance of roe deer and sheep. In our study, the estimated average lamb: roe deer density ratio inside lynx summer home ranges varied from 0.6:1 in the south to 4:1 in the northwest, and the northwestern area was the only area were sheep dominated the diet for both sexes.

In all our study areas male lynx killed sheep more frequently than females, given the same ecological settings. There is little comparable data on systematically sampled kill rates on livestock from a substantial sample of individual predators, but a few other studies of depredation by carnivores including leopard *Panthera pardus*, cougar *Puma concolor*, and black bear *Ursus americanus*, support this pattern of males killing more livestock than females [55–57]. The best comparable data comes from studies on coyotes *Canis latrans*, which have shown that most depredation is due to breeding adults and especially males [13,58,59]. However, telemetry studies of lynx in France, jaguars *Panthera onca* and grizzly bears *Ursus arctos* did not find any apparent effect of sex [7,60,61] on predation. The fact that male lynx killed sheep more frequently than females can partly be explained by the larger home ranges of males compared to females [42,62]. Males move faster and over larger areas, and thereby encounter sheep more often than females, but perhaps the most important sex difference is that males were responsible for all recorded cases of multiple killing in our study. Multiple killing of livestock is quite common among carnivores [4], and is typically associated with unusual environmental conditions that restrict the prey's ability to escape [63]. When faced with an abundant, vulnerable prey like sheep, there does not seem to be an adaptive reason why a lynx should limit killing, as the threat of injury is low [4]. The extent of multiple killing was disproportional to the greater energetic needs of males, and both sexes of lynx would tend to encounter sheep in the same setting, therefore these results probably reflect some intrinsic aspects of male behaviour, akin to their greater willingness to take risks [64].

Conclusions

This study confirmed our earlier findings [35] that a high proportion of lynx, especially males, will kill sheep at some stage as long as unguarded sheep are found at high densities throughout the natural habitats exploited by lynx. When depredation is not due to a few specific problem individuals, selective removal is impossible and lethal control will only reduce depredation if it reduces the overall number of lynx in the population [65]. Comparative studies of lynx from France and Sweden have shown that confining sheep in fenced fields or on alpine pastures (out of the forest) dramatically reduces depredation losses per lynx [7,53,66].

The present day Norwegian *ex post facto* compensation system creates widespread social conflicts because only a small fraction of the compensated sheep losses are documented through a formal examination of a carcass [38,67]. Combined with accurate lynx monitoring data, our findings on how lynx depredation rates vary with different factors can be used to evaluate current compensation levels based on empirical data, instead of a qualified guess of estimated losses by regional managers as it is today. This may be the first step towards a compensation system based on objective, accurate, and area specific estimation of sheep losses likely to be due to lynx. This would also facilitate transition to a risk based incentive system [26,68] as such systems are believed to encourage depredation prevention rather than damage documentation [69].

Acknowledgments

We would like to thank the hundreds of people who have helped us in the field during the years of this study. The study was conducted within the frames of the Scandinavian Lynx Project, Scandlynx (http://scandlynx. nina.no/). Hege Berg Henriksen, Morten Odden and Jenny Mattisson provided valuable comments on the manuscript. Reidar Andersen initiated this study and provided guidance throughout the process.

Author Contributions

Conceived and designed the experiments: JO JDCL. Performed the experiments: JO. Analyzed the data: JO EBN. Contributed reagents/materials/analysis tools: JO JDCL EBN. Wrote the paper: JO JDCL EBN.

References

1. Linnell JDC, Swenson J, Andersen R (2001) Predators and people: conservation of large carnivores is possible at high human densities if management policy is favourable. Anim. Conserv. 4: 345–350.

2. Baker PJ, Boitani L, Harris S, Saunders G, White PCL (2008) Terrestrial carnivores and human food production: impact and management. Mamm Rev 38: 123–166.

3. Inskip C, Zimmermann A (2009) Review Human-felid conflict: a review of patterns and priorities worldwide. Oryx 43: 18–34.

4. Linnell JDC, Odden J, Smith ME, Aanes R, Swenson JE (1999) Large carnivores that kill livestock: do "problem individuals" really exist? Wild. Soc. Bull. 27: 698–705.

5. Zimmermann A, Baker N, Linnell JDC, Inskip C, Marchini S, et al. (2010) Contemporary views of human-carnivore conflicts on wild rangelands. In: Du Toit JT, Kock R, Deutsch J, editors. Wild rangelands - conserving wildlife while maintaining livestock in semi-arid ecosystems. UK: Blackwells. pp. 129–151.

6. Meriggi A, Lovari S (1996) A review of wolf predation in southern Europe: does the wolf prefer wild prey to livestock? J. Appl. Ecol. 33: 1561–1571.

7. Stahl P, Vandel JM, Herrenschmidt V, Migot P (2001) Predation on livestock by an expanding reintroduced lynx population: long term trend and spatial variability. J. Appl. Ecol. 38: 674–687.

8. Treves A, Naughton-Treves L, Harper EK, Mladenoff DJ, Rose RA, et al. (2004) Predicting carnivore-human conflict: a spatial model derived from 25 years of data on wolf predation on livestock. Conserv. Biol. 18: 114–125.

9. Moa PF, Herfindal I, Linnell JDC, Overskaug K, Kvam T, et al. (2006) Does the spatiotemporal distribution of livestock influence forage patch selection in Eurasian lynx? Wildlife Biol. 12: 63–70.

10. Odden J, Herfindal I, Linnell JDC, Andersen R (2008) Vulnerability of domestic sheep to lynx depredation in relation to roe deer density. J. Wildl. Manage. 72: 276–282.

11. Mech LD, Fritts SH, Paul WJ (1988) Relationship between winter severity and wolf depredations on domestic animals in Minnesota. Wildl. Soc. Bull. 16: 269–272.

12. Sacks BN, Neale JCC (2007) Coyote abundance, sheep predation, and wild prey correlates illuminate Mediterranean trophic dynamics. J. Wildl. Manage. 71: 2404–2411.

13. Sacks BN, Neale JCC (2002) Foraging strategy of a generalist predator towards a special prey: coyote predation on sheep. Ecol Appl 12: 299–306.

14. Sidorovich VE, Tikhomirova LL, Jedrzejewska B (2003) Wolf Canis lupus numbers, diet and damage to livestock in relation to hunting and ungulate abundance in northeastern Belarus during 1990–2000. Wildlife Biol. 9: 103–111.

15. Pearson EW, Caroline M (1981) Predator control in relation to livestock losses in central Texas. J. Range Manage. 34: 435–441.

16. Cozza K, Fico R, Battistini ML, Rogers E (1996) The damage-conservation interface illustrated by predation on domestic livestock in central Italy. Biol. Conserv. 78: 329–336.

17. Stoddart LC, Griffiths RE, Knowlton FF (2001) Coyote responses to changing jackrabbit abundance affect sheep predation. J. Range Manage. 54: 15–20.

18. Patterson BD, Kasiki SM, Selempo E, Kays RW (2004) Livestock predation by lions (Panthera leo) and other carnivores on ranches neighboring Tsavo National Parks, Kenya. Biol. Conserv. 119: 507–516.

19. Woodroffe R, Lindsey P, Romanach S, Stein A, ole Ranah SMK (2005) Livestock predation by endangered African wild dogs (Lycaon pictus) in northern Kenya. Biol. Conserv. 124: 225–234.

20. Kolowski JM, Holekamp KE (2006) Spatial, temporal, and physical characteristics of livestock depredations by large carnivores along a Kenyan reserve border. Biol. Conserv. 128: 529–541.

21. Johnson A, Vongkhamheng C, Hedemark M, Saithongdam T (2006) Effects of human-carnivore conflict on tiger (Panthera tigris) and prey populations in Lao PDR. Anim. Conserv. 9: 421–430.

22. Bagchi S, Mishra C (2006) Living with large carnivores: predation on livestock by the snow leopard (Uncia uncia). J. Zool. (Lond.) 268: 217–224.

23. Meriggi A, Brangi A, Matteucci C, Sacchi O (1996) The feeding habits of wolves in relation to large prey availability in northern Italy. Ecography 19: 287–295.

24. Arnemo JM, Linnell JDC, Wedul SJ, Ranheim B, Odden J, et al. (1999) Use of intraperitoneal radio-transmitters in lynx Lynx lynx kittens: anaesthesia, surgery and behaviour. Wildlife Biol. 5: 245–250.

25. Arnemo JM, Ahlqvist P, Andersen R, Berntsen F, Ericsson G, et al. (2006) Risk of capture-related mortality in large free-ranging mammals: experiences from Scandinavia. Wildlife Biol. 12: 109–113.

26. Arnemo JM, Evans A, Fahlman Å (2012) Biomedical protocol for free-ranging brown bears, gray wolves, wolverines and lynx. Hedmark University College, Evenstad, Norway. Swedish University of Agricultural Sciences, Umeå, Sweden.

27. Kaczensky P (1996) Livestock-carnivore conflicts in Europe. Munich Wildlife Society.

28. Swenson JE, Andrén H (2005) A tale of two countries: large carnivore depredation and compensation schemes in Sweden and Norway. In: Woodroffe R, Thirgood S, Rabinowitz A, editors. People and wildlife: conflict or coexistence? Cambridge: Cambridge University Press. pp. 323–339.

29. Schwerdtner K, Gruber B (2007) A conceptual framework for damage compensation schemes. Biol. Conserv. 134: 354–360.

30. The Norwegian Directorate for Nature Management (2012) ROVBASE 3.0. pp. Open Access database on sheep and reindeer compensation and losses.

31. Linnell JDC, Broseth H, Odden J, Nilsen EB (2010) Sustainably Harvesting a Large Carnivore? Development of Eurasian Lynx Populations in Norway During 160 Years of Shifting Policy. Environ Manage 45: 1142–1154.

32. Warren JT, Mysterud I (1995) Mortality of domestic sheep in free-ranging flocks in southeastern Norway. J. Anim. Sci. 73: 1012–1018.

33. Warren I, Mysterud I (2001) Mortality of lambs in free-ranging domestic sheep (Ovis aries) in northern Norway. J. Zool. (Lond.), London 254: 195–202.

34. Knarrum V, Sørensen OJ, Eggen T, Kvam T, Opseth O, et al. (2006) Brown bear predation on domestic sheep in central Norway. Ursus 17: 67–74.

35. Odden J, Linnell JDC, Moa PF, Herfindal I, Kvam T, et al. (2002) Lynx depredation on domestic sheep in Norway. J. Wildl. Manage. 66: 98–105.

36. Odden J, Linnell JDC, Andersen R (2006) Diet of Eurasian lynx, Lynx lynx, in the boreal forest of southeastern Norway: the relative importance of livestock and hares at low roe deer density. Eur. J. Wildl. Res. 52: 237–244.

37. Mejlgaard T, Loe LE, Odden J, Linnell JDC, Nilsen EB (2013) Lynx prey selection for age and sex classes of roe deer varies with season. J. Zool. (Lond.) 289: 222–228.

38. Odden J, Linnell JDC, Andersen R (2006) Diet of Eurasian lynx, Lynx lynx, in the boreal forest of southeastern Norway: the relative importance of livestock and hares at low roe deer density. Eur. J. Wildl. Res. 52: 237–244.

39. Mattisson J, Odden J, Nilsen EB, Linnell JDC, Persson J, et al. (2011) Factors affecting Eurasian lynx kill rates on semi-domestic reindeer in northern Scandinavia: Can ecological research contribute to the development of a fair compensation system? Biol. Conserv. 144: 3009–3017.

40. Pedersen VA, Linnell JDC, Andersen R, Andrén H, Linden M, et al. (1999) Winter lynx Lynx lynx predation on semi-domestic reindeer Rangifer tarandus in northern Sweden. Wildl. Biol. 5: 203–211.

41. Grøtan V, Sæther B-E, Engen S, Solberg EJ, Linnell JDC, et al. (2005) Climate causes large scale spatial synchrony in population fluctuations of a temperate herbivore. Ecology 86: 1472–1482.

42. Herfindal I, Linnell JDC, Odden J, Nilsen EB, Andersen R (2005) Prey density, environmental productivity, and home range size in the Eurasian lynx (Lynx lynx). J. Zool. (Lond.), London 265: 63–71.

43. Melis C, Basille M, Herfindal I, Linnell JDC, Odden J, et al. (2010) Roe deer population growth and lynx predation along a gradient of environmental productivity and climate in Norway. Ecoscience 17: 166–174.

44. Melis C, Nilsen EB, Panzacchi M, Linnell JDC, Odden J (2013) Between the frying pan and the fire: roe deer face competing risks between predators along a gradient in abundance. Ecosphere 4:art 111. http://dx.doi.org/10.1890/ES13-00099.1.

45. Zuur AF, Ieno EN, Walker NJ, Saveliev AA, Smith GM (2009) Mixed effects models and extensions in ecology with R. New York, NY: Springer-Verlag New York.

46. Hurlbert SH (1984) Pseudoreplication and the Design of Ecological Field Experiments. Ecol Monogr 54: 187–211.

47. Burnham KP, Anderson DR (2002) Model selection and multimodel inference : a practical information-theoretic approach. New York: Springer. 488 p.

48. R Development Core Team (2010) R: a language and environment for statistical computing. Vienna, Austria: R Foundation for Statistical Computing.

49. Zeileis A, Kleiber C, Jackman S (2008) Regression models for count data in R. J Stat Softw 27: 1–25.

50. Zeileis A, Hothorn T (2002) Diagnostic checking in regression relationship. R News 2: 7–10.

51. Polisar J, Maxit I, Scognamillo D, Farrell L, Sunquist ME, et al. (2003) Jaguars, pumas, their prey base, and cattle ranching: ecological interpretations of a management problem. Biol. Conserv. 109: 297–310.

52. Melville H, Bothma JD (2006) Using spoor counts to analyse the effect of small stock farming in Namibia on caracal density in the neighbouring Kgalagadi Transfrontier Park. J. Arid Environ. 64: 436–447.

53. Stahl P, Vandel JM, Ruette S, Coat L, Coat Y, et al. (2002) Factors affecting lynx predation on sheep in the French Jura. J. Appl. Ecol. 39: 204–216.

54. MacArthur RH, Pianka ER (1966) On Optimal Use of a Patchy Environment. Am. Nat. 100: 603–609.

55. Jorgensen CJ (1983) Bear-sheep interactions, Targhee National Forest. Ursus 5: 191–200.

56. Mizutani F (1993) Home range of leopards and their impact on livestock on Kenyan ranches. Symp. Zool. Soc. London 65: 425–439.

57. Cunningham SC, Haynes LA, Gustavson C, Haywood DD (1995) Evaluation of the interaction between mountain lions and cattle in the Aravaipa-Klondyke area of southeast Arizona. Arizona Game and Fish Department Technical Report 17: 1–64.

58. Sacks BN, Jaeger MM, Neale JCC, McCullough DR (1999) Territoriality and breeding status of coyotes relative to sheep predation. J. Wildl. Manage. 63: 593–605.

59. Blejwas KM, Williams CL, Shin GT, McCullough DR, Jaeger MM (2006) Salivary DNA evidence convicts breeding male coyotes of killing sheep. J. Wildl. Manage. 70: 1087–1093.

60. Knight RR, Judd SL (1983) Grizzly bears that kill livestock. Ursus 5: 186–190.

61. Cavalcanti SMC, Gese EM (2010) Kill rates and predation patterns of jaguars (Panthera onca) in the southern Pantanal, Brazil. J. Mammal. 91: 722–736.

62. Linnell JDC, Andersen R, Kvam T, Andrén H, Liberg O, et al. (2001) Home range size and choice of management strategy for lynx in Scandinavia. Environ Manage 27: 869–879.

63. Kruuk H (1972) Surplus killing by carnivores. J. Zool. (Lond.) 166: 233–244.

64. Bunnefeld N, Linnell JDC, Odden J, van Duijn MAJ, Andersen R (2006) Risk-taking by Eurasian lynx in a human-dominated landscape: effects of sex and reproductive status. J. Zool. (Lond.), London 270: 31–39.

65. Herfindal I, Linnell JDC, Moa PF, Odden J, Austmo LB, et al. (2005) Does recreational hunting of lynx reduce depredation losses of domestic sheep? J. Wildl. Manage. 69: 1034–1042.

66. Karlsson J, Johansson O (2010) Predictability of repeated carnivore attacks on livestock favours reactive use of mitigation measures. J. Appl. Ecol. 47: 166–171.

67. Linnell JDC, Odden J, Mertens A (2012) Mitigation methods for conflicts associated with carnivore depredation on livestock. In: Boitani L, Powell R, editors. Carnivore Ecology and Conservation: A Handbook of Techniques. London, UK: Oxford University Press.

68. Zabel A, Holm-Müller K (2008) Conservation performance payments for carnivore conservation in Sweden. Conserv. Biol. 22: 247–251.

69. Ferraro PJ, Kiss A (2002) Direct payments to conserve biodiversity. 2981718–1719. Science 298: 1718–1719.

14

Transmission of *Mycobacterium tuberculosis* between Farmers and Cattle in Central Ethiopia

Gobena Ameni[1,2]*, **Konjit Tadesse**[1], **Elena Hailu**[2], **Yohannes Deresse**[2], **Girmay Medhin**[1], **Abraham Aseffa**[2], **Glyn Hewinson**[3], **Martin Vordermeier**[3], **Stefan Berg**[3]

1 Animal Health and Zoonotic Research Unit, Aklilu Lemma Institute of Pathobiology, Addis Ababa University, Addis Ababa, Ethiopia, **2** TB Research Team, Armauer Hansen Research Institute, Addis Ababa, Ethiopia, **3** TB Research Group, Animal Health and Veterinary Laboratories Agency, New Haw, Addlestone, Surrey, United Kingdom

Abstract

Background: Transmission of *Mycobacterium tuberculosis* (*M. tuberculosis*) complex could be possible between farmers and their cattle in Ethiopia.

Methodology/Principal Findings: A study was conducted in mixed type multi-purposes cattle raising region of Ethiopia on 287 households (146 households with case of pulmonary tuberculosis (TB) and 141 free of TB) and 287 herds consisting of 2,033 cattle belonging to these households to evaluate transmission of TB between cattle and farmers. Interview, bacteriological examinations and molecular typing were used for human subjects while comparative intradermal tuberculin (CIDT) test, post mortem and bacteriological examinations, and molecular typing were used for animal studies. Herd prevalence of CIDT reactors was 9.4% and was higher (p<0.01) in herds owned by households with TB than in herds owned by TB free households. Animal prevalence was 1.8% and also higher (p<0.01) in cattle owned by households with TB case than in those owned by TB free households. All mycobacteria (141) isolated from farmers were *M. tuberculosis*, while only five of the 16 isolates from cattle were members of the *M. tuberculosis* complex (MTC) while the remaining 11 were members of non-tuberculosis mycobacteria (NTM). Further speciation of the five MTC isolates showed that three of the isolates were *M. bovis* (strain SB1176), while the remaining two were *M. tuberculosis* strains (SIT149 and SIT53). Pathology scoring method described by "Vordermeier *et al.* (2002)" was applied and the average severity of pathology in two cattle infected with *M. bovis*, in 11 infected with NTM and two infected with *M. tuberculosis* were 5.5, 2.1 and 0.5, respectively.

Conclusions/Significance: The results showed that transmission of TB from farmers to cattle by the airborne route sensitizes the cows but rarely leads to TB. Similarly, low transmission of *M. bovis* between farmers and their cattle was found, suggesting requirement of ingestion of contaminated milk from cows with tuberculous mastitis.

Editor: Bernhard Kaltenboeck, Auburn University, United States of America

Funding: The authors are grateful to the Wellcome Trust for financially supporting this project through the programme "Animal diseases in the developing world". The funders had no role in study design, data collection and analysis, decision to publish, or preparation of the manuscript.

Competing Interests: The authors declared that no competing interest exists.

* E-mail: gobenachimidi2009@yahoo.co.uk

Introduction

M. tuberculosis and *M. bovis* are amongst the most important pathogens from the MTC, a highly related group of mycobacteria that cause TB in humans and other mammals [2]. *M. tuberculosis* is mainly considered as a human pathogen causing active TB in approximately eight million people every year [3], whereas *M. bovis* has a broader host range responsible for TB in domestic and wild animals [4]. It is well established that *M. bovis* also infects humans, causing zoonotic TB in humans [5].

According to the World Health Organization (WHO) Ethiopia ranks seventh in the world among the 22 high-burden TB countries, with an estimated incidence rate of 378 cases per 100,000 population of which ~1/3 are reported as new smear positive cases [6]. Smear positive cases are infectious and could be the sources of infection for healthy humans and animals. Although a recent comprehensive study on the proportion of human TB caused by *M. bovis* in Ethiopia suggested a rate less than 1% [7], smaller studies have reported higher isolation rates of *M. bovis* from humans [8]. The main routes of *M. bovis* transmission from infected animal to humans are believed to be through ingestion of raw milk and/or inhalation of aerosol from diseased animal, mainly in settings where pasteurization of milk is not widely established. Different studies have reported isolation of *M. tuberculosis* from domestic and wildlife animals [9–19]. The source of *M. tuberculosis* in animals is most frequently considered to be active TB patients expelling *M. tuberculosis* through sputum primarily, less often through urine or feces [11–14].

In developing countries, the dietary habit of people, close physical contact between humans and animals, rise in the incidence of immunosuppressive diseases, and inadequate disease control measures in animals and humans facilitate the transmission of the disease between animals and humans [5]. The present study took place in central Ethiopia where the majority of inhabitants in this area engaged in agriculture. As the climate condition is suitable for cattle production, these farmers practice mixed cattle farming in addition to crop cultivation. Previous studies conducted on bovine TB in herds in the study area have indicated high prevalence [20–22] of the disease. Thus, MTC prevalence in both human and cattle in the area suggest the possibility of existence of transmission between farmers and their cattle. The transmission of TB between farmers and their cattle could be associated with habits of farmers such as use of chewing tobacco for worming which is widely practiced in the area [19]. Therefore, this study was conducted to investigate the transmission of MTC between cattle and their owners in the area. To achieve this study, farmers with active TB were recruited from health institutions. TB free farmers living in the same village were identified to serve as control. This was followed by tracing and investigating cattle herds owned both groups of farmers.

Materials and Methods

Study Design

We hypothesize that the transmission of MTC between farmers and their cattle is prevalent in central Ethiopia. To prove or disprove this hypothesis, a case-control study (human TB cases and TB free controls) with two human and two linked animal cohort was conducted. The human risk factor study was a retrospective cohort study while the cattle study aspect was a prospective cohort study. The cases were farmers who visited health institutions in the area and were diagnosed as smear positive pulmonary TB patients by the health personnel. Households in which TB cases live were considered to be TB positive households. The controls were farmers residing in the households located in same village with TB positive households but did not have history of TB for the last decade. Households in which TB cases did not occur for the last decade were considered as TB negative households. Identification of the cases was done at health institutions while identification of controls was made in the villages after tracing the resident cases. Cases were defined on the basis of clinical and laboratory examination while controls were defined on the basis of interview and clinical examination. A total of 146 cases and 141 controls were recruited in the study. After identification of cases and controls, the herds were tested by CIDT for bovine TB. Thus, 146 "case" herds and 141 "control" herds were tested.

Study Subjects

The study was conducted in central Ethiopia (Figure 1). Fiche Hospital and district health centers were used for the identification of households with TB cases. Farmers with active TB were identified in the health institutions, consented and requested for submission of sputum samples before treatment. The sputa samples were collected and examined for acid fast bacilli (AFB) as routine diagnostic procedure. Leftover sputa of the AFB positive farmers were transported to TB laboratory of the Aklilu Lemma Institute of Pathobiology for mycobacterial culturing. This was followed by tracing the herds of households from where AFB positive farmers come and testing using CIDT. Side by side, herds belonging to households free from TB were identified and tested with CIDT for bovine TB. Using this procedure, 287 households (146 households with pulmonary TB and 141 TB-free households) and 2033 cattle owned by both groups were investigated. Human TB positivity was defined by routine diagnostic procedures including clinical examination and acid-fast staining. History and clinical examination were used for screening of TB free farmers residing in the same village with farmers with pulmonary TB. Both TB positive and TB negative households keep cattle for multi-purpose. Female cattle were kept for reproduction and milk production while male cattle were kept for crop production. Both female and male cattle were kept mixed together and as such there was no specialization of cattle production either as dairy or as beef cattle. CIDT was performed on both female and male cattle. During testing parameters such as age, sex, body condition score were recorded for each of the study cattle.

Interview of Cattle Owners

A total of 287 individuals from the same number of households (usually heads of the households) were interviewed. Of these 287 individuals, 146 were TB positive while the remaining 141 were TB negative. These individuals were interviewed by their local language. The interview consisted of closed and open questions which address the knowledge, attitude and practices for the farmers in relation to TB in humans and cattle (see attached annex).

Culturing of Sputum

A total of 146 sputum samples were collected in sterile universal tubes from smear positive pulmonary TB patients using sterile leak proof disposable plastic material. The sputum samples were processed (decontaminated and neutralized) for mycobacterial culturing according the standard operating procedure described earlier by WHO [23]. Thereafter, 100 μl neutralized sample was inoculated onto two slants of Lowenstein-Jensen (LJ) media, one supplemented with pyruvate and one with glycerol, and incubated at 37°C for up to 5–8 weeks. Growth of mycobacteria was monitored every week for up to 8 weeks. Sample negative for AFB after 8 weeks of growth were considered negative.

Comparative Intradermal Tuberculin (CIDT) Testing of Cattle

A total of 287 herds consisting of 2033 cattle (1063 owned by TB positive households and 970 owned by TB negative households) were tested by CIDT test [24]. Two sites, separated by 12 cm, on the middle the left neck were shaved and skin thickness was measure with a caliper. Aliquots of 0.1 ml of 2500 IU per ml of avian purified protein derivative (PPD) and 0.1 ml of 2000 IU per ml of bovine PPD (Lelystad, The Netherlands; Animal Health and Veterinary Laboratories Agency, Weighbridge, UK) were injected to the dermis of the study animals. The skin thickness at each injection site was measured again after 72 hours. The interpretation of the result was made on the basis of the difference in skin thickness at the bovine and avian PPD injection sites. Two cut-off values were used to determine BTB status of an animal; the animal was considered as positive for BTB if the skin thickness at bovine PPD injection site minus the skin thickness at avian PPD injection site was >4 mm [23] or if the difference was >2 mm [20].

Pathological Examination

Of the total 2,033 cattle subjected to CIDT test, 36 were found to be strong reactors and thus slaughtered for investigation. The lungs and lymph nodes of these 36 CIDT positive cattle were removed and examined for the presence of gross TB lesions. Each of the seven lobes of the lungs were thoroughly inspected and palpated for any suspicious gross TB lesions. Similarly, mandibular, retropharyngeal, cranial and caudal mediastinal, left and right bronchial, hepatic, and mesenteric lymph nodes were sliced into thin sections and inspected for the presence of visible lesions according to the protocol described earlier [1,25]. When gross lesions suggestive of BTB were found in any of the tissues, the

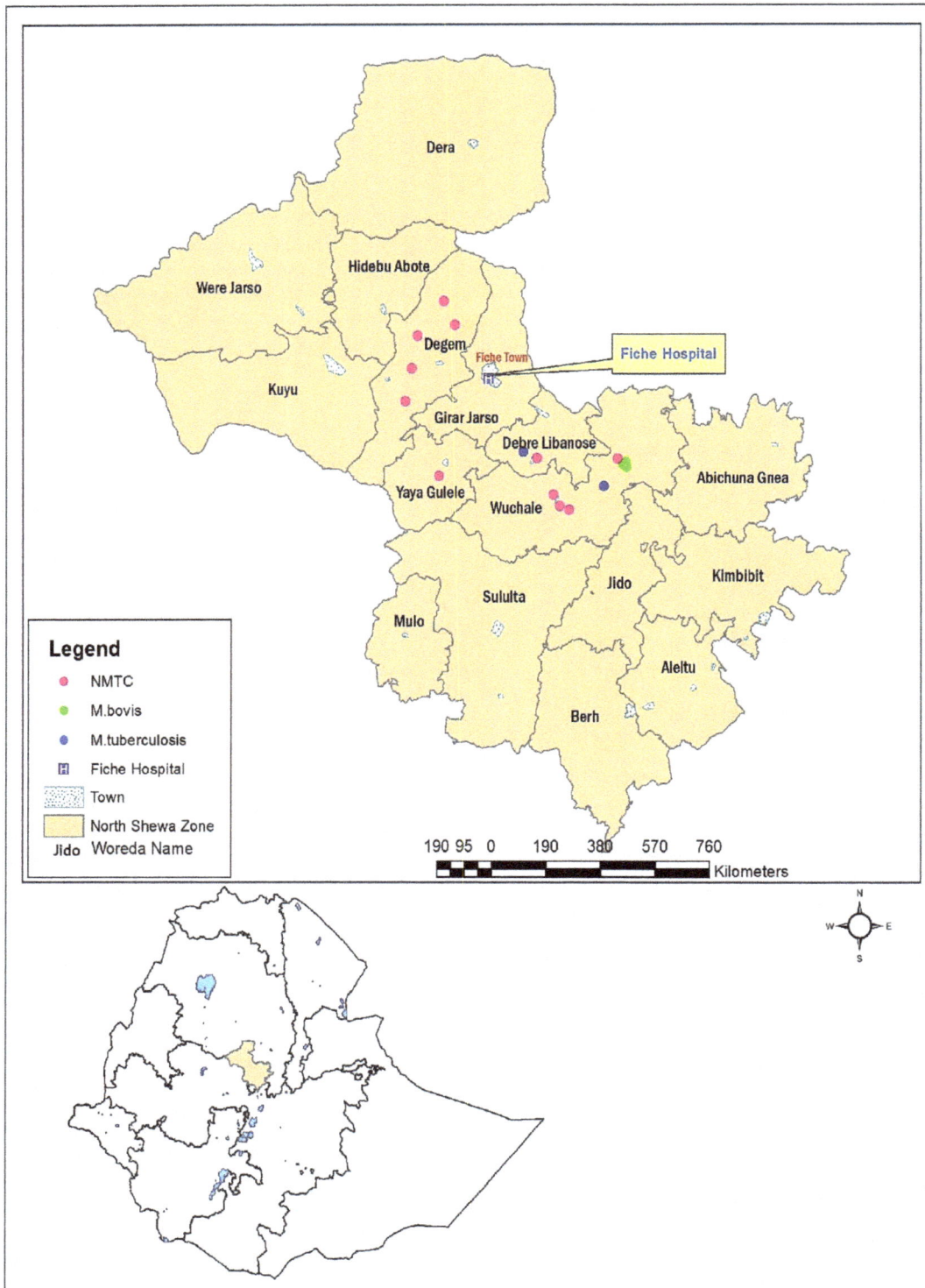

Figure 1. Distribution and frequencies of mycobacteria isolated from tissues of skin test positive cattle in central Ethiopia. Majority (68.8%) of the isolates were NTMs (pink dots on the map) while 18.8% were *M. bovis* (green dots) and 12.6% were *M. tuberculosis* (blue dots).

animal was classified as having lesions. Tissues with visible lesions were collected and processed for bacterial isolation using culture. Tissues with non-visible lesions were not cultured.

Culturing of Suspicious Tissue Lesion

A total of 33 tissues with suspicious lesions were collected from 36 necropsied cattle into universal bottles containing 5 ml

Table 1. Bovine tuberculosis prevalence in herds and animals owned by households with TB patients and TB free households in the same village.

Household status	>4 mm cut off			>2 mm cut-off		
	TB positive household	TB negative household	Total	TB positive household	TB negative household	Total
Herd prevalence						
Number of herds examined	146	141	287	146	141	287
Number of herds positive	21	6	27	39	19	58
Prevalence (95%CI)	14.4% (9.1,21.1)	4.3% (1.9, 9.0)	9.4% (6.3, 13.4)	26.7% (19.7, 34.7)	13.5% (8.3, 20.2)	20.2% (15.7, 25.3)
p-value	**<0.01**			**<0.01**		
Animal prevalence						
Number cattle examined	1,063	970	2,033	1,063	970	2,033
Number cattle positive	30	6	36	75	21	96
Prevalence (95%CI)	2.8% (1.9, 4.0)	0.6% (0.2, 1.3)	1.8% (1.2, 2.4)	7.1% (5.6,8.8)	2.2% (1.3,3.3)	4.7% (3.8, 5.7)
p-value	**<0.001**			**<0.001**		

of sterile 0.9% saline and transported in cold chain into the laboboratories for culturing. Each tissue was divided into two and processed for culturing in two laboratories, namely the Aklilu Lemma Institute of Pathobiology (ALIPB) and the Armauer Hansen Research Institute. In the laboratories, the specimens were sectioned using sterile blades and then homogenized with a mortar and pestle. The homogenate samples were processed (decontaminated and neutralized) for mycobacterial culturing according the OIE standard operating procedure [24]. Thereafter, 100 µl neutralized sample was inoculated onto two slants of Lowenstein-Jensen (LJ) media, one supplemented with pyruvate and one with glycerol, and incubated at 37°C for up to 5–12 weeks. Growth of mycobacteria was monitored every week for up to 12 weeks. Sample negative for AFB after 12 weeks of growth were classified as negative.

Molecular Typing of Mycobacteria

A total of 24 acid-fast bacilli (AFB) positive cattle isolates were heat-killed by mixing 2 loop-full of colonies in 400 µl distilled H_2O followed by incubation at 80°C for 1 h. Thereafter, they were subjected Genus typing using multiplex polymerase chain reaction (PCR) to detect the presence of the Genus *Mycobacterium* in the isolate and to differentiate *M. tuberculosis* complex from the other mycobacterial species [26]. Region of difference-based RD4 and RD9 deletion typing were used for identification of *M. tuberculosis* and *M. bovis* according to previous protocol [18]. Spoligotyping was performed on 139 human and five cattle isolates following the procedure as described previously [27].

Ethical Consideration

Both the human and animal components of the project were approved by the Ethical Clearance committee of the Aklilu Lemma Institute of Pathobiology (ALIPB/IRB/03/2009–10). All owners gave written consent to participate in the study and permission for their cattle to be included. All cattle that were recruited in the study were de-wormed after the PPD testing on free basis. TB positive cattle were slaughtered humanely in the Fitche Abattoir following the routine ante mortem and post mortem procedures.

Data Analysis

Cattle owners were classified in two categories; as households identified as TB positive (case group) and those that were TB negative (control group). The two categories were compared in terms of selected household and animal characteristics using chi-square. Crude and adjusted effects of TB positivity of the owner on the tuberculin test result of cattle were investigated using logistic regression. The clustering effect that could result from the fact that many animals being owned by the same owner was considered in logistic regression modeling.

Results

Herd and Animal Prevalence

A total of 2,033 cattle, 1063 owned by case households with TB and 970 owned by control households not diagnosed with TB, were CIDT tested in 287 herds from five districts. Two cut-off values of skin test result (>4 mm and >2 mm) were employed for the estimation of the prevalence at household/herd and animal levels. Table 1 shows the herd prevalence of BTB using these two cut-off values in herds owned by households with TB patients and in herds owned by TB free households from the same village. The overall herd prevalence was 9.4% at cut-off >4 mm and 20.8% at cut-off >2 mm. At both cut-off values the CIDT-positive herd prevalence was significantly greater in herds owned by households with TB patients than in herds owned by TB free households living in the same village. The overall animal prevalence was 1.8% (36/2033) at a cut-off >4 mm while it was 4.7% (96/2033) at a cut-off >2 mm. As for the herd prevalence, animal prevalence was significantly greater at both cut-off points in cattle owned by households with TB patients than in cattle owned by TB free individuals households from the same village (Table 1).

Risk Factors in Cattle Owners

A comparative assessment of different risk factors was made between TB positive (case) and TB negative (control) farmers and the results of selected characteristics of study participants are summarized in Table 2. Cattle owner who was TB positive was more likely to share house with the animals, more likely to have PPD positive animals, and more likely to have awareness about TB than the TB free control group living in the same village.

Table 2. Comparison of risk factors of tuberculosis between farmers with pulmonary tuberculosis and tuberculosis free farmers in central Ethiopia.

Characteristics of respondent	Number (%)		χ²	p-value
	TB cases (n = 146)	Controls (n = 141)		
Sex			35.3	<0.001
Male	74(50.7)	118(83.7)		
Female	72(49.3)	23(16.3)		
Age [mean(sd)]	34.7(1.3)	44.0(1.3)	5.07	<0.001
Consumption of raw milk	121(82.9)	115(82.1)	0.03	>0.05
Taking medication is a means to cure from TB	4(4.1)	27(19.3)	16.12	<0.001
Tobacco chewing			1.14	>0.05
Yes	25(17.1)	31(22.1)		
No	21(82.9)	109(77.9)		
Tobacco could transmit TB from human to animal	60(41.1)	59(42.1)	0.03	>0.05
Share house with animals			7.93	<0.01
Yes	94(64.4)	67(47.9)		
No	52(35.6)	73(52.1)		
Owns zebu breed	130(89.0)	123(87.2)	0.22	>0.05
Owns cross breed	96(65.8)	106(75.7)	3.42	>0.05
Owns Holstein breed	7(4.8)	10(7.2)	0.73	>0.05
Animals grazing on field			0.32	>0.05
Yes	137(93.8)	129(92.1)		
No	9(6.2)	11(7.9)		
Some of his/her animals are positive for TB	21(14.4)	6(4.3)	8.63	<0.01
Thinks that he/she know TB	90(61.6)	48(34.3)	21.4	<0.001
Has TB positive family member	58(39.7)	10(7.1)	41.9	<0.001
Know symptoms of TB	115(78.8)	63(45.0)	34.7	<0.001
Know cattle can be infected with TB	29(19.9)	59(42.1)	16.7	<0.001
Thinks that TB can be transmitted from cattle to human	40(27.4)	49(35.0)	1.93	>0.05
Thinks that animals can acquire TB from humans	49(33.6)	36(25.7)	2.12	>0.05

The result of logistic regression taking skin test as a binary outcome and TB status of the owner (confirmed positive versus negative as main exposure variable) is summarized in Table 3. The risk for an animal being positive for TB as measured by skin test result was significantly higher (crude OR = 4.52; 95%CI, 1.80–11.36 at >4 mm cut-off and crude OR = 3.32; 95%CI, 1.88–5.87 at 2 mm cut-off) in cattle owned by confirmed TB positive owner than in cattle owned by TB negative owner. These associations were not confounded by pre-specified characteristics of the owner (age, sex, tobacco treatment of cattle, sharing shelter with animals) and pre-specified characteristics of animals (sex, age, breed, body condition and field grazing practice) (Table 3).

After adjusting for TB status of the cattle owner (TB case or control), the risk of TB positivity (i.e. skin test result) was not significantly associated with pre-specified characteristics of the owner (age, sex, tobacco treatment of cattle, sharing shelter with animals, level of knowledge about TB) and animals (sex, age, bread, body condition and field grazing practice) (data not shown).

Post Mortem and Bacteriological Findings

General data on each of the 36 slaughtered cattle are presented in Table 4. The selection of the slaughtered cattle was on the basis of reaction to tuberculin, mainly reaction to bovine PPD. The proportion of reactors with visible lesion was 69% (25/36). Culture positivity was recorded for 16 (44%) of the 36 reactors. The average severity of pathology in two cattle infected with *M. bovis* was 5.5 while the average severity of pathology in two cattle infected with *M. tuberculosis* was 0.5 (Table 4). On the other hand, the average severity of pathology in 14 cattle infected with NTM was 2.1. Results of the skin test, post mortem, culture and molecular typing are presented. Most of the isolates (11/16) were NTM. Only five of the isolates were members of *M. tuberculosis* complex.

Molecular Characterization of Cattle Isolates

For identification of mycobacteria from farmers and their cattle molecular typing was used. Of the 24 cattle isolates (obtained from tissues of 16 cattle), 16 were positive for the Genus Mycobacterium. Out of these 16 isolates, five were members of the *M. tuberculosis* complex (MTC) while the remaining 11 were NTMs. Figure 1 shows the distribution of the different isolates in the study area. The five members of the MTC complex were further characterized and classified into three *M. bovis* isolates and two *M. tuberculosis* isolates.

The *M. tuberculosis* and *M. bovis* isolates from cattle tissues were further characterized at strain level using spoligotyping (Figure 2).

Table 3. The effect of being owned by households with confirmed TB positive patients on skin test result of cattle in central Ethiopia.

Character	Cut-off >4 mm	Cut-off >2 mm
	OR(95%CI)	OR (95%CI)
Unadjusted effect of being owned by confirmed case compared to being owned by control	4.52(1.80, 11.36)	3.32(1.88, 5.87)
Effect of being owned by confirmed case compared to being owned by control adjusted for owner characteristics:		
Sex of respondent	4.89(1.79, 13.36)	3.34(1.78, 6.29)
Age	4.57(1.71, 12.21)	3.50(1.86, 6.58)
Tobacco chewing	4.47(1.78, 11.21)	3.25(1.86, 5.70)
If they share house With animals	4.09(1.66, 10.09)	3.15(1.76, 5.63)
Knowledge level About TB	4.76(0.51, 44.28)	3.46(0.89, 13.39)
Effect of being owned by confirmed case compared to being owned by control adjusted for cattle characteristics:		
Sex	4.52(1.80, 11.31)	3.33(1.89, 5.89)
Age	4.44(1.79, 11.01)	3.37(1.92, 5.92)
Breed	4.62(1.82, 11.72)	3.35(1.89, 5.92)
Body condition	4.45(1.77, 11.19)	3.37(1.93, 5.90)
Grazing on field	4.79(1.94, 11.84)	3.40(1.95, 5.94)

The spoligotype patterns of the two *M. tuberculosis* isolates were SIT149 and SIT53 (SpolDB4 database). On the other hand, the three *M. bovis* isolates exhibited the same pattern of *M. bovis* and they were clustered as a single type SB1176, which is a very dominant strain in central Ethiopia. The two cattle isolates of *M. tuberculosis* were members of the Euro-American lineage.

Isolation and Molecular Characterization of *M. tuberculosis* Complex Isolates from Farmers

Culture positivity was observed in 97% (141/146) of the farmers with active pulmonary TB and confirmed AFB positive with Ziehl Neelsen staining. Of the 141 culture positive samples, 139 were confirmed to be *M. tuberculosis* isolates using RD9 deletion typing while the remaining two isolates did not give signal. These 139 isolates were characterized to the strain level by spoligotyping. Up on spoligotyping, 130 isolates gave good and interpretable patterns while the patterns of the remaining 9 isolates were poor and could not be interpreted.

The patterns of these 130 isolates are presented in Figure 3. The result of spoligotyping of the 130 isolates produced 49 distinct spoligotypes; 37 (30%) of them had a unique pattern. The remaining 93(71.5%) isolates were grouped into 12 clusters of strains possessing at least two isolates. The cluster size varied from 2 to 34 patients. Eighteen patterns of the 49 patterns were not previously reported. The most commonly occurring patterns were Spoligo International Type Number (SIT) 149, SIT53, and SIT37 each consisting of 34, 15, and 9 isolates, respectively, and these three strains accounted for 44.6% of the isolates. The most predominant lineage was Euro-American (lineage 4) consisting of 78.5% of the isolates while the lineages of 17.7% of the isolates were East African Indian (EAI) lineage (lineage 3).

Discussion

In the present study, we evaluated the presence and magnitude of transmission of TB between Ethiopian farmers and their cattle.

To achieve this study, TB-positive households and TB-negative households were included. In addition, cattle owned by both groups were investigated.

The result of this study indicated that neither transmission of *M. tuberculosis* from man to cattle nor of *M bovis* from cattle to man could be demonstrated. The main route of transmission of *M. bovis* is generally accepted to be through the milk which requires shedding of the organism by cow with tuberculous mastitis, which is a rare occurrence. Reports of *M. tuberculosis* among cattle exist [9–19] and cattle are likely to be exposed through inhalation of droplets of cough from active pulmonary TB cases of farmer and by ingestion of pasture contaminated with urine and sputum from infected farmer [11–14] but the exposure may not lead to disease. In general, it seems to be accepted that *M bovis* is substantially less virulent in humans than *M. tuberculosis* [28], conversely, *M. tuberculosis* is much less virulent in cattle than *M bovis* [29], although there is a need to confirm if transmission results in TB in cattle or only exposure and sensitization to tuberculin test. Our result showed larger number of skin test reactors in cattle owned by households with active TB.

In the present study, two *M. tuberculosis* strains of types SIT149 and SIT53 were isolated from five cattle. These two strains were the most commonly isolated strains from farmers who possess cattle. Together with isolates of SIT37, these made up 44.6% of the 130 isolates collected from patients keeping these cattle. This implies that these strains are present in the study area, and can be transmitted to cattle through different routes including ingestion of feed contaminated with infected sputum and/or urine from infected farmers. Similarly, the traditional animal husbandry practice of spitting tobacco juice into the oral cavity of cattle [19] could also be considered a means of transmission of *M. tuberculosis* from farmers to their cattle. The two *M. tuberculosis* cattle isolates were compared with the isolates from their respective owners. One of the owners (identification number 1180) could not provide sufficient sputum sample and was culture negative while the strain from his cattle was SIT53. The strain obtained from the second

Table 4. Overall characteristics of reactor cattle slaughtered for pathological and bacterial studies.

ID	District	Sex	Age (years)	Breed	Status of owner	BCS	PPDA	PPDB	PPD (B–A)	Pathology	Tissues with lesion	Culture (AFB+)	Species
C97c7	G.Jarso	M	10	Zebu	TB free	Lean	Reactor	Reactor	5	2	MS	Negative	Negative
P16c2	Wuchale	M	3	Cross	TB case	Medium	None	Reactor	6	9	MS, LB, RPH & RCL	Positive	2M. bovis
P16C4	Wuchale	M	2	Cross	TB case	Lean	None	Reactor	5	2	CRMD	Positive	M. bovis
P4C1	Wuchale	M	0.6	Cross	TB case	Lean	Reactor	Reactor	6	5	HP, MS	Positive	NTM
P4C6	Wuchale	F	7	Cross	TB case	Lean	Reactor	Reactor	3	1	RPH	Positive	NTM
P4C7	Wuchale	F	3	Cross	TB case	Lean	Reactor	Reactor	4	0	NVL	No result	Negative
629C3	Degem	M	6	Zebu	TB case	Lean	Reactor	Reactor	2	4	CAMD, MS	Positive	NTM
409C2	Wuchale	M	1.2	Zebu	TB case	Lean	None	Reactor	7	1	LDL	Positive	M. tuberculosis
409C4	Wuchale	F	1.5	Zebu	TB case	Lean	Reactor	Reactor	3	1	LB	Positive	NTM
409C10	Wuchale	M	5	Zebu	TB case	Medium	Reactor	Reactor	1	2	MS	Positive	NTM
520C1	Y. Gulele	M	4	Zebu	TB case	Lean	None	Reactor	6	0	NVL	No result	No result
1180C8	Y. Gulele	M	1	Zebu	TB case	Lean	Reactor	Reactor	6	0	NVL	No result	No result
1180C9	Y. Gulele	M	1	Zebu	TB case	Lean	Reactor	Reactor	4	1	CAMD	Positive	M. tuberculosis
723C1	Wuchale	M	4	Zebu	TB case	Lean	Reactor	Reactor	2	3	RCL	Negative	Negative
778C1	Degem	M	11	Cross	TB case	Lean	Reactor	Reactor	4	2	LB	Positive	NTM
1124C5	Y. Gulele	F	3	Zebu	TB case	Medium	Reactor	Reactor	7	3	MS	Negative	Negative
1124C6	Y. Gulele	F	7	Zebu	TB case	Medium	Reactor	Reactor	3	2	LDL	Positive	Negative
P44C5	Y. Gulele	M	8	Zebu	TB case	Medium	None	Reactor	4	2	MS	Positive	NTM
1300C 3	Y. Gulele	M	8	Zebu	TB case	Lean	None	Reactor	4	0	NVL	No result	No result
C13C1	Wuchale	M	7	Zebu	TB free	Medium	None	Reactor	5	3	MS	Negative	Negative
C16C7	Degem	M	5	Zebu	TB free	Lean	Reactor	Reactor	3	1	MS	Positive	NTM
CP25C5	Degem	M	12	Zebu	TB free	Lean	None	Reactor	7	2	HP, MS	Positive	NTM
CP301	Degem	M	5	Zebu	TB free	Medium	None	Reactor	5	4	CAMD, MS	Positive	NTM
Tag N06	D.libanos	F	7.8	Cross	TB case	Medium	None	Reactor	11	1	CAMD	Positive	NTM
Tag N14	D.libanos	F	4	Cross	TB case	Medium	Reactor	Reactor	5	0	NVL	No result	No result
Tag N03	D.libanos	F	1.5	Cross	TB case	Medium	Reactor	Reactor	5	0	NVL	No result	No result
Tag N04	D.libanos	F	1.7	Cross	TB case	Lean	Reactor	Reactor	3	0	NVL	No result	No result
Tag N07	D.libanos	F	2.5	Cross	TB case	Lean	None	Reactor	4	0	NVL	No result	No result
Tag N15	D.libanos	F	6	Cross	TB case	Medium	None	Reactor	6	2	MS	Negative	Negative
Tag N12	D.libanos	F	5	Cross	TB case	Lean	None	Reactor	3	0	NVL	No result	No result
35C5	Wuchale	F	6	Zebu	TB case	Lean	Reactor	Reactor	3	6	LCL, RDL, MS	Negative	Negative
P13C1	Wuchale	M	5	Zebu	TB case	Lean	Reactor	Reactor	4	0	NVL	No result	No result
P13C4	Wuchale	M	0.7	Cross	TB case	Lean	Reactor	Reactor	4	0	NVL	No result	No result
P13C6	Wuchale	F	9	Cross	TB case	Lean	Reactor	Reactor	4	3	MS	Negative	Negative
395C5	G.Jarso	F	6	Cross	TB case	Lean	Reactor	Reactor	1	1	MS	Negative	Negative
395C12	G.Jarso	M	10	Cross	TB case	Medium	Reactor	Reactor	3	3	CAMD	Negative	Negative

G. Jarso = Girar Jarso, D. libanos = Debre Libanos, Y. Gulele = Yaya Gulele, MS = Mesenteric lymph node (LN), LB = left bronchial LN, RPH = retropharyngeal LN, RCL = right cardiac lobe, CRMD = cranial mediastinal LN, HP = hepatic LN, CAMD = caudal mediastinal LN, LDL = left diaphragmatic lobe, LCL = left cardiac lobe, RDL = right diaphragmatic lobe, NTM, none-*M. tuberculosis* complex, BCS = body condition, NVL, =Non-visible lesion.

farmer (identification number 409) was different from the strain isolated from his cattle. The cattle strain was SIT149 while the strain isolated from the owner was new – octal number 777737377720771 and SIT number which matches to this strain was not found in the spoligotype database (spolDB4). Both two *M. tuberculosis* isolates were obtained from visible lesions with milder severity as compared to lesions caused by *M. bovis*. This implies that there could be non-visible lesions caused by *M. tuberculosis* which were not collected as they were not visible at post mortem examination. Of the 36 CIDT-positive cattle 11 had no visible

lesions. These cattle might have been infected with *M. tuberculosis* or MTC. If we cultured lymph nodes with non-visible lesions, we might have isolated larger number of isolates of *M. tuberculosis* from cattle.

The most common spoligotype identified from farmers was the T family and the predominant lineage was the Euro-American. Similar to the present study, previous studies in Ethiopia showed that T and CAS genotypes were the dominant families [30,31]. Nonetheless, no *M. bovis* was isolated from the TB positive farmers in this study. In agreement with result of the finding of the present

Binary format	Octal format	Spoligotype	Isolate in cluster No.	Lineage
▓▓□□□□□▓□▓▓▓▓▓□▓▓▓▓▓▓□□▓□□□□□□□□□□▓□□□□	602773761000200	SB1176	3	
▓▓▓▓▓▓▓▓□□□□□□□□□□□□▓▓▓▓▓▓▓▓▓▓▓□□□▓▓▓▓▓▓	777000377760771	SIT-149	1	Euro-American
▓▓▓▓▓▓▓▓▓▓▓▓▓▓▓▓▓▓▓▓▓▓▓▓▓▓▓▓▓□□□▓▓▓▓▓▓	777777777760771	SIT-53	1	Euro-American

Figure 2. Spoligotype patterns of *M. tuberculosis* complex species isolated from cattle owned by farmers with active pulmonary tuberculosis in central Ethiopia. Three isolates of *M. bovis* were isolated from two oxen of a farmer with active pulmonary tuberculosis. These three isolates had the same spoligotype pattern and were SB1176. The other two isolates were *M. tuberculosis* and were from cattle owned by farmers with active tuberculosis. These isolates had different spoligotype pattern and were SIT149 and SIT53.

study, recent studies in Ethiopia have reported transmission of *M. tuberculosis* from humans to cattle [18,19].

From the interview, it was observed that the majority of study participant in both TB cases and TB free farmers consumed raw milk, and there was no association between consumption of raw milk and occurrence of human TB case. This was different from earlier reports, which associate raw milk consumption with extra pulmonary TB [32–34]. As indicated earlier, only 1% of the cows with TB excrete tubercle bacilli in their milk [35], which decreases probability of milk transmitting *M. bovis* to humans. In the present study, the milk was consumed at individual household, and thus has minimal role in transmitting the tubercle bacilli to people outside the farm in question. Inhalation could be an important route of transmission between farmers and cattle, further exacerbated by the low level of awareness of the farmers on the route of transmission and prevention of TB [36,37]. The study participants were living in rural area and had poor housing condition, minimal access to health facilities, low awareness about disease, and usually shared house with their cattle. All these factors promote the transmission of TB between cattle and their owners.

Binary format	Octal format	SIT	Isolate in Cluster (N)	SpolDB4 Lineage	SpotCLUST	Lineage
	777737377720771	New	3		T1	Euro-American
	777765777760731	New	1		T1	Euro-American
	677777777420731	New	1		H3	Euro-American
	774777717420771	New	1		H3	Euro-American
	776637777760771	New	1		T1	Euro-American
	767676777760771	New	1		X1	Euro-American
	776727777760771	New	1		T1	Euro-American
	777777777760661	New	1		T1	Euro-American
	776737407760771	New	1		LAM	Euro-American
	006737777760771	New	1		T1	Euro-American
	337737777760771	New	1		T1	Euro-American
	777757777760700	New	1		T2	Euro-American
	777737677700011	New	1		T2	Euro-American
	777000377760771	SIT-149	34	T3_ETH		Euro-American
	777777777760771	SIT-53	15	T1		Euro-American
	777737377760771	SIT-37	9	T3		Euro-American
	777776777760731	SIT-336	5	X1		Euro-American
	777777404760771	SIT-41	5	LAM7_TUR		Euro-American
	777777777720771	SIT-50	2	H3		Euro-American
	777777777760731	SIT-52	2	T2		Euro-American
	777736077760771	SIT-93	2	LAM5		Euro-American
	777777775720771	SIT-121	1	H3		Euro-American
	777777777560771	SIT-462	1	T1		Euro-American
	777777403760771	SIT-1688	1	T1		Euro-American
	000000007760771	SIT-4	1	LAM3 and S/convergent		Euro-American
	767777777760771	SIT-1129	1	T1		Euro-American
	777776777760771	SIT-119	1	X1		Euro-American
	777767777760771	SIT-118	1	T2		Euro-American
	777777774020631	SIT-1552	1	H1		Euro-American
	777743777760771	SIT-913	1	T1		Euro-American
	777777777720631	SIT-134	1	H3		Euro-American
	677777777720571	SIT-699	1	H3		Euro-American
	777777777420771	SIT-777	1	H4		Euro-American
	777775777760731	SIT-584	1	T2		Euro-American
	777777770000771	SIT-602	1	U		Euro-American
	777737770000771	New	1		H1	Euro-American
	777767770000771	New	1		H1	Euro-American
	700000007177771	SIT-910	1	U		New
	777777777563771	SIT-1094	1	MANU2		Indo-Oceanic
	503767740003571	New	1		CAS	East African Indian
	703777740001631	New	1		CAS	East African Indian
	701377400007771	New	1		CAS	East African Indian
	703777740003171	SIT-25	6	CAS1_DELHI		East African Indian
	703777740003771	SIT-26	6	CAS1_DELHI		East African Indian
	703377400001771	SIT-21	4	CAS1_KILI		East African Indian
	703367400001771	SIT-1675	1	CAS1_KILI		East African Indian
	703777740003571	SIT-289	1	CAS1_DELHI		East African Indian
	703777740000771	SIT-357	1	CAS		East African Indian
	703777740000000	SIT-1264	1	CAS		East African Indian

Figure 3. Spoligotype patterns of 130 *M. tuberculosis* isolated from pulmonary tuberculosis cases of farmers in central Ethiopia. The filled boxes represent the presence of spacers, and the empty boxes represent the absence of spacers. Spoligotyping of the 130 isolates produced 49 distinct patterns: Of these, 37 of had a unique pattern while the remaining 93 (71.5%) isolates were grouped into 12 clusters of strains possessing at least two isolates. The cluster size varied from 2 to 34 patients. Eighteen patterns of the 49 unique patterns were not previously reported. The most commonly patterns were SIT149, SIT53, and SIT37 each consisting of 34, 15, and 9 isolates, respectively. The most predominant lineage was Euro-American consisting of 78.5% of the isolates while 17.7% of the isolates were East African Indian.

It was learnt that the local custom of spitting chewed tobacco or tobacco juice into the mouths of the cattle has animal husbandry significance. According to the respondents, animals fed on tobacco have good body condition and in a better health as compared to animals not fed on tobacco. As a result, this custom is widely been accepted and practiced by the community in the study area. Both men and women chew tobacco for this purpose.

The current study is not without limitations and the limitation of this study were (1) milk samples were not collected nor tested for the presence of mycobacterial pathogens; (2) the household survey did not collect data regarding consumption of soured milk vs. unpasteurized raw milk; (3) contamination of pasture with *M. tuberculosis* in sputum and tobacco juice (spit) was not demonstrated and (4) controls were required to be TB-negative at enrollment and have no history of TB+ household members for the previous 10 years. The 10 controls reporting "a TB+ family member" should not have been enrolled or should have been disqualified.

In conclusion, the present study did not identify any transmission of *M. tuberculosis* between humans and cattle. Similarly, no transmission of *M. bovis* between farmers and their cattle was found, even with 82% of households reporting consumption of unpasteurized milk produced by their animals. Herds and cattle belonging to 146 TB+ farmers had statistically significant increased rates of skin test reaction to the CIDT (14.4% vs. 4.3%; $\chi^2 = 8.63$, p<0,01) when compared to the 141 TB-negative control farmers and their herds, suggesting increased exposure and sensitization to mycobacteria. The practice of acidifying or "souring" milk prior to consumption can eliminate the risk of transmission if acidified to a pH <4.2, and may explain the absence of *M. bovis* infections in the TB+ farmers. When compared to controls, TB+ farmers were younger (34.7 years vs. 44.0 years; $\chi^2 = 5.07$, p<0.001), more likely to be female (49.3% vs. 16.3%; $\chi^2 = 35.3$; p<0,001), to share housing with their cattle (64% vs. 48%; $\chi^2 = 7.93$, p<0,01); and to have a TB+ family member (39.7% vs. 7.1%; $\chi^2 = 41.9$, p<0.001). Further study is needed to address the implications of these findings.

Acknowledgments

The authors would like also to thank the study farmers for their willingness to be involved in this study.

Author Contributions

Analyzed the data: GM GA. Wrote the paper: GA SB AA GH. Conceived the idea and designed the study: GA SB MV. Performed the study: GA KT EH YD.

References

1. Vordermeier HM, Chambers AM, Cockle JP, Whelan OA, Simmons J, et al. (2002) Correlation of ESAT-6-specific gamma interferon production with pathology in cattle following Mycobacterium bovis BCG vaccination against experimental bovine tuberculosis. Infec Immun 70: 3026–3032.
2. Brosch R, Gordon SV, Marmiesse M, Brodin P, Buchrieser C, et al. (2002) A new evolutionary scenario for the *Mycobacteriumtuberculosis* complex. Proc Natl Acad Sci USA 99: 3684–9.
3. WHO (2011) Global tuberculosis control: World Health Organization report 2011. 1–258.
4. Radostits SM, Blood DC, Gray CC (1994) Veterinary medicine, A textbook of disease of cattle, sheep, pigs, goats and horse, 8th ed, Baillière Tindal, London, United Kingdom, 748–785.
5. Cosivi O, Grange JM, Daborn CJ, Raviglione MC, Fujikura T, et al. (1998) Zoonotic tuberculosis due to *Mycobacterium bovis* in developing countries. Emerg Infect Dis 4: 59–70.
6. WHO (2009) Global tuberculosis Control: A short update to The 2009 Report. WHO/HTM/TB/2009.426. WHO, Geneva, Switzerland.
7. Firdessa R, Berg S, Hailu E, Schelling E, Gumi B, et al. (2013) Mycobacterial Lineages Causing Pulmonary and Extrapulmonary Tuberculosis in Ethiopia. Emerg Infect Dis 19 (3): 460–463.
8. Kidane D, Olobo JO, Habte A, Negesse Y, Aseffa A, et al. (2002) Identification of the causative organism of tuberculous lymphadenitis in Ethiopia by PCR. J Clin Microbiol 40: 4230–4.
9. Boulahbal F, Benelmouffok A, Brahimi K (1978) Role of *Mycobacterium tuberculosis* in bovine tuberculosis. Arch Inst Pasteur Alger 53: 155–64.
10. Michalak K, Austin C, Diesel S, Bacon MJ, Zimmerman P, Maslow JN (1998) *Mycobacterium tuberculosis* infection as a zoonotic disease: transmission between humans and elephants. Emerg Infect Dis 4: 283–7.
11. Alexander KA. Pleydell E, Williams MC, Lane EP, Nyange JF, et al. (2002) Mycobacterium tuberculosis: an emerging disease of free-ranging wildlife. Emerg Infect Dis 8: 598–601.
12. Sternberg S, Bernodt K, Holmstrom A, Roken B (2002) Survey of tuberculin testing in Swedish zoos. J Zoo Wildl Med 33: 378–80.
13. Michel AL, Venter L, Espie IW, Coetzee ML (2003) *Mycobacterium tuberculosis* infections in eight species at the National Zoological Gardens of South Africa 1991–2001. J Zoo Wildl Med 34: 364–70.
14. Pavlik I, Ayele WY, Parmova I, Melicharek I, Hanzlikova M, et al. (2003) *Mycobacterium tuberculosis* in animal and human populations in six Central European countries during 1990–1999. Czech Vet Med 48: 83–89.
15. Ocepek M, Pate M, Zolnir-Dovc M, Poljak M (2005) Transmission of *Mycobacterium tuberculosis* from human to cattle. J Clin Microbiol 43: 3555–7.
16. Pavlik I, Trcka I, Parmova I, Svobodova J, Melicharek I, et al. (2005) Detection of bovine and human tuberculosis in cattle and other animals in six Central European countries during the years 2000–2004. Czech Vet Med 7: 291–299.
17. Une Y, Mori T (2007) Tuberculosis as a zoonosis from a veterinary perspective. Comp Immunol Microbiol Infect Dis 30: 415–25.
18. Berg S, Firdessa R, Habtamu M, Gadisa E, Mengistu A, et al. (2009) The burden of mycobacterial disease in Ethiopian cattle: implications for public health. PLoS One 4: e5068.
19. Ameni G, Vordermeier M, Firdessa R, Aseffa A, Hewinson G, et al. (2011) *Mycobacterium tuberculosis* in grazing cattle in central Ethiopia. Vet J 188 (3): 359–361.
20. Ameni G, Amenu K, Tibbo M (2003) Prevalence and risk factor assessment in cattle and cattle owners in Wuchale-Jida District, Central Ethiopia. Int J Appl Res Vet Med 1: 17–25.
21. Ameni G, Hewinson GR, Aseffa A, Young DB, Vordermeier MH (2008) Appraisal of interpretation criteria for the comparative intradermal tuberculin test for the diagnosis of bovine tuberculosis in central Ethiopia. Clin. Vaccine Immunol 15 (8): 1272–1276.
22. Regassa A, Medhin G, Ameni G (2008) Bovine tuberculosis is more prevalent in cattle owned by farmers with active tuberculosis in central Ethiopia. Vet J 178: 119–25.
23. WHO (2008) World Health Organization. Global tuberculosis control: surveillance, planning and financing. WHO Report. WHO/HTM/TB/2008.393. Geneva, Switzerland:
24. OIE (2009) Terrestrial Manual 2009. Office Internationale des Epizooties. OIE:1–16.
25. Ameni G, Aseffa A, Engers H, Young D B, Hewinson GR, et al. (2006) Cattle husbandry is a predominant factor affecting the pathology of bovine tuberculosis and IFN-γ responses to mycobacterial antigens. Clin. Vaccine Immunol 13: 1030–1036.
26. Wilton S, Cousins D (1992) Detection and identification of multiple mycobacterial pathogens by DNA amplification in a single tube. PCR Methods Appl 1: 269–273.
27. Kamerbeek J, Schouls L, Kolk A, van Agterveld M, van Soolingen D, et al. (1997) Simultaneous detection and strain differentiation of *Mycobacterium tuberculosis* for diagnosis and epidemiology. J Clin Microbiol 35: 907–14.
28. Magnus K (1966) Epidemiological basis of tuberculosis eradication 3 Risk of pulmonary tuberculosis after human and bovine infection. Bull World Hlth Org. (35): 483–508.
29. Acha PN, Szyfres B (2001) Zoonotic tuberculosis In: Zoonoses and communicable diseases common to man and animals, 3rd ed Vol I, Bacterioses and mycoses. Pan American Health Organization/World Health Organization, Washington. Pp: 283–397.
30. Bruchfeld J, Aderaye G, Palme IB, Bjorvatn B, Ghebremichael S, et al (2002) Molecular Epidemiology and Drug Resistance of Mycobacterium tuberculosis isolates from Ethiopian Pulmonary Tuberculosis Patients with and without Human Immunodeficiency Virus Infection. J Clin Microbiol 40: 1636–1643.
31. Agonafir M, Lemma E, Wolde Meskel D, Goshu D, Santhanam A, et al. (2010) Phenotypic and genotypic analysis of multidrug-resistant tuberculosis in Ethiopia. Int J Tuberc Lung Dis 14(10): 1259–1265.
32. Kazwala RR, Daborn CJ, Kusiluka LJ, Jiwa SF, Sharp JM, et al. (1998) Isolation of Mycobacterium species from raw milk of pastoral cattle of the Southern Highlands of Tanzania. Trop Anim Health Prod 30: 233–9.
33. O'Reilly LM, Daborn CJ (1995) The epidemiology of Mycobacterium bovis infections in animals and man: a review. Tuber Lung Dis 76 (1): 1–46.
34. Shitaye JE, Tsegaye W, Pavlik I (2007) Bovine tuberculosis infection in animal and human population in Ethiopia: a review. Vet Med (8): 317–332.

35. Grange JM, Yates MD (1994) Zoonotic aspects of *Mycobacterium bovis* infection. Vet Microbiol 40: 137–51.

36. Legesse M, Ameni G, Mamo G, Medhin G, Shawel D, et al. (2010) Knowledge and perception of pulmonary tuberculosis in pastoral communities in the middle and Lower Awash Valley of Afar region, Ethiopia. BMC Public Hlth 10(187): 1–11.

37. Munyeme M, Muma JB, Munang'andu HM, Kankya C, Skjerve E, et al. (2010) Cattle owners' awareness of bovine tuberculosis in high and low prevalence settings of the wildlife-livestock interface areas in Zambia. BMC Vet Res 6(21): 1–9.

Diversity and Community Composition of Methanogenic Archaea in the Rumen of Scottish Upland Sheep Assessed by Different Methods

Timothy J. Snelling[1], Buğra Genç[2], Nest McKain[1], Mick Watson[3], Sinéad M. Waters[4], Christopher J. Creevey[5], R. John Wallace[1]*

1 Rowett Institute of Nutrition and Health, University of Aberdeen, Bucksburn, Aberdeen, United Kingdom, **2** Department of Animal Nutrition and Nutritional Diseases, Faculty of Veterinary Medicine, Ondokuz Mayis University, Samsun, Turkey, **3** ARK Genomics, The Roslin Institute, Easter Bush, Midlothian, United Kingdom, **4** Animal and Bioscience Research Department, Animal and Grassland Research and Innovation Centre, Teagasc, Grange, Dunsany, Co. Meath, Ireland, **5** Institute of Biological, Environmental and Rural Sciences, Aberystwyth University, Aberystwyth, Ceredigion, United Kingdom

Abstract

Ruminal archaeomes of two mature sheep grazing in the Scottish uplands were analysed by different sequencing and analysis methods in order to compare the apparent archaeal communities. All methods revealed that the majority of methanogens belonged to the Methanobacteriales order containing the *Methanobrevibacter, Methanosphaera and Methanobacteria* genera. Sanger sequenced 1.3 kb 16S rRNA gene amplicons identified the main species of *Methanobrevibacter* present to be a SGMT Clade member *Mbb. millerae* ($\geq$91% of OTUs); *Methanosphaera* comprised the remainder of the OTUs. The primers did not amplify ruminal Thermoplasmatales-related 16S rRNA genes. Illumina sequenced V6–V8 16S rRNA gene amplicons identified similar *Methanobrevibacter* spp. and *Methanosphaera* clades and also identified the Thermoplasmatales-related order as 13% of total archaea. Unusually, both methods concluded that *Mbb. ruminantium* and relatives from the same clade (RO) were almost absent. Sequences mapping to rumen 16S rRNA and *mcrA* gene references were extracted from Illumina metagenome data. Mapping of the metagenome data to 16S rRNA gene references produced taxonomic identification to Order level including 2–3% Thermoplasmatales, but was unable to discriminate to species level. Mapping of the metagenome data to *mcrA* gene references resolved 69% to unclassified Methanobacteriales. Only 30% of sequences were assigned to species level clades: of the sequences assigned to *Methanobrevibacter*, most mapped to SGMT (16%) and RO (10%) clades. The Sanger 16S amplicon and Illumina metagenome *mcrA* analyses showed similar species richness (Chao1 Index 19–35), while Illumina metagenome and amplicon 16S rRNA analysis gave lower richness estimates (10–18). The values of the Shannon Index were low in all methods, indicating low richness and uneven species distribution. Thus, although much information may be extracted from the other methods, Illumina amplicon sequencing of the V6–V8 16S rRNA gene would be the method of choice for studying rumen archaeal communities.

Editor: Bryan A. White, University of Illinois, United States of America

Funding: The Rowett Institute of Nutrition and Health is funded by the Rural and Environment Science and Analytical Services Division (RESAS) of the Scottish Government. Research at The Roslin Institute was supported by the Biotechnology and Biological Sciences Research Council (BBSRC; BB/J004243/1, BB/J004235/1), and by the Technology Strategy Board (TS/J000108/1, TS/J000116/1). BG thanks the Turkish Higher Education Council for the award of a travelling fellowship. CJC was supported by the BBSRC Institute Strategic Programme Grant, Rumen Systems Biology (BB/J004413/1). The funders had no role in study design, data collection and analysis, decision to publish, or preparation of the manuscript.

Competing Interests: The authors have declared that no competing interests exist.

* Email: john.wallace@abdn.ac.uk

Introduction

Methanogenic archaea are part of the anaerobic microbial community of the rumen. Though less abundant than the ruminal bacteria they have received a great deal of attention due to their ability to synthesise methane. Methanogenesis from the rumen can occur via three known metabolic pathways although the hydrogenotrophic reduction of CO_2 by H_2 predominates [1]. The production of methane by ruminants represents a loss of energy to the animal [2], and with 3.6 billion domestic ruminants and 1.1 billion sheep globally enteric methane is also believed to be a

significant contributor to anthropogenic global greenhouse gas (GHG) emissions [3]. In Scotland, sheep farming represents an important sector of the agricultural industry, with approximately 2.75 million currently registered breeding ewes [4]. Even at a national level, methane production from sheep has been recognised as a significant challenge to meeting proposed targets of lowering GHG emissions by 20% by 2020 [5]. Assessment of the archaeal community is a prerequisite for rational manipulation of the ruminal microbiota to lower methane emissions.

Molecular analyses of the ruminal archaea have been based mainly on 16S rRNA gene amplicons, revealing the methanogen diversity and phylogeny in a number of ruminant species, including cattle [6–9], alpaca [10], reindeer [11,12], domesticated red deer [13], and water buffalo [14,15]. Archaeal communities in sheep have been assessed previously in New Zealand [13], France [16], Japan [17], Australia [18,19] and Venezuela [20]. The methyl-coenzyme reductase A (*mcrA*) gene involved in the methanogenesis pathway has provided an alternative marker to identify ruminal archaea in lambs [21] and cattle [22], providing good correlation of the different phylogenetic analyses. In almost all studies, the rumen methanogens have been predominantly assigned to the order Methanobacteriales. These methanogens have been divided into two major and correlated groups of Methanobrevibacter species (*Mbb.*) [23], *Mbb. smithii, gottschalkii, millerae* and *thaueri*, referred to as the SGMT clade, and *Mbb. ruminantium* and *olleyae*, referred to as the RO clade. With amplicon based methods, there have been concerns highlighting the comparability of the different community analyses due to primer bias [24] and also a failure to appreciate the role of the Rumen Cluster C (RCC) clade, related to the order Thermoplasmatales [25].

Microbial diversity can also be assessed using metagenomic methods where function and taxonomy can be obtained from a single dataset. Moreover, using extracted genomic DNA as a starting material avoids potential amplicon sequencing biases. In studies comparing amplicon and metagenomic methods, the metagenomic analysis has compared well to synthetic reference archaeal and bacterial communities [26].

The aim of the present study was to characterise by different methods the community of methanogenic archaea in the rumen of the economically and environmentally important sheep grazing on Scottish upland pastures. Although the main properties of the community were consistent across different methods, important differences emerged in relative abundance and more detailed taxonomic identification.

Materials and Methods

Animals and sampling

All the animal experimentation for this study was carried out under the conditions set out by a UK Home Office licence no. 604028, procedure reference number 8. Samples of digesta were taken from the rumens of two mature Finn-Dorset cross sheep (Sheep A and Sheep B), each fitted with a ruminal cannula. The animals were grazing a mixed pasture at Glensaugh, Scotland (altitude 300 m, mean annual temperature 7.5°C, rainfall 1130 mm) in June 2011. Approximately 50 ml of digesta were taken from each animal via a 20-mm diameter plastic tube, homogenized to detach the fibre adherent microbes and strained through two layers of gauze. Aliquots of 20 ml of the filtrate were transferred into sterile plastic containers and placed on dry ice for transportation and then stored in a freezer at −20°C.

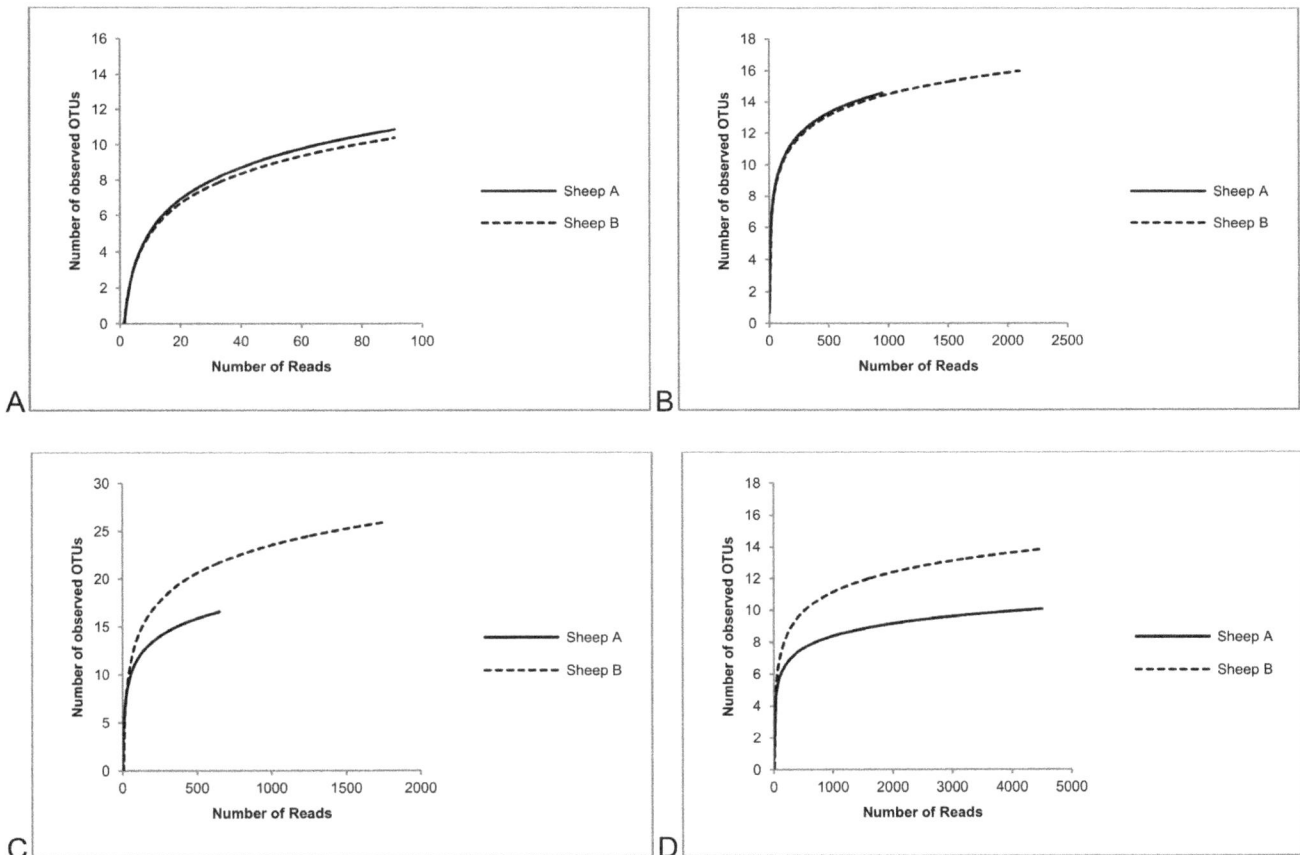

Figure 1. Multiple rarefaction collectors curves. Observed number of OTUs for different sequencing and analysis methods: A. SA *rrn*. B. IM *rrn*. C. IM *mcrA*. D. IA *rrn*.

Table 1. Methanogenic archaea of Sheep A and Sheep B.

16S rRNA OTU	Length (nt)	Reads Sheep Ag2	Reads Sheep Bg2	Nearest valid species	Acc No.	% sequence identity
Sanger						
RINH01	1030	84	63	Methanobrevibacter millerae	NR042785	98
RINH02	1031	1	0	Methanobrevibacter millerae	NR042785	99
RINH03	1029	4	14	Methanobrevibacter millerae	NR042785	99
RINH04	1027	3	1	Methanosphaera stadtmanae	CP000102	96
RINH05	1027	3	2	Methanosphaera stadtmanae	CP000102	97
RINH06	1083	2	8	Methanobrevibacter millerae	NR042785	99
RINH07	1030	2	1	Methanobrevibacter smithii	CP000678	95
RINH08	1027	2	0	Methanosphaera stadtmanae	CP000102	97
RINH09	1028	0	1	Methanosphaera stadtmanae	CP000102	96
RINH10	1036	0	1	Methanobrevibacter millerae	NR042785	99
RINH11	1028	0	1	Methanosphaera stadtmanae	CP000102	97
RINH12	1104	0	1	Methanosphaera stadtmanae	CP000102	94
RINH13	1032	0	1	Methanobrevibacter millerae	NR042785	98
RINH14	1034	0	1	Methanobrevibacter millerae	NR042785	97
RINH15	1027	1	0	Methanosphaera stadtmanae	CP000102	96
RINH16	1038	1	0	Methanobrevibacter millerae	NR042785	97
RINH17	1125	1	0	Methanobrevibacter millerae	NR042785	94
RINH18	1027	1	0	Methanosphaera stadtmanae	CP000102	95
RINH19	1029	1	0	Methanosphaera stadtmanae	CP000102	94
RINH20	1037	1	0	Methanobrevibacter millerae	NR042785	98
RINH21	1055	0	1	Methanobrevibacter millerae	NR042785	97
Illumina						
T01	492	3655	4630	Methanobrevibacter millerae	NR042785	99
T02	488	559	664	Aciduliprofundum boonei	NR074217	84
T03	488	345	545	Methanosphaera stadtmanae	CP000102	97
T04	486	153	30	Aciduliprofundum boonei	NR074217	82
T05	490	10	147	Methanobrevibacter ruminantium	NR074117	98
T06	486	90	40	Methanosphaera stadtmanae	CP000102	96
T07	486	27	28	Methanosphaera stadtmanae	CP000102	95
T08	488	0	27	Picrophilus torridus	NR074187	83
T09	478	5	9	Methanosarcina barkeri	JQ346756	100
T10	477	0	6	Methanoculleus palmolei	NR028253	99

Sanger and Illumina 16S rRNA amplicon OTUs clustered at 97% sequence identity. Taxonomic classification to the nearest valid species by BLASTn search of representative sequences to the GenBank nucleotide database.

Sanger amplicon sequencing of 16S rRNA gene (SA *rrn*)

Separate clone libraries were constructed for each of the two sheep from the respective digesta samples. DNA extraction was carried out using a method based on repeated bead beating using a Mini-Beadbeater (Biospec Products) plus column filtration (RBB+ C) [27]. Column filtration was carried out using the reagents and spin filter column provided with the QIAamp DNA Stool Mini Kit according the manufacturer's instructions (Qiagen, GmbH).

Methanogenic archaeal 16S rRNA genes (*rrn*) were amplified by PCR using the universal primers Arch f364 and Arch r1386 designed by Skillman *et al.* [28]. The amplicons were ligated into TOPO TA pCR 2.1 cloning vector (Life Technologies) and transformed into TOP10 chemically competent *Escherichia coli*. Positive transformed colonies were selected at random and the recombinant plasmids sequenced on a Beckman Coulter CEQ8000 platform following clonal amplification using a

QuickStart dye terminator master mix (Beckman Coulter Inc.) with M13 forward and reverse sequencing primers and the universal archaeal primers Met448F, Met448R, Met1027F and Met1027R [29].

Contigs were assembled by initially mapping the sequence fragments against a reference sequence obtained from the Ribosomal Database Project [30] of a type strain of *Methanobrevibacter ruminantium* (Acc. No. AY196666) [29]. The overlapping regions were inspected for mismatches or gaps and corrected to generate consensus sequences with the minimum number of ambiguities. Vector contamination was identified and removed after comparing the sequences to the UniVec database using VecScreen (NCBI).

Sequences were checked for possible chimeras using Bellerophon [31] and a non-redundant set of operational taxonomic units (OTUs) was generated *de novo* from a distance matrix of the

Table 2. Comparison of methanogenic archaeal diversity and depth of coverage for Sheep A and Sheep B.

	Sanger 16S		Illumina Metagenome 16S		Illumina Metagenome mcrA		Illumina 16S Amplicon	
	Sheep A	Sheep B	Sheep A	Sheep B	Sheep A	Sheep B	Sheep A	Sheep B
Shannon (H')	1.5	1.8	2.5	2.4	1.9	2.1	1.28	1.29
Chao1	19	31	18	17	20	35	10	18
Good (C)	0.67	0.57	0.83	0.88	0.92	1.0	1.0	0.75

Shannon index (H'), Chao1 estimated number of species and Good's statistic (C).

sequences constructed using the PHYLIP package dnadist [32] and clustered at values >98% sequence similarity using mothur) at values >98% sequence similarity [33]. Representative sequences for each OTU were entered as queries and searched using BLASTn against the NCBI GenBank nucleotide database [34] to assign taxonomy to the nearest valid species.

Illumina amplicon sequencing of 16S rRNA gene (IA *rrn*)

16S rRNA gene amplicons were generated using DNA Free Sensitive Taq polymerase (Bioron, GMbH) and primers Ar915aF and Ar1386R by Kittelmann *et al.*, (2013) [23]. Cycling conditions were: 94°C (2 min), then 30 cycles of 94°C (10 s), 68°C (20 s), 72°C (1 min). Amplicons were purified with a Qiaquick PCR purification kit (Qiagen, GmbH) according to manufacturer's instructions. 500 ng of each purified amplicon was then end repaired using the NEBNext End Repair Module (New England Biolabs Inc.). End repaired amplicons were purified with a Qiaquick PCR purification kit (Qiagen, GmbH) and a single adenine was added to the 3' ends using the NEBNext dA-Tailing Module (New England Biolabs Inc.). Partial Truseq standard paired end Illumina adapters with 6 bp barcodes (Integrated DNA Technologies) were ligated to the adenylated amplicons using the Quick Ligation Kit (New England Biolabs Inc.) and resulting adapter-ligated amplicons were purified with a Qiaquick PCR purification kit (Qiagen, GmbH). Full length adapter-ligated libraries were then generated by 7 cycles of PCR using Truseq paired end PCR primers (Integrated DNA Technologies) and Kapa HiFi Hotstart ready mix (Kapa Biosystems Inc.), then purified with Qiaquick PCR purification kit (Qiagen, GmbH). Cycling conditions were 95°C (2 min), then 7 cycles of 98°C (20 s), 58°C (15 s), 72°C (20 s) and a final extension step of 72°C (5 min). The resulting full-length libraries were denatured and diluted to 6 pM and spiked with 30% v/v denatured 12 pM PhiX control library (Illumina Inc.). The spiked 16S rRNA gene amplicon library was run on an Illumina MiSeq with 500 cycle Miseq Reagent Kit v2 sequencing chemistry (Illumina Inc.). Using the proprietary Illumina software suite, the reads were de-multiplexed and filtered to retain only those containing both 5' and 3' primer sequences. Paired-end reads were joined using the "fastq-join" command from ea-utils (http://code.google.com/p/ea-utils), specifying no difference between the two reads in the overlapping region (-p 0) and a minimum overlap of 10 nt (-m 10).

Clustering and determination of Operational Taxonomic Units (OTUs) were performed using CD-HIT-OTU [35] with a 97% identity cut-off. Taxonomic classification was carried out by submitting the representative sequence from each OTU to the RDP classifier [30] and individually queried using BLASTn against GenBank to assign taxonomy to the nearest valid species.

Illumina metagenome sequencing (IM *rrn*, IM *mcrA*)

Metagenomic sequences were generated from genomic DNA prepared in the form of 101-nucleotide (nt) paired-end reads using a HiSeq 2000 instrument (Illumina Inc.) at ARK-Genomics (University of Edinburgh). DNA from each sample was sheared randomly using an ultrasonicator (Covaris Inc.) and libraries for sequencing were constructed using a TruSeq DNA sample preparation kit (Illumina Inc.). Illumina metagenome 16S rRNA gene sequence dataset (IM *rrn*) was produced as follows: a rumen 16S rRNA gene database was constructed from Bacteria and Archaea references downloaded from the Ribosomal Database Project (RDP) website [30]. Sequences were selected from ≥ 1200 bp in length and with the quality tag 'Good' and with the keyword 'rumen'. Illumina metagenome sequences were clustered and aligned to this database using Novoalign (www.novocraft.com)

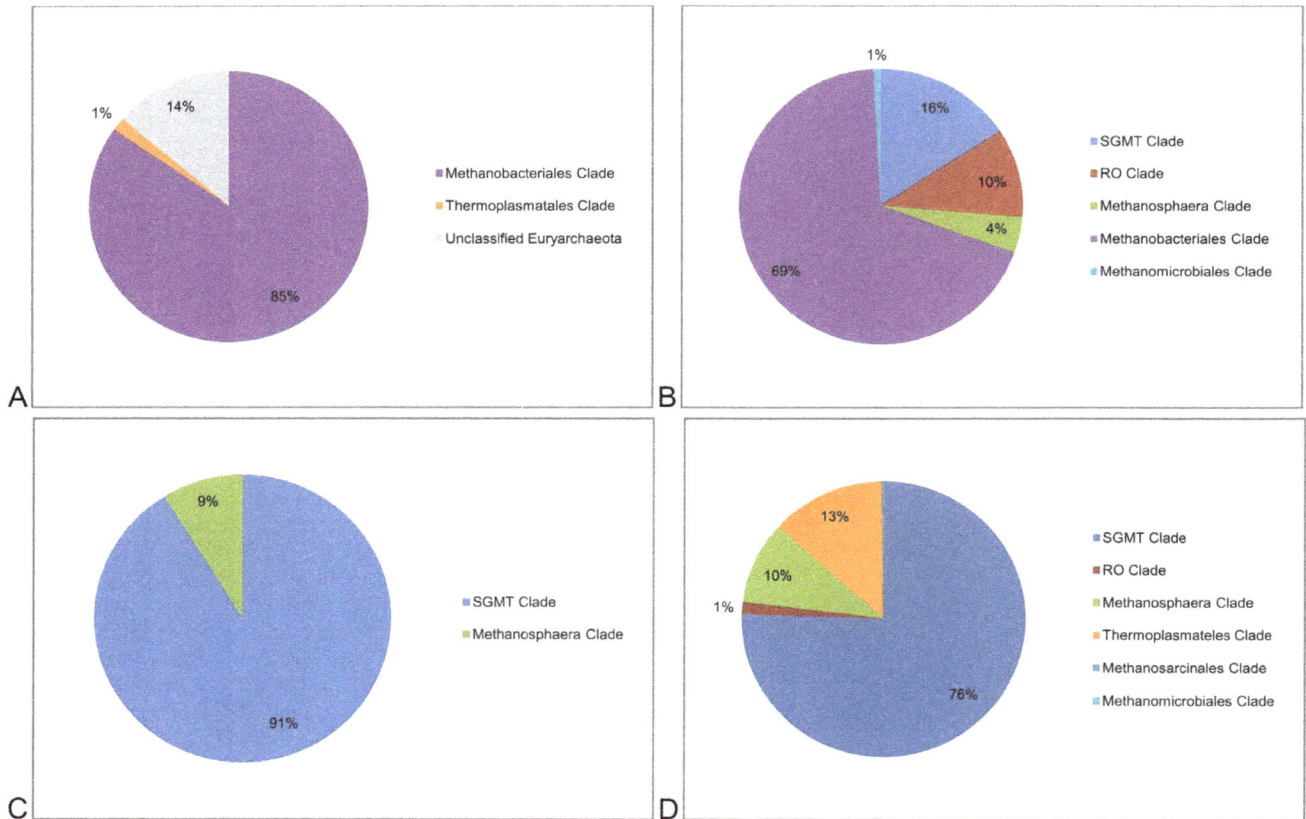

Figure 2. Relative distribution of methanogenic archaeal clades for different sequencing and analysis methods: A. SA *rrn*. B. IM *rrn*. C. IM *mcrA*. D. IA *rrn*.

using the '-r All' parameter setting to report all alignments. Where a fragment pair aligned to a single reference, the full taxon was reported. Where a fragment pair aligned equally well to multiple references, the lowest common taxon was reported.

The Illumina metagenome *mcrA* gene sequence dataset (IM *mcrA*) was prepared with a similar database, containing a complete set of *mcrA* genes, downloaded from the FunGene repository (http://fungene.cme.msu.edu/hmm_details.spr?hmm_id=16). OTU picking and taxonomic identification followed the same analysis method used previously for the IM *rrn* dataset.

Community diversity analysis

Good's depth of coverage (C) was assessed using the formula $C = 1-(n/N)$ where n is the number of singletons and N the total number of clones sequenced [36]. The diversity of the OTUs for each sheep was calculated using the Shannon index (H') as a summary measurement of species richness and evenness [37] and Chao1 indices as an estimate of the likely number of species [38]. Multiple rarefaction curves were also generated to assess depth of sequencing and species richness.

Phylogenetic analysis

Full multiple alignment and pairwise comparison of the OTU representative sequences was carried out using ClustalW [39]. A phylogenetic tree was constructed with MEGA5 [40] using the Neighbor-Joining method [41] with Jukes Cantor nucleotide substitution model and bootstrap resampling 1000 times.

Results

Sequencing and taxonomic identification

The Illumina HiSeq sequencing effort generated 307 million reads for Sheep A and Sheep B with a total of 216,038 reads mapped to prokaryote 16S rRNA references. Illumina metagenome *mcrA* gene (IM *mcrA*) and Illumina metagenome 16S rRNA gene (IM *rrn*) sequencing methods mapped 2442 reads and 3107 reads to methanogenic archaeal references and assigned them to 31 OTUs and 18 OTUs respectively. The assembly of the Sanger amplicon sequenced 16S rRNA amplicon (SA *rrn*) sequences produced 203 contigs of sequence length >1000 nt after vector removal, and clustering produced a set of 21 OTUs (RINH01-RINH21; Table 1). Representative sequences were submitted to EMBL, returning accession numbers HE858590 to HE858610. The Illumina amplicon sequencing of 16S rRNA genes (IA *rrn*) method produced 10982 reads with average sequence length 483 nt and clustered to 16 OTUs. The representative sequences of OTUs containing more than five reads per OTU were retained for phylogenetic analysis (T01–T10; Table 1). Alpha diversity statistics for each method including multiple rarefaction curves, Shannon index, Chao1 estimated number of species and Good's depth of coverage (C) are presented in Figure 1 and Table 2.

Taxonomic classification in all sequencing and analysis methods assigned all OTUs to the methanogenic archaea phylum Euryarchaeota. The taxonomic summaries for each method are presented in Table 1 and Figure 2. The majority of OTUs (85%–100%) obtained in all methods were assigned to the order

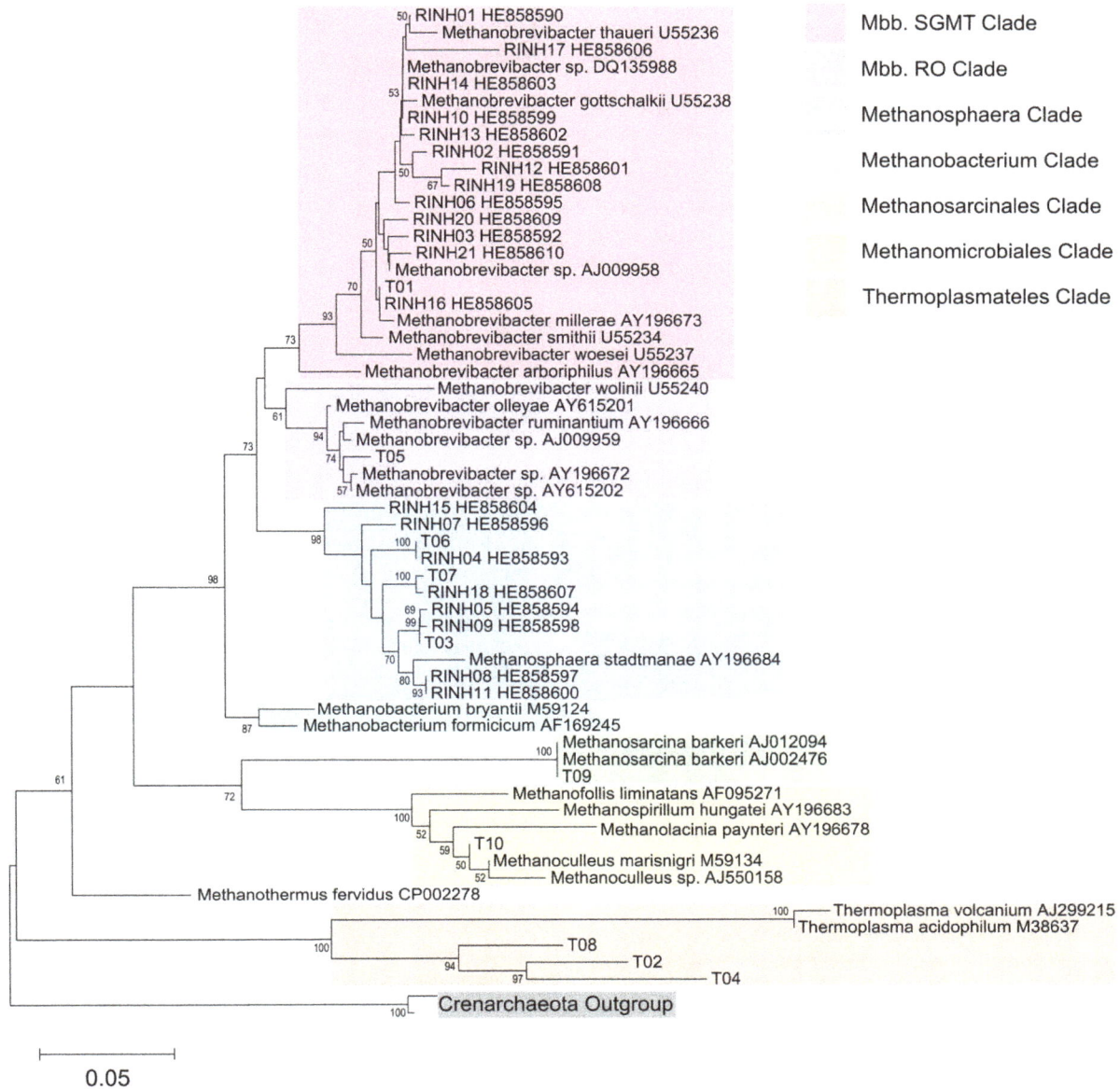

Figure 3. Phylogenetic analysis of SA *rrn* OTUs (RINH01–RINH21) and IA *rrn* OTUs (T01–T10). Placement of representative sequences of the present study in clades indicated with additional reference sequences obtained from GenBank. A sequence related to the Crenarchaeota phylum (Acc. No. AF418935) was used as an outgroup. Full multiple alignment using ClustalW and a consensus tree was constructed using the Neighbor-Joining method with the Jukes-Cantor substitution model. The trees were bootstrap resampled 1000 times with branch values ≥50% shown. Scale shows 0.05 nucleotide substitutions per nucleotide position.

Methanobacteriales. In detail, the resolution and relative abundance of the various clades varied between methods. However, a dominance of members of the genus *Methanobrevibacter*, and in particular the SGMT clade including the species *Mbb. smithii*, *Mbb. gottschalkii*, *Mbb. millerae* and *Mbb. thauerii*, was apparent (Figure 2). Notable differences were seen in the relative abundance of members of the RO clade (*Mbb. ruminantium* and *Mbb. olleyae*), with 10% assigned by the IM *mcrA* method, 1% by the IA *rrn* method and none from the SA *rrn* method. Taxonomic identification to species level was not possible by mapping the Illumina Metagenome to 16S rRNA gene references (IM *rrn* method). Therefore, the proportion of SGMT and RO clades within the Methanobacteriales could not be determined with this

method. The proportion of the Thermoplasmatales also varied between methods, from none detected using SA rrn and IM *mcrA* methods to 1% with the IM *rrn* method and 13% with the IA *rrn* method. The *Methanosphaera* clade was somewhat less variable between methods, where identified, with relative abundance of 4% detected with IM *mcrA*, 9% SA *rrn* and 10% IA *rrn* respectively.

Phylogenetic analysis

Phylogenetic analysis was carried out on the data where representative sequences were available and included data from the SA *rrn* and IA *rrn* methods. Full pairwise alignment of both the IM rrn and IM mcrA datasets was confounded by the fragmented nature of the reads. Sequences were mapped to random positions on the 16S rRNA gene and overlapping regions were variable or absent. Taxonomic identification was confirmed by placement in a branch containing valid species from the respective clades (Figure 3).

Phylogenetic analysis placed the most abundant SA *rrn* OTU (RINH01) in a group containing the methanogenic archaeal species *Mbb. thaueri* (U55236). This and a further 12 OTUs containing 92% of the sequences were all placed within the monophyletic SGMT clade [10,15]. The remaining nine OTUs containing 8% sequences were assigned to a clade containing the single valid species *Methanosphaera stadtmanae* (AY196684) (Figure 3). Similarly, the most abundant IA *rrn* OTU (T01) was placed near *Mbb. millerae*. This was the only IA *rrn* OTU placed within the SGMT clade although it accounted for over 75% of the relative abundance of methanogens detected using this method. The apparent richness was also greater with 1% OTU representatives assigned to the *Mbb*. RO clade and the Thermoplasmatales, Methanosarcinales and Methanobacteriales orders.

Discussion

Characterisation and measurements of diversity of methanogenic archaea were made from the ruminal digesta of two sheep kept on Scottish upland grazing. Different sequencing and analysis methodologies were applied to assess sequencing coverage, detection of the different taxa and effectiveness for calculating species diversity in each case. The methods used here were divided into two broad categories: an untargeted approach, where species are inferred by mining metagenomic data and mapping onto a reference database, and a targeted approach using PCR to amplify of a marker gene, alignment and clustering at specified sequences identity. The metagenome is the total gene content of an environmental microbiota at a given point in time [42]. Therefore, mapping the sequences to a specially curated reference gene database can provide a direct and representative measurement of the microbiota. The limitations are that metagenomic data can contain sequence fragments from any region of the target gene. This means that the information contained may not extend to cover the important hypervariable regions needed to make detailed taxonomic assignation.

Measuring the diversity of an environmental microbiota using amplicon sequencing whether based on a Sanger or NGS platform has also been subject to criticism. With this approach, results can be influenced by PCR amplification bias [43,44], choice of primers [24] and data analysis method [45]. However, it remains an established approach for measuring microbial communities with abundant resources of target gene databases [30,46,47] and analysis software [33,48].

To the best of our knowledge, methanogenic archaea of the phylum Crenarchaeota have been found in the rumen in one reported instance [49]. Otherwise the taxonomic richness of the ruminal archaeome is relatively poor compared to the bacteria [50] and can be summarised by a single phylum, the Euryarchaeota and four orders: Methanomicrobiales, Methanosarcinales, Methanobacteriales and the Thermoplasmatales (RCC). The latter, recently renamed Methanoplasmatales [51], is implicated in methane emissions in the rumen, possibly from methylamines [25]. Typically, the Methanobacteriales have been shown to be dominant in the rumen of sheep or lambs in a number of studies with members of the SGMT, RO, *Methanosphaera* spp. and *Methanobacterium* spp. clades found in varying relative abundance [20,21]. The exception to this has been methanogens from the rumen of sheep from Queensland, Australia where the Thermoplasmatales were the major order of methanogens [19].

The results presented here showed a clear majority of the Methanobacteriales using all methods. Mapping of the Illumina metagenome sequences to the rumen 16S rRNA database (IM *rrn*) identified 85% to this order with a further 1% was also mapped to the Thermoplasmatales. This method was severely limited in identifying OTUs with no detailed genus or species clades identified. The lack of resolution would be due in part to the absence of hypervariable regions offering the necessary taxonomic resolution in the sequence fragments.

This issue was also apparent after mapping the Illumina metagenome sequences to the *mcrA* reference database (IM *mcrA*). Here, 69% of the reads were assigned to Methanobacteriales. This method also detected a small proportion of Methanomicrobiales and was also able to identify some genera and species level clades. The failure to detect Thermoplasmatales may have been a result of the limited size and scope of the *mcrA* gene reference database [52].

The Sanger 16S rRNA amplicon sequencing methodology (SA *rrn*) benefited from high quality sequences, long read length and high resolution for taxonomic classification. OTUs for the entire archaeome were mapped to species level clades with members of the *Mbb*. SGMT clade predominating, with the highest number of reads related to *Mbb. millerae* and the remainder members of the *Methanosphaera* clade. SA *rrn* did not detect OTUs from the Thermoplasmatales related order. With any amplicon based method, detection of all the representative members of a microbial community depends on genuinely universal primers. Even a single base mismatch, particularly at the 3′ region can seriously affect primer annealing and bias the measurement of the microbial community [24]. This method in general is also limited by the number of sequences that can be produced in a single sequencing run and the calculation of Good's coverage (C) of 57%–67% indicated the need for increased sequencing effort. While falling well short of the upper limit of what can be achieved using this method, the 203 near full length SA *rrn* sequences presented here represented a reasonable sequencing effort for a single study. Characterisation of the methanogens in the gut of pigs and humans have yielded 763 and 1524 sequences respectively [53,54], the latter being a subset of 13355 prokaryotic *rrn* sequences. At this scale, the demands on time and cost begin to have an influence on practicality of Sanger sequencing compared to next generation sequencing methods.

The shorter average read length (483 nt) of the IA *rrn* method did not seem to affect the identification to genus and species level clades with SGMT dominant and the highest number of reads related to *Mbb. millerae*. *Mbb. ruminantium*, a member of the RO clade that was missed by the Sanger method was also detected albeit in small proportion (1%). In comparable studies, the proportions of these major methanogen clades between animals vary inversely [23]. However, the extreme bias toward SGMT was unusual to the Scottish sheep sampled here. The Thermoplasmatales clade was detected at 14% the highest proportion of all methods and the high coverage extended to detecting a few reads assigned to the Methanosarcinales and Methanomicrobiales clades, albeit making up less than 1% of the total archaeome.

An effective molecular method used to characterise the rumen methanogen community must have sufficient resolution to separate taxa at the minimum of genus and with sufficient depth to determine the presence and relative abundance of rare taxa in a population that is unevenly distributed and dominated by a few species. With appropriate primers, a high throughput amplicon sequencing strategy is currently the best way to assess the rumen methanogen community and in this study Illumina paired-end amplicon sequencing (IA rrn) effectively represented the subtle diversity of the ruminal archaeome.

The assessment and validation of the different methods presented here will serve as a guide to selecting the best approach for characterising methanogenic archaea in the rumen. Both microbiome and metagenomic methodologies will be essential tools as part of the investigation of the role of rumen microbiota in methane emissions and global climate change. Methane has been identified as a potent greenhouse gas with 27 times warming potential than CO_2 [55], and enteric methane emissions derived from the ruminal archaea have been estimated at 20–25%, making it the largest anthropogenic source [56]. Therefore, establishing an accurate and reliable method to characterise the methanogenic archaea in the rumen is an important step in the efforts to help mitigate the environmental impact of global livestock production.

Acknowledgments

We thank Bob Mayes and Dave Hamilton of the James Hutton Institute for their permission and help in sampling the sheep digesta. Gillian Campbell and Pauline Young provided an excellent DNA sequencing service. We also thank Dr Matthew McCabe for preparing V6–V8 amplicon libraries.

Author Contributions

Conceived and designed the experiments: RJW. Performed the experiments: RJW BG NM SMW CJC. Analyzed the data: TJS MW SMW CJC RJW. Contributed reagents/materials/analysis tools: NM RJW MW SMW CJC. Contributed to the writing of the manuscript: TJS MW SMW CJC RJW.

References

1. Cavicchioli R (2011) Archaea - timeline of the third domain. Nature Rev Microbiol 9: 51–61.
2. Czerkawski JW (1969) Methane production in ruminants and its significance. World Rev Nutr Diet 11: 240–282.
3. Moss AR, Jouany JP, Newbold J (2000) Methane production by ruminants: Its contribution to global warming. Ann Zootech 49: 231–253.
4. Scottish government (2010) Abstract of Scottish agricultural statistics 1982 to 2009. Available: http://www.scotland.gov.uk/Publications/2010/03/16160036/24. Accessed 2014 March 17.
5. Scottish government (2009) Climate change (Scotland) Act 2009 (asp 12). Available: www.legislation.gov.uk/asp/2009/12/contents. Accessed 2014 March 17.
6. Skillman LC, Evans PN, Strompl C, Joblin KN (2006) 16S rDNA directed PCR primers and detection of methanogens in the bovine rumen. Lett Appl Microbiol 42: 222–228.
7. King EE, Smith RP, St-Pierre B, Wright ADG (2011) Differences in the rumen methanogen populations of lactating Jersey and Holstein dairy cows under the same diet regimen. Appl Environ Microbiol 77: 5682–5687.
8. Wright ADG, Auckland CH, Lynn DH (2007) Molecular diversity of methanogens in feedlot cattle from Ontario and prince Edward island, Canada. Appl Environ Microbiol 73: 4206–4210.
9. Whitford MF, Teather RM, Forster RJ (2001) Phylogenetic analysis of methanogens from the bovine rumen. BMC Microbiol 1: 5. doi:10.1186/1471-2180-1-5.
10. St-Pierre B, Wright ADG (2012) Molecular analysis of methanogenic archaea in the forestomach of the alpaca (*Vicugna pacos*). BMC Microbiol 12: 1. doi:10.1186/1471-2180-12-1.
11. Sundset MA, Edwards JE, Cheng YF, Senosiain RS, Fraile MN, et al. (2009) Rumen microbial diversity in Svalbard reindeer, with particular emphasis on methanogenic archaea. FEMS Microbiol Ecol 70: 553–562.
12. Sundset MA, Edwards JE, Cheng YF, Senosiain RS, Fraile MN, et al. (2009) Molecular diversity of the rumen microbiome of Norwegian reindeer on natural summer pasture. Microb Ecol 57: 335–348.
13. Jeyanathan J, Kirs M, Ronimus RS, Hoskin SO, Janssen PH (2011) Methanogen community structure in the rumens of farmed sheep, cattle and red deer fed different diets. FEMS Microbiol Ecol 76: 311–326.
14. Singh KM, Tripathi AK, Pandya PR, Parnerkar S, Rank DN, et al. (2012) Methanogen diversity in the rumen of Indian Surti buffalo (*Bubalus bubalis*), assessed by 16S rDNA analysis. Res Vet Sci 92: 451–455.
15. Franzolin R, St-Pierre B, Northwood K, Wright AG (2012) Analysis of rumen methanogen diversity in water buffaloes (*Bubalus bubalis*) under three different diets. Microb Ecol 64: 131–139.
16. Mosoni P, Martin C, Forano E, Morgavi DP (2011) Long-term defaunation increases the abundance of cellulolytic *Ruminococci* and methanogens but does not affect the bacterial and methanogen diversity in the rumen of sheep. J Anim Sci 89: 783–791.
17. Yanagita K, Kamagata Y, Kawaharasaki M, Suzuki T, Nakamura Y, et al. (2000) Phylogenetic analysis of methanogens in sheep rumen ecosystem and detection of *Methanomicrobium mobile* by fluorescence *in situ* hybridization. Biosci Biotechnol Biochem 64: 1737–1742.
18. Wright ADG, Williams AJ, Winder B, Christophersen CT, Rodgers SL, et al. (2004) Molecular diversity of rumen methanogens from sheep in Western Australia. Appl Environ Microbiol 70: 1263–1270.
19. Wright ADG, Toovey AF, Pimm CL (2006) Molecular identification of methanogenic archaea from sheep in Queensland, Australia reveal more uncultured novel archaea. Anaerobe 12: 134–139.
20. Wright ADG, Ma X, Obispo NE (2008) Methanobrevibacter phylotypes are the dominant methanogens in sheep from Venezuela. Microb Ecol 56: 390–394.
21. Popova M, Morgavi DP, Martin C (2013) Methanogens and methanogenesis in the rumens and ceca of lambs fed two different high-grain-content diets. Appl Environ Microbiol 79: 1777–1786.

22. Ozutsumi Y, Tajima K, Takenaka A, Itabashi H (2012) The *mcrA* gene and 16S rRNA gene in the phylogenetic analysis of methanogens in the rumen of faunated and unfaunated cattle. Anim Sci J 83: 727–734.
23. Kittelmann S, Seedorf H, Walters WA, Clemente JC, Knight R, et al. (2013) Simultaneous amplicon sequencing to explore co-occurrence patterns of bacterial, archaeal and eukaryotic microorganisms in rumen microbial communities. PloS One 8: e47879.
24. Tymensen LD, McAllister TA (2012) Community structure analysis of methanogens associated with rumen protozoa reveals bias in universal archaeal primers. Appl Environ Microbiol 78: 4051–4056.
25. Poulsen M, Schwab C, Jensen BB, Engberg RM, Spang A, et al. (2013) Methylotrophic methanogenic Thermoplasmata implicated in reduced methane emissions from bovine rumen. Nature Commun 4: 1428.
26. Shakya M, Quince C, Campbell JH, Yang ZK, Schadt CW, et al. (2013) Comparative metagenomic and rRNA microbial diversity characterization using archaeal and bacterial synthetic communities. Environ Microbiol 15: 1882–1899.
27. Yu ZT, Morrison M (2004) Improved extraction of PCR-quality community DNA from digesta and fecal samples. BioTechniques 36: 808–812.
28. Skillman LC, Evans PN, Naylor GE, Morvan B, Jarvis GN, et al. (2004) 16S ribosomal DNA-directed PCR primers for ruminal methanogens and identification of methanogens colonising young lambs. Anaerobe 10: 277–285.
29. Wright ADG, Pimm C (2003) Improved strategy for presumptive identification of methanogens using 16S riboprinting. J Microbiol Meth 55: 337–349.
30. Cole JR, Wang Q, Cardenas E, Fish J, Chai B, et al. (2009) The ribosomal database project: Improved alignments and new tools for rRNA analysis. Nucleic Acids Res 37: D141–D145.
31. Huber T, Faulkner G, Hugenholtz P (2004) Bellerophon: A program to detect chimeric sequences in multiple sequence alignments. Bioinformatics 20.
32. Felsenstein J (1989) Phylogeny inference package (version 3.2). Cladistics 5: 164–166.
33. Schloss PD, Westcott SL, Ryabin T, Hall JR, Hartmann M, et al. (2009) Introducing MOTHUR: Open-source, platform-independent, community-supported software for describing and comparing microbial communities. Appl Environ Microbiol 75: 7537–7541.
34. Altschul SF, Gish W, Miller W, Myers EW, Lipman DJ (1990) Basic Local Alignment Search Tool. J Mol Biol 215: 403–410.
35. Li W, Godzik A (2006) Cd-hit: A fast program for clustering and comparing large sets of protein or nucleotide sequences. Bioinformatics 22: 1658–1659.
36. Good IJ (1953) The population frequencies of species and the estimation of population parameters. Biometrika 40: 237–264.
37. Shannon CE (1948) A mathematical theory of communication. Bell System Technical Journal 27: 379–423.
38. Chao A (1984) Nonparametric-estimation of the number of classes in a population. Scand J Stat 11: 265–270.
39. Larkin MA, Blackshields G, Brown NP, Chenna R, McGettigan PA, et al. (2007) Clustal W and clustal X version 2.0. Bioinformatics 23: 2947–2948.
40. Tamura K, Peterson D, Peterson N, Stecher G, Nei M, et al. (2011) MEGA5: Molecular evolutionary genetics analysis using maximum likelihood, evolutionary distance, and maximum parsimony methods. Mol Biol Evol 28: 2731–2739.
41. Saitou N, Nei M (1987) The neighbor-joining method - a new method for reconstructing phylogenetic trees. Mol Biol Evol 4: 406–425.
42. Handelsman J, Rondon MR, Brady SF, Clardy J, Goodman RM (1998) Molecular biological access to the chemistry of unknown soil microbes: A new frontier for natural products. Chem Biol 5: R245–R249.
43. Aird D, Ross MG, Chen W, Danielsson M, Fennell T, et al. (2011) Analyzing and minimizing PCR amplification bias in Illumina sequencing libraries. Genome Biol 12: R18.
44. Schloss PD, Gevers D, Westcott SL (2011) Reducing the effects of PCR amplification and sequencing artifacts on 16S rRNA-based studies. PloS One 6: e27310.

45. Schloss PD (2010) The effects of alignment quality, distance calculation method, sequence filtering, and region on the analysis of 16S rRNA gene-based studies. PLoS Comp Biol 6: e1000844.

46. DeSantis TZ, Hugenholtz P, Larsen N, Rojas M, Brodie EL, et al. (2006) Greengenes, a chimera-checked 16S rRNA gene database and workbench compatible with ARB. Appl Environ Microbiol 72: 5069–5072.

47. Pruesse E, Quast C, Knittel K, Fuchs BM, Ludwig W, et al. (2007) SILVA: A comprehensive online resource for quality checked and aligned ribosomal RNA sequence data compatible with ARB. Nucleic Acids Res 35: 7188–7196.

48. Caporaso JG, Kuczynski J, Stombaugh J, Bittinger K, Bushman FD, et al. (2010) QIIME allows analysis of high-throughput community sequencing data. Nature Methods 7: 335–336.

49. Shin EC, Choi BR, Lim WJ, Hong SY, An CL, et al. (2004) Phylogenetic analysis of archaea in three fractions of cow rumen based on the 16S rDNA sequence. Anaerobe 10: 313–319.

50. Edwards JE, McEwan NR, Travis AJ, Wallace RJ (2004) 16S rDNA library-based analysis of ruminal bacterial diversity. Anton v Leeuwen 86: 263–281.

51. Paul K, Nonoh JO, Mikulski L, Brune A (2012) "Methanoplasmatales," Thermoplasmatales-related archaea in termite guts and other environments, are the seventh order of methanogens. Appl Environ Microbiol 78: 8245–8253.

52. Santamaria M, Fosso B, Consiglio A, De Caro G, Grillo G, et al. (2012) Reference databases for taxonomic assignment in metagenomics. Briefings in Bioinformatics 13: 682–695.

53. Luo Y, Su Y, Wright AG, Zhang L, Smidt H, et al. (2012) Lean breed landrace pigs harbor fecal methanogens at higher diversity and density than obese breed Erhualian pigs. Archaea 2012: 605289. doi: 10.1155/2012/605289.

54. Eckburg PB, Bik EM, Bernstein CN, Purdom E, Dethlefsen L et al. (2005) Diversity of the human intestinal microbial flora. Science 308: 1635–1638.

55. Thorpe A (2009) Enteric fermentation and ruminant eructation: The role (and control?) of methane in the climate change debate. Clim Change 93: 407–431.

56. IPCC (2007) Climate Change 2007: Synthesis Report. Contribution of Working Groups I, II and III to the Fourth Assessment Report of the Intergovernmental Panel on Climate Change. Pachauri RK, Reisinger A, editors. IPCC, Geneva, Switzerland, 104 p.

The Effect of Different Adjuvants on Immune Parameters and Protection following Vaccination of Sheep with a Larval-Specific Antigen of the Gastrointestinal Nematode, *Haemonchus contortus*

David Piedrafita[1,2]*, Sarah Preston[1,2], Joanna Kemp[1], Michael de Veer[1], Jayne Sherrard[1], Troy Kraska[1], Martin Elhay[3], Els Meeusen[1,2]

1 School of Biomedical Sciences, Monash University, Clayton, Victoria, Australia, **2** The ARC Centre of Excellence in Structural and Functional Microbial Genomics, Monash University, Clayton, Victoria, Australia, **3** Veterinary Medicine Research and Development, Pfizer Animal Health, Parkville, Victoria, Australia

Abstract

It has recently been recognised that vaccine adjuvants play a critical role in directing the nature of a vaccine induced effector response. In the present study, several adjuvants were evaluated for their ability to protect sheep after field vaccination with the larval-specific *Haemonchus contortus* antigen, HcsL3. Using a suboptimal antigen dose, aluminium adjuvant was shown to reduce the cumulative faecal egg counts (cFEC) and worm burden by 23% and 25% respectively, in agreement with a previous study. The addition of Quil A to the aluminium-adjuvanted vaccine brought cFEC back to control levels. Vaccination with the adjuvant DEAE-dextran almost doubled the protection compared to the aluminium-adjuvanted vaccine resulting in 40% and 41% reduction in cFEC and worm counts compared to controls. Examination of skin responses following i.d. injection of exsheathed L3, revealed that cFEC was negatively correlated with wheal size and tissue eosinophils for the DEAE-dextran and aluminium-adjuvanted groups respectively. These studies have for the first time shown the potential of DEAE-dextran adjuvant for helminth vaccines, and discovered significant cellular correlates of vaccine-induced protection.

Editor: Paulo Lee Ho, Instituto Butantan, Brazil

Funding: Australian Research Council Linkage grant LP0668945 and the ARC-Centre of Excellence in Structural and Functional Microbial Genomics. The funders have no role in study design, data collection and analysis, decision to publish, or preparation of the manuscript.

Competing Interests: One of the co-authors, M Elhay, is an employee of Zoetis (formally Pfizer VMRD).

* E-mail: david.piedrafita@monash.edu

Introduction

The generation of natural immunity against gastrointestinal nematode (GIN) parasites, and helminth parasites in general, displays some unique characteristics compared to viral and bacterial infections, in particular in the recruitment and activation of 'allergic' or type-2 effector cells (mast cells and eosinophils) [1,2,3]. Attempts to generate subunit vaccines against GIN parasites have in the past relied heavily on successes achieved with microbial vaccines, including the use of potent vaccine adjuvants that generate high antibody responses, the major correlates of protection in most existing vaccines [4]. While helminth vaccines based on the 'hidden antigen' approach i.e. not boosted by natural immunity, may also rely on high antibody titres [5], it is likely that vaccination strategies aimed at mimicking and accelerating natural immunity will require the induction of both cellular and humoral immunity including the induction of a type-2 effector response.

Vaccine adjuvants have received increased attention in recent years with the realisation that they are the main drivers of both the magnitude and type of adaptive response generated after vaccination [6,7]. For helminth vaccines aimed at replicating natural immunity, a type-2 immune response may be essential to achieve protection. Indeed, in a previous small pen trial, we observed that immunization with a purified, larval-specific surface antigen of *H. contortus*, was only protective when administered with the type-2 adjuvants, aluminium hydroxide and cholera toxin, while addition of pertussis toxin increased antibody titres but abrogated protection [8].

In the present study, we performed a more extensive trial using 3 different adjuvants currently in use in veterinary vaccines, and determined both the levels of protection and the immune response generated. In order to be compatible with farm management practices, vaccinations were performed on pasture using only two subcutaneous immunizations. A preliminary dose-response trial established significant high levels (61–69%) of protection with the largest antigen dose and a significant but lower 27% protection in the second highest dose group using aluminium hydroxide as the adjuvant [9]. To assess the capacity of different adjuvants to improve protection and conserve antigen, the second highest antigen dose was used in the present study. The results of the study established that one rarely used adjuvant improved protection over aluminium adjuvant, while another more widely used type-1 adjuvant abrogated protection.

Assessment of vaccine efficacy based on infection trials is costly and labour intensive and the availability of more amenable immune correlates of protection is desirable for the development and validation of most vaccines [10]. Antibody levels are at present the only known correlates of vaccine-induced immunity, however no correlation with antibodies and protection has been observed with the current larval-specific vaccine [8,9]. In the present study, we also examined different immune parameters after vaccination and discovered significant immune correlates of protection in the two vaccinated groups of sheep that showed reduced egg counts and worm burdens.

Materials and Methods

Preparation and formulation of the vaccine

A surface extract was prepared from exsheathed L3 as described previously [9]. Briefly, L3 were exsheathed with $CO2$, resuspended in PBS and placed in a boiling water bath for 15 min. Larvae were pelleted by centrifugation and the supernatant, containing the surface extract, was depleted of small MW molecules and concentrated in one step using 50 kD cut-off Centricon Centrifugal Filter Units (Millipore). This resulted in one dominant, typically broad 75–90 kD band on PAGE, weakly staining with coomassie but reacting strongly with the HcsL3-specific mAb Hc22. As the antigen stained very poorly with coomassie (and not at all with silver stain) accurate weight estimation was not possible and vaccine doses were therefore expressed as L3-equivalents. Each antigen vaccinated sheep received an antigen dose extracted from 20,000 L3 equivalents, a suboptimal dose previously resulting in a small but still significant 27% reduction in cumulative faecal egg count (cFEC) using the adjuvant aluminium hydroxide [9].

Immunization setup is summarised in Table 1. Three groups of 10 sheep each were vaccinated with antigen added to aluminium adjuvant with (Group 2) or without (Group 4) addition of Quil A (Brenntag Nordic, Frederikssund, Denmark), or with antigen added to DEAE-dextran adjuvant (DD) (Sigma-Aldrich) (group 3). A separate group of sheep (Group 1, control) received the aluminium adjuvant without antigen. As preliminary *in vitro* experiments (not shown) established strongest binding of the antigen to aluminium phosphate ($AlPO_4$), this preparation was chosen instead of the aluminium hydroxide used in previous experiments [8,9], and prepared in house by mixing 4.73 mL of 35% w/v $AlCl_3$ with 17.04 mL of 25% w/v $Na_3PO_4.12H_2O$ and 0.47 mL of 30% w/v NaOH, adjusted to final volume with water. The aluminium phosphate was adjusted to a concentration of 1 mg/ml aluminium with sterile PBS (pH 7.2) in the final vaccination dose and mixed thoroughly with (Groups 2 &4) or without (control Group1) the antigen on an automated rotator for 1 h at room temperature (25°C). Quil A (2 mg/ml) was adjusted to a concentration of 1 mg/ml with sterile PBS (pH 7.2) in the final vaccination dose and added to the antigen/$AlPO_4$ preparation before mixing (Group 2). DEAE-dextran (DD) (20%w/v; pH 6.8) was adjusted to a concentration of 100 mg/ml with sterile PBS (pH 7.2) in the final vaccination dose and mixed thoroughly with the antigen as described above. Each vaccine dose was contained in 1 ml solution. Enough vaccine was prepared for 2 immunization doses, and the second dose was stored at 4°C until used.

Experimental animals, immunization and challenge protocol

Merino-cross wethers were raised and maintained on pasture at a Woodend farm (Northern Victoria). At 8 months of age (week -14), forty sheep were selected and treated with a long-acting anthelmintic, Cydectin®, to remove any existing parasites. Only low egg counts were detected throughout the grazing period (<100 eggs per gram) and no egg counts were detected after treatment. At week -8, sheep were randomised and allocated to 4 experimental groups (n = 10) based on stratified body weight ranking, bled for pre-vaccination serum and given their first immunization dose by subcutaneous injection in the neck region. Four weeks later (week – 4), they were given their second immunization and bled one week later (week -3) for the post-vaccination serum. Two weeks after the second immunization (week -2), they were transported and housed in a large indoor shed at the Monash University experimental Werribee farm. After two weeks acclimatization (week 0), they were infected twice with 7000 L3 on day 0 and day +3 using the *H. contortus* strain, Haecon-5. This strain was isolated in 2006 from the field by Novartis Animal Health, and shown to be more pathogenic than the previously used laboratory strains. Two sheep died on pasture and one indoors due to causes unrelated to the trial.

Parasitological, serological and haematological measurements

Faecal egg counts (FEC) were assessed between 21 to 56 days post infection (dpi) according to the modified McMaster method and expressed as eggs per gram (EPG) faeces. Egg counts were determined for each collection day, and mean cumulative faecal egg counts (cFEC) were calculated by adding all EPG values for each sheep over the whole collection period. Worm burdens were collected from Groups 1, 3 and 4 and worm recovery was performed as described previously [9]. Briefly, the stomach (abomasum) was removed and cut along the greater curvature. Abomasal contents and washings were collected and made up to a volume of 2 L with tap water containing 1% formalin. The solution was vigorously bubbled with air and 10% transferred to glass trays for manual counting of parasites.

Serum samples were collected before vaccination (week -8), one week after the second immunization (week -3), before challenge (week 0) and one week after challenge (week +1). For determining packed cell volume (PCV), blood was collected before (day 0) and 1, 4 and 7 weeks after challenge in 10 ml EDTA tubes and spun in a Haematocrit centrifuge.

Serum anti-HcsL3 antibodies were determined by enzyme linked immunosorbant assay (ELISA) as described previously [8,9]. Briefly, ELISA plates (Nunc, Denmark) were coated overnight with L3 surface extract in 100 μl carbonate buffer pH 9.6 and incubated with various dilutions of sheep sera. Specific antibody isotype responses were determined by incubating with mAbs against ovine IgG1, IgG2 [11] and IgA (Serotec, Bicester, UK) followed by HRP-conjugated rabbit anti-mouse reagent (DAKO, Denmark) and developed with 3′, 3′, 5′, 5′- tetramethyl-benzidine dihydrochloride hydrate (TMB; Invitrogen, VIC, Australia). Antigen-specific IgE was determined by ELISA of ammonium sulphate-treated serum samples as described previously [12]. Antibody levels were also compared to a separate control group of 8 sheep that had been kept in indoor pens, worm free for 2–3 months (Penned group).

Table 1. Sheep numbers, immunization protocols and levels of protection against *H.contortus* infection.

Group #	Sheep numbers (n)	Adjuvant	Antigen (*Hc*L3) L3 equivalents	cFEC; % protection relative to control	Mean worm burdens; % protection relative to control
1 (Control)	10	AlPO₄	0	-	-
2	10	AlPO₄ + Quil A	20 K	3	ND
3	10	DEAE-dextran(DD)	20 K	40*	41**
4	10	AlPO₄	20 K	23	25

ND: not done; *p<0.05, **p<0.001.

Intradermal injections and skin responses

Two days before kill, cutaneous hypersensitivity reactions were performed by two intradermal injections of 100 *H. contortus* exsheathed L3 in 100 μl saline in the right inside back leg of the animal. Two injections of saline were administered as a control at two adjacent sites. The skin temperature and wheal size were measured at 2 hours post injection and at 48 hrs, just prior to euthanasia. Skin temperature at the injection site was recorded with a digital infra-red temperature gauge (Kelly supply company, Australia) and wheal size was measured with digital calipers. After gross removal of the injected skin area, a hollow punch was used to cut skin samples of approximately 1 cm² in size which was sufficient to remove the majority of the inflamed tissue surrounding the injection site, and this was divided in 2 equal parts for histology and future RNA extraction. Histology samples were placed in 10% neutral buffered formalin and then embedded in paraffin, sectioned and stained with haematoxylin and eosin by the Monash University Histology Laboratory. Eosinophils in the entire biopsy sections were counted and expressed as the mean number of cells per 1 mm² tissue (+/- SEM).

Ethics Statement

Handling of animals and experimental procedures were approved by the Monash University Animal Ethics Committee (Ethics # SOBSA/P/2009/44).

Statistical Analysis

Data were analysed using GraphPad Prism5. Each vaccinated group was compared against the adjuvant control group using Student's t-test with significance set at p<0.05. Values were log transformed before analysis if variations were significantly different. One way ANOVA was used to compare values between all groups. The Spearman's rank correlation coefficient was calculated to determine significant dependence between two variables.

Results

Protection levels after challenge infection of sheep vaccinated with HcsL3 administered with different adjuvants.

Eggs per gram faeces (EPG) and cumulative faecal egg counts (cFEC) were reduced in group 4, immunized with antigen and aluminium adjuvant, compared to the control group 1 (Fig.1A&B). The degree of protection (23%, Table 1) was similar to that observed previously with this antigen dose (27%, [9]), although levels were not significant in this case. Addition of Quil A to the same vaccine (group 2) increased egg counts to control levels (Fig.1A&B). Group 3 which received DD as the adjuvant showed

the best protection, which was almost double that of group 4 (40%) and reached significance compared to the control group 1 (Table 1). To ensure that the increased protection in group 3 was not due to the DD adjuvant only, in a separate infection trial (not shown) a group of 10 sheep injected with DD alone were compared with a non-injected control group; no difference was observed in the cFEC of these groups, with cFEC of 23893 (SEM ±7628) and 23420 (SEM ±6693), respectively.

Post-mortem worm counts were performed on the control group 1 and the two vaccinated groups 3 and 4 (Fig. 1C). Mean worm counts in group 3 were significantly reduced compared to the control group (p = 0.008). Individual worm counts in group 4 showed wide segregation, indicative of responder and non-responder sheep resulting in no significant difference of the mean compared to group 1. Protection levels for worm counts were similar to the cFEC at 41% and 25% for groups 3 and group 4, respectively (Table 1).

All groups showed a decrease in PCV values after infection but the mean PCV values at the end of the experiment were similar in groups 3 and 4 and, although not significant, higher than those in groups 1 and 2 (Fig. 2).

Antibody responses during vaccination and challenge

As observed previously [9], antibody levels overall were elevated in pasture reared sheep compared to penned sheep even before the start of vaccination, and levels decreased in the control and most vaccinated groups when transferred to indoor pens (Fig. 3). Only the DD group showed significantly increased antibody levels 1 week after the second vaccination compared to the unvaccinated control group for the IgG2 and IgE isotypes (Fig. 3). No increase in antibody levels was detectable after the challenge infection.

Skin responses after intradermal injection of saline or L3 larvae

Skin temperatures at 2 h were consistently lower in the L3 compared to the saline injected sites (Fig.4A), and this was significant for the two aluminium adjuvant-vaccinated groups (Groups 2 and 4).

Wheal sizes were elevated in all groups at 2 h with no significant difference between groups (Fig.4B). Wheal sizes had decreased significantly by 48 hrs in groups 1, 2 and 4 but remained at a higher level in group 2 compared to group 4. Group 3 showed only a small decrease in wheal size from 2 to 48 h, and its 48 h wheal size was significantly higher than groups 1 and 4.

There was no difference in eosinophil counts between the control and vaccinated groups in biopsies taken 48 h after saline injection, with a combined mean of 21 ± 9.3 eos/mm². After injection of L3, skin eosinophil counts increased in all groups compared to the saline injections (Fig.4C). This was most

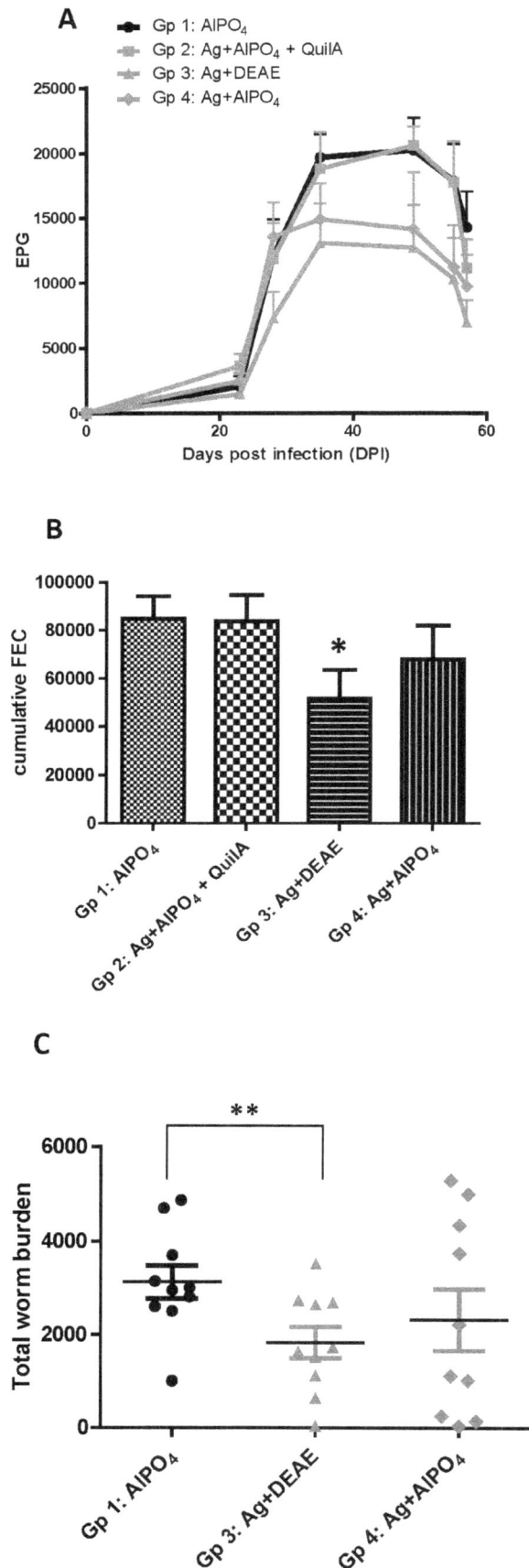

Figure 1. Faecal egg counts and worm burden after challenge of vaccinated sheep. A: Mean (+SEM) daily eggs/gram faeces (EPG); B: Mean (+SEM) cumulative faecal egg count (FEC); C: individual and mean (+/−SEM) total worm counts. For details of experimental groups, refer to Table 1. Significantly different from the control group at $p < 0.05$ (*) or $p < 0.01$ (**).

pronounced in the DD group 3, where numerous eosinophils were present, often aggregated in large granulomas (Fig. 5) which were not included in the counts. One sheep in the aluminium adjuvant vaccinated group 4 showed a much higher eosinophil count (2969 ± 801) then the other sheep in this group, and with this outlier excluded, mean eosinophil counts in group 4 were significantly below the control group 1 (Fig. 4C).

Correlations between protection, antibody levels and skin responses

The Spearman's rank correlation coefficients were calculated for all parasitological and immunological measurements, and correlations that showed significance detailed in Table 2. cFEC were significantly correlated with worm counts across all groups examined, and negatively correlated with PCV values. There was no significant correlation between cFEC and any of the antibody isotype levels, except for a slight negative correlation with IgE when all groups were combined. There were however significant but distinct correlations between skin responses and cFEC in the two vaccinated groups with lowest egg counts, groups 3 and 4. Group 3 showed a highly significant negative correlation of cFEC with the 2 h wheal response, while cFEC in group 4 were highly positively correlated with changes in skin temperature at 2 h and negatively correlated with eosinophil counts 48 h after i.d. injection of L3 larvae. The vaccinated group 2 (AlPO$_4$+Quil A) also showed a lower but significant negative correlation between cFEC and L3 eosinophil counts.

IgG2, but not IgE, correlated significantly with wheal size at 2 h

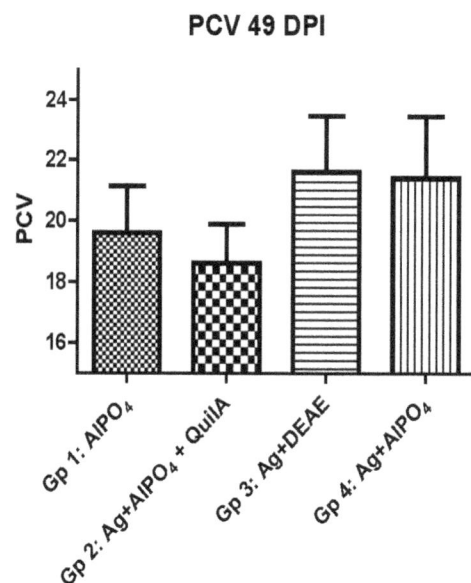

Figure 2. Packed cell volume (PCV) of peripheral blood after challenge of vaccinated sheep. Mean (+SEM) PCV of groups at 49 days post infection (DPI). The mean PCV of all sheep before infection (day 0) was 34.2 (SD 2.3). For details of experimental groups, refer to Table 1.

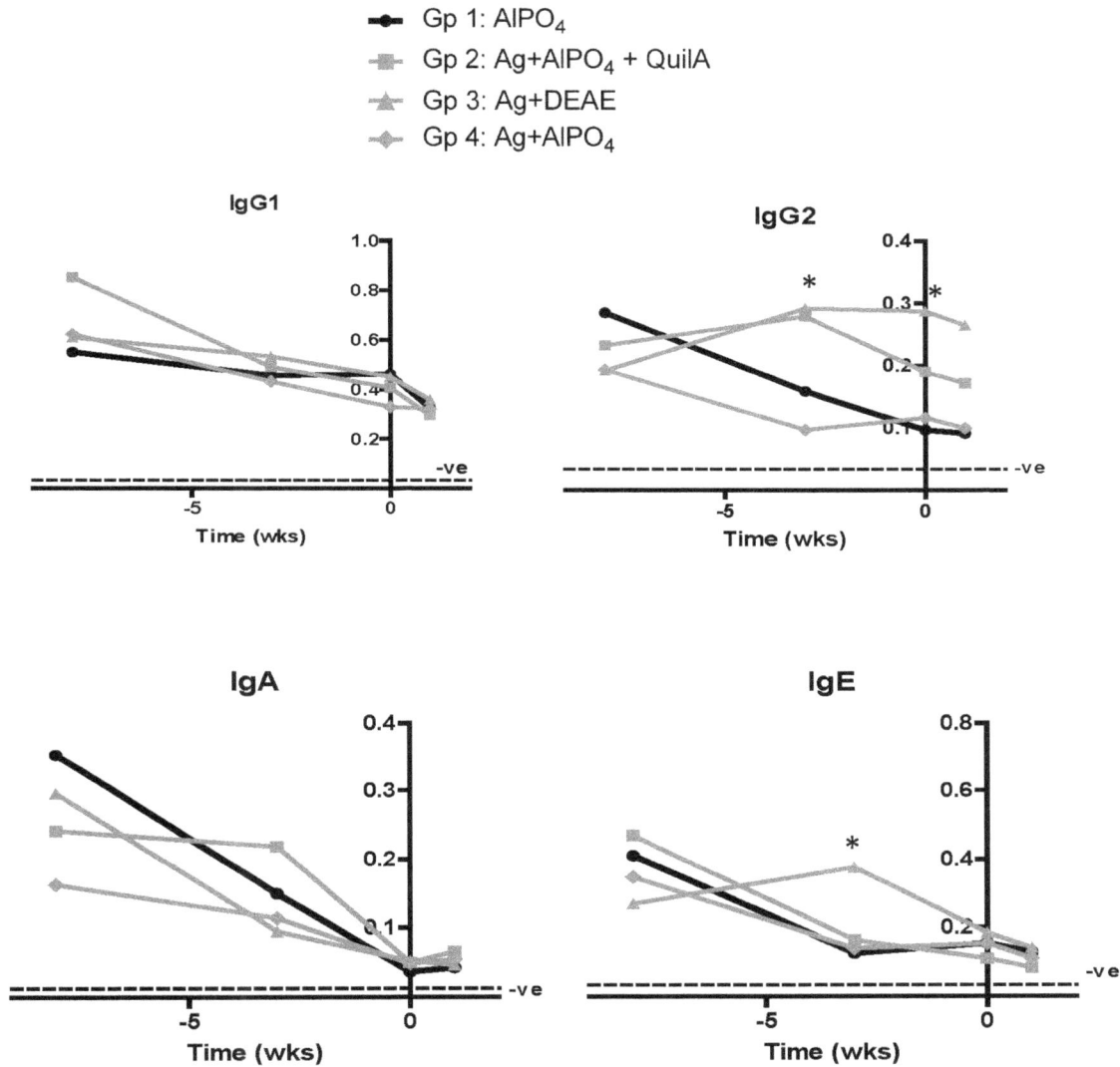

Figure 3. Serum antibody responses after challenge of vaccinated sheep. Mean antibody isotype responses of sheep before vaccination (wk -8), 1 week after vaccination (wk -3), at day of challenge (wk 0) and 1 week after challenge (wk +1). Serum diluted 1/5000 (IgG1), 1/500 (IgG2), 1/20 (IgA) or 1/32 ammonium sulphate cut (IgE). For details of experimental groups, refer to Table 1. Dashed line represents the mean of a separate group of 8 sheep kept in indoor pens for 2–3 months. *Significant difference between group 3 and control group 1.

in group 3 and when all groups were combined. At 48 h, this correlation was lost and replaced by significant correlations with IgE and IgG1, again only in group 3 or when all groups were combined.

Discussion

Recent advances in innate immunity have revealed the critical role adjuvants and vaccine delivery systems play in directing the strength and nature of a vaccine response [6,7]. In particular, innate receptors, including Toll-like receptors (TLRs) have received wide attention as potential new targets for incorporation into vaccine formulations [13]. However, in most cases of currently used adjuvant systems, the exact innate stimulation pathways are unknown and may not involve TLR activation [6]. This is the case for the most commonly used aluminium adjuvants [14], as well as the two other adjuvants used in the present study, DEAE-dextran (DD) and Quil A. As adjuvants may act differently

in different species [15], it is critical to test each experimental vaccine with different adjuvants in the target species.

Vaccination in the current study was performed with a larval-specific antigen and the protective effect is therefore likely to manifest against the early L3 stage during the first 1–2 days after infection, before it moults to an L4. The local response to challenge infections in the gastrointestinal tract is however difficult to study without sacrificing the animals. An attempt was therefore made to replicate the vaccine-induced response against L3 larvae by intradermal injections of exsheathed L3 and subsequent examination of the injection sites. This was done after establishment of the challenge infection so as not to compromise the vaccine efficacy study. Inflammation is generally associated with an increase in temperature and it was therefore surprising that temperature measurements at the L3 injected sites were consistently lower than in the saline injected sites, and this difference was significant in the two groups vaccinated with aluminium adjuvant. This may be related to the findings that type-2 immune responses

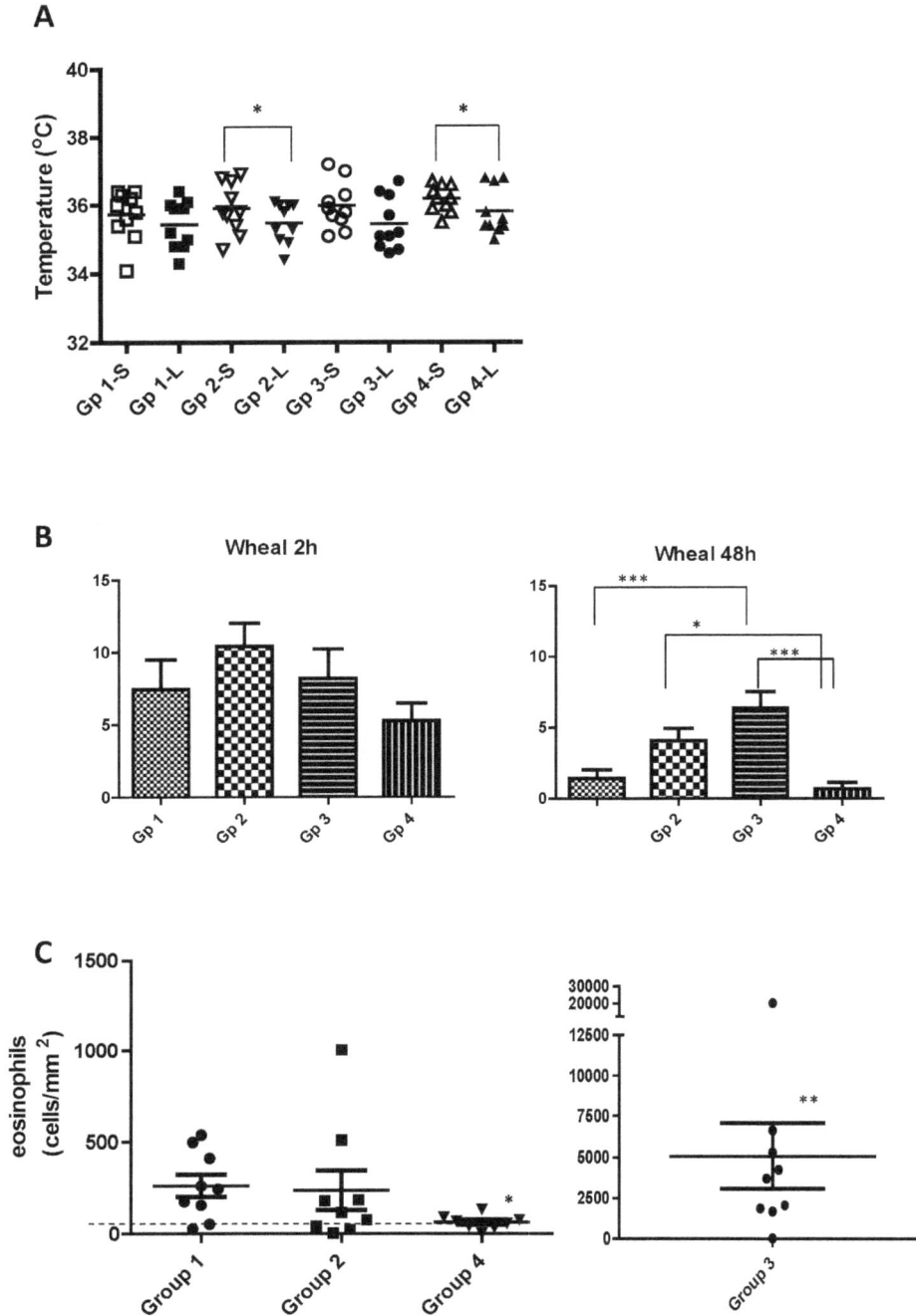

Figure 4. Skin responses after intradermal injection of saline or exsheathed L3 *H. contortus* larvae (xL3). A: temperature of the injected skin sites measured 2 h after intradermal injection of saline (S) or xL3 (L). B: wheal responses measured 2 h or 48 h after intradermal injection of xL3. C: Number of eosinophils/mm² tissue section of xL3 injected sites. Dashed line delineates the upper 95% confidence interval from the mean of the combined saline injected sites. For details of experimental groups, refer to Table 1. Asterisk denotes significance levels at $p<0.05$ (*), $p<0.01$ (**) or $p<0.001$ (***).

induced by helminth extracts or aluminium adjuvants dampen type-1 immune responses and pro-inflammatory cytokines [16,17].

Aluminium adjuvants have been used in most vaccines since the early 1900 and have a proven immune boosting and safety record [6,14]. The exact action of aluminium adjuvant has not been completely elucidated but seems to be associated with its unique interaction with dendritic cells both *in vitro* [18] and *in vivo* [19].

Aluminium adjuvants are known to bias the immune response towards a type-2 phenotype, including the recruitment of eosinophils, which is generally not considered favourable for bacterial or viral vaccines [14,17]. However, for protection against helminth parasites, a type-2 response may be advantageous, as eosinophil killing of helminth larvae is a prominent feature of natural immunity in these infections [20]. In particular, eosinophil-mediated killing of

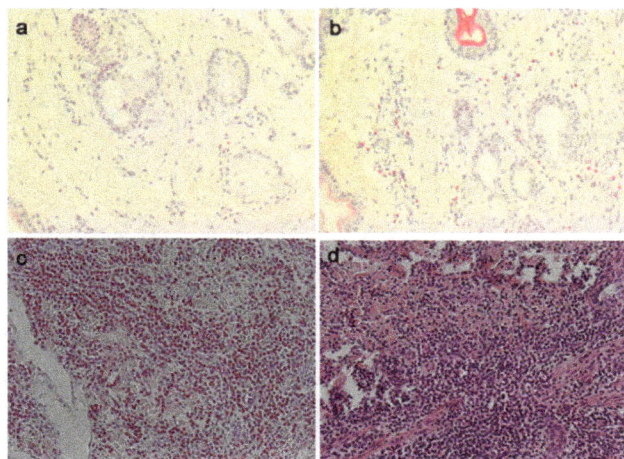

Figure 5. Eosinophil infiltration after intradermal injection of saline or exsheathed L3 *H. contortus* larvae (xL3). Representative H&E stained sections of skin tissue injected with saline (a) or xL3 (b) of group 1,2&4 sheep. Group 3 sheep (Ag+DEAE-dextran) showed much higher eosinophil infiltration (c) and occasional granulomas (d).

L3 *H. contortus* larvae has been shown both *in vitro* [21] and *in vivo* [22]. Surprisingly, eosinophil numbers were not increased but slightly reduced in skin biopsies after injection with L3 in the aluminium vaccinated sheep compared to the adjuvant control group. Eosinophil numbers did however correlate strongly with protection in the aluminium vaccine group, suggesting that they are involved in larval killing as observed during natural infection [20,22]. *In vitro* studies, have shown that activation of eosinophils is required for effective killing [21] and this may occur in the aluminium vaccinated but not in the control sheep. In addition, as eosinophils in immune sheep are strongly attracted to the tissue larvae [20,22], it is possible that cell counts of surrounding tissues may not adequately reflect the true level of eosinophil recruitment.

Quil A is a quintessential type-1 adjuvant currently incorporated into several veterinary vaccines [23]. In agreement with the type-2 hypothesis above and its known downregulation by type-1

cytokines, any protection observed in the aluminium adjuvant group was abrogated by the addition of Quil A to the vaccine. Previous studies using pertussis toxin or Freund's complete adjuvant in the HcsL3 vaccine, while increasing the antibody responses, have also resulted in a lack of protection or even exacerbation of infection [8,24]. Together, these studies emphasise the importance of choosing the right adjuvant for a particular disease or a desired immune effector response.

DEAE-dextran (DD) is a high molecular weight positively charged polymer that has received renewed interest as a vaccine adjuvant due to its antibody enhancing properties in anti-fertility vaccines [25]. The mode of action and the type of immune response elicited with this adjuvant have however not been studied in detail. In the present study, DD was the only adjuvant that elicited a detectable increase in antigen-specific antibodies, and these were significant for the IgG2 and IgE isotypes. In addition, skin biopsied after i.d. L3 injection revealed a massive increase in eosinophils, orders of magnitude above the control and aluminium injected sheep, indicating a predominant type-2 immune response induced by this adjuvant. Eosinophil numbers in skin biopsies taken 48 h after i.d. L3 injection, did not however correlate with protective immunity in this group, but were difficult to accurately enumerate due to their large numbers and the presence of numerous eosinophil-rich granulomas. Protection in the DD group of sheep was significantly correlated with the 2 h wheal responses suggesting that the protection observed in this group was manifest at this early time period and eosinophil numbers at this time may show correlation with protection. It is also possible that other cell types such as neutrophils and monocytes may have been recruited and contributed to protection and this requires further detailed studies at the earlier time points.

The present study has confirmed the critical dependence of effective vaccines against infections on the nature of the immune response induced by a particular adjuvant system. For helminth vaccines that are based on mimicking natural immunity against invading larvae, this may involve the use of adjuvants that induce a type-2 immune response, most likely involving eosinophil-mediated killing. In contrast to most bacterial and viral vaccines where protection is based on antibodies only, the present vaccine has shown no correlation with antibody levels, but significant correlations with cellular responses, in particular eosinophil recruitment

Table 2. Correlations between parasitological and immunological parameters in control and vaccinated sheep.

	Group1: AlPO$_4$; n = 10	Group 2: Ag+AlPO$_4$+QuilA; n = 10	Group 3: Ag+DEAE-dextran n = 10	Group 4: Ag +AlPO n = 10	All Groups n = 30/40
Correlations with cumulative faecal egg counts (cFEC)					
cFEC vs. worm burden	**0.78* (0.01)**	ND	**0.66 (0.04)**	**0.78 (0.01)**	**0.77 (<0.001)**
cFEC vs. PCV	**−0.68 (0.03)**	−0.60 (0.07)	**−0.76 (0.01)**	**−0.83 (0.003)**	**−0.74 (<0.001)**
cFEC vs. IgE (wk 0)	−0.048 (0.90)	−0.57 (0.09)	−0.09 (0.80)	−0.09 (0.80)	**−0.40 (0.01)**
cFEC vs. wheal (2 h)	−0.003 (0.99)	0.22 (0.54)	**−0.82 (0.004)**	0.032 (0.93)	−0.13 (0.41)
cFEC vs. eos/mm2 (L3)	−0.28 (0.46)	**−0.78 (0.02)**	0.37 (0.34)	**−0.82 (0.01)**	−0.30 (0.13)
cFEC vs. temp change(saline/L3)	0.34 (0.34)	0.05 (0.90)	0.16 (0.66)	**0.77 (0.01)**	0.29 (0.07)
Correlations of antibody isotypes 1 week after 2nd vaccination (wk -3) and skin responses after i.d. injection of L3					
wheal 2 h vs. IgE	0.08 (0.83)	0.38 (0.28)	0.28 (0.43)	0.32 (0.37)	0.22 (0.17)
wheal 2 h vs. IgG2	−0.20 (0.58)	0.59 (0.07)	**0.64 (0.047)**	0.47 (0.17)	**0.37 (0.02)**
wheal 48 h vs. IgE	0.04 (0.90)	0.33 (0.35)	**0.83 (0.003)**	0.63 (0.067)	**0.70 (<0.001)**
wheal 48 h vs. IgG1	−0.43 (0.22)	0.48 (0.16)	**0.65 (0.04)**	−0.10 (0.80)	**0.39 (0.01)**

*Spearman rank correlation coefficient (P value); bold numbers indicate significance at P<0.05. ND: not done.

and/or activation. It is however likely that specific antibodies will still be required for effective antibody-dependent cell mediated killing of invading larvae [20]. In addition to the protection observed previously with aluminium adjuvant, the results of this study have also provided a more effective adjuvant system for helminth vaccines and revealed significant correlates of vaccine-induced protection.

Acknowledgments

We thank Jason Sedgfield and staff at the Woodend farm for assistance with the sheep trial.

Author Contributions

Conceived and designed the experiments: DP ME EM. Performed the experiments: DP SP JK MdV JS TK EM. Analyzed the data: DP SP JK EM. Contributed reagents/materials/analysis tools: MdV ME. Wrote the paper: DP SP EM.

References

1. Anthony RM, Rutitzky LI, Urban JF Jr., Stadecker MJ, Gause WC (2007) Protective immune mechanisms in helminth infection. Nat Rev Immunol 7: 975–987.
2. Hein WR, Pernthaner A, Piedrafita D, Meeusen EN (2010) Immune mechanisms of resistance to gastrointestinal nematode infections in sheep. Parasite Immunol 32: 541–548.
3. Meeusen EN, Balic A, Bowles V (2005) Cells, cytokines and other molecules associated with rejection of gastrointestinal nematode parasites. Vet Immunol Immunopathol 108: 121–125.
4. Zinkernagel RM, Hengartner H (2006) Protective 'immunity' by pre-existent neutralizing antibody titers and preactivated T cells but not by so-called 'immunological memory'. Immunol Rev 211: 310–319.
5. Bethony JM, Loukas A, Hotez PJ, Knox DP (2006) Vaccines against blood-feeding nematodes of humans and livestock. Parasitology 133 Suppl: S63–79.
6. de Veer M, Meeusen E (2011) New developments in vaccine research—unveiling the secret of vaccine adjuvants. Discov Med 12: 195–204.
7. Pulendran B, Ahmed R (2011) Immunological mechanisms of vaccination. Nat Immunol 12: 509–517.
8. Jacobs HJ, Wiltshire C, Ashman K, Meeusen ENT (1999) Vaccination against the gastrointestinal nematode, Haemonchus contortus, using a purified larval surface antigen. Vaccine 17: 362–368.
9. Piedrafita D, de Veer MJ, Sherrard J, Kraska T, Elhay M, et al. (2012) Field vaccination of sheep with a larval-specific antigen of the gastrointestinal nematode, *Haemonchus contortus*. Vaccine 30: 7199–71204.
10. Thakur A, Pedersen LE, Jungersen G (2012) Immune markers and correlates of protection for vaccine induced immune responses. Vaccine 30: 4907–4920.
11. Beh KJ (1987) Production and characterization of monoclonal antibodies specific for sheep IgG subclasses IgG1 or IgG2. Vet Immunol Immunopathol 14: 187–196.
12. Bischof RJ, Snibson K, Shaw R, Meeusen EN (2003) Induction of allergic inflammation in the lungs of sensitized sheep after local challenge with house dust mite. Clin Exp Allergy 33: 367–375.
13. Duthie MS, Windish HP, Fox CB, Reed SG (2011) Use of defined TLR ligands as adjuvants within human vaccines. Immunol Rev 239: 178–196.
14. Marrack P, McKee AS, Munks MW (2009) Towards an understanding of the adjuvant action of aluminium. Nat Rev Immunol 9: 287–293.
15. Alving CR (2002) Design and selection of vaccine adjuvants: animal models and human trials. Vaccine 20 Suppl 3: S56–64.
16. Jankovic D, Kullberg MC, Caspar P, Sher A (2004) Parasite-induced Th2 polarization is associated with down-regulated dendritic cell responsiveness to Th1 stimuli and a transient delay in T lymphocyte cycling. J Immunol 173: 2419–2427.
17. Korsholm KS, Petersen RV, Agger EM, Andersen P (2010) T-helper 1 and T-helper 2 adjuvants induce distinct differences in the magnitude, quality and kinetics of the early inflammatory response at the site of injection. Immunology 129: 75–86.
18. Flach TL, Ng G, Hari A, Desrosiers MD, Zhang P, et al. (2011) Alum interaction with dendritic cell membrane lipids is essential for its adjuvanticity. Nat Med 17: 479–487.
19. de Veer M, Kemp J, Chatelier J, Elhay MJ, Meeusen EN (2010) The kinetics of soluble and particulate antigen trafficking in the afferent lymph, and its modulation by aluminum-based adjuvant. Vaccine 28: 6597–6602.
20. Meeusen EN, Balic A (2000) Do eosinophils have a role in the killing of helminth parasites? Parasitol Today 16: 95–101.
21. Rainbird MA, Macmillan D, Meeusen ENT (1998) Eosinophil-mediated killing of Haemonchus contortus larvae: Effect of eosinophil activation and role of antibody, complement and interleukin-5. Parasite Immunology 20: 93–103.
22. Balic A, Cunningham CP, Meeusen EN (2006) Eosinophil interactions with Haemonchus contortus larvae in the ovine gastrointestinal tract. Parasite Immunol 28: 107–115.
23. Sun HX, Xie Y, Ye YP (2009) Advances in saponin-based adjuvants. Vaccine 27: 1787–1796.
24. Turnbull IF, Bowles VM, Wiltshire CJ, Brandon MR, Meeusen EN (1992) Systemic immunization of sheep with surface antigens from Haemonchus contortus larvae. Int J Parasitol 22: 537–540.
25. Vizcarra JA, Karges SL, Wettemann RP (2012) Immunization of beef heifers against gonadotropin-releasing hormone prevents luteal activity and pregnancy: Effect of conjugation to different proteins and effectiveness of adjuvants. J Anim Sci 90: 1479–1488.

17

Functional Vascular Changes of the Kidney during Pregnancy in Animals

Joris van Drongelen[1]*, Rob de Vries[2], Frederik K. Lotgering[1], Paul Smits[3], Marc E. A. Spaanderman[1,4]

1 Department of Obstetrics and Gynecology, Radboudumc, the Netherlands, **2** Systematic Review Centre for Laboratory animal Experimentation, Radboudumc, the Netherlands, **3** Department of Pharmacology and Toxicology, Radboudumc, the Netherlands, **4** Department of Obstetrics and Gynecology, Research School GROW, Maastricht University Medical Centre, Maastricht, the Netherlands

Abstract

Renal vascular responses to pregnancy have frequently been studied, by investigating renal vascular resistance (RVR), renal flow, glomerular filtration rate (GFR), and renal artery responses to stimuli. Nonetheless, several questions remain: 1. Which vasodilator pathways are activated and to what extent do they affect RVR, renal flow and GFR across species, strains and gestational ages, 2. Are these changes dependent on renal artery adaptation, 3. At which cellular level does pregnancy affect the involved pathways? In an attempt to answer the questions raised, we performed a systematic review and meta-analysis on animal data. We included 37 studies (116 responses). At mid-gestation, RVR and GFR change to a similar degree across species and strains, accompanied by variable change in renal flow. At least in rats, changes depend on NO activation. At late gestation, changes in RVR, renal flow and GFR vary between species and strains. In rats, these changes are effectuated by sympathetic stimulation. Overall, renal artery responsiveness to stimuli is unaffected by pregnancy, except for Sprague Dawley rats in which pregnancy enhances renal artery vascular compliance and reduces renal artery myogenic reactivity. Our meta-analysis shows that: 1. Pregnancy changes RVR, renal flow and GFR dependent on NO-activation and sympathetic de-activation, but adjustments are different among species, strains and gestational ages; 2. These changes do not depend on adaptation of renal artery responsiveness; 3. It remains unknown at which cellular level pregnancy affects the pathways. Our meta-analysis suggests that renal changes during pregnancy in animals are qualitatively similar, even in comparison to humans, but quantitatively different.

Editor: Jean-Claude Dussaule, INSERM, France

Funding: The authors have no support or funding to report.

Competing Interests: The authors have declared that no competing interests exist.

* Email: joris.vandrongelen@radboudumc.nl

Introduction

Profound vasodilation takes place in early pregnancy, and results in a major reduction in systemic vascular resistance. Vasodilation seems to be maximal at mid-gestation and slowly decreases towards term pregnancy in both humans [1] and animals [2]. Several local vascular responses, affecting the endothelial cell (EC), smooth muscle cell (SMC) and extracellular matrix, are thought to be responsible for gestational vasodilation by: 1. Up regulation of the endothelium-dependent nitric oxide (NO) pathway [3], 2. Reduction of the responsiveness to vasoconstrictor stimuli [4], 3. Reduction in myogenic reactivity [5], and 4. Increase in arterial compliance [5]. It remains to be determined to what extent each factor contributes to gestational vasodilation at the various stages of pregnancy among different species and strains.

Many studies on the mechanisms of vasodilation in pregnancy have focused on the kidney and its renal arteries. The kidney is a prominent vascular region and major contributor to systemic vascular resistance in the non-pregnant and pregnant condition [6,7]. Vessel tone has been studied by investigating several local vascular responses. These responses can be divided by the stimulus involved: 1. Mechanical stimuli (flow-mediated vasodilation, myogenic reactivity, arterial compliance), and 2. Pharmacological stimuli, which can be subdivided by: a. The agent used (including acetylcholine, norepinephrine, phenylephrine, etc.) or b. The mediating receptor complex involved [8]. Most studies have concentrated on agents that affect the G-protein coupled receptors. These can be divided into three common G-protein coupled receptor pathways: the vasodilator Gq_{EC}-pathway which mediates NO-dependent vasodilation, the vasoconstrictor Gq_{SMC}-pathway which induces a calcium rise in the SMC, and the vasodilator Gs_{SMC}-pathway which induces potassium influx in the SMC [8–15].

Various species, strains and gestational age periods have been investigated and often the results have been extrapolated to pregnancy-induced renal adaptation in general and to responsible mechanisms in humans. One may question the validity of such extrapolations. A previous meta-analysis showed that in mesenteric arteries vascular adaptation to pregnancy is strongly dependent on species, strains and gestational ages [8]. This may also be the case for renal arteries.

To investigate which mechanisms are responsible for renal vascular changes in pregnancy and whether they are dependent on species, strains and gestational age periods, we performed a systematic review and meta-analysis on the functional vascular changes of the kidney during pregnancy. Questions to be answered included: 1. Which specific renal vasodilator pathways are activated in pregnancy and to what extent do they affect RVR, renal flow and GFR across species, strains and gestational ages, 2. Are the pregnancy-induced changes in RVR, renal flow and GFR dependent on renal artery adaptation and which specific vasodilator pathways affect the renal artery during pregnancy, and 3. At which cellular level does pregnancy affect the involved pathways?

Materials and Methods

Retrieving the literature

In an attempt to identify all original studies on renal vasoactive responses in animals and humans during first pregnancy, we searched the Pubmed (from 1948) and Embase (from 1980) databases until April 2012. We used a three phase search strategy. Tables 1 and 2 depict the search-strategy for Pubmed and Embase, respectively. No language restriction was used in the primary literature search; later in the process non-English studies were excluded. We also searched the reference lists of included studies to identify additional studies. The search strategy was developed in cooperation with an information specialist from the Medical Library of the Radboud University Nijmegen.

Selection of studies and data extraction

The selection and data extraction process was divided into three phases. In the first phase, a. one investigator (first author) selected studies on the basis of title and abstract, and b. two investigators (first and second author) independently screened all remaining abstracts for inclusion criteria. Differences were resolved by mutual agreement. Studies were selected if they met the following criteria: 1. renal vasoconstrictor/vasodilator responses, 2. first

pregnancy versus nulliparous non-pregnant control, 3. original data, 4.) healthy animals or humans, 5. age- or weight-matched subjects, and 6. for ex-vivo experiments: experiments performed in standard medium (defined as a generally accepted medium for the type of experiments). In the second phase of selection, non-English papers were excluded from further analysis and full articles were analyzed by two investigators (first and second author) independently using the same inclusion criteria as described above. Differences were resolved by mutual agreement. In the third phase, the first author extracted baseline measurements and responses to specific blockade from *in vivo* and whole organ perfusion experiments that related to renal function changes and the involved pathways. In addition, all vasodilator and vasoconstrictor responses were extracted both in the absence and presence of blockade of NO, prostaglandin (PGI_2), renin-angiotensin-aldosterone (RAAS), endothelin, vasopressin or the sympathetic system, or combinations of blockade, with and without denuded endothelium. Responses had to contain ≥ 5 measurements to be included in further analysis, as analysis of sensitivity to stimuli is not accurate with fewer data points.

From each article, study data were extracted and recorded for species, strain, and gestational age (early, mid and late pregnancy, defined as first, second and third trimester, respectively), estrous/cycle stage, and experimental setup (*in vivo*/whole organ perfusion, myograph type used). We extracted all reported vascular responses and recorded for each of them the number of subjects, effect size, and standard deviation or standard error of the mean. For *in vivo* and whole organ perfusion experiments we detected baseline renal vascular resistance (RVR), renal flow (combined renal plasma, blood and perfusion flow), glomerular filtration rate (GFR), autoregulatory threshold (lower pressure limit for stable renal flow), and the response to specific blockades. For pharmacological responses, we extracted EC_{50} (concentration inducing 50% of the maximum effect) and E_{max} (maximum response to stimulus) to indicate sensitivity and maximum reactivity, respectively. For the mechanical responses, we determined the direction of the effect (favoring vasodilation, vasoconstriction, or no effect) and used it for descriptive analysis. Parametric comparison of these responses was not possible, as

Table 1. Literature search-strategy for Pubmed.

Component	Description
Pregnancy	*"pregnancy" [MeSH Terms] OR "pregnancy" [tiab] OR "pregnancies" [tiab] OR "gestation" [tiab] OR "pregnant" [tiab] OR "maternal-fetal relations" [tiab]*
Renal circulation	*"renal artery" [MeSH] OR "renal artery" [tiab] OR "renal arteries" [tiab] OR "kidney artery" [tiab] OR "kidney arteries" [tiab] OR "renal blood vessel" [tiab] OR "renal blood vessels" [tiab] OR "kidney blood vessel" [tiab] OR "kidney blood vessels" [tiab] OR "renal vessel" [tiab] OR "renal vessels" [tiab] OR "kidney vessel" [tiab] OR "kidney vessels" [tiab] OR "arteria renalis" [tiab] OR "arteria renis" [tiab] OR "kidney/blood supply" [MeSH] OR "renal circulation" [MeSH] OR "renal circulation" [tiab] OR "kidney circulation" [tiab] OR "renal blood circulation" [tiab] OR "kidney blood circulation" [tiab] OR "renal blood supply" [tiab] OR "kidney blood supply" [tiab]*
Vasoconstrictor and vasodilator responses	*"vasoconstriction" [MeSH Terms] OR "vasoconstriction" [tiab] OR "vasoconstrictions" [tiab] OR "vasoconstrictor agents" [MeSH Terms] OR "vasoconstrictor agents" [Pharmacological Action] OR "vascular resistance" [MeSH Terms] OR "vascular resistance" [tiab] OR "vascular capacitance" [MeSH Terms] OR ("vascular" [tiab] AND "capacitance" [tiab]) OR "vasoconstrictor" [tiab] OR "vasoconstrictors" [tiab] OR "vasopressor" [tiab] OR "vasoactive agonist" [tiab] OR "vasoactive agonists" [tiab] OR "vasopressors" [tiab] OR "vasomotor system" [MeSH Terms] OR "vasomotor system" [tiab] OR "peripheral resistance" [tiab] OR "artery constriction" [tiab] OR "vessel constriction" [tiab] OR "vasoconstrictive" [tiab] OR "vasoconstricting" [tiab] OR "vasoconstricted" [tiab] OR "vasodilation" [MeSH Terms] OR "vasodilation" [tiab] OR "vasodilatation" [tiab] OR "vasodilatating" [tiab] OR "vasodilating" [tiab] OR "vasodilative" [tiab] OR "vasodilatative" [tiab] OR "artery dilation" [tiab] OR "vessel dilation" [tiab] OR "artery dilatation" [tiab] OR "vessel dilatation" [tiab] OR "vasodilator agents" [MeSH Terms] OR "vasodilator agents" [Pharmacological Action] OR "vasodilator" [tiab] OR "vasodilators" [tiab] OR "vasorelaxation" [tiab] OR "Vascular Endothelium Dependent Relaxation" [tiab] OR "Endothelium Dependent Relaxation" [tiab] OR "Vascular Endothelium-Dependent Relaxation" [tiab] OR "Endothelium-Dependent-Relaxation" [tiab] OR "hemodynamics" [MeSH Terms] OR "hemodynamics" [tiab] OR "hemodynamic" [tiab] OR "vasodilated" [tiab] OR "vasoactive agent" [tiab] OR "vasoactive drug" [tiab] OR "vasoactive drugs" [tiab] OR "dilation" [tiab] OR "dilatation" [tiab] OR "contraction" [tiab] OR "relaxation" [tiab]*

Table 2. Literature search-strategy for Embase.

Component	Description
Pregnancy	*exp pregnancy/OR (pregnancy or pregnant).ti,ab. OR (pregnacies or gestation).ti,ab. OR (pregnancy or pregnant or pregnacies or gestation).ti,ab.*
Renal circulation	*exp renal artery/OR renal artery.ti,ab. OR renal arteries.ti,ab. OR arteria renalis.ti,ab. OR exp kidney artery/OR kidney artery.ti,ab. OR kidney arteries.ti,ab. OR arteria renis.ti,ab. OR kidney blood vessel/OR renal blood vessel.ti,ab. OR renal blood vessels.ti,ab. OR kidney blood vessel.ti,ab. OR kidney blood vessels.ti,ab. OR intrarenal vessel.ti,ab. OR intrarenal vessels.ti,ab. OR renal vessel.ti,ab. OR renal vessels.ti,ab. OR kidney vessel.ti,ab. OR kidney vessels.ti,ab. OR exp kidney blood flow/OR kidney blood flow.ti,ab. OR intrarenal blood flow.ti,ab. OR kidney bloodflow.ti,ab. OR renal blood flow.ti,ab. OR kidney blood supply.ti,ab. OR renal blood supply.ti,ab. OR exp kidney circulation/OR kidney circulation.ti,ab. OR intrarenal circulation.ti,ab. OR kidney blood circulation.ti,ab. OR renal circulation.ti,ab. OR renal blood circulation.ti,ab.*
Vasoconstrictor and vasodilator responses	*exp vasoconstriction/OR exp vasoconstrictor agent/OR vasoconstrict*.ti,ab. OR exp vascular resistance/OR blood flow resistance.ti,ab. OR blood vessel resistance.ti,ab. OR (peripheral adj1 resistance).ti,ab. OR vascular vessel resistance.ti,ab. OR artery resistance.ti,ab. OR exp hemodynamics/OR exp blood vessel capacitance/OR blood vessel capacitance.ti,ab. OR vascular capacitance.ti,ab. OR (hemodynamic or hemodynamics).ti,ab. OR (haemodynamic or haemodynamics).ti,ab. OR vasopressor.ti,ab. OR exp vasoactive agent/ OR (vasoactive agonist OR vasoactive agonists or vasoactive agent or vasoactive agents or vasoactive drug or vasoactive drugs).ti,ab. OR (vessel contriction or vessel contrictions or artery contriction or artery constrictions).ti,ab. OR exp vasodilatation/OR (vasodilatation or vasodilation or vasodilator or vasodilatator or vasodilating or vasodilative or vasodilatative).ti,ab. OR (vaso dilatation or vaso dilation or vaso dilator or vaso dilatator or vaso dilating or vaso dilative or vaso dilatative).ti,ab. OR vaso relaxation.ti,ab. OR vascular endothelium dependent relaxation.ti,ab. OR vascular endothelium-dependent relaxation.ti,ab. OR endothelium dependent relaxation.ti,ab. OR endothelium-dependent relaxation.ti,ab. OR (vessel dilatation* or vessel dilation* or vascular dilation* or vascular dilatation* or vasorelaxation* or artery dilatation* or artery dilation*).ti,ab. OR (dilatation* or dilation* or relaxation* or contraction*).ti,ab.*

analyses (mostly ANOVA statistics) did not allow calculating overall effect sizes and effect measures were reported in variable ways (for myogenic reactivity and arterial compliance (absolute values, percentages of diameter changes, myogenic tone in relation to passive tone)). In case of missing, incomplete or indeterminate data, we approached the authors by email (response rate 33%).

Quality assessment

The methodological quality of the articles was assessed independently by two reviewers (first and second author). Data were scored on presence or absence of randomization of allocation to groups, blinding of outcome assessment, clearness on nulliparity at entry of study, age or weight of subjects described, and number of animals accounted for at the start and the end of the study. The quality of the responses reported was ranked for the clarity of the number of animals used for analyses, presence of a response graph containing ≥ 5 measurements, and achievement of E_{max} (defined as the presence of at least two measurements without any further increasing effect). The response quality was expressed as percentage of the number of responses that complied with these items divided by the total number of responses described in the study. It should be noted that the quality assessment determines the quality of the methodology required for comparison of responses to the predefined items and that it should not be interpreted as judgment of the value of the experiment per se.

Quantitative data synthesis and statistical analysis

Extracted data in RVR, renal flow and GFR at baseline, changes induced by specific blockades, and EC_{50} and E_{max} for pharmacological stimuli were analyzed for each gestational age period, stratified by species and strain, and displayed in forest plots, using Review Manager 5 (The Nordic Cochrane Centre, The Cochrane Collaboration, 2008). Pharmacological stimuli were analyzed by the type of G-protein-coupled pathway (Gq_{EC}, Gq_{SMC} and Gs_{SMC}) involved or by type of stimulus (NO-donors and potassium), when applicable. Combined effect sizes were calculated as weighed mean difference (MD) for RVR, renal flow, GFR and EC_{50}, and as standardized mean difference (SMD) for E_{max}. Both MD and SMD represent the effect sizes in

comparison to control values within the same study. In the forest plots data were presented as mean effect sizes with 95% confidence intervals (CI), as provided by Review Manager 5. Combined effect sizes were calculated by using the random effects model (Review Manager 5). In the forest plots, the size of the boxes represents the weight of the studies, the gray diamonds depict the overall effect size per species and strain, and the black diamonds illustrate the overall effect sizes. We used a significance level of alpha $= 0.05$ to identify statistically significant results. In the random effects model some heterogeneity beyond sampling errors is allowed to account for anticipated heterogeneity. Heterogeneity was presented as I^2; $I^2 < 60\%$ represents moderate heterogeneity, $I^2 > 60\%$ represents considerable to substantial heterogeneity [16]. Publication bias was assessed by linear regression analysis as proposed by Egger et al. (for ≥ 10 studies) [17] or subjective assessment of funnel plot asymmetry (for ≤ 10 studies).

For mechanical stimuli, we used the direction of the measured effects for qualitative analysis, and determined the overall weighed direction for each mechanical response, based on the total number of animals used.

We performed sensitivity analysis to assess the robustness of our findings, as is recommended for systematic reviews [16]. This type of analysis shows the influence of the used inclusion and exclusion criteria on the results. In the analysis we repeated all analyses in the presence of the studies that were excluded on the basis of the criterion first pregnancy versus nulliparous non-pregnant control.

References used for analysis included the following: [18–54].

Results

Our systematic literature search identified 1008 studies concerning the effects of pregnancy on vascular changes of the kidney (Figure 1). Based on title and abstract 917 studies were excluded, leaving 91 papers for full article evaluation. Thirty-seven studies did not meet the inclusion criteria. Seventeen studies were excluded based on their non-English language. Finally, 37 studies with 116 vascular responses met the inclusion criteria for meta-analysis.

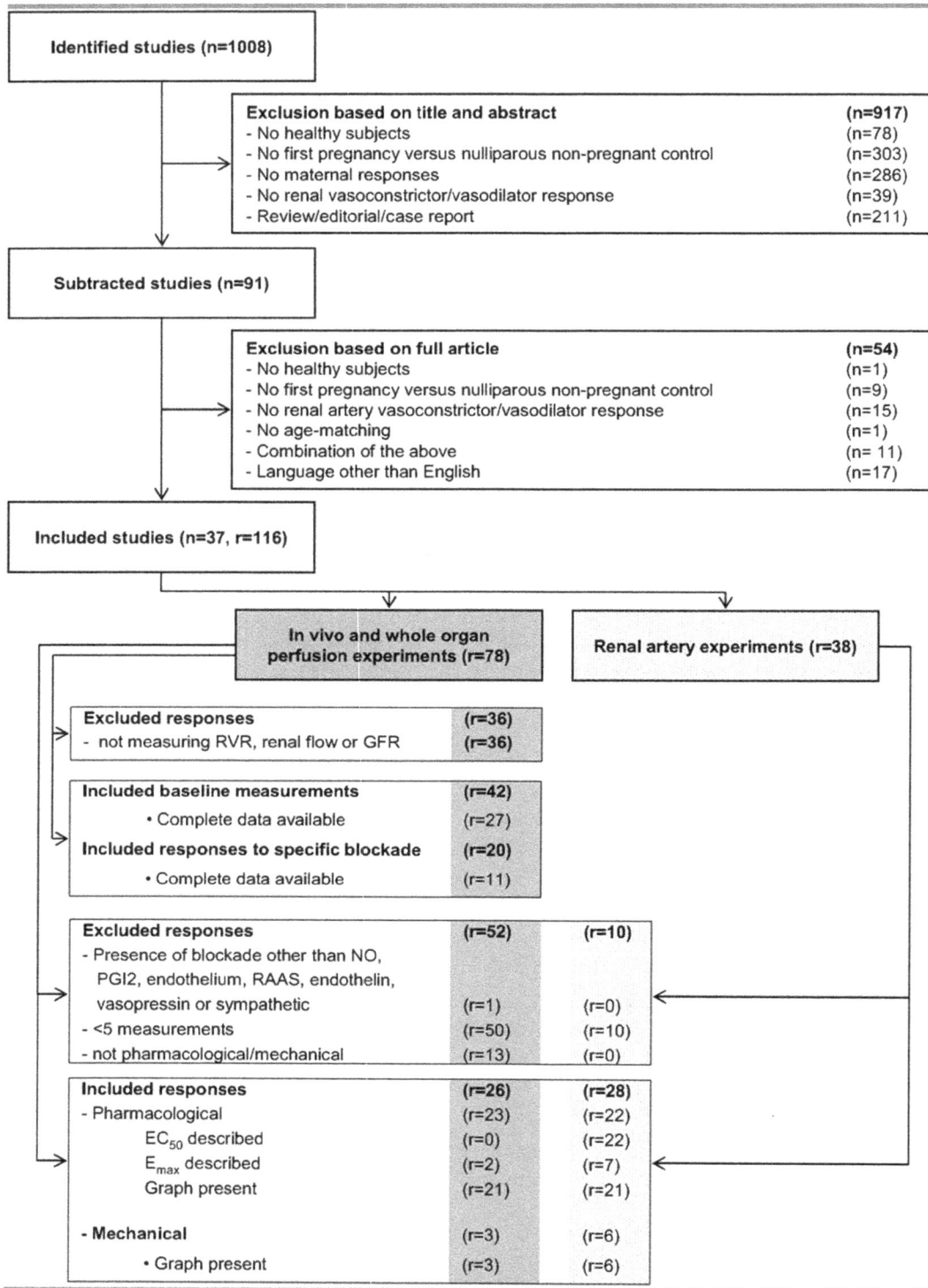

n= number of studies; r= number of responses (gray filling for in vivo and whole organ perfusion experiments, dotted filling for renal artery responses).

Figure 1. Flow chart for selection, inclusion and exclusion of studies and responses on renal vascular adaptation to pregnancy. n = number of studies; r= number of responses (gray filling for in vivo and whole organ perfusion experiments, dotted filling for renal artery responses).

The study characteristics are presented in Table 3. Most studies (n = 28) used rats (Long Evans rats (LER), Sprague Dawley rats (SDR), various strains of Wistar rats (Wistar Kyoto (WKR), Munich Wistar (MWR) and Wistar Hannover (WHR)); other species were used less frequently (sheep = 6, rabbits = 2 and guinea pigs (GP) = 1). Most studies contained *in vivo* experiments

Table 3. Characteristics of included studies, arranged by author and year.

Study	Species (strain)	Estrous stage control group	Gestation period	Experiment setup	Number of described baseline measurements	Number of described responses
Annibale et al., 1989	sheep	-	late	WM	n.a.	4
Baylis et al., 1986	rat (MWR)	-	late	IV	4	10
Baylis et al., 1988	rat (MWR)	-	mid	IV	3	0
Baylis et al., 1993	rat (SDR and MWR)	-	mid/late	IV	4	4
Baylis et al., 1995	rat (SDR)[#]	-	mid/late	IV	2	2
Bobadilla et al., 2001	rat (WR)	-	?	WOP	1	4
Bobadilla et al., 2003	rat (WR)	-	mid/late	WOP	2	16
Bobadilla et al., 2005	rat (WR)	-	late	WOP	1	2
Cha et al., 1993	sheep	-	late	IV	1	0
Cha et al., 1993	sheep	-	late	IV	1	0
Chu et al., 1997	rat (WKY)	-	late	WOP	0	0
Conrad et al., 1989	rat (LE)	-	late	IV	1	1
Conrad et al., 1999	rat (LE)	-	mid	IV	1	2
Danielson et al., 1995	rat (LE)	-	mid	IV	1	3
Danielson et al., 1996	rat (LE)	-	mid	IV	1	2
Fan et al., 1996	sheep	-	late	IV	1	0
Ferris et al., 1983	rabbit	-	late	IV	2	0
Gandley et al., 1997	rat (SDR)	-	late	PM	n.a.	4
Gandley et al., 2001	rat (LE)	-	mid	PM	n.a.	9
Greenberg et al., 1999	sheep	-	late	IV	1	0
Griggs et al., 1993	rat (SDR)	-	late	PM	n.a.	6
Hines et al., 1992	rat (SDR)	-	mid/late	IV	2	4
Hines et al., 2000	rat (SDR)	-	late	IV	1	1
Kim et al., 1994	guinea pig	-	Late	WM	n.a.	12
Knight et al., 2007	rat (SDR)	-	late	IV	0	0
Masilamani et al., 1994	rat (SDR)	-	mid/late	IV	1	0
McElvy et al., 2001	sheep	-	late	IV	1	0
Novak et al., 1997	rat (SDR)	-	mid	IV	1	0
Omer et al., 1999	rat (SDR)	-	late	IV	1	1
Patel et al., 1993	rat (SDR)	-	late	IV	2	0
Reckelhoff et al., 1992	rat (WR)	-	mid	IV	2	2
Sen et al., 1997	rat (?)	-	mid/late	IV	2	0
Sicinska et al., 1971	rat (SDR)	-	late	IV	1	0
Van Drongelen et al., 2011	rat (WR)	-	mid	WOP	1	1
Van Eijndhoven et al., 2003	rat (WR)	-	mid	WM	n.a.	1
Van Eijndhoven et al., 2007	rat (?)	-	mid	WM	n.a.	2
Woods et al., 1987	rabbit	-	late	IV	1	1

WM = wire myograph, PM = pressure myograph, PPM = pressure-perfusion myograph, WOP = whole organ perfusion, IV = *in vivo*. n.a. = not applicable.
[#]data received by email.

(n = 25), while whole organ (perfusion), perfusion myograph and wire myograph experiments were less common (n = 5, 3 and 4 respectively). Methodological limitations were present in a substantial portion of the included studies (Table 4): 43% (16/37) did not present the number of animals used, 32% (12/37) did not present parity, 97% (36/37) did not report having performed randomization, and 0% reported blinding of the outcome assessment.

The 116 subtracted vascular responses could be divided into two categories: 1. *in vivo* and whole organ perfusion studies measuring changes in RVR, renal flow and/or GFR, and 2. renal artery experiments measuring changes in vascular tone. Experiments were performed mainly in mid (34%, 40/116) and late (63%, 73/116) gestation; for four responses the gestation period was not defined. Early gestation was not investigated by any of the included studies.

Table 4. Quality assessment of the included studies and subsequent responses.

Study	Study quality						Response quality			
	(1)	(2)	(3)	(4)	(5)	Score	(6)	(7)	(8)	Score
Annibale et al., 1989	-	-	?	-	+	1	1.0	1.0	1.0	3.0
Baylis et al., 1986	-	+	+	+	+	3	1.0	0.0	0.0	1.0
Baylis et al., 1988	-	-	+	+	-	2	1.0	0.0	0.0	1.0
Baylis et al., 1993	-	-	+	-	+	2	1.0#	0.0	0.0	1.0#
Baylis et al., 1995	-	-	+	+	-	2	1.0	0.0	0.0	1.0
Bobadilla et al., 2001	-	-	+	+	-	2	1.0	1.0	0.0	2.0
Bobadilla et al., 2003	-	-	+	+	+	3	1.0	1.0	0.0	2.0
Bobadilla et al., 2005	-	-	+	+	-	2	1.0	1.0	0.5	2.5
Cha et al., 1993[a]	-	-	?	+	+	2	0.0	0.0	0.0	0.0
Cha et al., 1993[b]	-	-	?	-	+	1	0.0	1.0	1.0	2.0
Chu et al., 1997	-	-	?	+	-	1	1.0	0.0	0.0	1.0
Conrad et al., 1989	n.a.	-	+	+	+	3	1.0	0.0	0.0	1.0
Conrad et al., 1999	-	-	+	+	-	2	0.0	0.0	0.0	0.0
Danielson et al., 1995	-	-	+	+	+	3	0.0	0.0	0.0	0.0
Danielson et al., 1996	+	-	+	+	-	3	1.0	0.0	0.0	1.0
Fan et al., 1996	-	-	?	-	+	1	1.0	0.0	0.0	1.0
Ferris et al., 1983	-	-	?	-	+	1	1.0	0.0	0.0	1.0
Gandley et al., 1997	-	-	+	+	-	2	1.0	1.0	0.75	2.75
Gandley et al., 2001	-	-	+	-	-	1	0.9	0.2	0.2	1.3
Greenberg et al., 1999	-	-	?	+	+	2	0.0	0.0	0.0	0.0
Griggs et al., 1993	-	-	+	+	-	2	1.0	0.16	0.33	1.49
Hines et al., 1992	-	-	+	-	-	1	1.0	0.0	0.0	1.0
Hines et al., 2000	-	-	+	-	+	2	0.33	0.66	0.0	1.0
Kim et al., 1994	-	-	?	-	-	0	1.0	1.0	1.0	3.0
Knight et al., 2007	-	-	+	+	+	3	1.0	0.0	0.0	1.0
Masilamani et al., 1994	-	-	+	+	+	3	1.0	0.0	0.0	1.0
Masilamani et al., 1994	-	-	+	+	+	3	1.0	0.0	0.0	1.0
McElvy et al., 2001	-	-	?	+	+	2	1.0	1.0	0.0	2.0
Novak et al., 1997	-	-	?	-	-	0	0.0	0.0	0.0	0.0
Omer et al., 1999	-	-	+	+	-	2	0.0	0.0	0.0	0.0
Patel et al., 1993	-	-	+	-	+	2	1.0	0.0	0.0	1.0
Reckelhoff et al., 1992	-	-	+	+	-	2	1.0	1.0	1.0	3.0
Sen et al., 1997	-	-	?	-	-	0	1.0	0.0	0.0	1.0
Sicinska et al., 1971	-	-	?	+	+	2	0.0	0.0	0.0	0.0

Table 4. Cont.

| Study | Study quality | | | | | | Response quality | | | |
	(1)	(2)	(3)	(4)	(5)	Score	(6)	(7)	(8)	Score
Van Drongelen et al., 2011	-	-	+	+	+	3	1.0	0.0	1.0	2.0
Van Eijndhoven et al., 2003	-	-	+	+	+	3	0.0	0.0	0.0	0.0
Van Eijndhoven et al., 2007	-	-	+	+	+	3	0.5	0.0	0.0	0.5
Woods et al., 1987	-	-	+	-	+	2	1.0	1.0	1.0	3.0

(1) randomization, (2) blinding of outcome assessor, (3) virgin/nulliparous at entry study, (4) age or weight of animals at entry study, (5) number of used animals described, (6) fraction of responses with clear number of animals used for statistical analyses (as a percentage of the number of described responses), (7) fraction of responses with dose-response curve containing ≥5 measurements, (8) fraction of responses with accomplished E_{max}. n.a. = not applicable.
data received by email.

We identified 78 *in vivo* and whole organ perfusion experiments. Two types of experiments could be distinguished based on their objective to determine the effects of pregnancy on: 1) RVR, renal flow and GFR, or 2) the response of the kidney to vaso-active substances. For the first type of experiments we extracted 42 baseline measurements and 20 responses to specific blockades. Thirty-six responses were excluded as they did not measure RVR, renal flow or GFR. After exclusion of incomplete data, 27 baseline measurements and 11 responses to specific blockades were suitable for meta-analysis. For the second type of experiments we identified 23 pharmacological and 3 mechanical responses. For pharmacological stimuli, EC_{50} was reported in 0% of the responses, E_{max} was described in 9% and a graph was present in 91%. For mechanical stimuli, all responses were presented graphically without quantified effects.

From the 38 renal artery responses, 28 met the criteria for inclusion. They consisted of 22 pharmacological stimuli (EC_{50}, E_{max} and a graph present in 100%, 32% and 95% of the cases, respectively) and six mechanical stimuli, all with a graph without quantified effects.

In vivo and whole organ perfusion experiments

The influence of mid pregnancy on kidney function is presented in Figure 2. In mid pregnancy, *overall* (across species and strains) RVR decreases by 27%, varying from 18% in SDR to 31% in MWR. Renal flow is enhanced to a variable degree, ranging from 13% in WHR to 31% in LER. GFR increases by 17% in SDR to 29% in LER. In LER, blockade of the NO system completely normalizes the pregnancy-induced change in RVR and GFR, while pregnancy-induced increase in renal flow is reduced from 31% to 15%. In WR, NO-blockade normalizes renal flow completely. Sympathetic or vasopressin blockade has only been studied in SDR and does not affect the mid pregnancy-induced changes in RVR, renal flow and GFR. It appears that mid pregnancy changes in RVR, renal flow and GFR are dependent on up regulation of the NO system and not on sympathetic tone or vasopressin.

The magnitude of pregnancy-induced renal changes in rats decreases from mid pregnancy towards term. Figures 3 and 4 show renal function in late gestation compared to non-pregnant controls. Late pregnancy affects RVR, renal flow and GFR differently among species and strains. In LER and MWR, pregnancy does not affect any of the renal parameters. Blockade of the renin-angiotensin-aldosteron system (RAAS) did not show any effect in LER, but in MWR reduced RVR by 26% and increased renal flow by 26% without changing in GFR. In SDR, pregnancy decreases RVR by 19% and increases both renal flow and GFR, by 17% and 21% respectively. NO blockade in this strain returns renal flow to non-pregnant values with a persistently increased GFR. Under sympathetic blockade RVR, renal flow and GFR return to non-pregnant values in these rats, while vasopressin blockade has no effect on any of the parameters. In rabbits, late pregnancy does not significantly affect RVR, while renal flow and GFR are increased by 29% and 39% compared to non-pregnant values. These changes are not affected by RAAS-blockade. For late pregnant sheep, no data are available on RVR, but renal flow is 21% higher than in non-pregnant sheep without a significant difference in GFR. In short, there is no consistent pattern of renal changes in late pregnancy across species and strains.

In both mid and late pregnancy, the *overall* RVR, renal flow and GFR shows moderate to considerable heterogeneity. This cannot be attributed completely to species or strain specific effects, as stratification by species and/or strain does not reduce heterogeneity to a minimum in all groups. Heterogeneity depends

Substance / study	RVR (%) MD [95% CI]	I²	MD [95% CI]	Renal flow (%) MD [95% CI]	I²	MD [95% CI]	GFR (%) MD [95% CI]	I²	MD [95% CI]
MID GESTATION *(without blockade)*									
Danielson LER,3	-31.58 [-36.19, -26.97]			40.41 [34.92, 45.90]			29.83 [24.03, 35.63]		
Danielson LER,2	-23.26 [-34.65, -11.87]			21.59 [11.95, 31.23]			25.11 [14.28, 35.94]		
Baylis MWR,a,2,*	-25.00 [-46.91, -3.09]			25.00 [-22.62, 72.62]			25.66 [16.73, 34.59]		
Baylis MWR,b,2,*	-32.95 [-47.04, -18.86]			23.08 [1.76, 44.40]			27.78 [12.96, 42.60]		
Baylis SDR,1	-20.00 [-40.34, 0.34]			34.12 [7.83, 60.41]			16.67 [7.44, 25.90]		
Baylis SDR,2	-19.28 [-33.69, -4.87]			47.37 [34.48, 60.26]			19.05 [6.48, 31.62]		
Masilamani SDR,1	-13.13 [-32.78, 6.52]			21.79 [-0.69, 44.27]			14.29 [1.09, 27.49]		
Sen SDR,a,1,*	-			16.12 [-9.87, 42.11]			-		
Sen SDR,b,1,*	-			-1.74 [-21.64, 18.16]			-		
van Drongelen WHR,a,2	-			12.50 [10.33, 14.67]			-		
Overall LER	-29.12 [-36.56, -21.68]	43%		31.43 [13.01, 49.86]	91%		28.78 [23.67, 33.89]	0%	
Overall MWR	-30.63 [-42.47, -18.78]	0%		23.40 [3.94, 42.86]	0%		26.22 [18.58, 33.87]	0%	
Overall SDR	-17.84 [-27.93, -7.75]	0%		24.14 [4.34, 43.93]	78%		16.73 [10.25, 23.21]	0%	
Overall WHR	-			12.50 [10.33, 14.67]	n.a.		-		
Overall in vivo	-26.84 [-32.06, -21.62]	21%		26.85 [16.23, 37.47]	75%		23.58 [18.87, 28.29]	35%	
Overall all	-26.84 [-32.06, -21.62]	21%		24.45 [12.61, 36.29]	92%		23.58 [18.87, 28.29]	35%	
(NO-blockade)									
Danielson LER,a,3	-5.00 [-26.86, 16.86]			10.57 [-5.57, 26.71]			7.95 [-11.30, 27.20]		
Danielson LER,b,2	-15.33 [-34.01, 3.35]			23.91 [0.10, 47.72]			10.52 [-12.91, 33.95]		
van Drongelen WHR,a	-			3.23 [-1.24, 7.70]			-		
Overall LER	-10.97 [-25.17, 3.23]	0%		14.77 [1.41, 28.13]	0%		8.99 [-5.89, 23.86]	0%	
Overall WHR	-			3.23 [-1.24, 7.70]	n.a.		-		
Overall all	-10.97 [-25.17, 3.23]	0%		7.86 [-1.97, 17.69]	41%		8.99 [-5.89, 23.86]	0%	
(sympathetic blockade)									
Baylis SDR,b,2	-37.04 [-54.91, -19.17]			53.57 [31.16, 75.98]			27.92 [13.78, 42.06]		
Overall SDR	-37.04 [-54.91, -19.17]			53.57 [31.16, 75.98]			27.92 [13.78, 42.06]		
(vasopressin blockade)									
Baylis SDR,a,1	-27.06 [-48.32, -5.80]			33.33 [13.82, 52.84]			24.42 [16.65, 32.19]		
Overall SDR	-27.06 [-48.32, -5.80]			33.33 [13.82, 52.84]			24.42 [16.65, 32.19]		

-100 -50 0 50 100
Favors decreased / Favors increased RVR | Favors decreased / Favors increased renal flow | Favors decreased / Favors increased GFR

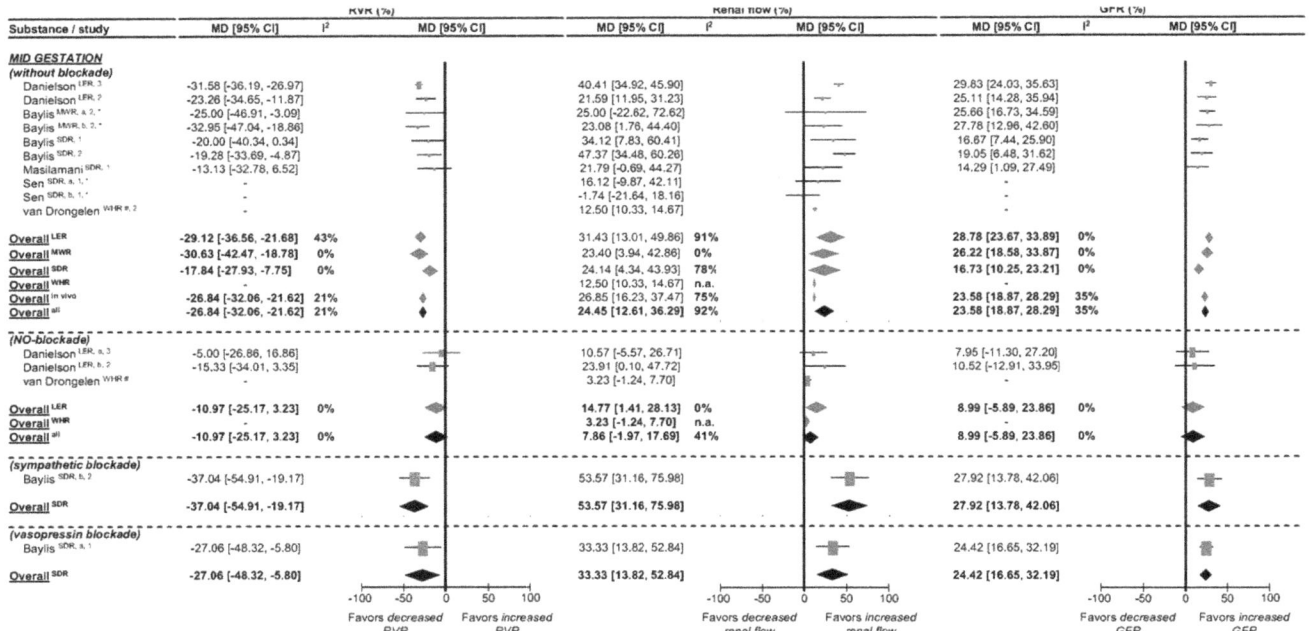

Figure 2. Effect of pregnancy at mid gestation on in vivo and whole organ perfusion# renal vascular resistance (RVR), renal flow and glomerular filtration rate (GFR) in absence and presence of NO and sympathetic blockade. The effect of pregnancy on RVR, renal flow (effects of renal plasma/blood/perfusion flow (RPF/RBF/RPPF) combined) and GFR is presented as percentage mean difference (MD) and its 95% CI. Studies and totals based on Long Evans rats (LER), Munich Wistar rats (MWR), Sprague Dawley rats (SDR) and Wistar Hannover rats (WHR). 1, 2, 3 represents first, second and third part of mid gestation. # Whole organ perfusion experiments, excluded in the "Overall in vivo" analysis. * Experiments performed under anesthesia. I² represents the amount of heterogeneity. n.a. = not applicable.

mainly on single responses with very small confidence intervals and responses with an estimated effect opposite to other studies. We could not identify any methodological issue that might explain these effects.

The influence of pregnancy on the autoregulatory threshold was analyzed qualitatively only, as quantitative effect measures of the lower pressure boundary for stable renal flow were not reported in the two relevant studies, which used WR and rabbits [21,54]. From these studies it appears that pregnancy does not affect the renal autoregulatory threshold in either mid or late pregnancy.

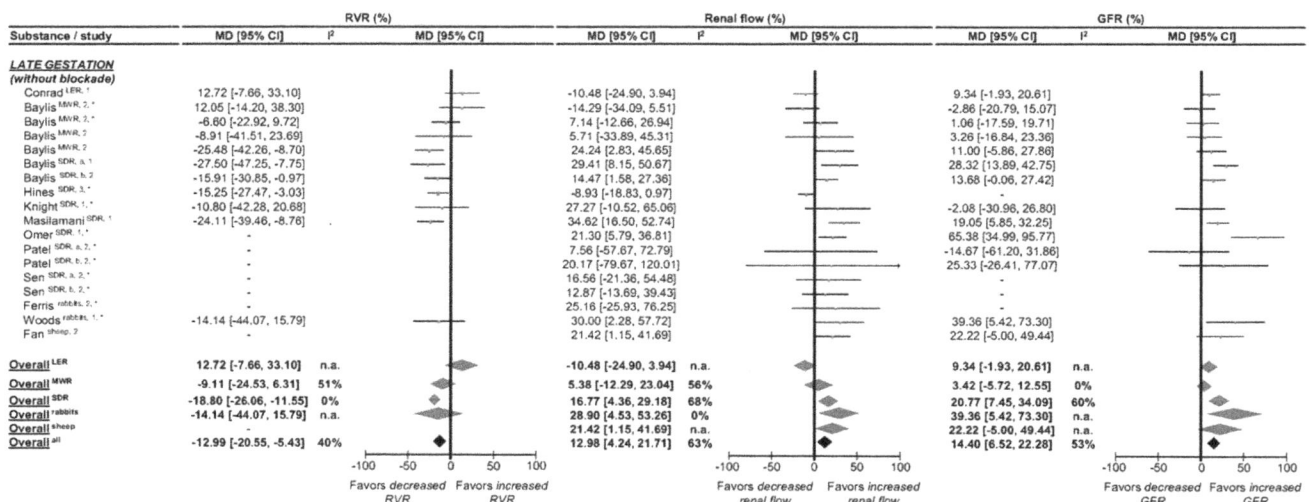

Substance / study	RVR (%) MD [95% CI]	I²	MD [95% CI]	Renal flow (%) MD [95% CI]	I²	MD [95% CI]	GFR (%) MD [95% CI]	I²	MD [95% CI]
LATE GESTATION *(without blockade)*									
Conrad LER,1	12.72 [-7.66, 33.10]			-10.48 [-24.90, 3.94]			9.34 [-1.93, 20.61]		
Baylis MWR,2,*	12.05 [-14.20, 38.30]			-14.29 [-34.09, 5.51]			-2.86 [-20.79, 15.07]		
Baylis MWR,2,*	-6.60 [-22.92, 9.72]			7.14 [-12.66, 26.94]			1.06 [-17.59, 19.71]		
Baylis MWR,2	-8.91 [-41.51, 23.69]			5.71 [-33.89, 45.31]			3.26 [-16.84, 23.36]		
Baylis MWR,2	-25.48 [-42.26, -8.70]			24.24 [2.83, 45.65]			11.00 [-5.86, 27.86]		
Baylis SDR,a,1	-27.50 [-47.25, -7.75]			29.41 [8.15, 50.67]			28.32 [13.89, 42.75]		
Baylis SDR,b,2	-15.91 [-30.85, -0.97]			14.47 [1.58, 27.36]			13.68 [-0.06, 27.42]		
Hines SDR,3,*	-15.25 [-27.47, -3.03]			-8.93 [-18.83, 0.97]					
Knight SDR,1,*	-10.80 [-42.28, 20.68]			27.27 [-10.52, 65.06]			-2.08 [-30.96, 26.80]		
Masilamani SDR,1	-24.11 [-39.46, -8.76]			34.62 [16.50, 52.74]			19.05 [5.85, 32.25]		
Omer SDR,1,*	-			21.30 [5.79, 36.81]			65.38 [34.99, 95.77]		
Patel SDR,a,2,*	-			7.56 [-57.67, 72.79]			-14.67 [-61.20, 31.86]		
Patel SDR,b,2,*	-			20.17 [-79.67, 120.01]			25.33 [-26.41, 77.07]		
Sen SDR,a,2,*	-			16.56 [-21.36, 54.48]			-		
Sen SDR,b,2,*	-			12.87 [-13.69, 39.43]			-		
Ferris rabbits,2,*	-			25.16 [-25.93, 76.25]			-		
Woods rabbits,1,*	-14.14 [-44.07, 15.79]			30.00 [2.28, 57.72]			39.36 [5.42, 73.30]		
Fan sheep,2	-			21.42 [1.15, 41.69]			22.22 [-5.00, 49.44]		
Overall LER	12.72 [-7.66, 33.10]	n.a.		-10.48 [-24.90, 3.94]	n.a.		9.34 [-1.93, 20.61]	n.a.	
Overall MWR	-9.11 [-24.53, 6.31]	51%		5.38 [-12.29, 23.04]	56%		3.42 [-5.72, 12.55]	0%	
Overall SDR	-18.80 [-26.06, -11.55]	0%		16.77 [4.36, 29.18]	68%		20.77 [7.45, 34.09]	60%	
Overall rabbits	-14.14 [-44.07, 15.79]	n.a.		28.90 [4.53, 53.26]	0%		39.36 [5.42, 73.30]	n.a.	
Overall sheep	-			21.42 [1.15, 41.69]	n.a.		22.22 [-5.00, 49.44]	n.a.	
Overall all	-12.99 [-20.55, -5.43]	40%		12.98 [4.24, 21.71]	63%		14.40 [6.52, 22.28]	53%	

-100 -50 0 50 100
Favors decreased / Favors increased RVR | Favors decreased / Favors increased renal flow | Favors decreased / Favors increased GFR

Figure 3. Effect of pregnancy at late gestation on in vivo renal vascular resistance (RVR), renal flow and glomerular filtration rate (GFR) in absence of blockade. The effect of pregnancy on RVR, renal flow (effects of renal plasma/blood/perfusion flow (RPF/RBF/RPPF) combined) and GFR is presented as percentage mean difference (MD) and its 95% CI. Studies and totals based on Long Evans rats (LER), Munich Wistar rats (MWR), Sprague Dawley rats (SDR), rabbits and sheep. 1, 2, 3 represents first, second and third part of late gestation. * Experiments performed under anesthesia. I² represents the amount of heterogeneity. n.a. = not applicable.

Substance / study	RVR (%) MD [95% CI]	I^2	Renal flow (%) MD [95% CI]	I^2	GFR (%) MD [95% CI]	I^2
LATE GESTATION						
(NO-blockade)						
Omer [SDR,1,*]	-		5.88 [-36.25, 48.01]		52.94 [16.17, 89.71]	
Overall [SDR]	-		5.88 [-36.25, 48.01]	n.a.	52.94 [16.17, 89.71]	n.a.
(RAAS-blockade)						
- ACE inhibition						
Conrad [LER,1]	8.83 [-14.62, 32.28]		-8.20 [-27.11, 10.71]		5.44 [-12.49, 23.37]	
Baylis [MWR,2,*]	-18.18 [-39.59, 3.23]		5.88 [-19.90, 31.66]		0.96 [-19.07, 20.99]	
Baylis [MWR,2,*]	-34.63 [-47.63, -21.63]		42.42 [29.14, 55.70]		9.80 [-5.70, 25.30]	
Ferris [rabbit,2,*]	-		55.17 [8.44, 101.90]		-	
Subtotal [LER]	8.83 [-14.62, 32.28]	n.a.	-8.20 [-27.11, 10.71]	n.a.	5.44 [-12.49, 23.37]	n.a.
Subtotal [MWR]	-28.70 [-44.18, -13.21]	40%	25.89 [-9.76, 61.54]	84%	6.49 [-5.77, 18.75]	0%
Subtotal [rabbits]	-		55.17 [8.44, 101.90]	n.a.	-	
Subtotal [all]	-16.10 [-41.11, 8.90]	81%	21.48 [-9.01, 51.97]	86%	6.16 [-3.96, 16.27]	0%
- Angiotensin-II						
Baylis [MWR,2,*]	-17.76 [-47.66, 12.14]		39.29 [9.59, 68.99]		8.33 [-14.77, 31.43]	
Baylis [MWR,2,*]	-9.73 [-44.69, 25.23]		3.23 [-37.26, 43.72]		4.71 [-21.38, 30.80]	
Subtotal [MWR]	-14.37 [-37.09, 8.36]	0%	23.99 [-10.94, 58.92]	50%	6.74 [-10.55, 24.03]	0%
Overall [LER]	8.83 [-14.62, 32.28]	n.a.	-8.20 [-27.11, 10.71]	n.a.	5.44 [-12.49, 23.37]	n.a.
Overall [MWR]	-26.34 [-37.22, -15.46]	8%	25.91 [4.59, 47.22]	64%	6.57 [-3.43, 16.57]	0%
Overall [rabbits]	-		55.17 [8.44, 101.90]	n.a.	-	
Overall [all]	-16.04 [-32.97, 0.89]	63%	21.73 [-0.93, 44.39]	79%	6.30 [-2.43, 15.04]	0%
(sympathetic blockade)						
Baylis [SDR,b,2]	-11.11 [-30.74, 8.52]		7.14 [-10.36, 24.64]		8.63 [-7.10, 24.36]	
Overall [SDR]	-11.11 [-30.74, 8.52]	n.a.	7.14 [-10.36, 24.64]	n.a.	8.63 [-7.10, 24.36]	n.a.
(vasopressin blockade)						
Baylis [SDR,a,1]	-30.59 [-51.85, -9.33]		33.33 [14.90, 51.76]		34.10 [21.55, 46.65]	
Overall [SDR]	-30.59 [-51.85, -9.33]		33.33 [14.90, 51.76]		34.10 [21.55, 46.65]	

RVR (%): Favors decreased RVR / Favors increased RVR (-100 -50 0 50 100)
Renal flow (%): Favors decreased renal flow / Favors increased renal flow (-100 -50 0 50 100)
GFR (%): Favors decreased GFR / Favors increased GFR (-100 -50 0 50 100)

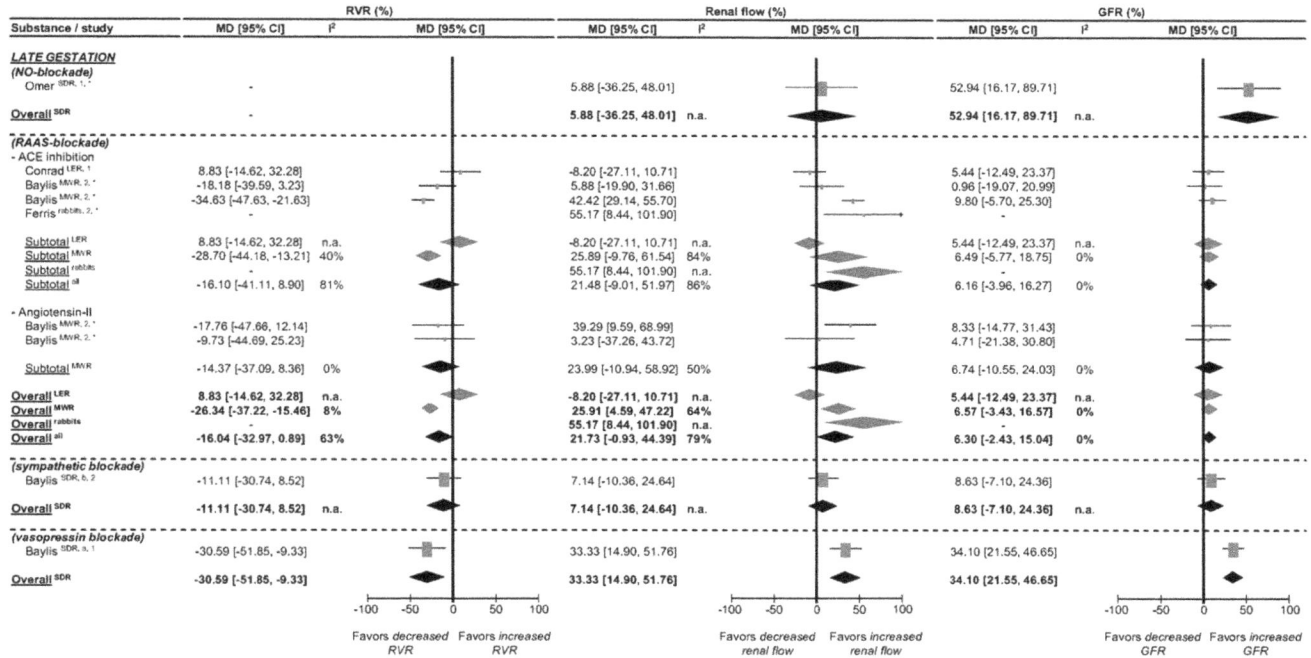

Figure 4. Effect of pregnancy at late gestation on in vivo renal vascular resistance (RVR), renal flow and glomerular filtration rate (GFR) in presence of NO, sympathetic and RAAS blockade. The effect of pregnancy on RVR, renal flow (effects of renal plasma/blood/perfusion flow (RPF/RBF/RPPF) combined) and GFR is presented as percentage mean difference (MD) and its 95% CI. Studies and totals based on Long Evans rats (LER), Munich Wistar rats (MWR), Sprague Dawley rats (SDR) and rabbits. 1, 2, 3 represents first, second and third part of late gestation. * Experiments performed under anesthesia. I^2 represents the amount of heterogeneity. n.a. = not applicable.

Renal artery experiments

The renal artery changes in response to vasodilator Gq_{EC}-coupled pharmacological stimuli in pregnancy are presented in Figure 5. Our search only detected responses during late gestation, and mainly in guinea pigs. In this species, pregnancy does not affect Gq_{EC}-coupled EC_{50} to acetylcholine and EC_{50} is unaffected by NO and PGI_2 blockade.

The effects of mid and late pregnancy on the vasoconstrictor Gq_{SMC}-coupled pathway in several species are shown in Figure 6; no data are available on early gestation. In both mid and late gestation, the Gq_{SMC}-coupled EC_{50} and E_{max} of phenylephrine and U46619 (thromboxane agonist) are unaffected by pregnancy for all species investigated. In guinea pigs under NO and PGI_2 blockade, pregnancy does not affect the response to U46619. In SDR in the absence of endothelium pregnancy increases EC_{50} in response to phenylephrine, which suggests that the pregnancy effect on the SMC is overruled by the endothelium. Apparently, pregnancy does not seem to affect the renal artery responsiveness to Gq_{SMC}-coupled vasoconstrictor stimuli.

The effect of pregnancy on the renal artery responses to NO and potassium are depicted in Figure 7. For NO, only single studies in rats and sheep were available. In mid and late gestation rats and late pregnant sheep pregnancy does not modify the response to NO or potassium. This suggests that in general pregnancy does not affect the properties of the renal SMC to NO and potassium.

The effect of pregnancy on the renal artery responses to mechanical stimuli was assessed qualitatively only, as shown in Figure 8. We did not find studies that investigated flow-mediated vasodilation of the renal artery during pregnancy, and only two studies that looked at myogenic reactivity and vascular compliance in late gestation. In SDR, late pregnancy reduces myogenic reactivity in an endothelium-dependent manner and enhances vascular compliance. In sheep, late pregnancy does not affect either myogenic reactivity or vascular compliance. Apparently, pregnancy-induced changes in myogenic reactivity and compliance depend on the species investigated.

Publication bias and sensitivity-analysis

Publication bias was assessed by subjective determination of funnel plot asymmetry, as all analyses consisted of less than 10 studies. We did not detect evidence suggestive for publication bias.

Sensitivity-analysis was performed to assess the influence of inclusion criteria on the results. We extended the inclusion for studies that did not match the criterium "nulliparity" (n = 3). The results were not affected by addition of these studies (data not shown). We did not extend the analysis for age-matching, because we had not excluded any study for that reason.

Discussion

Our meta-analysis confirms the commonly held view that pregnancy reduces RVR and enhances renal flow and GFR in animals, as it does in humans [1]. Our study has shown that the degree of change depends on the functional parameter, species, strain and gestational age investigated. Our meta-analysis did not detect changes in renal artery responsiveness. More importantly, the underlying mechanisms for the change in RVR, renal flow and GFR is not the same between species, strains and gestational age. Therefore each research question needs careful choice of the animal model and gestational age period of interest.

Across rat strains at mid-gestation, pregnancy decreases RVR and enhances renal flow and GFR, as shown in Figure 2. The effect of pregnancy on RVR and GFR is quite consistent, whereas

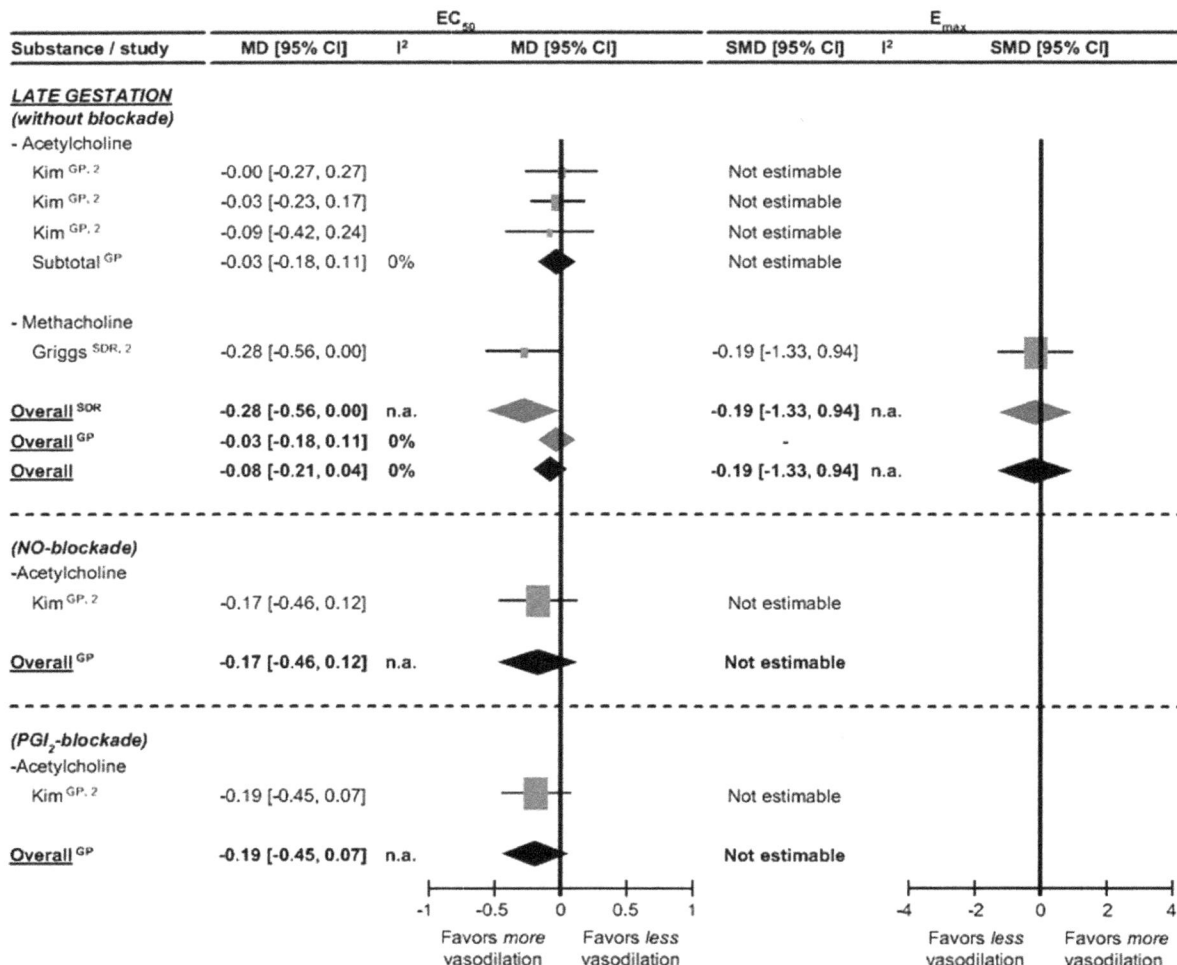

Figure 5. Effect of pregnancy at late gestation on renal artery responses to stimuli involved in vasodilation through the GqEC-coupled pathway in the presence and absence of blockade. The effect of pregnancy on EC_{50} (the dose of stimulus inducing 50% response) is depicted as mean difference (MD) and its 95% confidence interval (95% CI). E_{max} (maximum response) is presented as standardized mean difference (SMD) and its 95% CI. Studies and totals based on Sprague Dawley rats (SDR) and guinea pigs (GP). 1, 2, 3 represents first, second and third part of late gestation. I^2 represents the amount of heterogeneity. n.a. = not applicable.

there is considerable variation in renal flow. This may suggest differences in renal blood pressure between strains. We observed that the common changes in RVR, renal flow and GFR are caused by NO upregulation rather than by change in renal autoregulatory threshold to pressure, or sympathetic or vasopressin-regulation.

At late gestation, the pregnancy effects on RVR, renal flow and GFR are dependent on the species and strain investigated (Figure 3). In some species or strains, renal parameters are not different from non-pregnant values (LER and MWR), while in others late pregnancy affects RVR, renal flow and GFR (SDR, rabbits and sheep). The renal autoregulatory threshold to pressure changes was not affected in any species investigated. The NO, RAAS, sympathetic and vasopressin pathways are differently affected across species and strains. In SDR, NO activation affects renal flow, but not GFR. This implies a difference in response to NO activation between the afferent and efferent glomerular vasculature.

The pregnancy-induced reduction in RVR and the associated increase in renal flow and GFR across the species studied are dependent on gestational age, as shown in Figures 2 and 3. Our

analysis includes data on both mid- and late gestation in LER, MWR and SDR (Figure 9). In LER and MWR, all measured renal changes are maximal in mid-gestation and return towards non-pregnant levels in late gestation, whereas in SDR there is no difference between mid- and late pregnancy response. Responses in LER and MWR qualitatively correspond well with human pregnancy, in which renal flow increases in the first and second trimesters and decreases markedly towards term [1]. Quantitatively, in human pregnancy the renal flow increase (around 60%) is more pronounced than in the rat strains included in our meta-analysis (LER 31%, MWR 45% and SDR 14%). One may speculate on the underlying mechanism for this difference. Possibly the more mature human fetus poses a greater demand on the maternal vascular system, than the more immature rat fetus at comparable gestational age period. Our data suggest that renal changes during pregnancy in humans and rats are qualitatively similar but quantitatively different.

It is questionable if SDR is the right model for renal changes in healthy human pregnancy. In contrast to humans in which RVR, renal flow and GFR return towards non-pregnant values near

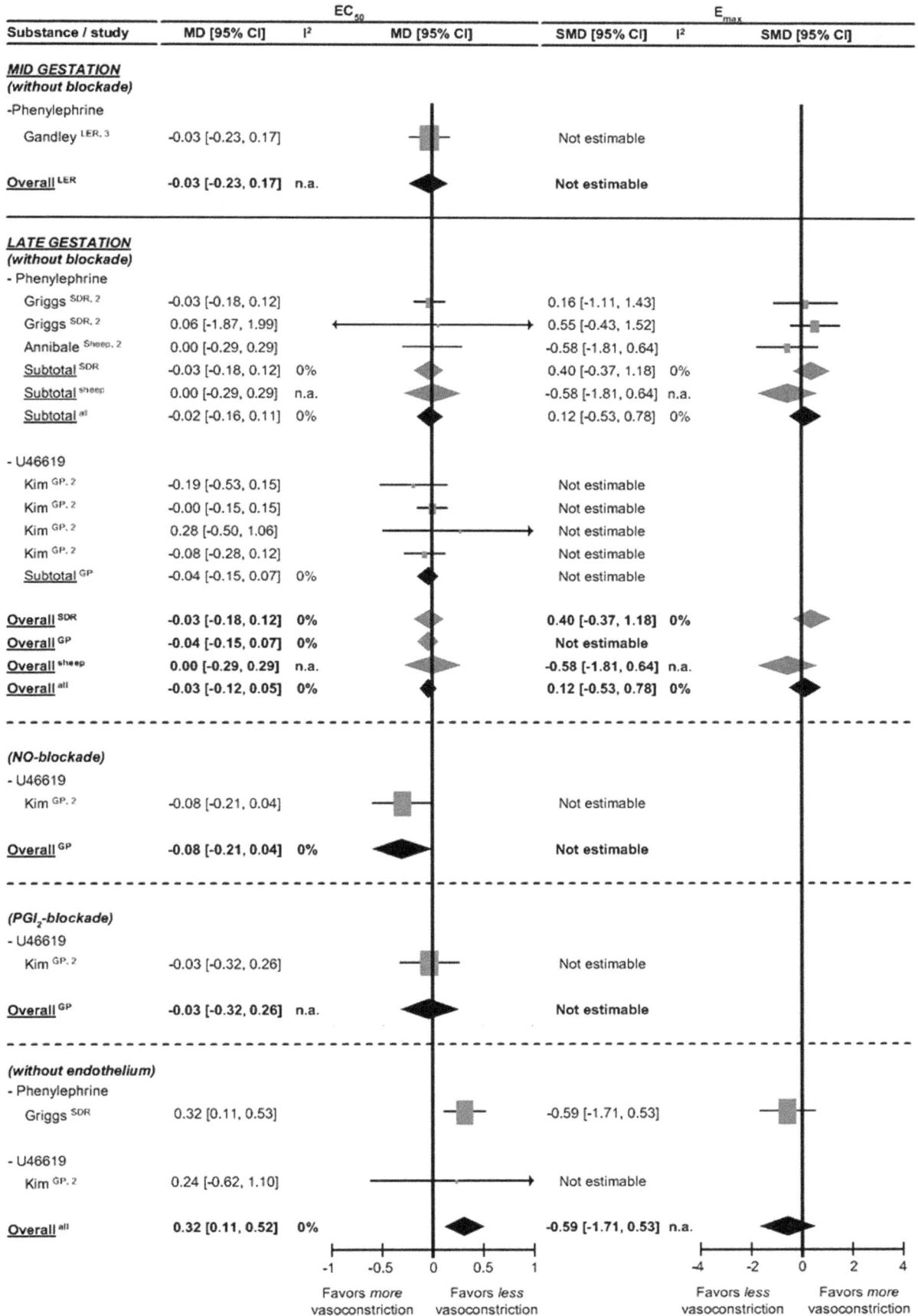

Substance / study	EC_{50} MD [95% CI]	I^2	MD [95% CI]	E_{max} SMD [95% CI]	I^2	SMD [95% CI]
MID GESTATION *(without blockade)*						
-Phenylephrine						
Gandley [LER, 3]	-0.03 [-0.23, 0.17]			Not estimable		
Overall [LER]	-0.03 [-0.23, 0.17]	n.a.		Not estimable		
LATE GESTATION *(without blockade)*						
- Phenylephrine						
Griggs [SDR, 2]	-0.03 [-0.18, 0.12]			0.16 [-1.11, 1.43]		
Griggs [SDR, 2]	0.06 [-1.87, 1.99]			0.55 [-0.43, 1.52]		
Annibale [Sheep, 2]	0.00 [-0.29, 0.29]			-0.58 [-1.81, 0.64]		
Subtotal [SDR]	-0.03 [-0.18, 0.12]	0%		0.40 [-0.37, 1.18]	0%	
Subtotal [sheep]	0.00 [-0.29, 0.29]	n.a.		-0.58 [-1.81, 0.64]	n.a.	
Subtotal [all]	-0.02 [-0.16, 0.11]	0%		0.12 [-0.53, 0.78]	0%	
- U46619						
Kim [GP, 2]	-0.19 [-0.53, 0.15]			Not estimable		
Kim [GP, 2]	-0.00 [-0.15, 0.15]			Not estimable		
Kim [GP, 2]	0.28 [-0.50, 1.06]			Not estimable		
Kim [GP, 2]	-0.08 [-0.28, 0.12]			Not estimable		
Subtotal [GP]	-0.04 [-0.15, 0.07]	0%		Not estimable		
Overall [SDR]	-0.03 [-0.18, 0.12]	0%		0.40 [-0.37, 1.18]	0%	
Overall [GP]	-0.04 [-0.15, 0.07]	0%		Not estimable		
Overall [sheep]	0.00 [-0.29, 0.29]	n.a.		-0.58 [-1.81, 0.64]	n.a.	
Overall [all]	-0.03 [-0.12, 0.05]	0%		0.12 [-0.53, 0.78]	0%	
(NO-blockade)						
- U46619						
Kim [GP, 2]	-0.08 [-0.21, 0.04]			Not estimable		
Overall [GP]	-0.08 [-0.21, 0.04]	0%		Not estimable		
(PGI₂-blockade)						
- U46619						
Kim [GP, 2]	-0.03 [-0.32, 0.26]			Not estimable		
Overall [GP]	-0.03 [-0.32, 0.26]	n.a.		Not estimable		
(without endothelium)						
- Phenylephrine						
Griggs [SDR]	0.32 [0.11, 0.53]			-0.59 [-1.71, 0.53]		
- U46619						
Kim [GP, 2]	0.24 [-0.62, 1.10]			Not estimable		
Overall [all]	0.32 [0.11, 0.52]	0%		-0.59 [-1.71, 0.53]	n.a.	

EC₅₀ scale: -1 -0.5 0 0.5 1

Favors *more* vasoconstriction | Favors *less* vasoconstriction

E_max scale: -4 -2 0 2 4

Favors *less* vasoconstriction | Favors *more* vasoconstriction

Figure 6. Effect of pregnancy at mid and late gestation on renal artery responses to stimuli involved in vasoconstriction through the GqSMC-coupled pathway in presence and absence of blockade. The effect of pregnancy on EC_{50} (the dose of stimulus inducing 50% response) is depicted as mean difference (MD) and its 95% confidence interval (95% CI). E_{max} (maximum response) is presented as standardized mean difference (SMD) and its 95% CI. Studies and totals based on Long Evans rats (LER), Sprague Dawley rats (SDR), sheep and guinea pigs (GP). 1, 2, 3 represents first, second and third part of mid and late gestation. I^2 represents the amount of heterogeneity. n.a. = not applicable.

term, the changes in SDR are maximal in mid-pregnancy and remain at maximum level throughout late pregnancy. As shown in Figure 9, this pattern is different from that in the other rat strains investigated (LER, MWR), that matches better with the human pattern. The persistent changes in SDR may reflect their relatively compromised vascular health. This strain is known to develop severe vascular dysfunction later in life, including chronic progressive nephropathy, peri-/vasculitis, and chronic cardiomyopathy [55]. SDR could be the better model for renal vascular maladaptation, whereas LER and MWR seem the more appropriate models for normal pregnancy-induced renal vasodilation.

Our meta-analysis detected considerable heterogeneity ($I^2>$ 60%) for overall renal flow and moderate heterogeneity ($I^2<60\%$) for overall GFR and RVR. This can only partially be attributed to differences between species and strains, as stratification reduced heterogeneity substantially only for some subgroups. Additionally, methodological differences in determining renal flow (flow probe

or para-aminohippurate clearance) may have contributed to the heterogeneity. The substantial degree of heterogeneity implies that data should not be quantitatively extrapolated to other species or strains.

Renal artery function is not affected by pregnancy, except for SDR. Pregnancy does not change the renal artery responsiveness to pharmacological stimuli (Gq$_{EC}$- and Gq$_{SMC}$-mediated stimuli, NO and potassium). One study in SDR showed an endothelium-dependent decrease in renal artery myogenic reactivity and increased vascular compliance in late pregnancy. Apparently SDR activate additional mechanisms to realize the same degree of vascular adaptation to pregnancy, as compared to other strains. This may represent a more generalized pattern, as it has been reported that SDR activate additional vascular adaptive pathways also in mesenteric arteries responses in pregnancy [8].

One may question what mechanisms are responsible for the changes in RVR, renal flow and GFR in pregnancy, given the lack of a role for renal artery adaptation. Our meta-analysis does not

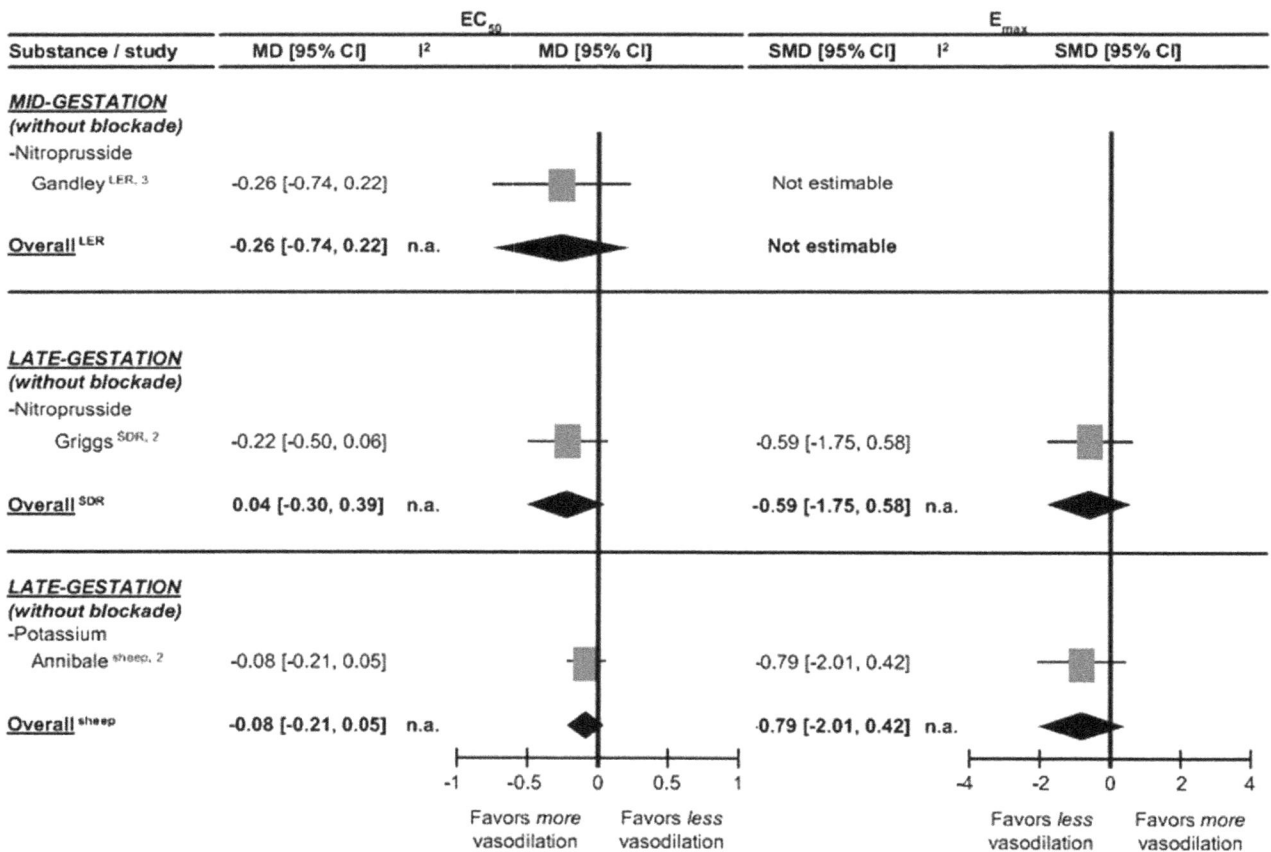

Figure 7. Effect of mid and late pregnancy on renal artery responses to nitric oxide (NO) and potassium. The effect of pregnancy on EC_{50} (the dose of stimulus inducing 50% response) is depicted as mean difference (MD) and its 95% confidence interval (95% CI). E_{max} (maximum response) is presented as standardized mean difference (SMD) and its 95% CI. Studies and totals based on Long Evans rats (LER), Sprague Dawley rats (SDR) and sheep. 1, 2, 3 represents first, second and third part of mid and late gestation. I^2 represents the amount of heterogeneity. n.a. = not applicable.

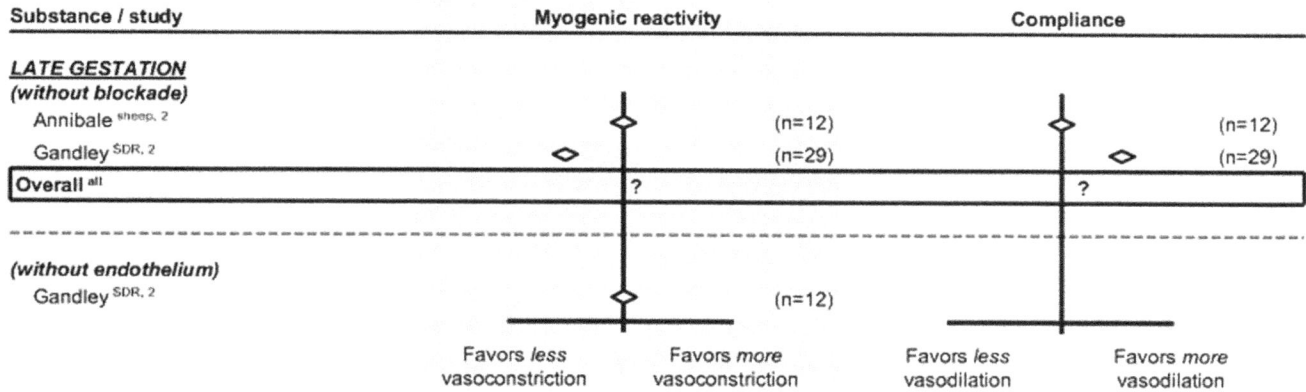

Figure 8. Effect of pregnancy at late gestation on renal artery responses to myogenic response and vascular compliance. The effect of pregnancy on the response to stimuli, depicted as the difference in direction.? = unknown or non-conclusive data. Studies and totals based on Sprague Dawley rats (SDR) and sheep. 1, 2, 3 represents first, second and third part of late gestation. n = number of subjects.

provide the answer. Several mechanisms can be considered. Reduced RVR and enhanced renal flow and GFR in pregnancy most likely result from regulation by the small resistance vessels of the kidney, rather than the renal artery [56]. The juxtaglomerular apparatus, which regulates afferent and efferent glomerular artery tone, may also be involved through resetting of the tubuloglomerular feedback system [57]. It seems likely that pregnancy-specific hormones, including relaxin or progesterone, play a role in these processes [58,59].

Several methodological aspects of our meta-analysis deserve discussion. First, the quality of all included studies was scored as poor, in terms of the reported number of animals, parity, randomization, and blinding the outcome assessment. This is a common finding in animal experiments [8], which are primarily concerned with generation and testing of hypotheses rather than with rigorous employment of randomized-controlled-trial methodology. Nonetheless, the quality of animal experimental work, and therefore the reliability of its findings, could benefit from application of strict methodological criteria, standardized procedures and reporting guidelines [60]. Second, our literature search

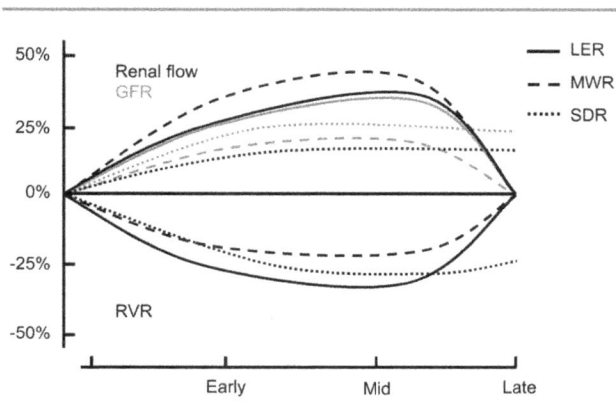

Figure 9. Summary of renal vascular resistance (RVR), renal flow and glomerular filtration rate (GFR) during pregnancy in Long Evens rats (LER), Munich Wistar rats (MWR) and Sprague Dawley rats (SDR). Graph design based on data at mid- and late gestation. No data were available on early pregnancy.

was not designed to identify all studies that reported on pregnancy-induced changes in RVR, renal flow and GFR. Because we were primarily interested in the underlying mechanisms, we restricted our search to studies investigating responses to vaso-active stimuli. Many studies in different animals have investigated RVR, RPF and GFR without using vaso-active stimuli. These studies were not included in our meta-analysis. Our data on pregnancy-induced changes imply species, strain and gestational age differences in RVR, RPF and GFR, without having the intention to be complete on this. Third, despite extensive searches, some of our observations are still based on a limited number of animals and strains of animals. Obviously, these results have to be interpreted with some restraint. Fourth, publication bias may have affected the results. Negative results tend to be underreported, which may therefore lead to overestimated effect sizes. Our funnel-plot analysis did not detect any such effect. Fifth, one may question whether it is legitimate to group together responses to pharmacological stimuli across species and strains. We observed considerable heterogeneity ($I^2 > 60\%$), which implies that one should not regard the responses as uniform across species and strains. Sixth, one may question the validity of combining the responses to different stimuli according to their assumed common G-protein coupled pathway. If heterogeneity would have been low, it would have been reasonable to group together the various stimuli to their common pathways, as observed in a former study of our group [61]. The moderate heterogeneity, observed for most responses in the present study, might imply that the assumption is not valid. However, it does not necessarily disprove the validity, because heterogeneity may be affected by many methodological aspects of the experiments. Given the moderate heterogeneity, one should mainly focus on the qualitative similarities.

In conclusion, our meta-analysis shows that pregnancy reduces RVR and increases renal flow and GFR through NO activation and sympathetic de-activation and not through a change in renal artery responsiveness. The cellular level at which pregnancy affects the respective pathways remains unknown. Quantitatively, renal vascular changes in pregnancy vary between species, strains and gestational age. Our meta-analysis suggests that renal changes during pregnancy are qualitatively similar in animals and even in comparison to humans, but quantitatively different.

Supporting Information

Checklist S1   Prisma Checklist.
(DOC)

Data S1   RevMan export data file.
(CSV)

Acknowledgments

The authors would like to thank Alice HJ Tillema, from the Medical Library of the Radboud University Nijmegen, for assistance with the search strategy.

Author Contributions

Conceived and designed the experiments: JVD RDV MS. Performed the experiments: JVD RDV. Analyzed the data: JVD. Wrote the paper: JVD RDV FL PS MS.

References

1. Sturgiss SN, Dunlop W, Davison JM (1994) Renal haemodynamics and tubular function in human pregnancy. Baillieres Clin Obstet Gynaecol 8: 209–234.
2. Slangen BF, Out IC, Verkeste CM, Peeters LL (1996) Hemodynamic changes in early pregnancy in chronically instrumented, conscious rats. Am J Physiol 270: H1779–H1784.
3. Sladek SM, Magness RR, Conrad KP (1997) Nitric oxide and pregnancy. Am J Physiol 272: R441–R463.
4. Chesley LC, Talledo E, Bohler CS, Zuspan FP (1965) Vascular Reactivity to Angiotensin II and Norepinephrine in Pregnant Woman. Am J Obstet Gynecol 91: 837–842.
5. McLaughlin MK, Keve TM (1986) Pregnancy-induced changes in resistance blood vessels. Am J Obstet Gynecol 155: 1296–1299.
6. Stanton BA, Koeppen BM (1998) The kidney; Elements of Renal Function. In: Berne RMaLMN, editors. Physiology. pp. 677–698.
7. Baylis C, Davison JM (1998) The Urinary System. In: Chamberlain G, Broughton Pipkin F, editors. Clinical Physiology in Obstetrics. Oxford, UK: Blackwell Science. pp. 263–307.
8. van Drongelen J, Hooijmans CR, Lotgering FK, Smits P, Spaanderman ME (2012) Adaptive changes of mesenteric arteries in pregnancy: a meta-analysis. Am J Physiol Heart Circ Physiol 303: H639–H657. ajpheart.00617.2011 [pii];10.1152/ajpheart.00617.2011 [doi].
9. Busse R, Edwards G, Feletou M, Fleming I, Vanhoutte PM, et al. (2002) EDHF: bringing the concepts together. Trends Pharmacol Sci 23: 374–380. S0165614702020503 [pii].
10. Davis MJ, Hill MA (1999) Signaling mechanisms underlying the vascular myogenic response. Physiol Rev 79: 387–423.
11. Feletou M (2011) Calcium Signaling in Vascular Cells and Cell-to-Cell Communications. In: Part 1: Multiple Functions of the Endothelial Cells-Focus on Endothelium-Derived Vasoactive Mediators. San Rafael (CA): Morgan & Claypool Life Sciences. pp. 19–43.
12. Feletou M (2011) Endothelium-Dependent Regulation of Vascular Tone. In: Part 1: Multiple Functions of the Endothelial Cells-Focus on Endothelium-Derived Vasoactive Mediators. San Rafael (CA): Morgan & Claypool Life Sciences. pp. 43–137.
13. Morello F, Perino A, Hirsch E (2009) Phosphoinositide 3-kinase signalling in the vascular system. Cardiovasc Res 82: 261–271. cvn325 [pii];10.1093/cvr/cvn325 [doi].
14. Ohno M, Gibbons GH, Dzau VJ, Cooke JP (1993) Shear stress elevates endothelial cGMP. Role of a potassium channel and G protein coupling. Circulation 88: 193–197.
15. Parkington HC, Coleman HA, Tare M (2004) Prostacyclin and endothelium-dependent hyperpolarization. Pharmacol Res 49: 509–514. 10.1016/j.phrs.2003.11.012 [doi];S1043661803004055 [pii].
16. Borenstein M, Hedges LV, Higgins JPT, Rothstein HR (2009) Introduction to Meta-Analysis. Chichester, UK: John Wiley and Sons Ltd.
17. Egger M, Davey SG, Schneider M, Minder C (1997) Bias in meta-analysis detected by a simple, graphical test. BMJ 315: 629–634.
18. Novak J, Reckelhoff J, Bumgarner L, Cockrell K, Kassab S, et al. (1997) Reduced sensitivity of the renal circulation to angiotensin II in pregnant rats. Hypertension 30: 580–584.
19. Omer S, Shan J, Varma DR, Mulay S (1999) Augmentation of diabetes-associated renal hyperfiltration and nitric oxide production by pregnancy in rats. J Endocrinol 161: 15–23.
20. Patel KP, Zhang PL (1993) Role of renal nerves in renal responses to acute volume expansion during pregnancy in rats. Proc Soc Exp Biol Med 203: 150–156.
21. Reckelhoff JF, Yokota SD, Baylis C (1992) Renal autoregulation in midterm and late-pregnant rats. Am J Obstet Gynecol 166: 1546–1550.
22. Sen AP, Dong Y, Gulati A (1997) Effect of diaspirin crosslinked hemoglobin on systemic and regional blood circulation in pregnant rats. Artif Cells Blood Substit Immobil Biotechnol 25: 275–288.
23. van Drongelen J, Pertijs J, Wouterse A, Hermsen R, Sweep FC, et al. (2011) Contribution of different local vascular responses to mid-gestational vasodilation. Am J Obstet Gynecol 205: 155–157. S0002-9378(11)00329-2 [pii];10.1016/j.ajog.2011.03.020 [doi].
24. van Eijndhoven HW, van der Heijden OW, Fazzi GE, Aardenburg R, Spaanderman ME, et al. (2003) Vasodilator reactivity to calcitonin gene-related peptide is increased in mesenteric arteries of rats during early pregnancy. J Vasc Res 40: 344–350.
25. van Eijndhoven HW, Aardenburg R, Spaanderman ME, De Mey JG, Peeters LL (2007) Pregnancy enhances the prejunctional vasodilator response to adrenomedullin in selective regions of the arterial bed of Wistar rats. Reprod Sci 14: 771–779.
26. Masilamani S, Castro L, Baylis C (1994) Pregnant rats are refractory to the natriuretic actions of atrial natriuretic peptide. Am J Physiol 267: R1611–R1616.
27. McElvy S, Greenberg SG, Mershon JL, Da Seng Y, Magill C, et al. (2001) Mechanism of uterine vascular refractoriness to endothelin-1 in pregnant sheep. American Journal of Physiology Heart and Circulatory Physiology 281: H804–H812. Article.
28. Hines T, Porter JP (1992) Pressor effect of posterior hypothalamic stimulation is enhanced in pregnant rats. Am J Physiol 262: R604–R609.
29. Hines T, Herzer WA (2000) Effect of cardiac receptor stimulation on renal vascular resistance in the pregnant rat. Am J Physiol Regul Integr Comp Physiol 278: R171–R178.
30. Kim TH, Weiner CP, Thompson LP (1994) Effect of pregnancy on contraction and endothelium-mediated relaxation of renal and mesenteric arteries. Am J Physiol 267: H41–H47.
31. Kuczynska-Sicinska J (1972) Effect of angiotensin II on arterial blood pressure, glomerular filtration rate and urinary sodium excretion in normal, pregnant and hypertensive rats. Pol Med J 11: 1113–1124.
32. Danielson LA, Conrad KP (1995) Acute blockade of nitric oxide synthase inhibits renal vasodilation and hyperfiltration during pregnancy in chronically instrumented conscious rats. J Clin Invest 96: 482–490.
33. Danielson LA, Conrad KP (1996) Prostaglandins maintain renal vasodilation and hyperfiltration during chronic nitric oxide synthase blockade in conscious pregnant rats. Circ Res 79: 1161–1166.
34. Fan L, Mukaddam-Daher S, Gutkowska J, Nuwayhid BS, Quillen EW Jr (1996) Enhanced natriuretic response to intrarenal infusion of atrial natriuretic factor during ovine pregnancy. Am J Physiol 270: R1132–R1140.
35. Ferris TF, Weir EK (1983) Effect of captopril on uterine blood flow and prostaglandin E synthesis in the pregnant rabbit. Journal of Clinical Investigation 71: 809–815. Article.
36. Gandley RE, Griggs KC, Conrad KP, McLaughlin MK (1997) Intrinsic tone and passive mechanics of isolated renal arteries from virgin and late-pregnant rats. Am J Physiol 273: R22–R27.
37. Gandley RE, Conrad KP, McLaughlin MK (2001) Endothelin and nitric oxide mediate reduced myogenic reactivity of small renal arteries from pregnant rats. Am J Physiol Regul Integr Comp Physiol 280: R1–R7.
38. Greenberg SG, Clark KE (1999) Hemodynamic effects of platelet-activating factor in nonpregnant and pregnant sheep. Am J Physiol 277: R996–R1001.
39. Cha SC, Aberdeen GW, Mukaddam-Daher S, Quillen EW Jr, Nuwayhid BS (1993) Tubular handling of fluid and electrolytes during ovine pregnancy. American Journal of Physiology Renal Fluid and Electrolyte Physiology 265: F278–F284. Article.
40. Cha SC, Aberdeen GW, Mukaddam-Daher S, Quillen EW Jr, Nuwayhid BS (1993) Autoregulation of renal blood flow during ovine pregnancy. Hypertension in Pregnancy 12: 71–83. Article.
41. Chu ZM, Beilin IJ (1997) Demonstration of the existence of nitric oxide-independent as well as nitric oxide-dependent vasodilator mechanisms in the in situ renal circulation in near term pregnant rats. Br J Pharmacol 122: 307–315.
42. Conrad KP, Morganelli PM, Brinck-Johnsen T, Colpoys MC (1989) The renin-angiotensin system during pregnancy in chronically instrumented, conscious rats. Am J Obstet Gynecol 161: 1065–1072.
43. Conrad KP, Gandley RE, Ogawa T, Nakanishi S, Danielson LA (1999) Endothelin mediates renal vasodilation and hyperfiltration during pregnancy in chronically instrumented conscious rats. Am J Physiol 276: F767–F776.
44. Baylis C, Collins RC (1986) Angiotensin II inhibition on blood pressure and renal hemodynamics in pregnant rats. Am J Physiol 250: F308–F314.
45. Baylis C (1988) Effect of amino acid infusion as an index of renal vasodilatory capacity in pregnant rats. Am J Physiol 254: F650–F656.
46. Baylis C (1993) Blood pressure and renal hemodynamic effects of acute blockade of the vascular actions of arginine vasopressin in normal pregnancy in the rat. Hypertension in Pregnancy 12: 93–102. Article.

47. Baylis C (1995) Acute blockade of alpha-1-adrenoreceptors has similar effects in pregnant and nonpregnant rats. Hypertension in Pregnancy 14: 17–25. Article.

48. Bobadilla LRA, Perez-Alvarez VM, Robledo LA, Sanchez PL (2005) Renal vascular responses in an experimental model of preeclampsia. Proceedings of the Western Pharmacology Society 48: 49–51. Conference Paper.

49. Bobadilla RA, Anguiano Robledo L, Perez-Alvarez VM, Valencia Hernandez I, Lopez Sanchez P (2001) Pregnancy influence on rat renal response to phenylephrine. Proc West Pharmacol Soc 44: 43–44.

50. Bobadilla RA, Anguiano L, Perez-Alvarez V, Lopez SP (2003) Pregnancy influence on the vascular interactions between nitric oxide and other endothelium-derived mediators in rat kidney. Can J Physiol Pharmacol 81: 1–8.

51. Annibale DJ, Rosenfeld CR, Kamm KE (1989) Alterations in vascular smooth muscle contractility during ovine pregnancy. Am J Physiol 256: H1282–H1288.

52. Griggs KC, Conrad KP, Mackey K, McLaughlin MK (1993) Endothelial modulation of renal interlobar arteries from pregnant rats. Am J Physiol 265: F309–F315.

53. Knight S, Snellen H, Humphreys M, Baylis C (2007) Increased renal phosphodiesterase-5 activity mediates the blunted natriuretic response to ANP in the pregnant rat. American Journal of Physiology Renal Physiology 292: F655–F659. Article.

54. Woods LL, Mizelle HL, Hall JE (1987) Autoregulation of renal blood flow and glomerular filtration rate in the pregnant rabbit. Am J Physiol 252: R69–R72.

55. Weber K (2009) Differences in Rat Strains - Where stands the RccHanWist™ Hannover Wistar Rat? Online.

56. Ren Y, Garvin JL, Liu R, Carretero OA (2009) Cross-talk between arterioles and tubules in the kidney. Pediatr Nephrol 24: 31–35. 10.1007/s00467-008-0852-8 [doi].

57. Baylis C, Blantz RC (1985) Tubuloglomerular feedback activity in virgin and 12-day-pregnant rats. Am J Physiol 249: F169–F173.

58. Novak J, Danielson LA, Kerchner LJ, Sherwood OD, Ramirez RJ, et al. (2001) Relaxin is essential for renal vasodilation during pregnancy in conscious rats. J Clin Invest 107: 1469–1475. 10.1172/JCI11975 [doi].

59. Ogueh O, Clough A, Hancock M, Johnson MR (2011) A longitudinal study of the control of renal and uterine hemodynamic changes of pregnancy. Hypertens Pregnancy 30: 243–259. 10.3109/10641955.2010.484079 [doi].

60. Hooijmans CR, Leenaars M, Ritskes-Hoitinga M (2010) A gold standard publication checklist to improve the quality of animal studies, to fully integrate the Three Rs, and to make systematic reviews more feasible. Altern Lab Anim 38: 167–182.

61. van Drongelen J, Hooijmans CR, Lotgering FK, Smits P, Spaanderman ME (2012) Adaptive changes of mesenteric arteries in pregnancy: a meta-analysis. Am J Physiol Heart Circ Physiol 303: H639–H657. ajpheart.00617.2011 [pii]; 10.1152/ajpheart.00617.2011 [doi].

18

Environmental Drivers of *Culicoides* Phenology: How Important Is Species-Specific Variation When Determining Disease Policy?

Kate R. Searle[1]*, **James Barber**[2], **Francesca Stubbins**[2], **Karien Labuschagne**[3], **Simon Carpenter**[2], **Adam Butler**[4], **Eric Denison**[2], **Christopher Sanders**[2], **Philip S. Mellor**[2], **Anthony Wilson**[2], **Noel Nelson**[5], **Simon Gubbins**[2], **Bethan V. Purse**[1]

1 Centre for Ecology and Hydrology, Bush Estate, Penicuik, Midlothian, United Kingdom, **2** The Pirbright Institute, Ash Road, Pirbright, Woking, United Kingdom, **3** PVVD, ARC-Onderstepoort Veterinary Institute, Private Bag X05, Onderstepoort, South Africa, **4** Biomathematics and Statistics Scotland, James Clerk Maxwell Building, The King's Buildings, Edinburgh, United Kingdom, **5** The Met Office, Based at The Pirbright Institute, Pirbright, Woking, Surrey, United Kingdom

Abstract

Since 2006, arboviruses transmitted by *Culicoides* biting midges (Diptera: Ceratopogonidae) have caused significant disruption to ruminant production in northern Europe. The most serious incursions involved strains of bluetongue virus (BTV), which cause bluetongue (BT) disease. To control spread of BTV, movement of susceptible livestock is restricted with economic and animal welfare impacts. The timing of BTV transmission in temperate regions is partly determined by the seasonal presence of adult *Culicoides* females. Legislative measures therefore allow for the relaxation of ruminant movement restrictions during winter, when nightly light-suction trap catches of *Culicoides* fall below a threshold (the 'seasonally vector free period': SVFP). We analysed five years of time-series surveillance data from light-suction trapping in the UK to investigate whether significant inter-specific and yearly variation in adult phenology exists, and whether the SVFP is predictable from environmental factors. Because female vector *Culicoides* are not easily morphologically separated, inter-specific comparisons in phenology were drawn from male populations. We demonstrate significant inter-specific differences in *Culicoides* adult phenology with the season of *Culicoides scoticus* approximately eight weeks shorter than *Culicoides obsoletus*. Species-specific differences in the length of the SVFP were related to host density and local variation in landscape habitat. When the *Avaritia Culicoides* females were modelled as a group (as utilised in the SFVP), we were unable to detect links between environmental drivers and phenological metrics. We conclude that the current treatment of *Avaritia Culicoides* as a single group inhibits understanding of environmentally-driven spatial variation in species phenology and hinders the development of models for predicting the SVFP from environmental factors. *Culicoides* surveillance methods should be adapted to focus on concentrated assessments of species-specific abundance during the start and end of seasonal activity in temperate regions to facilitate refinement of ruminant movement restrictions thereby reducing the impact of *Culicoides*-borne arboviruses.

Editor: Nikos T. Papadopoulos, University of Thessaly, Greece

Funding: This study was funded by EU grant FP7-261504 EDENext and is catalogued by the EDENext Steering Committee as EDENext268 (http://www.edenext.eu). The contents of this publication are the sole responsibility of the authors and don't necessarily reflect the views of the European Commission. The funders had no role in study design, data collection and analysis, decision to publish, or preparation of the manuscript.

Competing Interests: The authors have declared that no competing interests exist.

* Email: katrle@ceh.ac.uk

Introduction

Northern Europe is currently experiencing an unprecedented series of incursions of arboviruses transmitted between ruminants by *Culicoides* (Diptera: Ceratopogonidae) [1]. Five separate strains of bluetongue virus (BTV) have been recorded in the region since 2006 [2–4], and a novel *Culicoides*-borne arbovirus, provisionally named Schmallenberg virus (SBV), was discovered by metagenomic studies in Germany in 2011 [5], following unexplained clinical signs in dairy cattle. SBV has since spread rapidly across a large geographic area, and has been found to inflict foetal abnormalities in both cattle and sheep [6]. The route of entry of several of these arbovirus strains remains undefined [7]; hence the risk of further emergence of *Culicoides*-borne pathogens in this region cannot easily be estimated or mitigated.

In temperate ecosystems, the seasonal incidence and abundance of adult female *Culicoides* is a key parameter in determining the timing and intensity of arbovirus outbreaks and varies widely across geographical space [8]. It is thought that livestock-associated *Culicoides* in northern Europe overwinter in their final (fourth) larval instar and do not generally survive the winter as adults [9]. While live *Culicoides* adults have been recovered from animal housing in winter in northern Europe [10,11], their numbers seem to be insufficient to drive BTV outbreaks since new confirmed clinical cases of BT are only very rarely recorded in winter (December to March).

The response to incursion of livestock arboviruses in Europe is dependent on strain pathogenicity and whether legislation exists that defines control measures. For BTV, emergence is notifiable to the World Organisation for Animal Health (OIE). Movement and trade restrictions of susceptible stock in the surrounding area are imposed immediately to limit spread of BTV (defined under Council Directive 2000/75/EC of EU legislation). Although these measures reduce the speed and extent of spread of viruses, they also impose huge logistic and welfare costs on affected regions [12] that could be minimised with enhanced understanding of geographical and annual variation in adult vector seasons.

To date, the most damaging outbreak of bluetongue (BT) in northern Europe was inflicted by a serotype 8 strain of sub-Saharan origin [1,13]. In response to this incursion, movement restrictions were imposed across the region but vaccination to combat spread was not available until spring 2008, some eighteen months after the initial incursion was identified. In 2007, to alleviate the impact of movement restrictions, a 'seasonal vector free period' (SVFP) scheme was defined by the European Union council enabling movement of susceptible livestock during winter under what was hypothesised to be an extremely low risk of BTV transmission (defined in Annexe V of Commission Regulation (EC) No. 1266/2007). The declaration enabled movement of these livestock from farms in the affected zones to winter sites or markets, substantially easing the economic difficulties of farmers in the affected region and enabling some 85,000 animal movements in the UK alone during the winter of 2007 [14].

Maintenance of the SVFP is reliant upon the operation of a network of light-suction traps designed to monitor adult *Culicoides* activity [15]. In northern Europe, a threshold of catching less than five parous (abdominally pigmented) female *Culicoides* per trap was set as a limit for declaring freedom of adult activity. In some countries (e.g., the UK), this restriction was additionally underpinned using data concerning the thermal limits of BTV replication [16]. Due in part to fact that the primary vectors of BTV-8 in northern Europe were not convincingly identified to species level [1], no attempt was made to account for potential variation in phenology in *Culicoides* species that could be involved in transmission.

Unlike BTV-8, SBV has not been declared a notifiable disease by a majority of affected countries partly because substantial geographical spread had already occurred before the pathogen was first detected. For mitigating SBV impacts, it is even more imperative to understand *Culicoides* phenology, because the extent of clinical birth malformations in lambs and calves is governed by whether the ewe or dam receives an infectious *Culicoides* bite during a particular period of pregnancy [17,18], which is close to the last portion of the adult vector season. In the event that SBV persists in this region, or in the event of emergence of more pathogenic strains, this knowledge could be employed to make alterations in husbandry practises that would reduce the impact of the disease including changes in tupping schedules [19].

In this paper light-suction trap data collected over a five year period within the UK are analysed to quantify inter-specific differences in the phenology of *Culicoides*. Phenological metrics are related to remote environmental variables with the aim of understanding how variation in the SVFP may be produced under different climate, host and landscape conditions. The livestock-associated *Culicoides* fauna in northern Europe is dominated by species belonging to the subgenus *Avaritia* and four species of this group have been identified in the UK (*Culicoides obsoletus*, *Culicoides scoticus*, *Culicoides dewulfi* and *Culicoides chiopterus*).

Due to poor levels of discrimination during vector competence studies it remains unclear which of these species were involved in transmission of BTV-8 and to what degree [1]. We examine the potential policy impact of treating these individual species as a single group and then identify key environmental drivers of the timing of each phenological metric for both the group and its constituent species. More specifically, we hypothesise that the four species differ significantly in their phenology and that these responses can be explained by varying responses to environmental variables characterising different habitat and climatic requirements.

In a broader sense, we also address the statistical complications that arise routinely when analysing phenological data. Phenological variables, which refer to the date on which particular events occur, are inherently circular (with Day = 366, or Day = 367 in leap years, being equivalent to Day = 1). Standard statistical methods (such as linear regression, linear mixed models and GLMMs) fail to account for this circularity, and can therefore lead to biased results and potentially to the detection of spurious relationships. We present a hierarchical Bayesian method for dealing with the circular nature of phenological event data by using a natural extension of linear mixed modelling [20]. This method can be implemented using freely available software packages.

Materials and Methods

Trapping methods and locations

Trapping was conducted from 2006–2010 using standard 8w ultraviolet Onderstepoort Veterinary Institute (OVI) light-suction traps, as previously recommended for *Culicoides* surveillance purposes [21]. OVI traps were hung at approximately 1.5 m as close to livestock as logistically possible to allow permanent positioning and in all cases were ≤25 m distance from ruminant hosts throughout the trapping period. A total of 29 trap locations on private land were used for a variable period of time ranging from one to four years across the UK (Figure 1; for further information regarding the precise location of trapping sites please contact KRS). Access to private land was granted by the landowners. Sampling was carried out by volunteers on one night from dusk until dawn each week with no attempt to synchronise trapping day across the network. Collections were made into approximately 250 ml of water with a drop of detergent to reduce surface tension and then later transferred to 70% ethanol and transported to the Pirbright Institute for identification. Fieldwork did not involve endangered or protected species.

Following receipt, all non-*Culicoides* were removed from the samples based upon morphological features of wing pattern [22] and the absence of humeral pits diagnostic for *Culicoides*. The remaining portion of each sample was categorised into *Culicoides* species groups using wing patterns. Male members of the subgenus *Avaritia* were then identified to species level using genitalia morphology (hereafter referred to as '*Avaritia* males'), while females were treated as a single group (hereafter referred to as '*Avaritia* females'). Females of the subgenus *Avaritia* were characterised according to the appearance of the abdomen as pigmented, non-pigmented or gravid [23]. Large catches (≥1000 individuals) were subsampled using volumetric methods.

Timing metrics

From the raw time-series data of weekly trap catches at each site in each year, the date of first appearance (start), last appearance (end), and the length of the inactive overwinter period (length overwinter) were derived for *Avaritia* females and for the

Figure 1. Locations of UK trapping sites for *Culicoides* surveillance (2006–2010). Circle area is proportional to maximum catch ever recorded per site, where complete yearly data were available. Sites with crosses lacked complete yearly trapping profile.

individual species in the *Avaritia* male dataset. The start of the season for *Avaritia* females was defined as the first day of the year in which more than five pigmented females were caught (in accordance with the definition used by European Union Council to define the start and end of the SVFP in Europe). Site-by-year combinations were only included in the analysis if at least two trapping nights prior to this date had been recorded. The end of the season for *Avaritia* females was similarly determined as the last day of the year in which more than five pigmented females were caught (with the site-by-year combination being excluded from analysis if there were less than two subsequent trapping nights

following this date). For the *Avaritia* males, the start and end of the season were defined as the first (start) and final (end) day of the season when more than five males were caught. The length of the overwinter period was determined as the difference in days between the start of the season in one year, and the end of the season the previous year. Relationships between these three timing metrics and a suite of environmental variables that have previously been found to be important in influencing the seasonal dynamics and abundance of these species were then investigated [24–27].

Environmental covariates

We derived a series of meteorological variables for each site and year combination from the nearest UK Met Office weather station using raw data on total daily precipitation (mm), mean daily temperature (°C), and daily humidity (%) as measured at 15:00 hrs (Table 1). The mean distance to the nearest weather station was 3.9 km (range: 0.1 km−12.4 km). Photoperiod was calculated daily for each site using latitude and was expressed as the first day of the year at which nine hours of daylight were achieved in each year of observation. Percentage cover of two broad land-cover habitat classes – moorland and heathlands (*moors*) and broadleaf mixed woodland (*brdlf*) - in the surrounding 1 km around each trapping site were determined from the CEH Landcover Map 2007 [28]. The mean densities of cattle and sheep around each trapping site were determined for each year using the Edina AgriCensus data at a 2 km resolution (http://edina.ac.uk/agcensus/).

Statistical models

We used statistical models to analyse relationships between seasonal timing metrics and environmental drivers (see supplementary material for complete details, Section S1). Three different response variables were considered: length of the overwinter period (in days), start of the season (Julian day) and end of the season (Julian day). Analyses were performed for the entire group

of *C. obsoletus* complex parous females (*Avaritia* females) and, separately, for each of the individual species within this complex (*Avaritia* males). We constructed sets of environmental variables that were relevant for each seasonal timing metric. The five non-meteorological variables (photoperiod, percent cover of moorland (*moors*), percent cover of broadleaf woodland (*brdlf*), mean density of cattle, mean density of sheep) were assumed to be relevant to all three metrics, but distinct meteorological variables were identified for each seasonal timing metric (Table 1).

Models for length of the overwinter period (days) assumed that the response variable had a normal distribution, whilst models for the start and end of the season were based upon a Wrapped Normal distribution (WN). 'Start of season' and 'end of season' are circular variables. It would be inappropriate to assume that these variables have a normal distribution because this would imply that the difference between January 1st and December 31st (a difference in Julian days of 364) is twice as large as the difference between January 1st and July 1st (a difference in Julian days of 182). Distributions for circular data deal with this problem by treating the Julian date as a variable that lies on a circle rather than a line, so that the distance between December 31st and January 1st is equal to one rather than 364 [29]. Circular data also arise in other contexts – e.g. when modelling angles and directions. A range of distributions exist for modelling data of this form, but the Wrapped Normal (WN) is commonly used [29]. The WN

Table 1. Derived meteorological variables from UK Met Office weather station data for each of the trapping site by year combinations used in the timing analyses.

Seasonal metric	Meteorological variable	Definition	Notation
Start of season & length of overwinter	Mean winter temperature (°C)	Mean daily temperature over November 1st to February 28th	T_w
Start of season	Accumulated degree days over winter	Accumulated degree days greater than 10°C between November and May to capture temperature variation over the preceding winter and current spring $\sum_{\text{Nov 1st}}^{\text{May 31st}} [\text{mean temperature-10°C, 0}]$	DD_w
Start of season	Mean spring temperature (°C)	Mean daily temperature over March 21st to April 30th	T_{spr}
Start of season	Total spring precipitation (mm)	Summed daily precipitation over March 1st to May 31st	P_{spr}
Start of season	Mean spring relative humidity (%)	Mean daily (15:00 hrs) relative humidity over March 21st to April 30th.	RH_{spr}
End of season	Accumulated degree days over summer	Accumulated degree days greater than 10°C between June and September to capture temperature variation over the current summer and autumn $\sum_{\text{June 1st}}^{\text{Sept 30th}} [\text{mean temperature-10°C, 0}]$	
End of season	Mean summer temperature (°C)	Mean daily temperature over 1st to September 30th	T_{sum}
End of season	Total summer precipitation (mm)	Summed daily precipitation over June 1st to September 30th	P_{sum}
End of season	Mean summer relative humidity (%)	Mean daily (15:00 hrs) relative humidity over June 1st to September 30th	RH_{sum}

distribution is symmetric and unimodel, and is obtained by wrapping a normal distribution on the real line around a circle [20]. The WN distribution contains two unknown parameters, which are directly analogous to the mean and variance of the normal distribution. The WN distribution is defined as a distribution for angles, which lie between zero and 2π, and variables that are defined in Julian days therefore need to be multiplied by $2\pi/365$ before the distribution is applied.

We assumed that the mean of the normal distribution (for length of overwinter period) or WN distribution (for start of season and end of season) had a linear relationship with environmental drivers. For *Avaritia* males the linear predictor also included a categorical variable to account for differences between species (as a fixed effect), and allowed for interactions between 'species' and the environmental drivers (species-environment interactions). Random effects were also included in models for both males and females in order to account for the structure of the data (and thereby avoid pseudo-replication). When performing analyses for the *Avaritia* females normally distributed random effects were included to account for variation between years (unstructured temporal heterogeneity) and sites (unstructured spatial heterogeneity). The model for *Avaritia* males included a single random effect to capture both site and year effects because the data were highly unbalanced with respect to site and year and so contained little information from which to separate out the effects of site, year and site-by-year interaction (leading to convergence problems in models for the phenological metrics with least data). To check that the use of a single random effect for males was reasonable and did not lead to pseudo-replication, we re-ran the final best-fitting model for the most data-rich phenological metric with additional random effects included (site, year, site-by-species, year-by-species) and found that the inferences regarding environmental relationships and species differences did not change.

Statistical inference

All models were fitted via Bayesian inference using WinBUGS [30] and the package R2WinBUGS on the R platform [31]. Standard diffuse priors, Normal (0, 100000) for slope parameters and Uniform (0,100) for standard deviations, were assumed for all parameters other than the intercept of the WN model. That parameter is assumed to have a prior of the form Normal $(-\pi,\pi)$, because the use of a more diffuse prior may lead to problems with convergence (B. Reich, personal communication). The fitting of the WN model in WinBUGS is not trivial, and we follow the approach of Modlin *et al.* [20]: further details of this approach, along with the BUGS code, are given in Section S2.

Stepwise selection using Deviance Information Criterion [32] was used to select environmental variables (and, for males, species-by-environment interactions) in order to find the most parsimonious model for each seasonal timing metric. DIC is a generalisation of the Akaike Information Criterion (AIC), and is derived as the mean deviance adjusted for the estimated number of parameters in the model – DIC accounts for both model fit and complexity, and providing a relative measure of out-of-sample predictive performance [33]. DIC comparisons to 'null' models are presented for all best-fitting models, where the null model contained all fixed and random effects but no environmental variables.

Ideally, the regression models for the start, end and length of the overwinter period should include an estimate of the total abundance of each species at each site in each year. This is because more accurate estimates of the timing of emergence and disappearance are expected at sites with higher abundances of *Culicoides*. However, because of the unequal number and uneven

timing of trapping across the sites, we were unable to create a meaningful estimate of overall site abundance for each year within this dataset. This problem is confounded by a degree of circularity between *Culicoides* phenology and abundance. We expect that trapping sites with higher abundance will produce more reliable estimates of the timing of phenological events such as the start and end of the season. However, it is also feasible that the phenology of the species may directly influence abundance, such that sites that are environmentally suitable for earlier spring emergence may also support greater population abundance. Therefore, disentangling phenology from abundance for these species is inherently difficult. To try and reduce this confounding we conducted a second analysis (Section S3) to compare the proportions of the total population of each of the four species that emerged in different seasonal periods for which trapping data were available using the *Avaritia* males dataset (Fig. S3 in Section S3). This analysis allowed us to examine the influence of environmental drivers on the proportion of each species' population that was active at different periods of the year (Section S3), bypassing the issue of not having a reliable estimate of overall abundance to include in the model. In so doing, we were able to determine that the same environmental variables affected both our measure of phenology, and proportional abundance, thereby providing a measure of the consistency in our inference regarding species-specific phenology (full details in *Section S3: 'Multinomial analysis of Avaritia males phenology'*).

Results

Using only those sites and years for which trapping captured the full seasonal profile of *Culicoides* abundance (n = 69, 14 sites over four years), the maximum nightly catch of males per species was 123 (mean 44.1, s.d. 34.8) for *C. chiopterus*, 217 (mean 81.6, s.d. 68.3) for *C. dewulfi*, 460 (mean 86.4, s.d. 116.0) for *C. obsoletus,* and 120 (mean 44.6, s.d. 37.9) for *C. scoticus*. Of these sites, the *Avaritia* male data shows that *C. obsoletus* was caught most frequently (53% of the total maximum catch across all years), followed by *C. dewulfi* (24%), *C. scoticus* (13%) and *C. chiopterus* (10%).

Start of season

Across all sites and years first appearance of *Culicoides* females occurred on average in early May (Table 2, Fig. 2) with the first record in late March (2007, site 3 'Chobham', Fig. 1) and the latest in late May (2009, site 25 'Winterslow', Fig. 1). Average appearance across all sites was remarkably consistent from year to year (early May), with the exception of 2007 in which the *Avaritia* females emerged on average two weeks earlier (Table 2), probably due to the warmer temperatures in winter and spring during that year relative to the other four years of observation (Fig. 3).

Among the subgenus *Avaritia* males, all species emerged in late May, with *C. obsoletus* and *C. dewulfi* tending to emerge on average one week earlier than *C. chiopterus* and *C. scoticus* (Table 2, Fig. 2). The best-fitting model for this metric detected no significant species differences in timing of emergence, however, between *C. obsoletus* and the other three species (Table 3).

The start of the season for *Avaritia* females was significantly negatively influenced by spring temperature, and close to being significantly positively influenced by photoperiod (more than 94% of posterior density mass greater than zero; Table 4). The best fitting model included both of these variables and received considerably more support in the data than the null model – containing no environmental variables (ΔDIC 4.5, Table S1),

Table 2. Summary of number of site-by-year combinations observed for each complex or species by timing metric used in the analysis.

	observations	sites	years	Min	Max	Range	Mean	Mean by year				
								2006	2007	2008	2009	2010
START												
Subgenus Avaritia females	49	20	5	88	143	55	123.7	126.3	115.6	127.3	126.1	125.7
C. obsoletus s.s	42	18	5	100	236	136	145.4	154.2	140.5	141.8	116.7	165.5
C. scoticus	19	10	4 (2009)	114	219	105	150.8	153	150.8	141	-	169
C. dewulfi	20	12	5	107	207	100	146.1	163.3	149	133.3	125	129
C. chiopterus	17	8	5	107	271	164	150.2	173	158	134.8	139	129
END												
Subgenus Avaritia females	47	17	5	255	347	92	302	303.6	311.8	297.8	300.6	296
C. obsoletus s.s	47	18	5	183	362	179	270.7	275.3	256.6	259	288.8	272
C. scoticus	23	10	5	183	328	145	274.6	274.3	255.3	278.3	286.2	276.6
C. dewulfi	26	11	5	183	308	125	270.8	267.5	278.5	245.7	294.4	275.2
C. chiopterus	24	8	5	229	306	77	278.3	279.5	277.5	262.6	284.3	289.2
OVERWINTER												
Subgenus Avaritia females	35	16	4	106	221	115	185.4	178.4	184.3	199.8	190.2	-
C. obsoletus s.s	26	13	4	147	336	189	247.7	246.3	250.9	235	248	-
C. scoticus	8	4	3 (2008)	182	301	119	236.5	219.3	241.3	-	255	-
C. dewulfi	8	3	3 (2008)	175	239	64	205.9	201.3	207.3	-	210.5	-
C. chiopterus	11	6	4	175	268	93	226.4	234	226.4	235	210.5	-

Years span from 2006–2010, missing years noted in parentheses. Summary of seasonal metrics (minimum, maximum and mean day of the year) across all sites, and mean for each year for C. obsoletus group (Avaritia females), C. obsoletus (Avaritia males), C. scoticus (Avaritia males), C. dewulfi (Avaritia males), and C. chiopterus (Avaritia males). START: start of season (day of year), END: end of season (day of year), OVERWINTER: length of overwinter period (days).

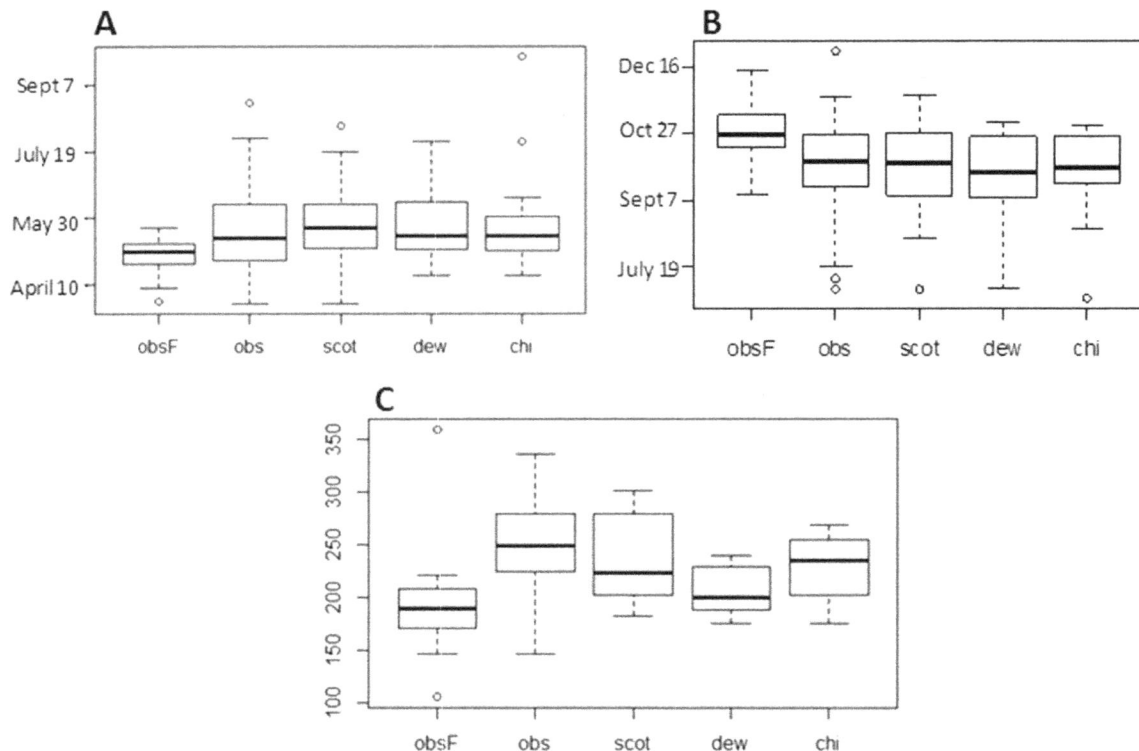

Figure 2. Raw data summarised for the start (A) and end (B) of seasonal activity (date), and length of overwinter period (C; days) derived from UK *Culicoides* surveillance data during 2006 to 2010. Box plots show the median (central line), box denotes 25th and 75th percentiles, error bars represent 10th and 90th percentiles, and dots are points outside the 10th and 90th percentiles. Data are shown for the subgenus *Avaritia* (*Avaritia* females; obsF), *C. obsoletus Avaritia* males (obs), *C. scoticus Avaritia* males (scot), *C. dewulfi Avaritia* males (dew), and *C. chiopterus Avaritia* males (chi).

although model fit to the data was poor (R^2 0.42, including both fixed and random effects).

The best-fitting model for the start of the season for the *Avaritia* males included a significant positive influence of cattle density on the start of seasonal activity for *C. obsoletus* with separate slopes for the other three species showing that start dates for *C. scoticus* and *C. dewulfi* were significantly less delayed by cattle density, and a non-significant difference for *C. chiopterus* (Table 3). A non-significant positive effect of spring humidity for *C. obsoletus* was also identified, with separate slopes for the other three species showing that start dates for *C. dewulfi* were significantly less affected by relative humidity than *C. obsoletus* (Table 3). In addition, a significant positive influence of spring temperature over all species was also determined and a non-significant positive influence of spring precipitation over all species occurred (best-fitting model: $R^2 = 0.73$; Table 3). An identical model without spring precipitation received essentially equal support in the data as the best-fitting model (ΔDIC 0.6; Table S2), indicating this variable is perhaps less important than the other three included in the best model. The null model received almost no support in the data in comparison to the best-fitting model (ΔDIC 37.9; Table S2).

End of the season

Averaged over all sites and years, the last capture date for the *Avaritia* females occurred at the end of October (Table 2, Fig. 2). The earliest final capture occurred in mid-September (2006, site 4 'Compton', Fig. 1), while the latest was in mid-December (2006, site 3 'Chobham', Fig. 1). The four constituent species of the

Avaritia males had mean dates of last capture within five days of each other, occurring in late September (Table 2, Fig. 2). *C. dewulfi* adults disappeared earliest, followed by *C. obsoletus*, *C. chiopterus* and *C. scoticus* (Table 2, Fig. 2). The best-fitting model for this metric detected no significant species differences between *C. obsoletus* and the other three species (Table 3). Overall, there was greater variability in the end dates for seasonal activity than for the start dates except for *C. chiopterus* (Table 2, Fig. 2).

No significant relationships were detected between environmental variables and the end of the season for *Avaritia* females (Table 4). Moreover, the null model received approximately similar support in the data as all other models, indicating that effects of climate and landcover on this phenological metric were not well captured when modelled as a complex (ΔDIC 1.3, Table S1), and model fit was poor (R^2 best-fitting model 0.42).

The best-fitting model for the end of seasonal activity for the *Avaritia* males contained a significant negative influence of mean summer humidity for *C. obsoletus* and separate and significantly different slopes for *C. dewulfi* and *C. chiopterus* showing that *C. obsoletus* was inhibited more by summer humidity than both these dung-breeding species (Table 3). A significant positive effect of sheep density was also identified across all species, along with a non-significant positive effect of photoperiod across all species and a non-significant positive effect of cattle density across all species ($R^2 = 0.85$; Table 3). Adding mean summer temperature to this model resulted in the same level of support in the data (ΔDIC 0.1, Table S2), as did dropping cattle density (ΔDIC 0.4, Table S2) or dropping photoperiod (ΔDIC 0.6, Table S2). The best-fitting model received considerably more support in the data than the

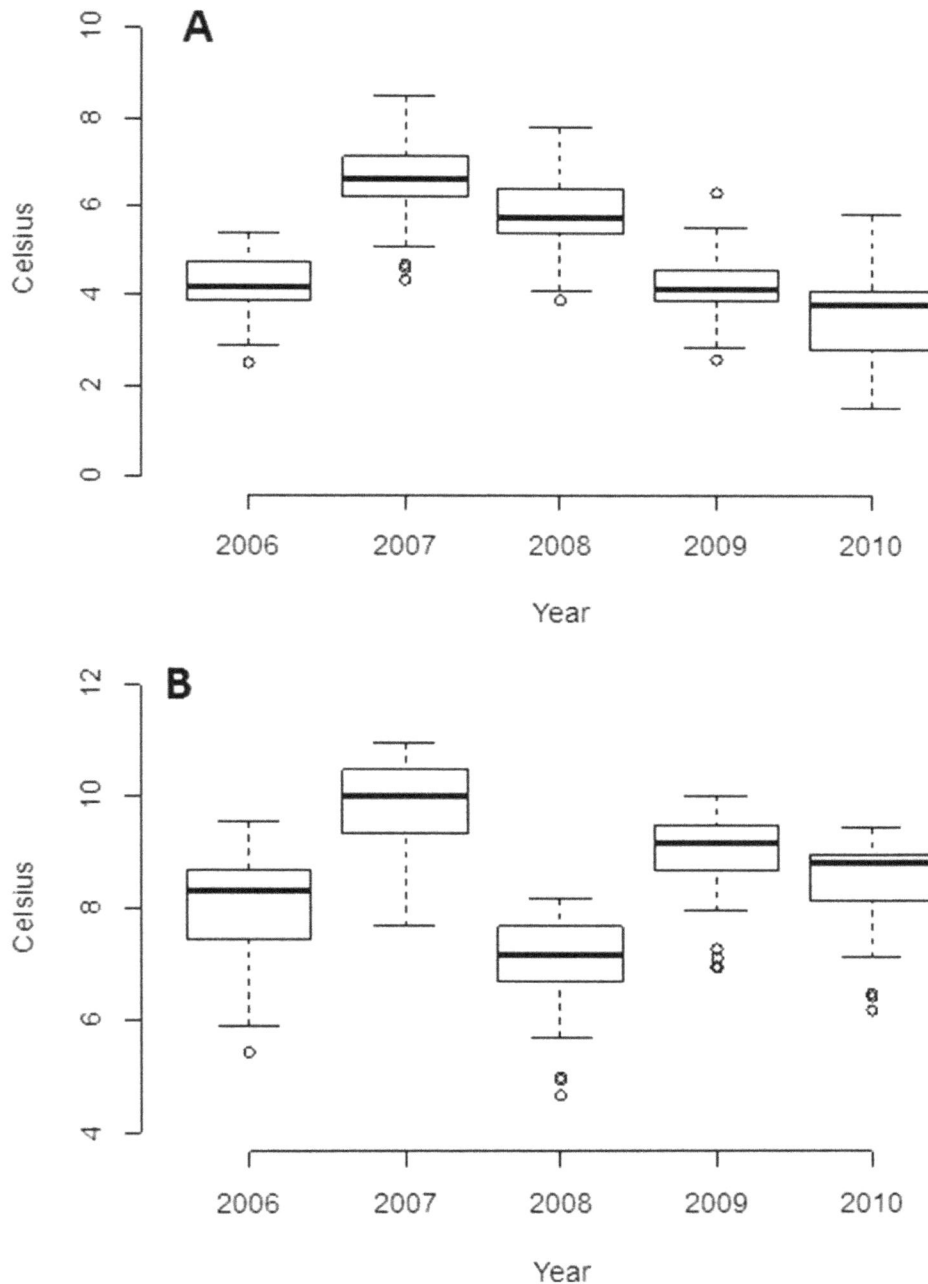

Figure 3. Mean winter (A) and spring (B) temperatures preceding *Culicoides* **activity season across all sites for each year of observation.** Box plots show the median (central line), box denotes 25th and 75th percentiles, error bars represent 10th and 90th percentiles, and dots are points outside the 10th and 90th percentiles.

null model (ΔDIC 4.28, Table S2). This suggests the predominant variables influencing the end of the season are summer relative humidity and sheep density; indeed, a model including only these two variables received very similar support in the data as the best fitting model (ΔDIC 0.7, Table S2).

Length of overwinter

The length of the overwinter period for the *Avaritia* females, when averaged over all sites and years, was approximately 185 days. The shortest overwinter period was 106 days (2006, site 3 'Chobham', Fig. 1), while the longest was 221 days (2006, site 4

'Compton', Fig. 1). The individual species making up the *Avaritia* males showed considerable variation in the length of the inactive overwinter period, with maximum overwinter lengths per species differing by up to 97 days (Table 2, Fig. 2). Averaged over all sites and years, *C. dewulfi* had the shortest overwinter period (206 days), followed by *C. chiopterus* (226 days), *C. scoticus* (237 days), and finally *C. obsoletus* (248 days) (Fig. 2). The best-fitting model for this metric strongly suggested that *C. scoticus* had a longer overwinter period (~60 days) than *C. obsoletus* (~93% of the posterior mass for this effect was greater than zero, Table 3), while

Table 3. Parameter estimates and 95% credible intervals for the fixed effects for the best-fitting model identified for each seasonal metric for individual subgenus *Avaritia Culicoides* males.

Variable	Estimate (95% credible interval)			
	Overall	*C. scoticus*	*C. dewulfi*	*C. chiopterus*
Start of season				
Intercept	149.5* (141.0, 158.7)			
Sp		0.22 (−12.82, 13.14)	1.91 (−10.84, 15.43)	−1.82 (−15.91, 12.54)
Cattle	0.32* (0.090, 0.56)			
Cattle:sp		−0.43* (−0.68, −0.18)	−0.43* (−0.68, −0.19)	−0.034 (−0.29, 0.23)
RH_{spr}	0.16 (−0.057, 0.38)			
RH_{spr}:sp		−0.068 (−0.30, 0.15)	−0.32* (−0.57, −0.059)	0.069 (−0.20, 0.33)
T_{spr}	1.94* (0.40, 3.53)			
P_{spr}	0.11 (−0.070, 0.29)			
End of season				
Intercept	273.7* (263.6, 283.9)			
Sp		−8.17 (−17.45, 1.56)	−6.31 (−16.14, 3.41)	−3.28 (−13.75, 7.72)
RH_{sum}	−2.76* (−5.23, −0.28)			
RH_{sum}:sp		1.43 (−0.084, 2.85)	1.62* (0.15, 3.33)	2.38* (0.00081, 4.85)
Sheep	0.024* (0.0058, 0.043)			
Photo	0.31 (−0.067, 0.68)			
Cattle	0.057 (−0.0041, 0.12)			
Length of overwinter period				
Intercept	262.5* (236.3, 286.8)			
Sp		60.09 (−18.53, 144.8)	−30.25 (−140.6, 88.21)	−12.27 (−67.10, 42.30)
Cattle	0.042 (−0.035, 0.12)			
Cattle:sp		−0.15* (−0.25, −0.046)	−0.11 (−0.25, 0.023)	−0.14* (−0.24, −0.029)
Sheep	−0.036* (−0.062, −0.0090)			
Sheep:sp		−0.0032 (−0.052, 0.041)	0.035 (−0.017, 0.085)	0.030 (−0.010, 0.071)
Moors	2.62 (−0.88, 6.23)			

Asterix denotes significance (95% credible interval does not bridge zero). Interactions with species are denoted '***:sp', and the parameter estimates associated with them refer to the differences in effect sizes relative to *C. obsoletus*. The 'overall' column contains parameter estimates for main effects (which refer to all species if the corresponding interaction is not present in the model, and refer to *C. obsoletus* if the corresponding interaction is present).

Table 4. Parameter estimates and 95% credible intervals for the fixed effects within the best-fitting model identified for each seasonal metric for the subgenus *Avaritia* female *Culicoides*.

Seasonal metric	Variable	Estimate (95% credible interval)
Start of season	Intercept	123.9* (118.8, 130.9)
	T_{spr}	−3.27* (−6.06, −0.44)
	Photoperiod	1.05 (−0.13, 2.22)
End of season	Intercept	296.4* (294.3, 302.3)
	DD_s	0.031 (−0.0045, 0.064)
Length of	Intercept	184.3* (165.5, 201.9)
overwintering	Photoperiod	2.02 (−1.81, 6.46)
period	Moors	2.83 (−0.54, 6.26)
	Cattle	−0.037 (−0.11, 0.039)

Asterix denotes significance (95% credible interval does not bridge zero).

both *C. chiopterus* and *C. dewulfi* showed no significant difference in overwinter length compared to *C. obsoletus* (Table 3).

The best-fitting model for the length of the inactive overwinter period for the *Avaritia* females included non-significant positive effects of photoperiod and percent cover of moorland and heathland, and a non-significant negative effect of cattle density (Table 4). However, model fit was relatively poor (R^2: 0.63), and the null model received essentially similar support in the data (ΔDIC 1.1, Table S1).

The best-fitting model for the length of the inactive overwinter period for the *Avaritia* males included a non-significant positive effect of cattle density on *C. obsoletus* with separate slopes for the other three species showing that the length of the overwinter period for all three species was less affected by cattle density than *C. obsoletus*, with both *C. scoticus* and *C. chiopterus* significantly less affected (Table 3). In addition, a significant negative influence of sheep density was demonstrated for *C. obsoletus* with separate, but non-significant slopes for the other three species, and a close to significant positive influence of percent cover of moorland and heathland on all four species (~93% of posterior density mass greater than zero) ($R^2 = 0.87$; Table 3). Several similar models received very close support in the data to the best-fitting model (ΔDIC<2 for next two best-fitting models, Table S2), however, the null model received no support in the data compared to the best-fitting models (ΔDIC 15.8, Table S2).

Variation across years and sites in the phenological models for *Avaritia* females were similar for the start of the season, although there was slightly more variation between years than sites for the end of the season, and slightly less variation between years than sites for the length of the overwinter period (Table S3). For the *Avaritia* males, there was greatest variation across sites and years (combined site*year random effect) for the end of the season, followed by the length of the overwinter and the start of the season (Table S3).

Discussion

This study has demonstrated that the phenology of *Culicoides* in the UK is species-specific, but also exhibits considerable intra-specific variation between sites and years. The most significant finding in relation to disease control was the documentation of large variations in the length of the inactive overwinter period amongst the constituent species of the subgenus Avaritia. When assessed from catches of males in light-suction traps, *C. dewulfi* had on average the shortest overwinter period (206 days), followed by *C. chiopterus* (~2 weeks longer), *C. obsoletus* (~4 weeks longer) and *C. scoticus* (~8 weeks longer). Importantly, this variation in the overwinter period of males was related to underlying land-cover and host density variables. As such, when defined using data for males, the overwinter SVFP differed by up to eight weeks between the four species. Moreover, evidence from this nationwide surveillance dataset showed that the end of the flight season (autumn) is considerably more variable than the start (spring) across the species comprising the subgenus Avaritia. This is perhaps due to the synchronisation of overwintering larvae into the fourth instar stage resulting in relatively synchronous emergence as adults in the spring. This result implies that accurately predicting the start of the SVFP using current monitoring methods aggregated across species will be difficult. Importantly, when the subgenus *Avaritia* females were modelled as a group, we were unable to detect underlying links with environmental drivers for most of the phenological metrics, indicating that accounting for species-specific variation within this group is important for both understanding phenology and

producing models that can be used to predict the SVFP from remote environmental variables.

Models describing the start and end of seasonal activity and the length of the overwinter period for subgenus *Avaritia* males were strongly supported at a species level. Species-specific drivers of phenology identified included increased cattle density (which led to a later start of season in *C. obsoletus*, but had less impact on *C. scoticus* or *C. dewulfi*) and percentage land-cover of moorland (with increasing cover tending to lead to longer overwinter across all species). Additionally, significantly shorter overwintering periods were documented at increased sheep density for *C. obsoletus*, and extended season end dates across all species. *Culicoides obsoletus* also exhibited significantly earlier season end dates at sites with higher mean summer humidity, but that both *C. dewulfi* and *C. chiopterus* were significantly less affected by this variable. *Culicoides dewulfi* was also significantly less influenced by relative humidity in the spring than *C. obsoletus* in relation to the timing of the start of seasonal activity. Warmer spring temperatures also resulted in significantly later start dates for all species using the *Avaritia* male data in direct contrast to the *Avaritia* females, which appeared earlier under warmer conditions.

The biological drivers of these relationships are challenging to interpret, although limited conclusions can be drawn given current knowledge of the ecology of each species. A key consideration is the contrasting types and availability of larval habitat used, with *C. obsoletus* and *C. scoticus* occupying a diverse range of development sites while *C. chiopterus* and *C. dewulfi* are restricted to cattle and horse dung [9,34]. Somewhat counter-intuitively from an ecological point of view, season start dates for both *C. scoticus* and *C. dewulfi* were significantly less influenced by cattle density than *C. obsoletus*. This is in contrast to our expectations because cattle are known to be an important host for *C. obsoletus*, *C. dewulfi* and *C. chiopterus* [11,35]. Cattle have been suggested as the most attractive hosts for Palaearctic *Culicoides* [36,37], though, because hosts have not been enumerated in field studies, a robust host preference has not been conclusively demonstrated at the species level [11]. This observation could arise from interspecific differences in both intrinsic sensitivity of males to UV light and the availability and localisation of mating sites and resting areas. Numbers of *Culicoides* males occurring in UV or incandescent light-suction trap collections usually constitute only a small proportion of the total catch, as seen in both the current study and those conducted across Europe [38–40]. No attempt has been made, however, to quantify population responses at a species level in the presence of competing mating cues (e.g. pheromones) or to map resting populations at a farm level scale.

While speculative, similar factors could also drive the observations that greater sheep density resulted in both shorter overwinter periods and significantly later end dates across male populations of all species, although this could also be related to the greater provision of overwinter livestock accommodation of a suitable type to allow later survival of *Culicoides* as documented in northern Europe [10,11]. Sheep are known to be important hosts for most of these species – *Culicoides scoticus*, *C. dewulfi* and *C. chiopterus* have been collected in abundance on the body of sheep [41], and all four species have been found to engorge on sheep in recent blood meal analysis studies in both farm and extensive pasture settings [35,37,42–45]. *Obsoletus* group catches in Wales were found to increase with the number of sheep on farms [46], a relationship also identified by a controlled study reporting a linear relationship between *C. obsoletus* trap catches and the number of sheep positioned beneath light traps [47]. It is likely that much finer resolution livestock density data and *C. chiopterus* population data from a passive trapping method would be needed to better understand the relationship and mechanisms whereby host density differentially affects the phenology of these species.

Biological factors underlying climatic drivers of populations similarly require further studies to confirm relationships. In this regard the differential response between dung-breeding species (*C. dewulfi* and *C. chiopterus*) and those utilising less localised and uniform larval development sites is of interest. The localisation of breeding habitats has the potential to lead to restrictions on suitable resting sites for males in order to maximise contact with gravid females. This close association with larval development sites may explain a reduced vulnerability to high humidity levels although adaptation to these conditions in dung breeding species may also play a role. Overall, our results suggest that the resolution and specificity of the freely available climate and ecological datasets we used were not sufficient to accurately decipher how temperature and precipitation interact to drive phenological events in these species, highlighting the need for further work at finer spatial scales.

Importantly, the evidence for considerable variation between species in the length of the inactive period overwinter revealed by analysing data for males in this study has significant implications for disease management via the SFVP in temperate zones. Transmission of vector-borne diseases is highly dependent on the host-vector ratio [48,49], therefore if these four species are found to be differentially competent for arboviruses, the species-specific variation documented here in terms of their temporal phenology could impact strongly on the length and infection risk of the transmission season for arboviruses in the UK and elsewhere in northern Europe. A key caveat, is that our analysis of male trap catches is an acceptable representation of the phenology of female *Culicoides*. This is important because only female *Culicoides* blood-feed and are responsible for the transmission of viruses. Demonstrating correlation and synchrony in the phenology of males and females of the *Avaritia* species is difficult, and cannot be accomplished without using recently developed high-throughput qPCR assays for pooled samples (which are yet to be proven with collections arising from surveillance). To address this concern with our dataset, we analysed the extent of seasonal correlation in weekly trap catch abundance between males and parous females for two related Palaearctic *Culicoides* species (*C. pulicaris* and *C. impunctatus*) for which we have data spanning the same time period and sites as used in this study (Figs S4a, b in Section S4). This analysis demonstrated that both *C. pulicaris* and *C. impunctatus* exhibited good correlation between the seasonal abundance of male and parous female trap catches with 13 (*C. pulicaris*) and 10 (*C. impunctatus*) of the 15 sites examined for each species showing a correlation of greater than 0.5. In summary, we believe that these data demonstrate good correlation between seasonal trap catches of males and parous females for these two related *Culicoides* species. Moreover, there is no biological reason we are aware of that would suggest a different relationship between male and female seasonal activity for *C. pulicaris* or *C. impunctatus* in comparison to the four *C. obsoletus* complex species used in our main analysis. All of these species require tight synchronisation between male and female seasonal activity to ensure successful reproduction, and this is particularly true in temperate zones such as the UK where multiple generations of these species occur within a single year. Finally, although our analysis cannot define the length of the SFVP for these species because correlation in male and female seasonal abundance and phenology is by no means absolute, it does demonstrate a clear potential for species-specific variation in this important disease management tool. If these findings can be replicated using data on

females the implications for disease management and spatio-temporal variation in risk are profound.

A key consideration for the current study and for future surveillance as part of the SVFP lies in the ability in the future to accurately differentiate female members of the subgenus *Avaritia* as these constitute the vast majority of light trap catches. The recent development of high-throughput real-time RT-PCR assays to differentiate species within pools of the subgenus *Avaritia* has great potential in offering a processing method that is sufficiently rapid to sustain surveillance trapping programmes [50]. While the current study has demonstrated that male populations of these species vary in their phenology, uptake of such techniques to examine such variation in females is likely to be determined by the ability to demarcate the role of specific species in the transmission of arboviruses and thereby provide a practical tool for estimating the risk of transmission. While this was not achieved for BTV-8 during the northern European incursion [1], results from the SBV outbreak strongly imply the presence of multiple vector species [51,52]. This is likely to significantly complicate future attempts to model the risk of transmission according to season. A key observation, however, lay in the observation that the beginning of the adult *Culicoides* flight season was significantly more straightforward to predict than the end. This may enable at least partial prediction of SVFP's without recourse to costly and time consuming direct surveillance methods.

Recommendations for management or policy

As part of a surveillance system designed to allow ruminant movements during incursion of *Culicoides*-borne arboviruses, it was suggested that a SVFP could be maintained during which animal movement restrictions could be relaxed. Our study demonstrates that active surveillance of haematophagous female *Culicoides* vector populations cannot currently be replaced using remote models of abundance. This failure was most likely related to the diverse ecology of species conflated within this taxonomic grouping and was partially resolved by the use for species level modelling based on collections of male *Culicoides*.

The differences identified in this study of around eight weeks in the length of the overwinter period for the four species are particularly relevant to disease policy in the UK in relation to defining the SVFP. For instance, the Schmallenberg virus is known to have its greatest impact on mammalian hosts when infection occurs at a particular point in the gestation cycle of the host [5,18], which coincides with the tail end of the *Culicoides* vector season in the UK. This coincidence in vector phenology and host susceptibility has been demonstrated to drive the extent and size of potential outbreaks of SBV in Scotland [19]. A key finding was that the timing of the end of the season may be more difficult to forecast, and should perhaps be treated with more caution by policy-makers than the beginning of the season, because it varies widely between species, years and locations in response to environmental heterogeneity. We recommend more intensive trapping across a range of climatic zones with species-level identification of *Culicoides* females wherever feasible to facilitate more accurate detection and understanding of the start of the SVFP in temperate zones.

Supporting Information

Table S1 Model fit statistics for top ten models identified using forwards and backwards selection with DIC for the start and end of season, and length of overwinter period for the subgenus *Avaritia* (*Avaritia* females). Null model includes only intercept and site and year random effects. The Null model, excluding all environmental effects, is provided for comparison. The difference in DIC between the best-fitting model and each other model is shown by ΔDIC, and *pD* is the effective number of parameters in each model. (DOC)

Table S2 Model fit statistics for top ten models identified using forwards and backwards selection with DIC for the start and end of the season, and length of overwinter period for the individual species of the subgenus Avaritia (*Avaritia* males). Null model includes intercept, fixed species effect and site*year random effect, and are provided for comparison. The difference in DIC between the best-fitting model and each other model is shown by ΔDIC, and *pD* is the effective number of parameters in each model. (DOCX)

Table S3 Parameter estimates and 95% credible intervals for the variances of the random effects within the best-fitting models identified for each seasonal metric for the subgenus Avaritia (*Avaritia* females and males). (DOCX)

Section S1 Mathematical description of the models used. (DOCX)

Section S2 Bayesian statistical inference for the Wrapped Normal (WN) model. (DOCX)

Section S3 Multinomial analysis of *Avaritia* males phenology. File includes: **Figure S3:** Maximum catch size and proportion of the population (based on mean trap catch) trapped in spring (spr), summer (sum) or autumn (aut) months for the four species comprising the subgenus Avaritia (*Avaritia* males) across the traps used in the multinomial analysis (chi: *C.*

chiopterus, dew: *C. dewulfi*, obs: *C. obsoletus*, scot: *C. scoticus*). **Table S3:** Multinomial results showing posterior means and 95% credible intervals for estimated parameters. Species effects (*sp* and *bᵢsp*) denoted by [2]: *C. scoticus*, [3]: *C. dewulfi*, and [4]: *C. chiopterus*. (DOCX)

Section S4 Examination of the correlation in seasonal abundance of male and parous females using data for *C. pulicaris* and *C. impunctatus*. File includes: **Figure S4a.** Correlation in seasonal trap catches for male and parous female *C. pulicaris* over 15 sites from the UK *Culicoides* surveillance dataset. These 15 site by year combinations had complete seasonal trapping coverage and represent the 15 most abundant site by year combinations in the dataset for this species. **Figure S4b.** Correlation in seasonal trap catches for male and parous female *C. impunctatus* over 15 sites from the UK *Culicoides* surveillance dataset. These 15 site by year combinations had complete seasonal trapping coverage and represent the 15 most abundant site by year combinations in the dataset for this species. (DOCX)

Acknowledgments

We thank the private landowners who allowed us access to their property to conduct the trapping. We particularly wish to thank Brian Reich for providing assistance in the development of the circular regression model in WinBUGS. We also thank two anonymous reviewers for their insights during the review process.

Author Contributions

Conceived and designed the experiments: KRS SC AB PSM SG BVP. Performed the experiments: JB FS KL ED CS AW NN. Analyzed the data: KRS SC BVP. Contributed to the writing of the manuscript: KRS BVP SC PSM SG.

References

1. Carpenter S, Wilson A, Mellor PS (2009) Culicoides and the emergence of bluetongue virus in northern Europe. Trends in Microbiology, 17: 172–178.
2. Vandenbussche F, De Leeuw I, Vandemeulebroucke E, De Clercq K (2009) Emergence of Bluetongue Serotypes in Europe, Part 1: Description and Validation of Four Real-Time RT-PCR Assays for the Serotyping of Bluetongue Viruses BTV-1, BTV-6, BTV-8 and BTV-11. Transboundary and Emerging Diseases, 56: 346–354.
3. van Rijn PA, Geurts Y, van der Spek AN, Veldman D, van Gennip RGP (2012) Bluetongue virus serotype 6 in Europe in 2008-Emergence and disappearance of an unexpected non-virulent BTV. Veterinary Microbiology, 158: 23–32.
4. Hofmann MA, Renzullo S, Mader M, Chaignat V, Worwa G, et al. (2008) Genetic Characterization of Toggenburg Orbivirus, a New Bluetongue Virus, from Goats, Switzerland. Emerging Infectious Diseases, 14: 1855–1861.
5. Hoffmann B, Scheuch M, Hoper D, Jungblut R, Holsteg M, et al. (2012) Novel Orthobunyavirus in Cattle, Europe, 2011. Emerging Infectious Diseases, 18: 469–472.
6. Davies I, Vellema P, Roger P (2012) Schmallenberg virus - an emerging novel pathogen. In Practice, 34: 598–+.
7. Mintiens K, Meroc E, Mellor PS, Staubach C, Gerbier G, et al. (2008) Possible routes of introduction of bluetongue virus serotype 8 into the epicentre of the 2006 epidemic in North-Western Europe. Preventive Veterinary Medicine, 87: 131–144.
8. Mellor PS, Boorman J, Baylis M (2000) Culicoides biting midges: Their role as arbovirus vectors. Annual Review of Entomology, 45: 307–340.
9. Kettle DS, Lawson JWH (1952) The early stages of British biting midges Culicoides Latreille (Diptera: Ceratopogonidae) and allied genera. Bulletin of Entomological Research, 43: 421–467.
10. Meiswinkel R, Goffredo M, Dliskstra EGM, van der Ven IJK, Baldet T, et al. (2008) Endophily in Culicoides associated with BTV-infected cattle in the province of Limburg, South-Eastern Netherlands, 2006. Preventive Veterinary Medicine, 87: 182–195.
11. Viennet E, Garros C, Rakotoarivony I, Allene X, Gardes L, et al. (2012) Host-Seeking Activity of Bluetongue Virus Vectors: Endo/Exophagy and Circadian Rhythm of Culicoides in Western Europe. PLoS ONE: 7.
12. Schley D, Gubbins S, Paton DJ (2009) Quantifying the Risk of Localised Animal Movement Bans for Foot-and-Mouth Disease. PLoS ONE: 4.
13. Maan S, Maan NS, Ross-Smith N, Batten CA, Shaw AE, et al. (2008) Sequence analysis of bluetongue virus serotype 8 from the Netherlands 2006 and comparison to other European strains. Virology, 377: 308–318.
14. Carpenter S, Wilson A, Mellor PS (2009a) Bluetongue virus and Culicoides in the UK: The impact of research on policy. Outlooks on Pest Management, 20: 161–164.
15. EFSA, *Annual Report* 2008: European Food Safety Authority, European Union. 108.
16. Wilson A, Carpenter S, Gloster J, Mellor P (2007) Re-emergence of bluetongue in northern Europe in 2007. Veterinary Record, 161: 487–489.
17. Gariglinany MM, Hoffmann B, Dive M, Sartelet A, Bayrou C, et al. (2012) Schmallenberg virus in calf born at term with porencephaly, Belgium [letter]. Emerg Infect Dis [serial on the Internet]. 2012 June. http://dx.doi.org/10.3201/eid1806.120104.
18. Lievaart-Peterson K, Luttikholt SJM, Van den Brom R, Vellema P (2012) Schmallenberg virus infection in small ruminants – First review of the situation and prospects in Northern Europe. Small ruminant research 106: 71–76.
19. Bessell PR, Searle KR, Auty HK, Handel IG, Purse BV, et al. (2013) Epidemic potential of an emerging vector borne disease in a marginal environment: Schmallenberg in Scotland. Scientific Reports: 3.
20. Modlin D, Fuentes M, Reich B (2012) *Circular conditional autoregressive modeling of vector fields.* Environmetrics, 2012. 23(1): p. 46–53.
21. Mellor PS, Tabachnick W, Baldet T, Baylis M, Bellis G, et al. (2004) Conclusions of working groups. Group 2. Vectors. Proceedings of the 3rd OIE Bluetongue International Symposium, 26–29 October 2003, Taormina, Italy. Veterinaria Italia: 40: 715–717.
22. Campbell JAP (1960) A taxonomic review of the British species of Culicoides Latreille (Diptera: Ceratopogonidae). Proceedings of the Royal Entomological Society of London: Series B. 67: 181–302.
23. Dyce AL (1969) The recognition of mulliparous and parous Culicoides (Diptera Ceratopogonidae) without dissection. Journal of the Australian Entomological Society, 8: 11–15.
24. Blackwell A (1997) *Diel flight periodicity of the biting midge Culicoides impunctatus and the effects of meteorological conditions.* Medical and Veterinary Entomology, 1997. 11(4): p. 361–367.

25. Purse BV, Falconer D, Sullivan MJ, Carpenter S, Mellor PS, et al. (2012) Impacts of climate, host and landscape factors on Culicoides species in Scotland. Medical and Veterinary Entomology, 26: 168–177.

26. Sanders CJ, Shortall CR, Gubbins S, Burgin L, Gloster J, et al. (2011) Influence of season and meteorological parameters on flight activity of Culicoides biting midges. Journal of Applied Ecology, 48: 1355–1364.

27. Searle KR, Purse BV, Blackwell S, Falconer D, Sullivan M, et al. (2013) Environmental drivers of insect phenology across space and time: Culicoides in Scotland as a case study. Bulletin of Entomological Research, 103: 155–170.

28. Morton D, Rowland C, Wood C, Meek L, Marston C, et al. (2011) Final Report for LCM2007 - the new UK Land Cover Map. Countryside Survey Technical Report No. 11/07 NERC/Centre for Ecology & Hydrology 112 pp. (CEH Project Number: C03259).

29. Fisher NI (1993) Statistical Analysis of Circular Data. Cambridge University Press, Cambridge.

30. Lunn D, Spiegelhalter D, Thomas A, Bes N (2011) The BUGS project: Evolution, critique and future directions. Statistics in Medicine, 28: 3049–3067.

31. R-Development_Core_Team, R: a language and environment for statistical computing, 2009, www.r-project.org/: Vienna, Austria.

32. Spiegelhalter DJ, Best NaG, Carlin BP, Van Der Linde A (2002) Bayesian measures of model complexity and fit. Journal of the Royal Statistical Society: Series B (Statistical Methodology), 64: 583–639.

33. Gelman A, Hill J (2007) Data Analysis Using Regression and Multilevel/Hierarchical Models. Cambridge University Press, Cambridge, UK.

34. Harrup LE, Purse BV, Golding N, Mellor PS, Carpenter S (2013) Larval development and emergence sites of farm-associated Culicoides in the United Kingdom. Medical and Veterinary Entomology, 2013. Online early.

35. Garros C, Gardèsa L, Allènea X, Rakotoarivonya I, Vienneta E, et al. (2011) Adaptation of a species-specific multiplex PCR assay for the identification of blood meal source in Culicoides (Ceratopogonidae: Diptera): applications on Palaearctic biting midge species, vectors of Orbiviruses. Infection, Genetics and Evolution, 11: 1103–1110.

36. Bartsch S, Bauer B, Wiemann A, Clausen P-H, Steuber S (2009) Feeding patterns of biting midges of the Culicoides obsoletus and Culicoides pulicaris groups on selected farms in Brandenburg, Germany. Parasitology Research, 105: 373–380.

37. Ninio C, Augot D, Delecolle J-C, Dufour B, Depaquit J (2001) Contribution to the knowledge of Culicoides (Diptera: Ceratopogonidae) host preferences in France. Parasitology Research, 108: 657–663.

38. Takken W, Verhulst N, Scholte E-J, Jacobs F, Jongema Y, et al. (2008) The phenology and population dynamics of Culicoides spp. in different ecosystems in The Netherlands. Preventive Veterinary Medicine, 2008. 87(1–2): p. 41–54.

39. Purse BV, Tatem AJ, Caracappa S, Rogers DJ, Mellor PS, et al. (2004) Modelling the distributions of Culicoides bluetongue virus vectors in Sicily in relation to satellite-derived climate variables. Medical and Veterinary Entomology, 2004. 18(2): p. 90–101.

40. Capela R, Purse BV, Pena I, Wittman EJ, Margarita Y, et al. (2003) Spatial distribution of Culicoides species in Portugal in relation to the transmission of African horse sickness and bluetongue viruses. Medical and Veterinary Entomology, 2003. 17(2): p. 165–177.

41. Carpenter S, Szmaragd C, Barber J, Labuschagne K, Gubbins S, et al. (2008) An assessment of Culicoides surveillance techniques in northern Europe: have we underestimated a potential bluetongue virus vector? Journal of Applied Ecology, 45: 1237–1245.

42. Lassen SB, Nielsen SA, Skovgard H, Kristensen M (2011) Molecular identification of bloodmeals from biting midges (Diptera: Ceratopogonidae: Culicoides Latreille) in Denmark. Parasitology Research, 2011. 108: p. 823–829.

43. Lassen SB, Nielsen SA, Kristensen M (2012) Identity and diversity of bloodmeal hosts of biting midges (Diptera: Ceratopogonidae: Culicoides Latreille) in Denmark. Parasites and Vectors, 5: p. 153–161.

44. Martínez-de la Puente, Martínez J, Ferraguti M, Morales-de la Nuez A, Castro N, et al. (2012) Genetic characterization and molecular identification of the bloodmeal sources of the potential bluetongue vector Culicoides obsoletus in the Canary Islands, Spain. Parasites and Vectors, 5: 147–155.

45. Viennet E, Garros C, Lancelot AXR, Gardes L, Rakotoarivony I, et al. (2011) Assessment of vector/host contact: comparison of animal-baited traps and UV-light/suction trap for collecting Culicoides biting midges (Diptera: Ceratopogonidae), vectors of Orbiviruses. Parasites and Vectors, 4: 119.

46. Kluiters G, Sugden D, Guis H, McIntyre KM, Labuschagne K, et al. (2012) Modelling the spatial distribution of Culicoides biting midges at the local scale. Journal of Applied Ecology. Doi:10.1111/1365-2664.12030.

47. Garcia-Saenz A, McCarter P, Baylis M (2010) The influence of host number on the attraction of biting midges, Culicoides spp., to light traps. Med Vet Entomol., 25: 113–5.

48. Hartemink NA, Purse BV, Meiswinkel R, Brown HE, de Koeijer A, et al. (2009) Mapping the basic reproduction number (R-0) for vector-borne diseases: A case study on bluetongue virus. EPIDEMICS, 1: 153–161.

49. Gubbins S, Carpenter S, Baylis M, Wood JLN, Mellor PS (2008) Assessing the risk of bluetongue to UK livestock: uncertainty and sensitivity analyses of a temperature-dependent model for the basic reproduction number. Journal of the Royal Society Interface, 5: 363–371.

50. Mathieu B, Delecolle JC, Garros C, Balenghien T, Setier-Rio ML, et al. (2011) Simultaneous quantification of the relative abundance of species complex members: Application to Culicoides obsoletus and Culicoides scoticus (Diptera: Ceratopogonidae), potential vectors of bluetongue virus (vol 182, pg 297, 2011). Veterinary Parasitology, 2012. 187(3–4): p. 578–578.

51. Elbers ARW, Meiswinkel R, van Weezep E, Sloet van Oldruitenborgh-Oosterbaan MM, et al. (2011) Schmallenberg virus in Culicoides spp. biting midges, the Netherlands, 2011. Emerging infectious diseases, 2013. 19(1): p. 106–9.

52. De Regge N, Deblauwe I, De Deken R, Vantieghem P, Madder M, et al. (2012) Detection of Schmallenberg virus in different Culicoides spp. by real-time RT-PCR. Transboundary and Emerging Diseases, 2012. 59(6): p. 471–475.

Antenatal Dexamethasone after Asphyxia Increases Neural Injury in Preterm Fetal Sheep

Miriam E. Koome, Joanne O. Davidson, Paul P. Drury, Sam Mathai, Lindsea C. Booth, Alistair Jan Gunn, Laura Bennet*

Department of Physiology, the University of Auckland, Auckland New Zealand

Abstract

Background and Purpose: Maternal glucocorticoid treatment for threatened premature delivery dramatically improves neonatal survival and short-term morbidity; however, its effects on neurodevelopmental outcome are variable. We investigated the effect of maternal glucocorticoid exposure after acute asphyxia on injury in the preterm brain.

Methods: Chronically instrumented singleton fetal sheep at 0.7 of gestation received asphyxia induced by complete umbilical cord occlusion for 25 minutes. 15 minutes after release of occlusion, ewes received a 3 ml i.m. injection of either dexamethasone (12 mg, n = 10) or saline (n = 10). Sheep were killed after 7 days recovery; survival of neurons in the hippocampus and basal ganglia, and oligodendrocytes in periventricular white matter were assessed using an unbiased stereological approach.

Results: Maternal dexamethasone after asphyxia was associated with more severe loss of neurons in the hippocampus (CA3 regions, 290±76 *vs* 484±98 neurons/mm², mean±SEM, P<0.05) and basal ganglia (putamen, 538±112 *vs* 814±34 neurons/mm², P<0.05) compared to asphyxia-saline, and with greater loss of both total (913±77 *vs* 1201±75/mm², P<0.05) and immature/mature myelinating oligodendrocytes in periventricular white matter (66±8 *vs* 114±12/mm², P<0.05, *vs* sham controls 165±10/mm², P<0.001). This was associated with transient hyperglycemia (peak 3.5±0.2 vs. 1.4±0.2 mmol/L at 6 h, P<0.05) and reduced suppression of EEG power in the first 24 h after occlusion (maximum −1.5±1.2 dB vs. −5.0±1.4 dB in saline controls, P<0.01), but later onset and fewer overt seizures.

Conclusions: In preterm fetal sheep, exposure to maternal dexamethasone during recovery from asphyxia exacerbated brain damage.

Editor: Kang Sun, Fudan University, China

Funding: This study was supported by public good grants from the Health Research Council of New Zealand, the Auckland Medical Research Foundation, and Lottery Health New Zealand. The funders had no role in study design, data collection and analysis, decision to publish, or preparation of the manuscript.

Competing Interests: The authors have declared that no competing interests exist.

* E-mail: l.bennet@auckland.ac.nz

Introduction

Fetuses at risk of premature delivery are now routinely exposed to maternal treatment with synthetic glucocorticoids. Preterm infants have a high rate of neuronal and white matter damage [1,2]. Clinically, maternal glucocorticoids substantially reduce acute neonatal morbidity and mortality and reduce intraventricular hemorrhage [3]. However, although they reduce the risk of white matter injury [4,5], the effect on later neurodevelopmental outcome is still unclear [6], and some studies report reduced head size [7].

Some of the apparent inconsistency may be related to the effect of glucocorticoids on the brain after exposure to hypoxia-ischemia (HI). For example, in preterm infants dying within 4 days after birth, exposure to maternal steroid therapy was associated with reduced hippocampal neuronal density [8]. Critically, there is now compelling evidence that early loss of vulnerable cells such as pre-oligodendrocytes is followed by chronic impairment of white and grey matter maturation [2,9,10]. Thus it is vital to understand how maternal glucocorticoids affect recovery from perinatal brain injury. Studies in newborn rodents suggest little effect of glucocorticoids after HI, as reviewed [11]. In near-term fetal sheep, at an age when brain maturity is broadly equivalent to term infants, maternal dexamethasone treatment 48 hours *before* cerebral ischemia did not modify the pattern of injury [12]. However, perhaps surprisingly, there is no information on how glucocorticoids after preterm HI affect the pervasive white and grey matter injury that underpins long-term impairment of neurodevelopment [1,2].

We recently reported that in normoxic preterm fetal sheep a standard clinical course of maternal dexamethasone was associated with significant transient EEG hyperactivity from approximately 3 hours after the first injection [13]. Although there was no cerebral injury after 5 days of recovery, loss of neural suppression after hypoxia-ischemia can be associated with increased neural injury [14,15].

Therefore, we examined the hypothesis that maternal dexamethasone therapy shortly after a period of severe asphyxia would increase fetal neural activity during the early recovery ('latent') phase, and increase neural injury. These studies were conducted in preterm fetal sheep at an age that is broadly equivalent in brain maturation to the 27–30 week human [16].

Materials and Methods

Animals and experimental procedures

All procedures were approved by the Animal Ethics Committee of the University of Auckland. Singleton Romney/Suffolk fetal sheep were surgically instrumented at 98–100 days of gestation (term = 147 days). Ewes were anesthetized by intravenous injection of propofol (5 mg/kg, AstraZeneca Limited, Auckland, New Zealand), followed by 2–3% isofluorane in oxygen. A midline incision was made to expose the uterus, and the fetus was partially exteriorized for instrumentation. Polyvinyl catheters were placed in the amniotic sac, left femoral artery and vein and right brachial artery to measure blood pressure and for pre-ductal blood sampling. Two pairs of electrodes (Cooner Wire, Chatsworth, CA, USA) were placed over the parietal cortex bilaterally, 10 mm lateral to bregma and 5 mm and 10 mm anterior to measure electroencephalographic (EEG) activity. A reference electrode was sewn over the occiput. A pair of electrodes was placed across the fetal chest to measure the fetal electrocardiogram (ECG). In addition, an inflatable silicone occluder was placed around the umbilical cord (In Vivo Metric, Healdsburg, CA, USA). All fetal leads were exteriorised through the maternal flank, and a maternal leg vein was catheterised for post-operative care and euthanasia.

Antibiotics were administered into the amniotic sac (80 mg Gentamicin, Pharmacia & Upjohn, Rydalmere, NSW, Australia) before the uterus was closed. Ewes were given 5 ml of Streptocin (Stockguard Labs Ltd., Hamilton, New Zealand) i.m. 30 min before surgery for prophylaxis. The maternal midline skin incision was infiltrated with a local analgesic, 10 ml 0.5% bupivacaine plus adrenaline (AstraZeneca Ltd., Auckland, New Zealand). After surgery, ewes were housed together in separate metabolic cages with *ad libitum* access to food and water. Rooms were temperature and humidity controlled (16 ± 1°C, humidity 50 ± 10%) with a 12 h light/dark cycle (light 0600 to 1800 h). Ewes were given daily i.v. antibiotics (600 mg Crystapen, Biochemie, Vienna, Austria and 80 mg Gentamicin) for 4 days after surgery. Fetal catheters were maintained patent with continuous infusion of heparinised saline (20 U/ml). Experiments began 4–5 days after surgery.

Recordings

Fetal MAP, ECG and EEG were recorded continuously. Fetal MAP was recorded using Novatrans II pressure transducers (MX860, Carslbad, USA) and corrected for maternal movement by subtraction of amniotic fluid pressure. The blood pressure signal was collected at 64 Hz and low-pass filtered at 30 Hz. The analogue fetal EEG signal was low-pass filtered with the cut-off frequency set with the −3 dB point at 30 Hz, and digitized at a sampling rate of 512 Hz. The intensity (power) and frequency were derived from the intensity spectrum signal between 0.5 and 20 Hz. For data presentation, the total EEG power was normalized by log transformation (dB, 20 × log intensity) and data were stored to disk as one min averages (Labview, National Instruments, Austin, TX, USA). Additionally, the unaveraged EEG signal was processed through a digital finite impulse response low-pass filter with a cut-off frequency of 30 Hz, and stored at a sampling rate of 64 Hz for analysis of seizures [13].

Experimental Protocol

Experiments were conducted at 103–104 days of gestation. Fetuses were randomly assigned to saline-sham asphyxia (sham control, n = 8), asphyxia-saline (n = 10) or asphyxia-dexamethasone (asphyxia-DEX, n = 10) groups. We have previously reported sham control data [13,17]. Fetal asphyxia was induced by complete umbilical cord occlusion for 25 minutes. Occlusions were started between 9:00 and 9:30 am. Fifteen minutes after the end of occlusion, ewes received a 3 ml i.m. injection of either dexamethasone (12 mg dexamethasone sodium phosphate, David Bull Laboratories, Mulgrave, Australia) or the equivalent volume of saline. The maternal weight was 60.5 ± 1.0 kg. Recordings were started from 24 hours before asphyxia and continued for seven days after release of occlusion. Fetal arterial blood samples (0.3 ml) were taken 15 min prior to occlusion, at 5 and 17 min during occlusion, and at 10 min, 1, 2, 4 and 6 h post-occlusion, then daily thereafter between 8:30 and 9:30 am. Blood samples were analyzed for pH and blood gases (Ciba-Corning Diagnostics 845 blood gas analyzer and co-oximeter, MA, USA) and glucose and lactate levels (YSI model 2300, Yellow Springs, OH, USA).

Seven days after occlusion, ewes and fetuses were killed by an overdose of sodium pentobarbital i.v. to the ewe (9 g Pentobarb 300, Chemstock International, Christchurch, NZ), and fetal brains were perfusion fixed by gravity feed, from 1.3 m above the fetus, with 10% neutral buffered formalin and left for fixation in 10% formalin for one week, cut into 5 mm thick slices and processed for paraffin embedding using standard techniques for histological evaluation.

Histological procedures

Histological analysis was undertaken using 8 μm thick coronal sections mounted on chrome alum-gelatin pre-coated slides, at the level of caudo-putamen and hippocampus (Figure 1). Oven dried (60 °C) slides were de-paraffinized in xylene and rehydrated in a series of ethanol solutions of decreasing concentrations, and then washed in 0.1 m PBS. Sections were stained with acid-fuchsin/thionin to assess gross morphological changes. On separate sections for immunohistochemical analysis, antigen retrieval was performed using citrate buffer (pH 6.0) using a pressure cooking method (2100 retriever, Prestige Medical Ltd, Blackburn, UK). After repeated washings in PBS the sections were treated with 1% H_2O_2 in methanol for 30 min in dark to quench endogenous peroxidase activity and again washed in PBS. 5% goat/horse serum in PBS was used for blocking. Antibodies were diluted with 2.5% goat/horse serum in PBS. Sections were incubated with following primary antibodies overnight at 4°C; Iba-1 (abcam; 1:200 dilution) for activated microglia, NeuN (Millipore, 1:200) for neuronal nuclei, Olig-2 (Millipore, 1:200.) a marker for oligodendrocytes at all stages of the lineage [18], CNPase (abcam,1:200), a marker for immature and mature oligodendrocytes and Ki-67 (Dako, 1.200) for proliferating cells. Washed sections were then incubated overnight with anti-goat secondary antibody for Iba-1, anti-mouse secondary antibody for NeuN and anti-rabbit secondary antibody for Olig-2 (Vector Laboratories, all 1:200). After repeated washing with PBS, sections were then incubated with ExtrAvidin ® (Sigma, 1:200) for 3 h at room temperature and finally treated with SIGMAFASTTM 3,3′ diamino benzidine (DAB) for brown colour development. Ki-67 labelled slides were counterstained with weak thionine solution. All slides were dehydrated in increasing alcohol concentrations and xylene, and finally mounted with DPX.

Qualitative and quantitative analysis of brain injury

Gross morphological changes to the brain and neuronal death were assessed from acid fuchsin-thionin stained slides by light microscopy. Dead cells were identified by the characteristic acidophilic cytoplasm and pyknotic nuclei. Activated microglia

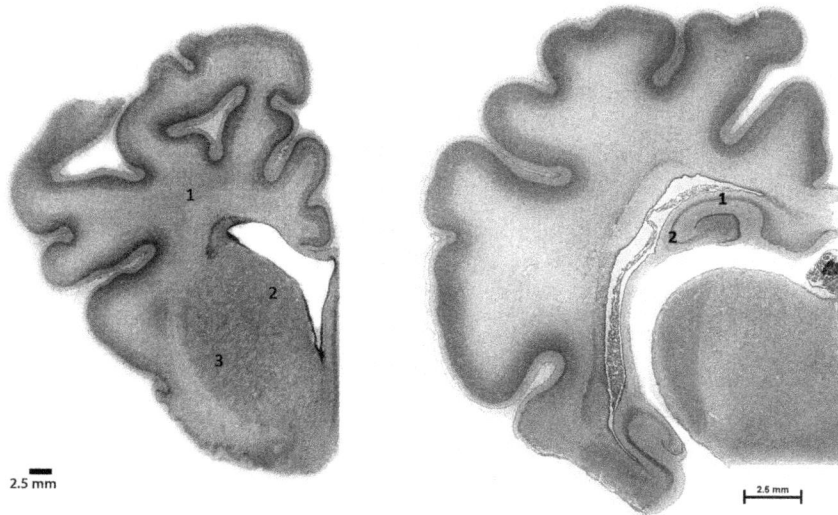

Figure 1. Photomicrographs of coronal sections of a preterm fetal sheep brain showing the fields used for analysis. Left panel: periventricular white matter (1) and Caudate (2) and putamen (3). Right panel: CA1-2 (1) and CA3 (2) regions of the hippocampus. Scale bars = 2.5 mm.

and Olig-2 positive cells were estimated in periventricular white matter (PVWM, Figure 1), intragyral white matter of the cingulate gyrus (IGWM) and deep white matter (DWM, the internal capsule within the caudoputamen). Numbers of NeuN positive cells were estimated in caudate and putamen of striatum and CA1-2 & CA3 areas of hippocampus as previously described [19].

Sampling was performed using stereological principles by first tracing around each region of interest at $2\times$ magnification, and then randomly translating a grid onto the sections and applying a fractionator probe consisting of a counting frame for object inclusion/exclusion at $40\times$ magnification. The grid and counting frame size for the CA1/2 region were 100×100 µm and for CA3 were 50×50 µm. The grid and counting frame sizes used for the caudate nucleus and putamen were 500×500 µm and 50×50 µm, respectively. Cells touching the bottom and right-hand boundaries were included, whereas those touching the top and left were excluded. Cell counts for each region were converted to density (cells/mm^2) by the following formula: (estimated total counts by fractionator/contour area $(\mu m^2)) \times 10^6$. For each animal, average scores across both hemispheres from two sections were calculated for each region. Counts were made by an assessor blinded to treatment group.

Data Analysis

Offline analysis of the physiological data was carried out using customized Labview programs (National Instruments). EEG power is expressed as change from baseline. Continuous EEG traces were analyzed for the presence of stereotypic evolving seizures defined as rhythmic, repetitive waveforms with a stereotypic evolving pattern and seizure-like activity [20]. Seizure-like events were classified as bursts of large amplitude events which had no evolving pattern or large amplitude bursts preceding or following rhythmic rolling waveforms (see Figure 2). Repetitive rhythmic slow-wave activity was quantified as% time/ hour. These waveforms were defined by a period of 200–350 ms from trough to peak, with waveforms forming consistent events lasting >20 sec. These waveforms were often seen in conjunction with sharp and fast wave epileptiform transients, characterised as

individual or multiple waveforms with a duration of >70 ms and <300 ms [21].

Statistics

For analysis of changes during recovery from asphyxia, data were calculated as hour averages from the end of occlusion until the end of the experiment (SPSS v15, SPSS Inc., Chicago, Illinois, USA). The baseline was taken as the mean of the 24 h before occlusion. Treatment effects were evaluated by analysis of variance with time as a repeated measure (ANOVA, SPSS v12, SPSS Inc., Chicago, Il., USA) followed by Fisher's protected least-significant difference (LSD) post-hoc test when a significant overall effect was found. Statistical significance was accepted when P<0.05. Data are mean ± S.E.M.

Results

Fetal biochemistry

Umbilical cord occlusion was associated with profound hypoxemia, hypercarbia and metabolic acidosis (Table 1), which resolved progressively after release of occlusion. DEX was associated with higher fetal PaO$_2$ at 1 and 2 days after occlusion, greater fetal plasma glucose from 4 to 48 h and higher plasma lactate at 4 and 6 h compared to asphyxia-saline. 4 asphyxia-DEX fetuses, but no asphyxia-saline or sham controls, went into labor within 48 h after asphyxia.

Fetal heart rate, blood pressure

Umbilical cord occlusion was associated with profound bradycardia and hypotension, that resolved rapidly after restoration of umbilical blood flow, with transient tachycardia in both asphyxia groups at 3–5 h (not shown, P<0.05). MAP was significantly elevated in both asphyxia groups for the first 48 h (Figure 3, P<0.05) and was transiently higher in the asphyxia-DEX group between 4–6 h (P<0.05). Thereafter there were no differences between groups.

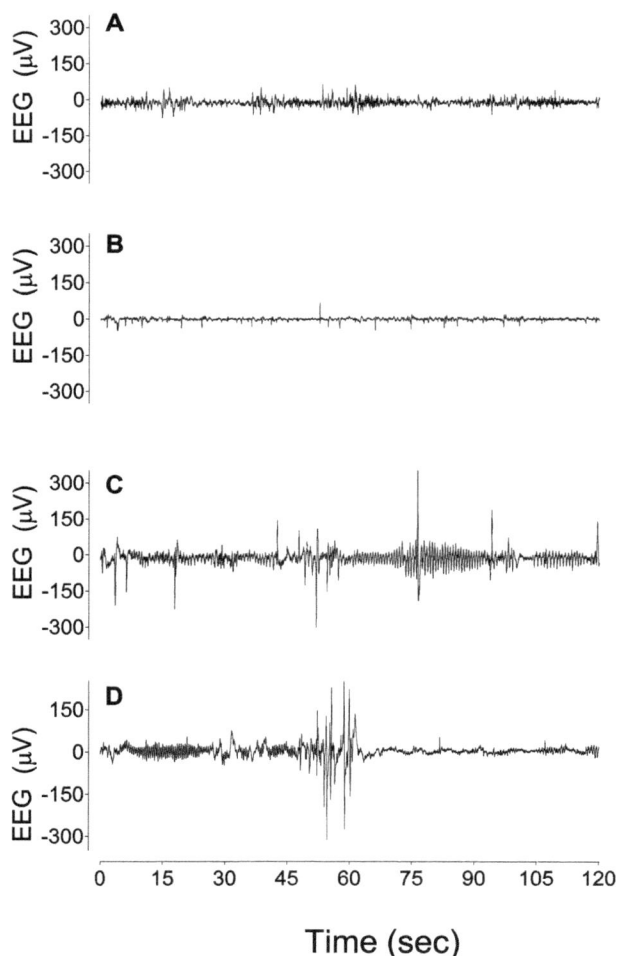

Figure 2. Examples of EEG patterns in preterm fetal sheep.
Panel A shows an example of normal EEG activity prior to asphyxia, showing the normal mixed EEG amplitude and frequency. Panel B shows an example from the asphyxia-saline group, 4 hours after asphyxia, showing significant suppression of EEG amplitude and the presence of repetitive sharp and fast wave transients [39]. Panel C shows an example from the asphyxia-DEX group, 4 hours after asphyxia, showing the presence of interictal high amplitude slow-waves and fast waves. Panel D shows an example of a seizure-like event in the asphyxia-DEX group, 3 hours after asphyxia, characterised as slow-wave activity ending in a high amplitude burst. All figures are continuous raw data from individual animals.

EEG power and seizures

After occlusion, EEG power was initially suppressed in both groups (Figure 3). EEG amplitude then increased, peaking at 9 h in the asphyxia-saline group, and 11 h in the asphyxia-DEX group. EEG power was significantly higher in the asphyxia-DEX group from 5–20 h ($P<0.005$). Thereafter EEG power in both groups rose progressively, but remained suppressed compared to baseline at day 7. Stereotypic evolving seizures were seen in both groups, and started earlier (7.5 ± 0.6 vs. 14.9 ± 4.0 h, $P<0.005$) and continued for longer (21.1 ± 3.9 vs.17.7 ± 6.4 h, $P<0.05$) in the asphyxia-saline group compared to the asphyxia-DEX group, but with no difference in the amplitude (171 ± 24 vs.169 ± 26 μV), or duration (81 ± 7 vs. 80 ± 5 sec/seizure) of stereotypic seizures.

Analysis of the continuous EEG recordings showed that the relative increase in EEG power noted above corresponded with an

increase in rhythmic slow-wave and high amplitude sharp wave transient activity (Figure 2). This activity was seen in a much greater proportion of the EEG recordings in asphyxia-DEX fetuses than asphyxia-saline fetuses during the first 24 hours after occlusion (Figure 4); post-hoc tests suggested that the groups were significantly different from 5–21 h ($P<0.01$). There were no differences between groups after 24 hours.

Brain histology and immunohistochemistry

No white or grey matter injury was seen in sham controls. Asphyxia was associated with marked induction of Iba-1 positive microglia in all white matter regions, which was not affected by DEX (Figure 5). In contrast, there was a marked reduction of Ki-67 labeled proliferation after asphyxia compared to sham controls. Asphyxia-DEX was associated with a significant increase in proliferation compared with asphyxia-saline, to sham control values. There was no significant change in numbers of Olig-2 positive oligodendrocytes in the PVWM in the asphyxia-saline group, in contrast with a significant reduction after asphyxia-DEX compared to both sham controls and asphyxia-saline.

There was a marked reduction in numbers of CNPase positive cells (a marker of immature and mature myelinating oligodendrocytes) in PVWM after asphyxia, with a further significant reduction in the asphyxia-DEX group. The percentage of CNPase positive oligodendrocytes in the periventricular white matter was markedly reduced after asphyxia (asphyxia-saline $9\pm2\%$ vs sham controls $16\pm2\%$, $P=0.005$), with no further change in the asphyxia-DEX group ($9\pm2\%$).

Asphyxia-saline was associated with marked neuronal loss in the hippocampal regions but not the caudoputamen (Figures 6 & 7). Asphyxia-DEX was associated with significantly greater neuronal loss overall compared with both sham controls and asphyxia-saline. Post-hoc analysis showed significantly greater neuronal loss in the putamen and the CA3 region of the hippocampus after asphyxia-DEX compared with asphyxia-saline ($P<0.05$).

Discussion

This study demonstrates, for the first time, that dexamethasone treatment of pregnant ewes shortly after *in utero* asphyxia was associated with greater subcortical neuronal loss and white matter injury as shown by loss of oligodendrocytes in the periventricular white matter tracts in preterm fetal sheep. During recovery from asphyxia, from 6 to 24 h after occlusion, maternal dexamethasone was associated with a marked transient increase in background fetal EEG power, reflecting mainly abnormal slow-wave activity, but delayed onset of seizures. Further, fetal plasma glucose levels were significantly greater during both the latent and secondary phases after maternal DEX treatment, with a transient increase in fetal lactate.

Maternal glucocorticoid therapy is now standard care for threatened premature delivery, including cases where the fetus is at risk of hypoxia. There is surprisingly little information on the potential impact of antenatal steroids on fetal or newborn brain injury [11]. In postnatal day 7 rats, dexamethasone after HI did not affect infarct size [22,23], although there was increased neuronal apoptosis raising the possibility of greater long-term injury. Similarly, studies in adult animals suggest little effect of dexamethasone given after cerebral ischemia [11], although in one study prolonged post-insult treatment reduced injury in the caudate nucleus, but no other regions [24]. Strikingly, there are no translational studies in fetal animals of steroids given after HI.

Periventricular white matter damage is highly characteristic of preterm brain injury. In this study we found no net loss of total

Table 1. Fetal blood composition data.

	Group	Baseline	5 min	17 min	+10 min	+1 h	+2 h	+4 h	+6 h	+1 day	+3 days	+5 days	+7 days
pH	C	7.37±0.0	7.36±0.0	7.36±0.0	7.37±0.0	7.36±0.0	7.38±0.0	7.40±0.0	7.40±0.0	7.36±0.0	7.38±0.0	7.39±0.0	7.38±0.0
	S	7.37±0.01	7.04±0.02#	6.86±0.02#	7.16±0.01#	7.29±0.01#	7.35±0.01	7.43±0.01	7.41±0.01	7.38±0.01	7.38±0.01	7.38±0.0	7.37±0.0
	D	7.38±0.00	7.05±0.02#	6.85±0.02#	7.14±0.01#	7.30±0.01#	7.34±0.03	7.37±0.02	7.38±0.02	7.38±0.02	7.35±0.1	7.36±0.0	7.37±0.0
pCO2 (mmHg)	C	48.0±1.5	45.0±1.4	46.1±1.3	46.1±1.3	48.5±1.6	49.4±1.6	51.3±1.2	50.3±1.2	48.8±1.5	46.1±1.9	46.2±1.6	47.0±0.8
	S	48.6±0.9	100.0±4.6#	138.2±3.2#	54.1±1.6#	43.1±1.3	45.0±1.5	43.7±0.9	46.8±0.7	47.5±0.6	47.5±0.9	47.9±1.4	48.7±1.9
	D	50.0±1.6	91.8±3.0#	131.3±5.0#	52.0±0.8#	41.2±1.3	41.0±1.0	43.0±0.8	44.4±2.1	42.1±1.1#*	48.5±1.3	46.4±1.4	49.2±0.7
pO2 (mmHg)	C	25.2±0.9	26.2±1.2	25.3±1.0	24.7±1.0	25.3±1.3	25.2±1.1	24.6±0.9	25.0±0.9	25.5±0.9	24.4±0.7	24.1±0.6	23.8±0.6
	S	24.1±1.2	7.1±0.9#	6.4±0.8#	35.4±1.0#	29.2±1.5	25.7±1.4	27.2±1.6	28.1±1.5#	28.3±0.8	27.1±0.9	26.3±0.9	25.2±1.0
	D	24.0±0.8	9.2±1.3#	10.0±2.1#	34.9±1.5#	32.0±2.2#	29.1±2.2#	28.5±1.0#	28.9±1.1#	31.9±0.4#*	30.7±0.8#	28.1±1.2#	29.2±1.2#
Lactate (mmol/L)	C	0.8±0.1	0.8±0.1	0.9±0.1	0.8±0.1	0.8±0.1	0.7±0.1	0.7±0.1	0.7±0.1	0.8±0.1	0.9±0.1	0.9±0.0	0.9±0.1
	S	0.9±0.2	4.2±0.2#	7.0±0.4#	6.4±0.3#	4.6±0.2#	3.3±0.3#	2.0±0.4#	2.2±0.4#	1.1±0.2	0.9±0.1	0.9±0.1	0.9±0.1
	D	0.8±0.1	4.1±0.3#	6.5±0.4#	6.4±0.4#	4.8±0.4#	4.1±0.7#	4.4±0.7#*	5.5±0.9#*	1.3±0.1#	0.6±0.3	0.7±0.1	0.7±0.1
Glucose (mmol/L)	C	1.0±0.1	0.9±0.1	1.0±0.1	1.0±0.1	1.0±0.1	1.0±0.1	1.1±0.1	0.9±0.0	1.0±0.0	0.9±0.0	0.9±0.0	0.9±0.0
	S	1.0±0.1	0.3±0.1#	0.6±0.1#	1.6±0.1#	1.3±0.1	1.3±0.1	1.3±0.2	1.4±0.2	1.1±0.1	1.0±0.1	1.1±0.0	1.0±0.1
	D	1.1±0.1	0.3±0.1#	0.6±0.1#	1.6±0.1#*	1.5±0.1	1.6±0.1#	2.5±0.1#*	3.5±0.2#*	2.1±0.2#*	1.3±0.1#	1.2±0.0	1.3±0.1#*

Data from sham controls (C), asphyxia-saline (S) and asphyxia-DEX (D) groups before occlusion (baseline), at 5 and 17 minutes during occlusion, at 10 minutes, 1, 2, 4 and 6 hours after occlusion, and daily thereafter. Days 2, 4 and 6 omitted for brevity. # P<0.05 asphyxia group compared to sham group, * P<0.05 asphyxia-saline compared to asphyxia-DEX.

(Olig-2 positive) oligodendrocytes in the periventricular white matter after asphyxia and only a minor reduction in numbers in fetuses exposed to maternal dexamethasone. In contrast, there was a marked reduction in both numbers and the proportion of immature/mature myelinating oligodendrocytes in the periventricular white matter after asphyxia-saline. There was a further reduction in numbers of myelinating oligodendrocytes after asphyxia-DEX, although interestingly, the reduction in their percentage of total oligodendrocytes in the PVWM was similar to asphyxia-saline. We have previously shown marked loss of pre-oligodendrocytes 3 days after asphyxia in the same paradigm as the present study [25]. This combination of findings is consistent with the seminal report of extensive restorative proliferation after

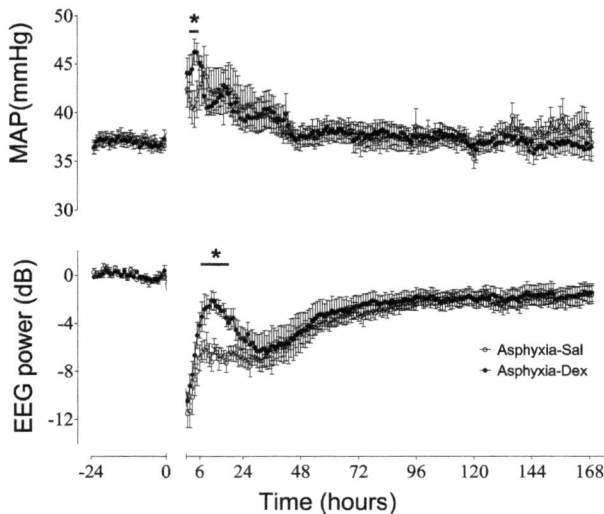

Figure 3. Time course of changes in mean arterial blood pressure (MAP, mmHg), and EEG power (dB) from 24 hours before, until 7 days after occlusion (occlusion data not shown) in the asphyxia-saline (open circles) and asphyxia-DEX (closed circles) groups. Data are one hour means ± SEM. *P<0.05 asphyxia-DEX vs. asphyxia-saline, # P<0.05, both occlusion groups vs. baseline.

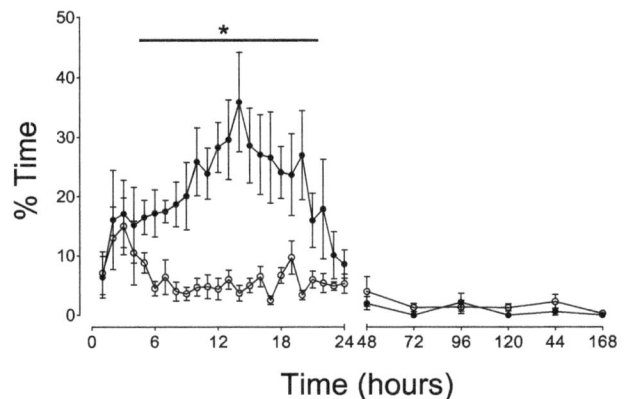

Figure 4. Time course of changes in % time/h of prolonged slow wave activity after asphyxia-saline (open circles) and asphyxia-DEX (closed circles). Data are one hour means ± SEM up to 24 h then 24 h averages until the end of experiment. *P<0.01 asphyxia-DEX vs. asphyxia-saline.

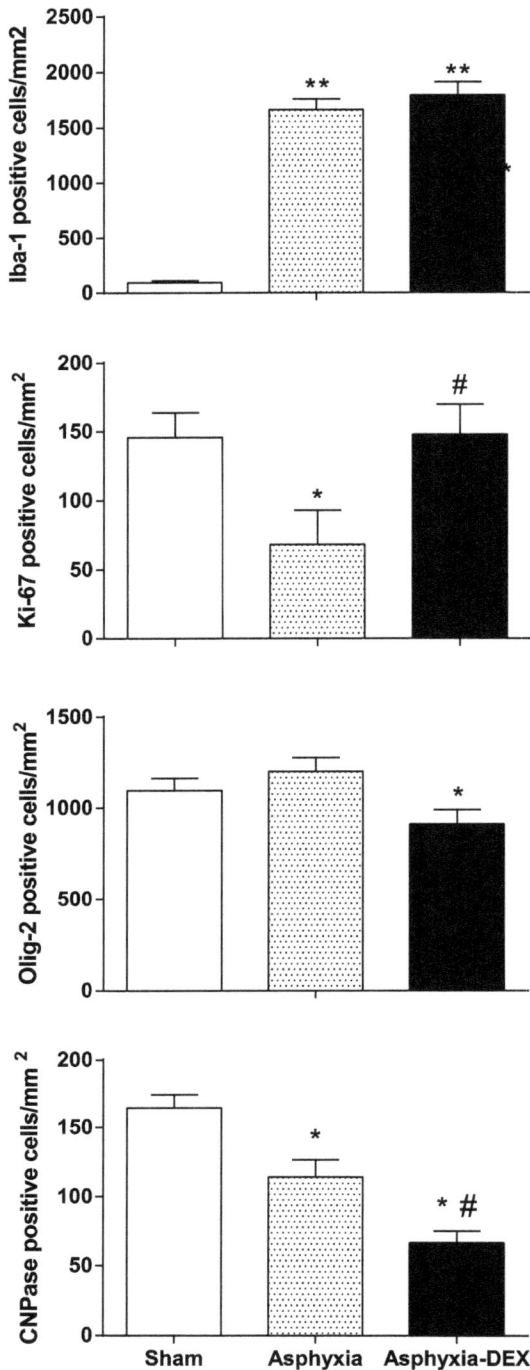

Figure 5. Cell counts seven days after occlusion in sham-control (white bars), asphyxia-saline (asphyxia, grey bars) and asphyxia-DEX groups (black bars), showing numbers of microglia (Iba-1), proliferating cells (Ki-67), total oligodendro-cytes (Olig-2) and immature to mature oligodendrocytes (CNPase positive) in the periventricular white matter. Data are mean ± SEM. * P<0.05, ** P<0.01 compared to sham controls, # P<0.05 compared to asphyxia-saline.

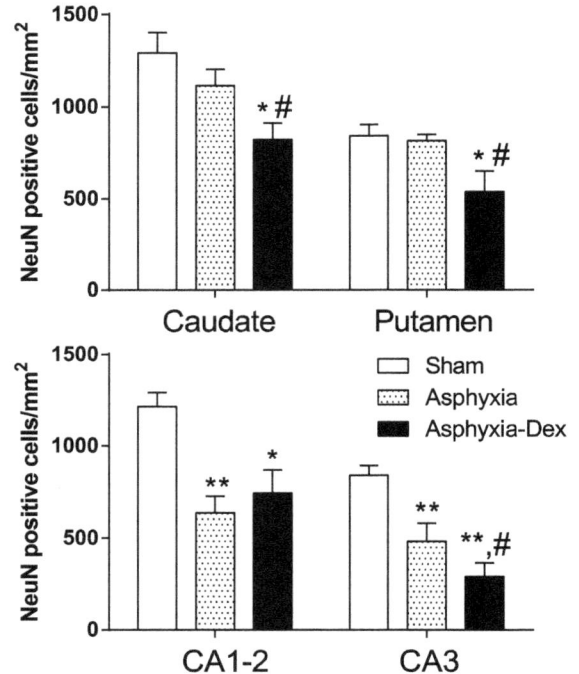

Figure 6. Numbers of surviving neurons (NeuN) per mm² in the Caudate nucleus and the Putamen (Top) and the Cornu Ammonis regions (CA1-2 and CA3) of the hippocampus. * P<0.05, ** P<0.01 compared to sham controls, # P<0.05 compared to asphyxia-saline.

loss of white matter cells, however, it is highly likely that it impaired restorative proliferation, consistent with the finding of suppressed fetal neural proliferation after maternal glucocorticoids in rats [26]. The increased Ki-67-labelled proliferation in white matter by day 7 in this study is consistent with the delayed compensatory rebound increase in proliferation seen up to three weeks after maternal glucocorticoid treatment in the rat [26]. Further long-term studies are needed to determine whether this impaired maturation will resolve with time.

The mechanisms by which maternal glucocorticoids increased fetal neural injury in the present study are likely multifactorial. Increased injury was not related to post-asphyxial hypotension. Although dexamethasone is a potent anti-inflammatory agent, there was no effect on microglial induction by day 7 [11]. As previously reviewed [27], even after surprisingly severe insults there can be transient normalization of oxidative metabolism in a critical 'latent' phase when cell survival can be modulated, followed by a secondary phase of progressive cell death hours to days after reperfusion. Loss of neural suppression in the latent phase can increase neural injury [14,15]. In this study, there was no change in EEG activity in the first 6 hours suggesting that neuronal recovery in the latent phase was not compromised. However, in the secondary phase maternal dexamethasone treatment was associated with a marked increase in background fetal EEG power compared to continued suppression of back-ground EEG power in the asphyxia-saline group. This was not due to increased seizures but to greater interictal activity. Since, glucocorticoids can enhance glutamate release [28], this suggests the hypothesis that dexamethasone increased delayed excitotox-icity in the present study. Conversely, the onset of stereotypic seizures was delayed. Clinically and experimentally, delayed onset of seizures is associated with more severe injury [29–31]. Thus,

hypoxic-ischemic death of pre-oligodendrocytes in the developing brain, followed by chronically impaired maturation [9]. At day 7 after injury we cannot determine whether DEX exacerbated acute

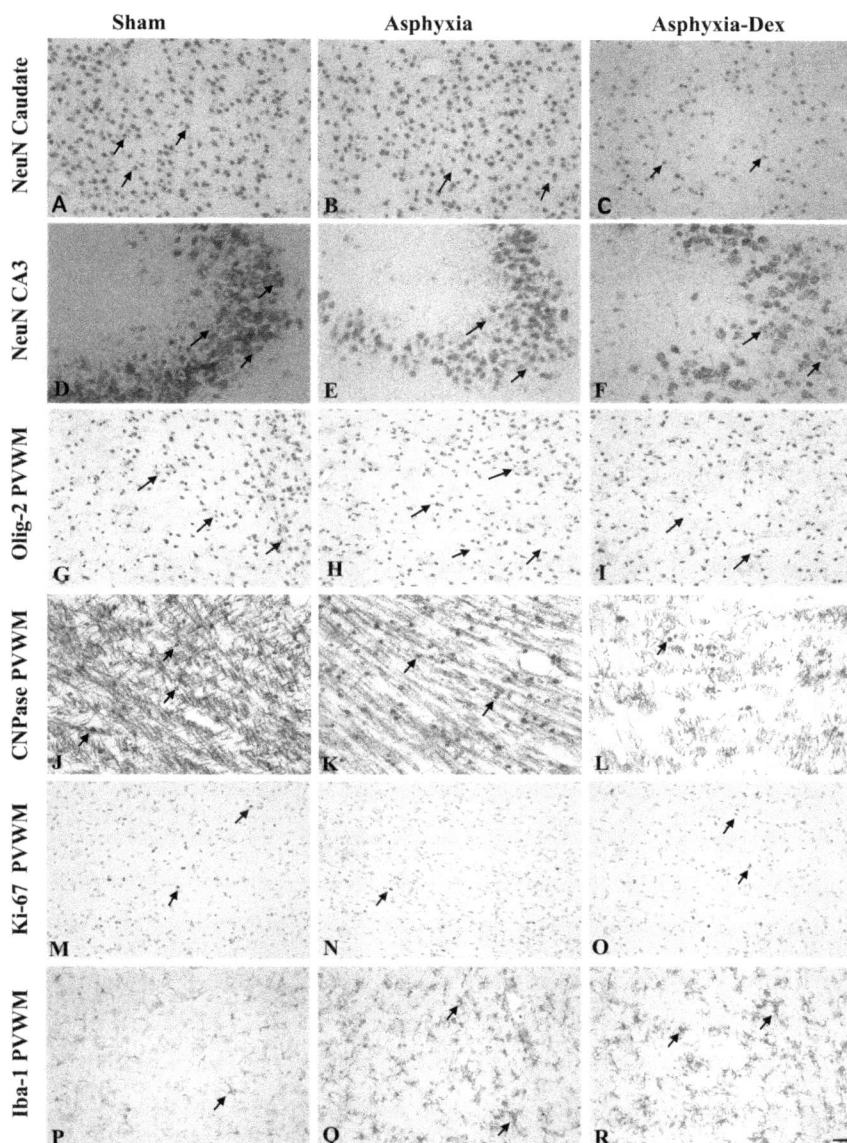

Figure 7. Photomicrographs showing examples of immunohistochemically stained neurons (NeuN) in the Caudate Nucleus (A-C), and in the CA3 region of the hippocampus (D-F). In Plates G to R show photomicrographs from periventricular white matter (PVWM), including all oligodendrocytes (Olig-2, G-I), immature to mature oligodendrocytes (CNPase, J-L), proliferating cells (Ki-67, M-O) and microglia (Iba-1, P-R) from sham-control, asphyxia-saline and asphyxia-DEX fetuses. Arrows show examples of labelled cells. Note the reduction in number of neurons and oligodendrocytes after asphyxia and further loss with DEX treatment, and marked induction of microglia in the PVWM. Scale bar is 40 μm.

this finding is consistent with the greater neural injury in the present study [11,32,33].

Finally, it is possible that the increase in plasma glucose levels in the latent phase and later may contribute to greater injury. As recently reviewed, the effect of blood sugar on ischemic injury is complex [11]. Clinically it is highly controversial whether controlling glucose affects outcomes after ischemia [34]; however, hyperglycemia in preterm infants is associated with increased morbidity and mortality [35]. The fetus is an obligatory glucose user and hyperglycemia increases glucose utilization in the fetus [36], potentially increasing metabolism. In newborn piglets, although hyperglycemia during HI exacerbated brain injury, hyperglycemia after HI similar to that seen in the present study did not affect the severity of neural injury [37].

The present study examined the effects of dexamethasone, one of the two major synthetic glucocorticoids used for antenatal treatment. As previously reviewed, it is unclear whether there is any material difference in neonatal outcomes between dexamethasone and betamethasone, despite these compounds differing only in the orientation of one methyl group [11,32]. Both are widely used [32]. Potentially, glucocorticoids may have sex specific effects in some situations [26,38]. The current study was not powered to evaluate this possibility, although the study groups had similar numbers of male and female fetuses (Table 2). Further, there is potential for much longer term effects of glucocorticoids than addressed in this study [33]. Given this controversy, it will be important to directly compare the effects of these glucocorticoids in future studies and to assess their long-term impact on brain

Table 2. Postmortem data.

	Body Weight (g)	Brain weight (g)	Sex (m:f)
C	1950±148	30.5±2.7	4:4
S	2065±128	28.6±1.4	5:5
D	1933±105	27.3±0.7	6:4

Sham controls (C), asphyxia-saline (S) and asphyxia-DEX (D) groups.

development and function and whether they may have sex specific effects.

In conclusion, the present study shows that a clinical dose of maternal dexamethasone given shortly after asphyxia was associated with increased periventricular white matter damage and neuronal loss in the putamen and the CA3 region of the hippocampus, and increased interictal fetal EEG activity during recovery. These findings suggest that maternal glucocorticoid therapy can adversely affect recovery of the immature brain after hypoxia-ischemia.

Author Contributions

Conceived and designed the experiments: AG LB MEK. Performed the experiments: MEK JOD PPD LCB LB. Analyzed the data: MEK LB JOD SM AG. Wrote the paper: MEK JOD PPD SM LCB AG LB.

References

1. Volpe JJ (2009) Brain injury in premature infants: a complex amalgam of destructive and developmental disturbances. Lancet Neurol 8: 110–124.
2. Buser JR, Maire J, Riddle A, Gong X, Nguyen T, et al. (2012) Arrested preoligodendrocyte maturation contributes to myelination failure in premature infants. Ann Neurol 71: 93–109.
3. Roberts D, Dalziel S (2006) Antenatal corticosteroids for accelerating fetal lung maturation for women at risk of preterm birth. Cochrane Database Syst Rev 3: CD004454.
4. Baud O, Foix-L'Helias L, Kaminski M, Audibert F, Jarreau PH, et al. (1999) Antenatal glucocorticoid treatment and cystic periventricular leukomalacia in very premature infants. N Engl J Med 341: 1190–1196.
5. Whitelaw A, Thoresen M (2000) Antenatal steroids and the developing brain. Arch Dis Child Fetal Neonatal Ed 83: F154–F157.
6. Shinwell ES, Eventov-Friedman S (2009) Impact of perinatal corticosteroids on neuromotor development and outcome: review of the literature and new meta-analysis. Semin Fetal Neonatal Med 14: 164–170.
7. Khan AA, Rodriguez A, Kaakinen M, Pouta A, Hartikainen AL, et al. (2011) Does in utero exposure to synthetic glucocorticoids influence birthweight, head circumference and birth length? A systematic review of current evidence in humans. Paediatr Perinat Epidemiol 25: 20–36.
8. Tijsseling D, Wijnberger LD, Derks JB, van Velthoven CT, de Vries WB, et al. (2012) Effects of antenatal glucocorticoid therapy on hippocampal histology of preterm infants. PLoS ONE 7: e33369.
9. Segovia KN, McClure M, Moravec M, Luo NL, Wan Y, et al. (2008) Arrested oligodendrocyte lineage maturation in chronic perinatal white matter injury. Ann Neurol 63: 520–530.
10. Dean JM, McClendon E, Hansen K, Azimi-Zonooz A, Chen K, et al. (2013) Prenatal cerebral ischemia disrupts MRI-defined cortical microstructure through disturbances in neuronal arborization. Sci Transl Med 5: 168ra167.
11. Bennet L, Davidson JO, Koome M, Gunn AJ (2012) Glucocorticoids and preterm hypoxic-ischemic brain injury: the good and the bad. J Pregnancy 2012: 751694.
12. Elitt CM, Sadowska GB, Stopa EG, Pinar H, Petersson KH, et al. (2003) Effects of antenatal steroids on ischemic brain injury in near-term ovine fetuses. Early Hum Dev 73: 1–15.
13. Davidson JO, Quaedackers JS, George SA, Gunn AJ, Bennet L (2011) Maternal dexamethasone and EEG hyperactivity in preterm fetal sheep. J Physiol 589: 3823–3835.
14. Dean JM, Gunn AJ, Wassink G, George S, Bennet L (2006) Endogenous alpha(2)-adrenergic receptor-mediated neuroprotection after severe hypoxia in preterm fetal sheep. Neuroscience 142: 615–628.
15. Hunter CJ, Bennet L, Power GG, Roelfsema V, Blood AB, et al. (2003) Key neuroprotective role for endogenous adenosine A1 receptor activation during asphyxia in the fetal sheep. Stroke 34: 2240–2245.
16. McIntosh GH, Baghurst KI, Potter BJ, Hetzel BS (1979) Foetal brain development in the sheep. Neuropathol Appl Neurobiol 5: 103–114.
17. Quaedackers JS, Roelfsema V, Fraser M, Gunn AJ, Bennet L (2005) Cardiovascular and endocrine effects of a single course of maternal dexamethasone treatment in preterm fetal sheep. Br J Obstet Gynaecol 112: 182–191.
18. Jakovcevski I, Filipovic R, Mo Z, Rakic S, Zecevic N (2009) Oligodendrocyte development and the onset of myelination in the human fetal brain. Front Neuroanat 3: 5.
19. George SA, Barrett RD, Bennet L, Mathai S, Jensen EC, et al. (2012) Nonadditive neuroprotection with early glutamate receptor blockade and delayed hypothermia after asphyxia in preterm fetal sheep. Stroke 43: 3114–3117.
20. Scher MS, Aso K, Beggarly ME, Hamid MY, Steppe DA, et al. (1993) Electrographic seizures in preterm and full-term neonates: clinical correlates, associated brain lesions, and risk for neurologic sequelae. Pediatrics 91: 128–134.
21. Davidson JO, Green CR, Nicholson LF, O'Carroll SJ, Fraser M, et al. (2012) Connexin hemichannel blockade improves outcomes in a model of fetal ischemia. Ann Neurol 71: 121–132.
22. Barks JD, Post M, Tuor UI (1991) Dexamethasone prevents hypoxic-ischemic brain damage in the neonatal rat. Pediatr Res 29: 558–563.
23. Charles MS, Ostrowski RP, Manaenko A, Duris K, Zhang JH, et al. (2012) Role of the pituitary-adrenal axis in granulocyte-colony stimulating factor-induced neuroprotection against hypoxia-ischemia in neonatal rats. Neurobiol Dis 47: 29–37.
24. Koide T, Wieloch TW, Siesjo BK (1986) Chronic dexamethasone pretreatment aggravates ischemic neuronal necrosis. J Cereb Blood Flow Metab 6: 395–404.
25. Barrett RD, Bennet L, Naylor A, George SA, Dean JM, et al. (2012) Effect of cerebral hypothermia and asphyxia on the subventricular zone and white matter tracts in preterm fetal sheep. Brain Res 1469: 35–42.
26. Scheepens A, van de Waarenburg M, van den Hove D, Blanco CE (2003) A single course of prenatal betamethasone in the rat alters postnatal brain cell proliferation but not apoptosis. J Physiol 552: 163–175.
27. Gunn AJ, Thoresen M (2006) Hypothermic neuroprotection. NeuroRx 3: 154–169.
28. Abraham I, Juhasz G, Kekesi KA, Kovacs KJ (1996) Effect of intrahippocampal dexamethasone on the levels of amino acid transmitters and neuronal excitability. Brain Res 733: 56–63.
29. Filan P, Boylan GB, Chorley G, Davies A, Fox GF, et al. (2005) The relationship between the onset of electrographic seizure activity after birth and the time of cerebral injury in utero. BJOG 112: 504–507.
30. Glass HC, Nash KB, Bonifacio SL, Barkovich AJ, Ferriero DM, et al. (2011) Seizures and magnetic resonance imaging-detected brain injury in newborns cooled for hypoxic-ischemic encephalopathy. J Pediatr 159: 731–735 e731.
31. Williams CE, Gunn AJ, Mallard C, Gluckman PD (1992) Outcome after ischemia in the developing sheep brain: an electroencephalographic and histological study. Ann Neurol 31: 14–21.
32. Carlo WA, McDonald SA, Fanaroff AA, Vohr BR, Stoll BJ, et al. (2011) Association of antenatal corticosteroids with mortality and neurodevelopmental outcomes among infants born at 22 to 25 weeks' gestation. JAMA 306: 2348–2358.
33. Scheepens A, Wassink G, Piersma MJ, Van de Berg WD, Blanco CE (2003) A delayed increase in hippocampal proliferation following global asphyxia in the neonatal rat. Dev Brain Res 142: 67–76.
34. Mehta S (2003) The glucose paradox of cerebral ischaemia. J Postgrad Med 49: 299–301.
35. Ogilvy-Stuart AL, Beardsall K (2010) Management of hyperglycaemia in the preterm infant. Arch Dis Child Fetal Neonatal Ed 95: F126–131.
36. Hay WW, Jr., Myers SA, Sparks JW, Wilkening RB, Meschia G, et al. (1983) Glucose and lactate oxidation rates in the fetal lamb. Proc Soc Exp Biol Med 173: 553–563.
37. LeBlanc MH, Huang M, Patel D, Smith EE, Devidas M (1994) Glucose given after hypoxic ischemia does not affect brain injury in piglets. Stroke 25: 1443–1447.
38. Singh RR, Cuffe JS, Moritz KM (2012) Short- and long-term effects of exposure to natural and synthetic glucocorticoids during development. Clin Exp Pharmacol Physiol 39: 979–989.
39. Bennet L, Booth L, Gunn AJ (2010) Potential biomarkers for hypoxic-ischemic encephalopathy. Semin Fetal Neonatal Med 15: 253–260.

Selenium Supplementation Restores Innate and Humoral Immune Responses in Footrot-Affected Sheep

Jean A. Hall[1]*, William R. Vorachek[1], Whitney C. Stewart[2,3], M. Elena Gorman[1], Wayne D. Mosher[2], Gene J. Pirelli[2], Gerd Bobe[2,4]

1 Department of Biomedical Sciences, College of Veterinary Medicine, Oregon State University, Corvallis, Oregon, United States of America, 2 Department of Animal and Rangeland Sciences, College of Agriculture, Oregon State University, Corvallis, Oregon, United States of America, 3 Current Address: Department of Animal and Range Sciences, New Mexico State University, Las Cruces, New Mexico, United States of America, 4 Linus Pauling Institute, Oregon State University, Corvallis, Oregon, United States of America

Abstract

Dietary selenium (Se) alters whole-blood Se concentrations in sheep, dependent upon Se source and dosage administered, but little is known about effects on immune function. We used footrot (FR) as a disease model to test the effects of supranutritional Se supplementation on immune function. To determine the effect of Se-source (organic Se-yeast, inorganic Na-selenite or Na-selenate) and Se-dosage (1, 3, 5 times FDA-permitted level) on FR severity, 120 ewes with and 120 ewes without FR were drenched weekly for 62 weeks with different Se sources and dosages (30 ewes/treatment group). Innate immunity was evaluated after 62 weeks of supplementation by measuring neutrophil bacterial killing ability. Adaptive immune function was evaluated by immunizing sheep with keyhole limpet hemocyanin (KLH). The antibody titer and delayed-type hypersensitivity skin test to KLH were used to assess humoral immunity and cell-mediated immunity, respectively. At baseline, FR-affected ewes had lower whole-blood and serum-Se concentrations; this difference was not observed after Se supplementation. Se supplementation increased neutrophil bacterial killing percentages in FR-affected sheep to percentages observed in supplemented and non-supplemented healthy sheep. Similarly, Se supplementation increased KLH antibody titers in FR-affected sheep to titers observed in healthy sheep. FR-affected sheep demonstrated suppressed cell-mediated immunity at 24 hours after intradermal KLH challenge, although there was no improvement with Se supplementation. We did not consistently prevent nor improve recovery from FR over the 62 week Se-treatment period. In conclusion, Se supplementation does not prevent FR, but does restore innate and humoral immune functions negatively affected by FR.

Editor: Dipshikha Chakravortty, Indian Institute of Science, India

Funding: The work presented in this study was performed at the Department of Biomedical Sciences, College of Veterinary Medicine, Oregon State University, Corvallis, Oregon 97331-4802, USA, and supported in part by USDA CSREES 2008-35204-04624, Agricultural Research Foundation, and Animal Health and Disease Project Formula Funds, Oregon State University, Corvallis, Oregon 97331-2219. The funders had no role in study design, data collection and analysis, decision to publish, or preparation of the manuscript.

Competing Interests: The authors have declared that no competing interests exist.

* E-mail: Jean.Hall@oregonstate.edu

Introduction

Dietary selenium (Se) alters whole-blood (WB) Se concentrations in sheep, depending upon the chemical source and dosage administered [1–5]. Less is known about how different chemical forms of Se (inorganic Na-selenate or Na-selenite, and organic Se-yeast) at comparative dosages alter immune functions. In domestic animals, including sheep, Se deficiency results in immunosuppression. Specifically, Se deficiency decreases resistance to bacterial and viral infections, and decreases neutrophil function, antibody production, proliferation of T and B lymphocytes in response to mitogens, and cytodestruction by T lymphocytes and natural killer cells (reviewed in [6–10]). The effect of supranutritional Se supplementation on specific immune functions has not been well studied. We hypothesize that the amount of Se required for optimum health is higher than the amount required for prevention of nutritional myodegeneration [11].

Current Food and Drug Administration (FDA) regulations limit the amount of dietary Se supplementation, regardless of chemical source, to 0.3 mg/kg (as fed), or 0.7 mg per sheep per day [12]. Concentrations that exceed 0.3 mg/kg but that are less than the maximum tolerable level (5 mg/kg of diet, as fed) are referred to as supranutritional. There is interest in supranutritional supplementation relative to health, performance, and disease prevention in animals and humans [13–15]. We recently reported that supranutritional Se supplementation improved response to vaccination with a J-5 *Escherichia coli* bacterin in adult beef cows [16]. Furthermore, we reported that supranutritional Se supplementation of ewes improves growth and survival of their offspring [17], which may be due in part to greater colostral IgG concentrations and greater calculated amounts of IgG transferred to their lambs [18], suggesting that supranutritional Se supplementation may enhance passive immunity.

A suitable model in sheep to test the effects of supranutritional Se supplementation on immune function is footrot (FR), an endemic bacterial disease of sheep feet that results in lameness and large production loses for the industry. Footrot is caused by infection with *Dichelobacter nodosus*, a gram negative, anaerobic and fastidious bacterium, in association with other bacteria, particularly *Fusobacterium necrophorum* [19–21]. If the interdigital skin of the foot is damaged or wet for prolonged periods of time, it may become infected by the ubiquitous soil and fecal bacterium *F. necrophorum*. *F. necrophorum* causes interdigital dermatitis and produces toxins that cause necrosis of the superficial layer of the skin allowing co-infection with other bacteria such as *D. nodosus*. *D. nodosus* contains surface fimbriae and stable extracellular proteases that allow it to colonize the interdigital epithelial tissue, digesting the living dermis, and feeding on collagen [22,23]. A foul smell is associated with the accumulation of grey pasty exudate between the dermis and epidermal horn, and ultimately separation of the hoof horn from the underlying dermis occurs [23]. A strict culling program during the hot, dry summer months (non-transmission period) has proven successful in eliminating FR in flocks in Western Australia [24]. However, this protocol is unfeasible in countries with cool, wet climates and widespread infection in flocks [23]. Instead, management programs to control rather than eliminate infection are more commonly employed. Strategies include parenteral antibiotic treatment, topical antibacterial sprays, trimming of horn hoof, vaccination, low stocking density, and genetic selection for sheep breeds less susceptible to FR [23,25]. We previously documented in a small-scale study that WB-Se concentrations are lower in FR-affected compared with healthy sheep, and that supranutritional Se-supplementation (inorganic sodium selenite administered parenterally once monthly) hastens FR recovery [26].

The role of the immune response and Se supplementation in the pathogenesis or recovery from FR is unclear. Adaptive immunity, including humoral and cell mediated immunity (CMI), likely play a role in protection against FR [27,28], yet infected or vaccinated sheep do not develop long-term immunity and may become re-infected over time [23,29,30]. Heritability of resistance to FR may be related to a specific range of MHC II haplotype that is required to generate a sufficient immune response to *D. nodosus* [28]. We recently showed that supranutritional Se-yeast supplementation in ewes increased WB-neutrophil expression of genes involved in innate immunity, including reversing those impacted by FR [31].

The objective of the current study was to determine the effects of Se-source (inorganic Na-selenate or Na-selenite versus organic Se-yeast) and Se-dosage (0, 1, 3, or 5 times the FDA-allowed Se supplementation rate) on immune function. We hypothesized that both arms of the immune response, i.e., innate immunity of neutrophils and adaptive immunity (both humoral and cell mediated immunity), would be enhanced by supplementation with supranutritional levels of Se, with a greater benefit from Se-yeast. Innate immunity was evaluated by measuring the bacterial killing ability of isolated neutrophils. Humoral immunity was evaluated by measuring an antibody titer response to a novel protein challenge (KLH, keyhole limpet hemocyanin). Cell mediated immunity was evaluated using a delayed hypersensitivity test (DTH) to KLH. Finally, we monitored the incidence and severity of FR across time in sheep receiving varying dosages of organic versus inorganic Se.

Materials and Methods

Animals, Experimental Design, and Selenium Treatments

Experimental procedures used in this study were approved by the Institutional Animal Care and Use Committees of Oregon State University (Permit Number: 3778) and have been described in detail previously [1]. As shown in **Figure 1**, mature breeding ewes that had been randomly assigned to 8 treatment groups (n = 30 each; 15 healthy and 15 foot rot affected) based on Se

supplementation rate and source were used: one group received water (no Se supplement group), one group received Na-selenate (8.95 mg Se/wk), 3 groups received Na-selenite (RETORTE Ulrich Scharrer GmbH, Röthenbach, Germany) at 4.9 (maximum FDA-allowed level), 14.7 (3 times maximum FDA-allowed level), or 24.5 mg Se/wk (5 times maximum FDA-allowed level), and 3 groups received Se-yeast (Prince Se Yeast 2000, Prince Agri Products Inc., Quincy, IL) at 4.9, 14.7, or 24.5 mg Se/wk. Sodium-selenate (Na_2SeO_4) was 418 g/kg Se or 41.8% Se, and Na-selenite (Na_2SeO_3) was 456 g/kg Se or 45.6% Se. The organic Se source had a guaranteed analysis of 2 g/kg of organically bound Se with 78% being selenomethionine (SeMet).

The treatment period started approximately 2 wk before breeding and lasted for 62.5 wk. All dosages were below the maximum tolerable level (5 mg/kg as fed, which is 16.7 times the FDA-allowed Se supplementation rate) for small ruminants [32]. Treatment groups were stratified for age of ewe and FR severity (scale 0 to V; with no FR = 0 being the lowest category). Ewes were from three genotypes (Polypay, Suffolk, and Suffolk by Polypay cross), and ranged in age and body weight (BW) from 2 to 6 yr, and 51 to 93 kg, respectively. The three genotypes were not completely balanced across treatments. The experiments were conducted at the Oregon State University Sheep Center, Corvallis, OR.

Ewe treatments were administered individually once per wk by oral drench (with the calculated weekly amount of Se supplement being equal to the summed daily intake). The weekly Se dose (4.9, 8.95, 14.7, or 24.5 mg Se/wk per ewe) was suspended in water (11, 11, 30, and 48 mL for the 4.9, 8.95, 14.7, or 24.5 mg Se solutions, respectively). Stock solutions were made up fresh each wk and administered with a dose syringe. To ensure a homogeneous dosage, stock solutions were stirred each time prior to being drawn into the dose syringe. The no-Se treatment group received 11 mL water.

Individual treatment aliquots were submitted for Se analysis to the Center for Nutrition, Diagnostic Center for Population and Animal Health, Michigan State University (East Lansing, MI), and Se was quantified according to previously described methods [1]. The 4.9, 14.7, and 24.5 mg weekly dosages of Na-selenite (4.9, 14.9, and 24.6 mg Se, respectively), and the weekly dosages of organic Se-yeast supplements (4.8, 14.4, and 24.0 mg of organically bound Se-yeast, respectively) were found to be within expected analytical variance of their targeted concentrations. The dosage of Na-selenate administered was determined to be 8.95 mg Se/wk per ewe. Concentrations of Se in the pasture forage from the sheep center pastures ranged from 0.12 to 0.14 mg/kg dry matter [1]. The Se concentrations of the grass hay, alfalfa hay, alfalfa pellets, whole corn, and custom-made mineral supplement (Oregon State University Sheep Mineral Premix, Wilbur-Ellis Company, Clackamas, OR) were 0.02, 0.05, 0.06, 0.01, and 0.44 mg Se/kg dry matter, respectively. Assuming pasture dry matter intake of 2% of BW, ewes would consume between 0.12 and 0.26 mg Se/d. For an average mineral intake of 8 g/d, an additional 3 μg of Se would be consumed. Other feed ingredients would contribute less than 20 μg Se/d. Thus, the majority of Se intake (4.9, 14.7, or 24.5 mg Se/wk) was provided by the oral Se drench [1].

Ewes were fed on pasture, except for a 3-mo period around lambing when ewes were housed in the barn. Ewes on pasture were supplemented with grass hay and later with alfalfa hay when grass was scarce. In the barn, sheep were fed alfalfa hay and shelled corn, except for 2 d in the lambing jug when ewes were fed alfalfa pellets. Ewe feed sources and management details have

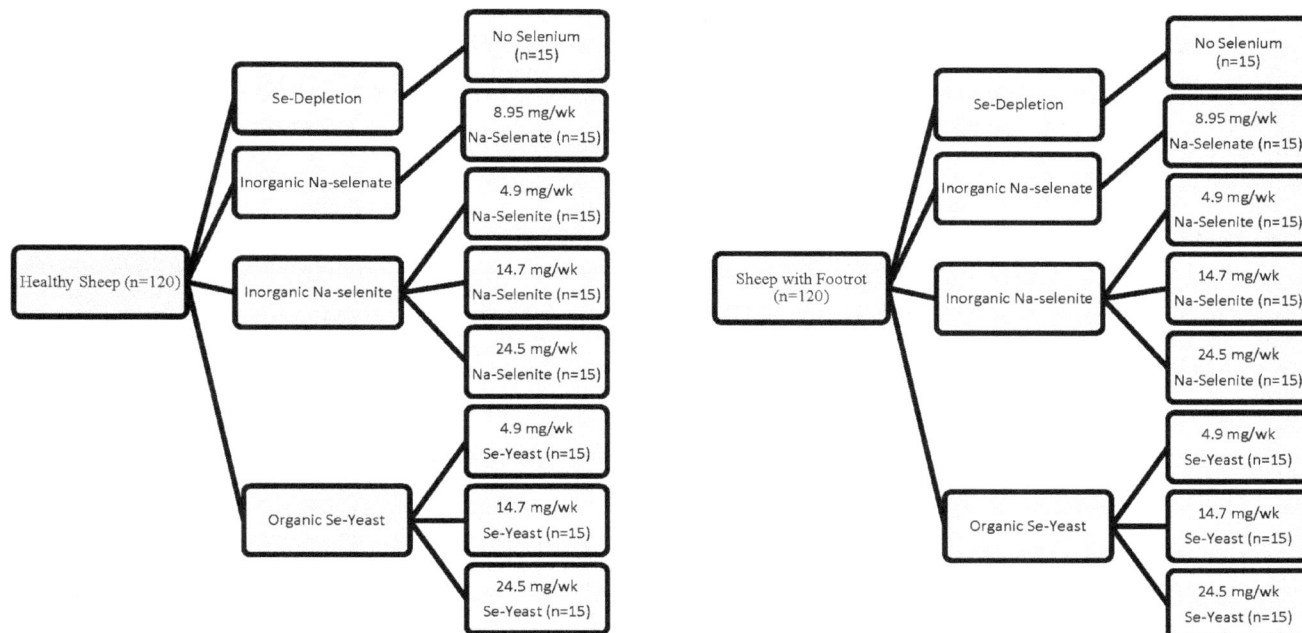

Figure 1. Study design. Mature breeding ewes were randomly assigned to 8 treatment groups (n = 30 each; 15 healthy and 15 foot rot affected) for 62 wk based on Se source (no Se, inorganice Na-selenate, inorganic Na-selenite, and organic Se-yeast) and Se supplementation rate (0, 4.9, 14.7, and 24.5 mg Se/wk; Na-selenatate only at 8.95 mg/wk Se/wk).

been previously described [1]. Sheep were fed to meet or exceed National Research Council [32] recommendations.

Foot Scores

At 0, 20, 28, 40, and 60 wk of Se supplementation, each sheep foot was examined, trimmed, and scored for FR using a scale of 0 to 4. This scoring system was based on pathologically defined criteria [33]. Definitions for the scores per foot are as follows: 0: normal hoof, no evidence of FR; 1: interdigital dermatitis, presence of heat, and characteristic FR odor; 2: initial underrunning of the hoof wall between the toes; 3: underrunning of the sole; and 4: extensive underrunning of the sole and lateral walls of the hoof. The same person assigned scores (blinded) at each time point for consistency of evaluations.

Based on the FR scores, sheep were classified into the following FR severity categories: Category 0: normal hoofs, no evidence of FR; Category I: single foot affected and FR score of 1 or 2; Category II: single foot affected and FR score of 3 or 4; Category III: at least 2 feet affected and at least one foot with low FR score of 1 or 2; Category IV: at least 2 feet affected with FR scores equal or greater than 3; and Category V: at least 3 feet affected, FR score greater than 3 in at least one foot, and FR score equal to or greater than 2 in other feet. For each treatment group (n = 30) sheep were randomly assigned as follows: Category 0 (n = 15), Category I (n = 4), Category II (n = 5), Category III (n = 3), Category IV (n = 2), and Category V (n = 1).

Sheep used in this study had naturally acquired FR. The FR control program [34] in this flock consisted of a walkthrough footbath containing 10% zinc sulfate solution (used monthly or less often). A FR vaccination program was not used. The guidelines for administration of parenteral oxytetracycline antibiotic (Liquamycin LA-200; Pfizer Animal Health, Exton, PA) at the 28- and 40-wk foot trimming sessions were if one foot had a score of 4, or one foot had a score of 3 and a second foot had a score of 2 or greater,

or if all 4 feet had FR, then 20 mg/kg oxytetracycline was administered subcutaneously.

Blood Collection and Se Analysis

Jugular venous blood was collected from ewes monthly. For this study, blood samples from 0, 14, 27, 40, and 54 wk of Se supplementation were used. Whole blood was collected into evacuated tubes without EDTA (10 mL; Becton Dickinson, Franklin Lakes, NJ). Tubes were centrifuged at $850 \times g$ for 10 min and serum was harvested, centrifuged again at $16,300 \times g$ for 1 min in a microcentrifuge to remove remaining red blood cells (RBC), and transferred to 2.0 mL screw cap self-standing microcentrifuge tubes and stored at $-20°C$ for a serum-Se assay. Jugular venous blood was collected into evacuated tubes with EDTA (2 mL; final EDTA concentration 2 g/L; Becton Dickinson) and stored on ice until it could be frozen at $-20°C$ for a WB-Se assay. Selenium concentrations in WB and serum were determined by a commercial laboratory (Center of Nutrition, Diagnostic Center for Population and Animal Health, Michigan State University) using ionized coupled plasma mass spectrometry (ICP-MS) method with modifications as previously described [1]. Finally, jugular venous blood was collected at 14 months into EDTA tubes (10 mL; final EDTA concentration 2 g/L, Becton Dickinson) and transported on ice to the lab for subsequent isolation of neutrophils.

Isolation of peripheral blood neutrophils

Neutrophils were isolated within 4 h of collection, using a Percoll (Sigma-Aldrich, St. Louis, MO) gradient technique, then resuspended in 1× Hank's balanced saline solution (HBSS; Life Technologies, Grand Island, NY) plus 0.5% fetal bovine serum (FBS; Life Technologies). Cells were counted using a Coulter counter (Beckman Coulter, Indianapolis, IN) to determine cell concentration. Briefly, 10 mL of anticoagulated blood was transferred into 50-mL tubes (Thermo Fisher Scientific, Waltham, MA) and centrifuged at $1000 \times g$ in a TJ-6 swinging bucket centrifuge (Beckman Coulter) for 20 min at 22 °C. The plasma, buffy coat, and one third of the RBC pack from each tube were aseptically removed. The remaining RBC pack and leukocytes were mixed with 34-mL ice-cold phosphate buffered saline (PBS; Life Technologies) and layered onto 10 mL of freshly prepared 1.084 g/mL Percoll. Tubes were centrifuged at $400 \times g$ for 40 min at 22 °C. After centrifugation, RBC and neutrophils pelleted at the bottom of the tube. The mononuclear cell band remained at the sample/medium interface, and was aspirated and discarded.

Neutrophils were isolated from RBC by lysing the RBC with 24-mL ice-cold hypotonic lysis buffer (10.56 mM Na_2HPO_4, 2.67 mM NaH_2PO_4, pH 7.3) for 90 s, and then isotonicity was restored by adding 12-mL ice-cold hypertonic restore buffer (10.56 mM Na_2HPO_4, 2.67 mM NaH_2PO_4, 0.43 M NaCl, pH 7.3) to stop lysis. Neutrophils were pelleted by centrifugation of tubes at $800 \times g$ for 5 min at 22 °C in a TJ-6 centrifuge. The lysis solution was decanted, and the neutrophils were resuspended and washed twice more with $1 \times$ HBSS plus 0.5% FBS. The neutrophils were then resuspended in 0.25 mL of $1 \times$ HBSS with 0.5% FBS and stored on ice until needed. A 20-μL aliquot of the cell suspension was used to determine cell concentration using a Coulter counter (Beckman). Another 5-μL aliquot was used to assess purity of neutrophil preparations (differential cell count) by microscopic examination after Wright-Giemsa staining (96±1% neutrophils; mean ± SEM).

Neutrophil bacterial killing of Lactococcus lactis or Escherichia coli

To evaluate innate immune function, we measured neutrophil bacterial killing after 14 months of Se supplementation using a cell proliferation assay (CellTiter 96® Non-Radioactive Cell Proliferation Assay; Promega Corp, Madison, WI). In this assay, living bacteria convert the tetrazolium component of the dye solution into a formazan product during the 4-h incubation. A solution is added to wells to solubilize the formazan product and absorbance at 570 nm is recorded using a 96-well plate reader. In short, Escherichia coli (E. coli) were grown overnight to stationary phase in 5-mL Luria Broth media (prepared using ingredients from Thermo Fisher Scientific) after which the bacteria were put on ice. The bacteria were washed twice with $1 \times$ HBSS (centrifuged at 4 °C in a TJ-6 centrifuge at $1300 \times g$ for 10 min), and then resuspended in $1 \times$ HBSS to approximately 2×10^5 colony forming units (CFU)/μL and stored on ice until needed. Next 1.2×10^6 CFU of E. coli were added to all experimental wells of a 96-well flat bottom plate (Corning Inc., Corning, NY). Neutrophils (1.2×10^5 cells) from each calf were seeded into triplicate wells containing bacteria, and $1 \times$ HBSS added such that the total volume was 50 μL. The multiplicity of infection (MOI) was 10 bacteria to 1 neutrophil. A standard curve was prepared from serial dilutions of bacteria in $1 \times$ HBSS (2.4×10^6 to 7.5×10^4 CFU); these bacterial dilutions were seeded into triplicate wells without neutrophils. Three wells without bacteria or neutrophils were used for background controls. The 96-well plate was covered and incubated at 37 °C in a 5% CO_2 humidified chamber for 2 h for neutrophil bacterial killing. After incubation, 50 mL of a 0.2% saponin solution (Sigma-Aldrich) was added to each well to lyse remaining neutrophils, and the plate was incubated at 37°C for 1 h in a 5% CO_2 humidified chamber. After incubation was complete, 15 μL of dye solution was added to each well and the plate was incubated at 37°C for 4 h in a 5% CO_2 humidified chamber. Then 100 μL of solubilization solution/stop mix was added to each well and the plate was incubated overnight at ambient

temperature in the dark. The 96-well plate was then scanned at 570 nM using a BioRad plate reader (BioRad Inc., Hercules, CA). Data was analyzed using Excel (Microsoft Inc., Bellevue, WA) to determine the number of live bacteria in each well based on the bacteria standard curve. Final data are expressed as % bacteria killed.

Delayed-type hypersensitivity (DTH) skin test with keyhole limpet hemocyanin (KLH)

Adaptive immune function was evaluated after 12 months of Se supplementation by immunizing all sheep twice 2-weeks apart with 0.5 mL of KLH (500 μg of KLH emulsified in 1.0 mg of T1501 adjuvant for a total volume of 0.5 mL; administered intramuscularly) as previously described [35]. Three weeks after the second KLH immunization, intradermal skin testing was performed. Swelling and induration following an intradermal challenge are measured in the DTH test to assess the CMI response by T cells. Individual disposable tuberculin syringes were filled with heat aggregated KLH and a 25-gauge needle was used to inject 0.05 mL of KLH intradermally on the ear tip. The 0.05-mL dose of heat-aggregated KLH consisted of approximately 3 mg of KLH, and was prepared as described previously [35]. Measurements of wheal diameter and ear thickness were made at 30 min (to test innate immune function) and at 24, 48, 72, and 96 h for the DTH test. Reactions were recorded according to the diameter of induration and erythema. The test was administered by the same person to all sheep. No chemical restraint was used.

KLH antibody titer

The KLH antibody titer was also used to assess adaptive immune function. The humoral immune response was evaluated after 12 months of Se supplementation by comparing the KLH antibody titer before vaccination with KLH (after 9 months of Se supplementation and 2 months before vaccination) to the titer obtained 3.5 weeks after the second KLH immunization.

Serum was assayed for KLH antibody titer by an indirect ELISA. Briefly, 96-well microtiter plates (Costar, Cambridge, MA, USA) were coated with 5 μg/mL of KLH (100 μL/well; Sigma Chemical Co., St. Louis, MO, USA) in a 0.1 M sodium bicarbonate buffer, sealed to prevent evaporation, and incubated at 4°C overnight. The next day the coating solution was removed and 100 μL of StabilCoat (SurModics Inc., Eden Praire, MN, USA) was added to each well to block nonspecific binding sites and plates were incubated for 30 min at room temperature. After incubation, the StabilCoat was removed; plates were resealed and stored at 4°C until used. Serum samples were serially diluted (1:100 to 1:1,024,000) in 0.05 M PBS with 0.05% Tween-20 (T-PBS; pH 8.0). Each dilution was added to the plate in duplicate, and incubated for 30 min at room temperature. Positive and negative control serums were included on each plate. After incubation, plates were washed eight times with T-PBS and then 100 μL of horseradish peroxidase conjugated to rec-Protein G (Zymed Laboratories Inc., San Francisco, CA, USA) was added to each well at a previously determined dilution (1:20,000). Rec-Protein G binds to IgG immunoglobulin through their Fc regions and was used instead of an anti-species conjugate as it resulted in equivalent results with a stronger signal and less background. After another 30 min incubation at room temperature, plates were again washed 5 times with T-PBS and 100 μL of 3, 3', 5, 5'-tetramethyl benzidine (TMB; Sigma, St Louis, MO) was added to each well. Plates were read at 655 nm until an absorbance of 1.000 O.D. was reached in the least dilute standard well. The TMB reaction was then stopped by adding 100 μL of 1M H_2SO_4 and the plate was read at 450 nm.

For standards, we used pooled sheep serum at each time point before and after immunization. The change in titer from pre-immunization to 3.5 weeks after the second immunization was determined using a fold multiple calculation based on the pooled sheep serum standards. Ten µL of serum from each ewe before immunization were pooled into a single tube and mixed, and aliquots of this pooled serum were placed into screw cap tubes for later use as a pre immunization-pooled standard. The same procedure was followed for the post-immunization-pooled standard. On every ELISA plate, a duplicate set of eight dilutions from the pooled standard was assayed along with a set of serum dilutions from each individual ewe. The dilutions of pooled serum provided a log linear standard curve that was used to normalize all ELISA plates. A log linear curve was then generated for each individual ewe sera. A linear equation was derived in Excel from the log linear portion of each curve and the quantity of serum that would have been needed to achieve an absorbance of 0.40 at 450 nm was calculated. The amount of serum from each individual ewe that would have been needed to achieve an absorbance of 0.40 was compared to the amount of pooled serum standard needed to achieve an absorbance of 0.40 and expressed as an inverse fold multiple. Values are inverse fold multiples of log transformed raw data.

Statistical analysis

Statistical analyses were performed using SAS, version 9.2 (SAS, Inc., Cary, NC, USA) software. For evaluating the effect of Se supplementation on WB-Se and serum-Se concentration, data were analyzed in PROC GLM. Fixed effects in the model were Se-source and dosage (no-Se, 8.95 mg Na-selenate/wk, 4.9 mg Na-selenite/wk, 14.7 mg Na-selenite/wk, 24.5 mg Na-selenite/wk, 4.9 mg Se-yeast/wk, 14.7 mg Se-yeast/wk, and 24.5 mg Se-yeast/wk), ewe FR status at the time of blood collection (no FR, FR), ewe breed (Polypay, Suffolk, or Suffolk x Polypay; the Suffolk and Suffolk x Polypay ewes were combined into one group because animal numbers for both groups were small and phenotypically the crossbreds more closely resembled the Suffolks in size; the reference group was Polypay ewes), ewe age at lambing (<5, 5, >5 yr), and the interaction between FR status at the time of blood collection and Se-source and dosage.

For evaluating the effect of Se source and dosage on FR prevalence and severity, data were analyzed in PROC GLIMMIX assuming a binomial distribution for prevalence of FR (% ewes with FR or % ewes with FR ≤ I) and assuming a negative binomial distribution for FR severity (0 to V). Fixed effects in the model were the same as for blood Se concentrations. We conducted a stratified analysis by FR status to evaluate whether Se supplementation improved FR prevalence and severity in ewes affected with FR. The effect of Se source and dosage on response to oxytetracycline treatment at 28 wk and 40 wk of supplementation (yes = FR severity 0 or I, no = FR severity ≥ II) was evaluated utilizing the same statistical model as for FR prevalence.

Indicators of innate and adaptive immunity were analyzed using PROC GLM (neutrophil bacterial killing assay, 30-min skin test response to intradermal KLH injection, and KLH antibody titers) for single measures in time and PROC MIXED (DTH skin test responses) for repeated measures in time within the same animals. Fixed effects in the model were the same as previously described. For DTH, additional fixed effects in the model were time of measurement (24, 48, 72, and 96 hr after KLH challenge) and the interactions of Se source and dosage by time, FR status by time, and Se source by FR status by time. A completely unrestricted variance-covariance structure was used to account for repeated measures taken on individual ewes across time. To obtain the

correct degrees of freedom, the KENWARDROGER option was invoked. The KENWARDROGER option consists of the Satterthwaite adjustment for degrees of freedom with a Kenward-Roger adjustment on standard errors, which can be used for repeated measures studies. Antibody titers for KLH were normalized by log 2 transformation.

The effect of Se-source and dosage were evaluated using preplanned comparisons. The effect of Se supplementation was evaluated by comparing the estimated values of the no Se group with the average of the estimated values of the seven Se-supplemented ewe groups. The effect of Se source was evaluated by comparing the three Se sources (no Se, Na-selenate, Na-selenite, and Se-yeast), for which the average of the estimated values of the three Na-selenite and three Se-yeast groups were used. The effect of Se dosage was evaluated by comparing the estimated values of the three Na-selenite and three Se-yeast dosage groups (4.5, 14.7, 24.5 mg/wk). The effect of supranutritional Se dosage was evaluated by comparing the average of the estimated values of the two higher dosages (14.7 and 24.5 mg/kg) with that of the maximal FDA-allowed dosage (4.5 mg/wk). Data are reported as least square means ± SEM except for FR lesions. All statistical tests were two-sided. Statistical significance was declared at P ≤ 0.05 and tendencies at 0.05 <P ≤ 0.10.

Results

Blood Se concentrations and ewe health

The effects of Se-source and Se-dosage on WB- and serum Se concentrations are shown in **Table 1** and **Table 2**, respectively. At baseline, healthy ewes had higher WB-Se concentrations than ewes with FR (250 versus 235 ng/mL; P = 0.01) and higher serum-Se concentrations (102 versus 90 ng/mL; P < 0.0001). After Se supplementation, no significant differences in WB- and serum-Se concentrations were observed between healthy and FR-affected ewes within the same treatment group; this effect was irrespective of Se-source and Se-dosage (**Figure 2** and **Figure 3**). Both healthy and FR-affected ewes that received no Se treatment had decreased WB-Se and serum-Se concentrations (both P < 0.0001). Supranutritional Se-yeast supplementation increased WB-Se and serum-Se concentrations linearly with dosage (P < 0.0001), whereas ewes receiving supranutritional Na-selenite supplementation at 14.7 and 24.5 mg Se/wk achieved similar WB-Se concentrations as ewes receiving 4.9 mg/wk of Se-yeast. Ewe FR status did not affect supplementation-induced changes in WB- or serum-Se concentrations.

None of the ewes in the no-Se treatment group showed clinical signs of nutritional myodegeneration from Se deficiency [11]. None of the ewes receiving supranutritional Se supplementation showed clinical signs of Se toxicity at any time during the study.

Foot lesions

The effects of supplementing ewes with Se on FR severity are shown in **Figure 4** (effects of Se dosage) and **Figure 5** (effects of Se source). The proportion of sheep (%) in each FR-severity category within each treatment group is shown at wk 0, 20, 28, 40, and 60. There were no consistent effects of Se source and dosage on FR prevalence and severity over the 62 week treatment period. Overall, FR severity and prevalence decreased from 0 to 20 wk (50±3% to 30±3%; P < 0.0001). The effect of Se treatment on FR prevalence was influenced by the presence of FR disease ($P_{Interaction}$ = 0.04). Selenium source and dosage did not influence the interaction. Ewes that were healthy at baseline and received Se treatment had lower FR prevalence at wk 20 (FR score II or greater) than ewes that were healthy at baseline and received no Se

($8\pm3\%$ versus $29\pm13\%$, respectively; $P = 0.03$). No differences in FR prevalence were observed in ewes that had FR at baseline and received Se treatment ($28\pm4\%$) or no Se ($20\pm11\%$).

After the ewes were moved from pasture into the barn for lambing at wk 20, the FR severity and prevalence increased in all ewes, to $82\pm2\%$ at wk 28 ($P < 0.0001$). Ewes receiving Se-yeast ($77\pm5\%$) or no Se ($72\pm8\%$) had lower FR prevalence than ewes receiving Na-selenite ($90\pm3\%$; both $P = 0.02$) (**Figure 5**). No dosage-associated differences were observed.

Once the ewes were returned to pasture at wk 30, the FR prevalence decreased from $82\pm2\%$ at wk 28 to $39\pm3\%$ at wk 40 ($P < 0.0001$). In ewes previously affected with FR at wk 28, the effect of Se treatment on FR prevalence (FR score II or greater) at wk 40 differed by Se source and dosage. Ewes receiving Na-selenite at increasing dosages responded differently than ewes receiving Se-yeast at increasing dosages ($P_{\text{Interaction}} = 0.02$). The FR prevalence was higher in previously FR-affected ewes receiving

the highest Na-selenite dosage compared with the two lower Na-selenite dosages (4.9 mg/wk: $21\pm8\%$; 14.7 mg/wk: $11\pm6\%$; 24.5 mg/wk: $52\pm10\%$; $P = 0.002$), but lower with supranutritional Se-yeast supplementation compared with the lowest Se-yeast dosage (4.5 mg/wk: $45\pm11\%$; 14.7 mg/wk: $20\pm9\%$; 24.5 mg/wk: $32\pm10\%$; $P = 0.04$), with values for ewes receiving no Se being $20\pm9\%$ and for ewes receiving Na-selenate $29\pm9\%$.

At wk 28, the more severely FR-affected ewes were treated with parenteral oxytetracycline (52% of all ewes). This contributed to the overall decrease in FR prevalence at wk 40, to $39\pm3\%$ ($P < 0.0001$). No significant differences in parental oxytetracycline treatment were observed for individual treatment groups at wk 28 (**Figure 6**); however; more ewes with supranutritional Na-selenite treatment were treated with oxytetracyline than ewes receiving the lowest Na-selenite dosage ($P = 0.05$). In oxytetracycline-treated ewes, the effect of Se treatment on FR prevalence at wk 40 (FR score II or greater) also differed by Se source and dosage. Again,

Table 1. Whole-blood Se concentrations (ng/mL) in ewes after weekly oral drenching with no Se, inorganic Se (Na-selenate or Na-selenite), and organic Se (Se-yeast) at varying supplementation rates (4.9, 14.7, and 24.5 mg Se/wk; Na-selenate only at 8.95 mg Se/wk) for 62 wk.[1]

Se-Source	No-Se	Na-Selenate	Na-Selenite			Se-Yeast				P-value
										Oral Selenium Drench
Se-Dosage, mg Se/wk	0	8.95	4.9	14.7	24.5	4.9	14.7	24.5	SEM[2]	Overall
Week 0										
Healthy N[3]	15	15	15	15	15	15	15	15		
LSMean[4]	259	258	241	253	249	264	240	240	12	0.75
Footrot N[5]	15	15	15	15	15	15	15	15		
LSMean[4]	241	232	227	253	235	242	227	225	12	0.71
Week 14										
Healthy N	19	23	20	21	20	23	19	20		
LSMean	176	307	282	359	360	358	485	645	13	<0.0001
Footrot N	10	7	10	9	9	7	9	9		
LSMean	180	318	293	355	368	383	509	636	22	<0.0001
Week 27										
Healthy N	8	4	4	2	3	6	7	7		
LSMean	84	364	292	388	350	349	507	737	41	<0.0001
Footrot N	21	26	26	28	26	23	21	22		
LSMean	109	316	297	361	383	388	529	691	13	<0.0001
Week 40										
Healthy N	22	14	16	22	11	15	21	17		
LSMean	50	289	277	364	369	342	485	656	16	<0.0001
Footrot N	6	14	12	8	17	12	6	11		
LSMean	61	292	275	331	341	341	519	612	23	<0.0001
Week 54										
Healthy N	18	11	15	21	12	13	10	10		
LSMean	53	296	271	353	371	361	550	768	15	<0.0001
Footrot N	9	13	13	9	15	13	17	14		
LSMean	62	311	282	364	356	369	586	775	15	<0.0001

[1]Adapted from [1]; numbers differ because all sheep are included here and classified as healthy or footrot-affected.
[2]The largest SEM of the 8 treatment groups is shown.
[3]Number of healthy sheep in each treatment group.
[4]Least squares means.
[5]Number of footrot-affected sheep in each treatment group.

Table 2. Serum-Se concentrations (ng/mL) in ewes after weekly oral drenching with no Se, inorganic Se (Na-selenate or Na-selenite), and organic Se (Se-yeast) at varying supplementation rates (4.9, 14.7, and 24.5 mg Se/wk; Na-selenate only at 8.95 mg Se/wk) for 62 wk.[1]

		Oral Selenium Drench								P-value
Se-Source	No-Se	Na-Selenate	Na-Selenite			Se-Yeast				
Se-Dosage, mg Se/wk	0	8.95	4.9	14.7	24.5	4.9	14.7	24.5	SEM[2]	Overall
Week 0										
Healthy N[3]	15	15	15	15	15	15	15	15		
LSMean[4]	108	106	106	94	104	104	99	97	6	0.62
Footrot N[5]	15	15	15	15	15	15	15	15		
LSMean[4]	90	92	87	89	94	89	92	83	6	0.92
Week 14										
Healthy N	19	23	20	21	20	23	19	20		
LSMean	62	142	123	162	165	168	217	250	6	<0.0001
Footrot N	10	7	10	9	9	7	9	9		
LSMean	66	154	127	160	173	189	205	260	9	<0.0001
Week 27										
Healthy N	8	4	4	2	3	6	7	7		
LSMean	48	128	129	162	162	142	182	226	15	<0.0001
Footrot N	21	26	26	28	26	23	21	22		
LSMean	58	136	120	151	165	147	185	218	4	<0.0001
Week 40										
Healthy N	22	14	16	22	11	15	21	17		
LSMean	34	139	125	154	172	151	199	236	6	<0.0001
Footrot N	6	14	12	8	17	12	6	11		
LSMean	31	141	127	145	160	144	204	243	8	<0.0001
Week 54										
Healthy N	18	11	15	21	12	13	10	10		
LSMean	35	139	124	156	176	172	221	296	9	<0.0001
Footrot N	9	13	13	9	15	13	17	14		
LSMean	46	150	141	166	179	159	221	293	10	<0.0001

[1]Adapted from [1]; numbers differ because all sheep are included here and classified as healthy or footrot-affected.
[2]The largest SEM of the 8 treatment groups is shown.
[3]Number of healthy sheep in each treatment group.
[4]Least squares means.
[5]Number of footrot-affected sheep in each treatment group.

ewes receiving Na-selenite at increasing dosages responded differently than ewes receiving Se-yeast at increasing dosages ($P_{Interaction} = 0.05$), although with fewer numbers of sheep in this oxytetracycline-treatment subgroup, significance was more difficult to demonstrate. In oxytetracyline-treated ewes, FR prevalence at wk 40 tended to be higher in ewes receiving the highest Na-selenite dosage compared with the two lower Na-selenite dosages (4.9 mg/wk: 9±9%; 14.7 mg/wk: 17±9%; 24.5 mg/wk: 39±12%; $P = 0.06$), but lower with supranutritional Se-yeast supplementation compared with the lowest Se-yeast dosage (4.5 mg/wk: 41±12%; 14.7 mg/wk: 31±13%; 24.5 mg/wk: 21±11%; $P = 0.29$). Ewes receiving no Se (33±14%) or Na-selenate (29±13%) had higher FR prevalence than the overall mean of oxytetracycline-treated ewes (28±4%).

The overall FR prevalence increased again between wk 40 and wk 60, to 48±3% ($P = 0.02$), despite a second oxytetracycline treatment of the more severely affected ewes at wk 40 (16% of all ewes), which dropped their FR prevalence (FR score II or greater) from 100% to 59±9%. More ewes receiving Na-selenite at the highest dosage had to be treated with oxytetracycline at wk 40 than ewes receiving no Se, Na-selenite at both lower dosages, or Se-yeast at 14.5 mg/wk (all $P \leq 0.05$; **Figure 6**). Across all groups, the lowest FR prevalences were observed in ewes receiving no Se (26±9%) and 14.7 mg Na-selenite/wk (27±8%), whereas FR prevalences for the other treatment groups (8.95 mg Na-selenate: 50±8%; 4.5 mg Na-selenite/wk: 43±10%; 24.5 mg Na-selenite/wk: 48±10%; 4.5 mg Se-yeast/wk: 50±10%; 14.7 mg Se-yeast/wk: 63±9%; 24.5 mg Se-yeast/wk: 52±10%) were similar or greater than the overall mean (48±3%). Ewes receiving Se-yeast had on average a greater FR prevalence than ewes receiving Na-selenite ($P = 0.03$) or no Se ($P = 0.01$). Whereas FR prevalence was similar in ewes receiving different dosages of Se-yeast, ewes receiving 14.7 mg Na-selenite/wk had a lower FR prevalence than ewes receiving 4.9 or 24.5 mg Na-selenite ($P = 0.05$).

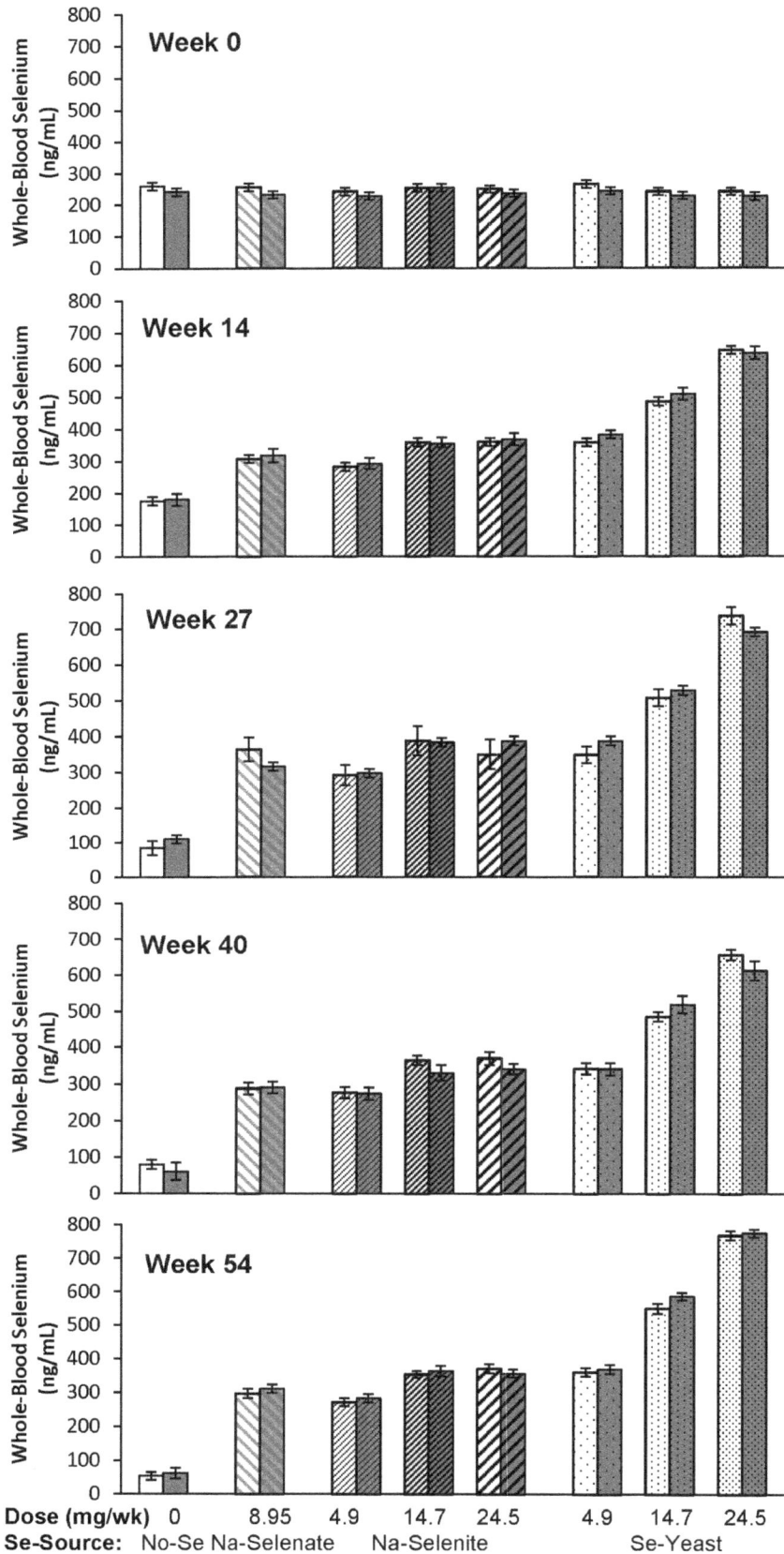

Figure 2. The effect of Se-source, Se-dosage, and foot rot status on whole-blood Se concentrations in sheep. Whole-blood Se concentrations were measured after 0, 14, 27, 40, and 54 wk of Se supplementation, and foot rot severity was assessed after 0, 20, 28, 40, and 60 wk of Se supplementation in ewes receiving no Se treatment, Na-selenate at a dosage rate of 8.95 mg Se/wk per ewe, or Na-selenite and Se-Yeast at 4.9, 14.7, or 24.5 mg Se/wk per ewe for 62 wk. Whole-blood Se concentrations are shown as separate bars for healthy sheep (lighter background) and for sheep with foot rot (darker background). At baseline (wk 0), no significant treatment group differences were observed; however, healthy ewes had higher WB-Se concentrations than ewes with FR ($P = 0.01$). After treatments started, group differences by foot-rot status subsided, whereas Se-source and Se-dosage affected WB-Se concentrations. Both healthy and FR-affected ewes that received no Se treatment had decreased WB-Se concentrations ($P < 0.0001$). Supranutritional Se-yeast supplementation increased WB-Se concentrations linearly with dosage ($P < 0.0001$), whereas ewes receiving supranutritional Na-selenite supplementation at 14.7 and 24.5 mg Se/wk achieved similar WB-Se concentrations as ewes receiving 4.9 mg/wk of Se-yeast.

Overall, the proportion of ewes that had a worse FR score at wk 60 compared with wk 40 did not differ significantly between groups. There was, however, an interaction between Se source and previous FR presence on change in FR score between wk 40 and wk 60 ($P = 0.02$). Whereas in healthy ewes the number with worse scores at wk 60 was lower in ewes receiving no Se and Na-selenate compared with ewes receiving Na-selenite and Se-yeast (no Se: 18±8%; Na-selenate: 14±10%; Na-selenite: 35±7%; Se-yeast: 42±7%; $P = 0.03$), the reverse was true in previously FR affected ewes (no Se: 50±22%; Na-selenate: 50±14%; Na-selenite: 24±7%; Se-yeast: 23±8%; $P = 0.04$). In other words, Na-selenite and Se-yeast treatments decreased FR severity at wk 60 in ewes affected with FR at wk 40 compared with ewes receiving no Se or Na-selenate.

In general, older ewes had a higher FR severity ($P < 0.0001$) and prevalence ($P < 0.0001$). For example, ewes 5 years and older at lambing had a higher FR severity score and FR prevalence than younger ewes at wk 60 (60±5% versus 39±4%, respectively).

Neutrophil bacterial killing and the 30 minute DTH skin test response are innate immunity measurements

The effect of Se treatment on neutrophil bacterial killing was influenced by the presence of FR disease ($P_{Interaction} = 0.02$; **Figure 7A**). Weekly oral Se drenching improved neutrophil bacterial killing in FR-affected ewes from 40±4% to 50±1% ($P = 0.007$), percentages which were similar to healthy ewes receiving no-Se (49±3%) or healthy ewes receiving Se treatment (49±1%). Compared with no Se treatment, neutrophil bacterial killing was greater in FR-affected ewes receiving Na-selenate, Na-selenite, and Se-yeast (all $P < 0.05$) and tended to be greater in FR-affected ewes receiving Se-yeast compared with Na-selenite ($P = 0.09$; **Figure 7B**). Supranutritional Se treatment did not provide additional benefits, regardless of Se source (data not shown).

Neutrophil bacterial killing decreased linearly in ewes with age (< 5 years: 54±1%; 5 years: 49±2%; >5 years: 43±2%; $P < 0.0001$). The age-associated decline in neutrophil bacterial killing was delayed with Se treatment (< 5 years and no-Se: 50±3%; < 5 years and Se treatment: 54±1%; 5 years and no-Se: 42±5%; 5 years and Se treatment: 50±2%; >5 years and no-Se: 39±4%; >5 years and Se: 44±2%).

The effect of FR status on 30 minute responses to KLH was not consistent. Ewes affected with FR had smaller 30 min ear-thickness responses to KLH intradermal injection than healthy ewes (4.11±0.08 versus 4.35±0.09 mm; $P = 0.05$), which was not influenced by Se treatment ($P_{Interaction} = 0.94$; **Figure 8A**). Such an effect was not observed for 30 min ear-wheal diameter responses to KLH injection ($P = 0.95$; **Figure 8B**). Selenium dosage, but not Se source affected 30 minute responses to KLH. Se supplementation at lower dosages (8.95 mg/wk Na-selenate, 4.9 mg/wk Na-selenite, and 4.9 mg Se-yeast/wk) had smaller ear-thickness responses than supranutritional Se supplementation (14.7 and 24.5 mg/wk Na-selenite and Se-yeast; $P = 0.01$) or no

Se supplementation ($P = 0.01$; **Figure 8C**). A similar effect was observed for ear-wheal diameter response, as Se supplementation at lower dosages had or tended to have smaller ear-wheal diameter responses than supranutritional Se supplementation ($P = 0.03$) or no Se supplementation ($P = 0.07$; **Figure 8D**). In addition, Polypay ewes had smaller ear-wheal diameter responses than Suffolk and Suffolk cross ewes (10.1±0.3 versus 13.0±0.4 mm; $P < 0.0001$).

Cell-mediated immunity as measured by the delayed-type hypersensitivity (DTH) skin test with KLH

FR-affected sheep demonstrated suppressed CMI at 24 h after intradermal KLH challenge. For the DTH ear thickness response, there was no significant effect of ewe FR status ($P = 0.15$; **Figure 9A**). Significance was achieved, however, for the DTH ear wheal diameter response as sheep affected with FR had an attenuated DTH response compared with healthy sheep (overall $P = 0.03$), which was significant only at 24 h (39.6±2.3 versus 48.2±2.2 mm, respectively; $P = 0.007$; $P_{Interaction} = 0.05$; **Figure 9B**). Selenium source and dosage did not significantly alter the ear thickness response (**Figure 9C**) or the ear wheal diameter response (**Figure 9D**).

Humoral immunity as measured by the KLH antibody titer

Similar to the results for neutrophil bacterial killing, the effect of Se treatment on KLH antibody titer tended to be influenced by the presence of FR disease ($P_{Interaction} = 0.09$; **Figure 10A**). Weekly oral Se drenching improved KLH antibody titer in FR-affected ewes from 13.15±0.31 to 13.97±0.10 ($P = 0.01$), titers which were similar to healthy ewes receiving no-Se (13.82±0.22) or healthy ewes receiving Se treatment (13.97±0.11). Compared with no Se treatment, KLH antibody titers were greater in all Se-treatment groups (all $P < 0.06$) except for the 24.5 mg/wk Na-selenite group (**Figure 10B**).

Discussion

Effect of dietary Se depletion, different dietary Se sources, and supranutritional Se-supplementation dosages on immune responses in healthy and FR-affected ewes

The immune system has two functional divisions: innate and adaptive immunity. Both divisions involve various blood-borne factors (e.g., complement, antibodies, and cytokines) and cells (e.g., neutrophils, lymphocytes, and macrophages). Neutrophils are the most numerous and important cellular component of innate immunity. Their primary functions are phagocytosis and destruction of microorganisms. They serve as the body's first line of defenses against invading microorganisms. Phagocytosed bacteria are rapidly killed by proteolytic enzymes (e.g., myeloperoxidase), antimicrobial proteins, and ROS when membrane-bound granules

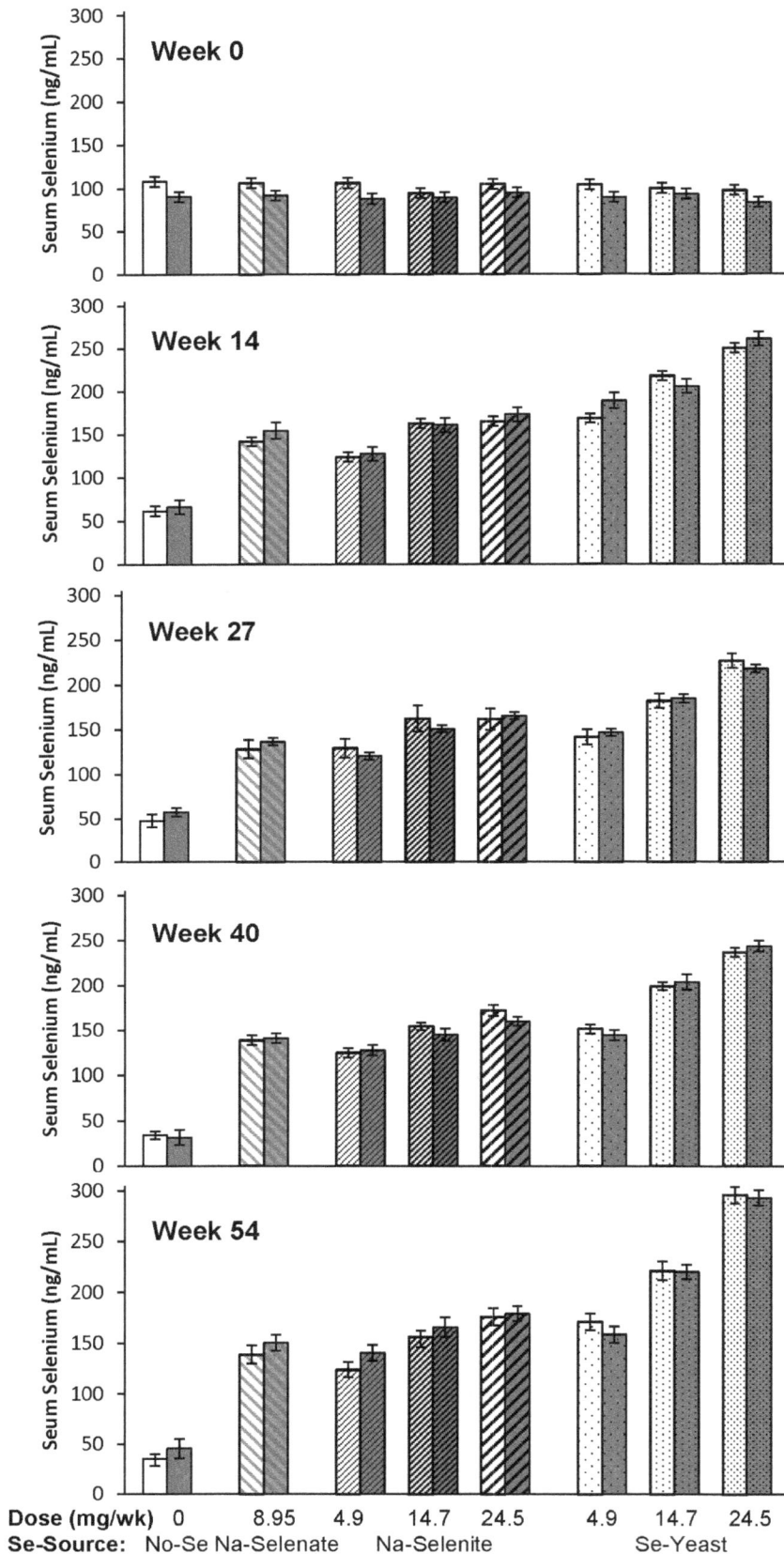

Figure 3. The effect of Se-source, Se-dosage, and foot rot status on serum-Se concentrations in sheep. Serum-Se concentrations were measured after 0, 14, 27, 40, and 54 wk of Se supplementation, and foot rot severity was assessed after 0, 20, 28, 40, and 60 wk of Se supplementation

in ewes receiving no Se treatment, Na-selenate at a dosage rate of 8.95 mg Se/wk per ewe, or Na-selenite and Se-Yeast at 4.9, 14.7, or 24.5 mg Se/wk per ewe for 62 wk. Serum-Se concentrations for each treatment group are shown as separate bars for healthy sheep (lighter background) and for sheep with foot rot (darker background). At baseline (wk 0), no significant treatment group differences were observed; however, healthy ewes had higher serum-Se concentrations than ewes with FR ($P = 0.01$). After treatments started, group differences by foot-rot status subsided, whereas Se-source and Se-dosage affected serum-Se concentrations. Both healthy and FR-affected ewes that received no Se treatment had decreased serum-Se concentrations ($P < 0.0001$). Supranutritional Se-yeast supplementation increased serum-Se concentrations linearly with dosage ($P < 0.0001$), whereas ewes receiving supranutritional Na-selenite supplementation at 14.7 and 24.5 mg Se/wk achieved similar serum-Se concentrations as ewes receiving 4.9 mg/wk of Se-yeast.

fuse with phagocytic vesicles. To assess innate immunity of neutrophils, an *ex vivo* biologic assay was performed using *E. coli* and measuring percent bacterial killing. Neutrophils from healthy sheep not receiving Se supplementation demonstrated higher percent bacterial killing compared with neutrophils from FR-affected sheep not receiving Se supplementation, consistent with our results published previously for another flock [26]. In the current study we were able to show that Se supplementation, regardless of source or dosage, restored neutrophil bacterial killing in FR-affected ewes back to percentages consistent with Se-supplemented or non-supplemented healthy ewes. The bacterial killing percentage for neutrophils tended to be greater in FR-affected ewes receiving Se-yeast compared with Na-selenite. We saw no clear benefit from supranutritional Se dosages in neutrophil bacterial killing.

In a companion paper [31], we reported on neutrophil-related gene expression profiles from these ewes and showed a U-shaped relationship with supranutritional Se-yeast supplementation and Se depletion both enhancing gene expression of L-Selectin (L-Sel), interleukin 8 receptor (IL-8R), and toll-like receptor 4 (TLR4). All three are essential for bacterial recognition and neutrophil migration, phagocytosis, and killing. In addition, expression of selenoprotein S (SEPS1) and glutathione peroxidase 4 (GPx4), which are both involved in controlling inflammation, was increased for both with supranutritional Se-yeast supplementation and Se depletion [31]. When we correlated (nonparametric spearman correlation) neutrophil bacterial killing activity in this study with the previously reported neutrophil gene expression profiles for ewes receiving 0, 4.9, 14.7, and 24.5 mg Se/wk, significant negative correlations were observed for GPx4 ($r = -0.24$; $P = 0.01$) and SEPS1 ($r = -0.26$; $P = 0.01$). Both genes act as anti-inflammatory agents: GPx4 promotes cell survival and blocks eicosanoid synthesis, including cyclooxygenase (COX) II [36,37], and SEPS1 protects immune cells from apoptosis and decreases the release of the proinflammatory cytokines IL-6 and TNF-α [38,39]. Others have shown in mice that moderate selenium deficiency down-regulates inflammation-related genes and reduces myeloperoxidase and lysozyme activities in Se-restricted leukocytes [40]. Myeloperoxidase is expressed in neutrophils and monocytes and generates ROS that are important for antimicrobial and cytotoxic effects. Thus, Se-supplementation may restore optimal neutrophil bacteria killing in the presence of FR disease, but healthy ewes may have different requirements for dietary Se.

Neutrophil bacterial killing also decreased linearly in ewes with age. Ewes six years and older (43%) had similar killing percentages as FR-affected ewes receiving no Se supplement (40%). Se supplementation delayed the age-associated decline in neutrophil killing ability, as ewes receiving Se had similar killing percentages as non supplemented ewes that were one age category younger. We have previously shown in aged Beagle dogs that older dogs have a significant decrease in neutrophil bacterial killing and, in addition, have lower levels of mRNA for neutrophil-related gene expression compared with younger dogs, including mRNA for myeloperoxidase [41], which may contribute to increased mor-

bidity and mortality with aging. These results are consistent with our finding in this study that older ewes had a higher FR severity and prevalence. Neutrophils from older humans have also been shown to have less phagocytic ability than those from younger adults, and the respiratory burst was altered in neutrophils from aged participants [42].

Tests used to assess adaptive immunity include measuring an antibody titer in response to sensitization/immunization (humoral immune response). The ewe is injected with a novel protein (e.g., KLH) that elicits an immune response. Following sensitization, antibody titers to KLH are measured. Consistent with our results in another flock [26] and results for bacterial killing by neutrophils in the current study, healthy ewes receiving no Se supplementation had higher KLH antibody titers compared with FR affected ewes receiving no Se supplementation. In the current study, we were also able to show that Se supplementation, regardless of source, restored KLH antibody titers in FR-affected ewes back to titers consistent with Se-supplemented and non-supplemented healthy ewes. Others have shown that Se supplementation increases antibody production in Se deficient sheep (reviewed in [16]). Our results suggest that Se supplementation may also improve antibody titers in response to a novel antigen in Se-replete yet FR-affected sheep, which is consistent with results we demonstrated in Se-replete adult beef cattle [16] and weaned beef calves (paper under review).

We did see a disadvantage of supranutritional Se treatment with Na-selenite (24.5 mg/wk dosage) in that KLH titers in FR-affected ewes receiving this Se source and dosage remained suppressed. We also observed at this Na-selenite dosage a higher propensity for FR lesions at several time points. Furthermore, WB-Se concentrations did not increase from the 14.7 mg/wk to the 24.5 mg/wk Na-selenite. In a companion paper, we reported a lower transfer of IgG from ewe colostrum to lamb serum if ewes received 24.5 versus 4.9 mg Se/wk as Na-selenite. [18]. These results suggest that Na-selenite may have potentially deleterious effects at higher dosages. In support, comparative toxicosis studies in sheep showed that oxidative effects were greater for Na-selenite than equivalent amounts of SeMet [43], which is the main selenocompound in Se-yeast [44]. Thus, the efficacy of supranutritional treatment with Na-selenite at 5 times the maximal FDA-permitted level requires further study.

The DTH test, which is also known as a type IV hypersensitivity reaction, is another test used to assess the adaptive immune response. This test provides a general measure of CMI. Professional antigen presenting cells, e.g., dendritic cells, present antigen to T lymphocytes. This results in antigen specific activation of T lymphocytes in local tissues. Inflammatory cytokines produced by these stimulated T lymphocytes cause other mononuclear cells (lymphocytes and macrophages) to migrate to the area and proliferate. To perform this test, foreign antigen is injected under the epidermis of the skin. The immune system responds to this antigen by producing a small raised wheal that can be measured 24 to 96 h after injection. The larger and thicker the wheal, the greater is the DTH response.

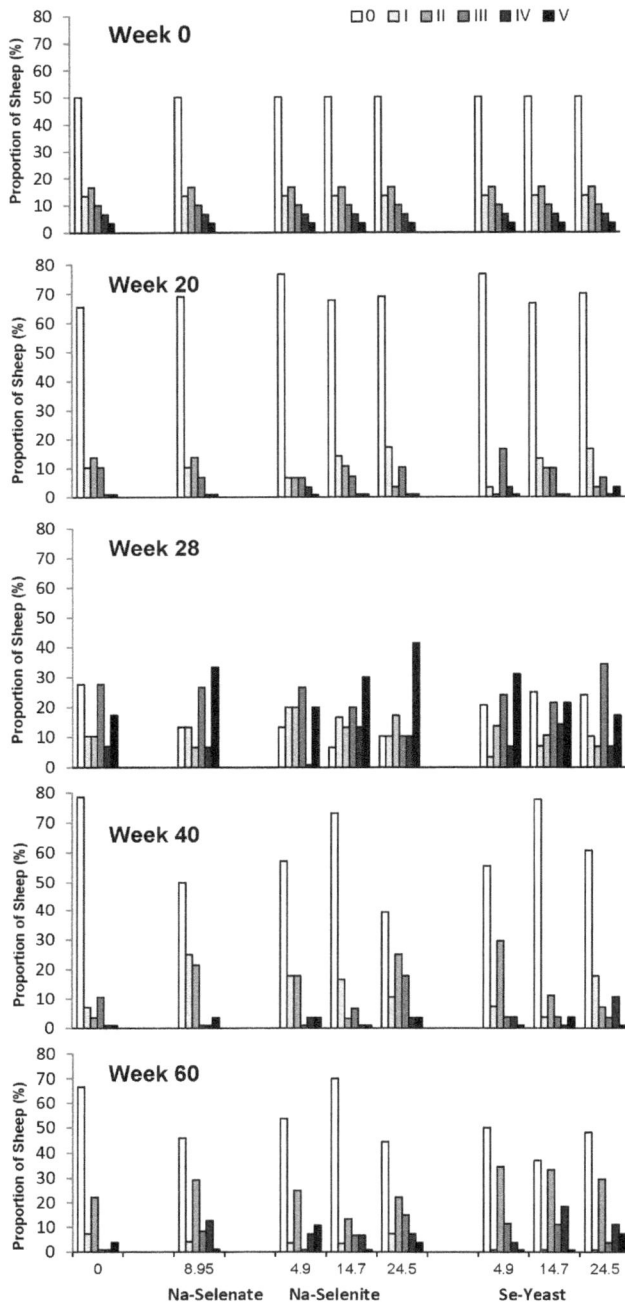

Figure 4. The effect of Se-source and dosage on foot rot (FR) severity in sheep. Foot rot severity was assessed after 0, 20, 28, 40, and 60 wk of Se supplementation in ewes receiving no Se treatment, Na-selenate at a dosage rate of 8.95 mg Se/wk per ewe, or Na-selenite and Se-Yeast at 4.9, 14.7, or 24.5 mg Se/wk per ewe for 62 wk. The proportion of sheep (%) in each FR-severity category within each treatment group is shown (scale 0 to V; with no FR = 0 being the lowest category). The scoring and categorization for FR is described in detail in the Materials and Methods. There were no consistent effects of Se source and dosage on FR prevalence and severity across time.

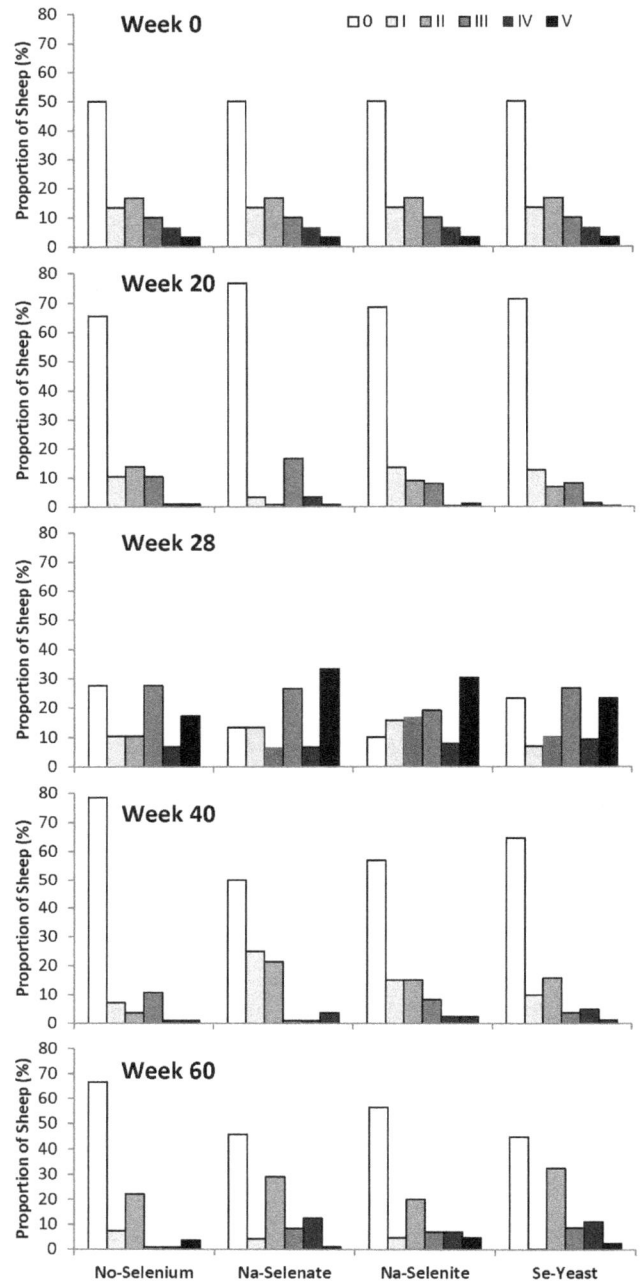

Figure 5. The effect of Se-source on foot rot (FR) severity in sheep. Foot rot severity was assessed after 0, 20, 28, 40, and 60 wk of Se supplementation in ewes receiving no Se treatment, Na-selenate, Na-selenite, and Se-Yeast for 62 wk. All dosage groups were combined for each chemical source of Se treatment. The proportion of sheep (%) in each FR-severity category within each treatment group is shown (scale 0 to V; with no FR = 0 being the lowest category). The scoring and categorization for FR is described in detail in the Materials and Methods. There were no consistent effects of Se source on FR prevalence and severity across time.

In our study, FR-affected sheep demonstrated suppressed CMI at 24 h after intradermal KLH challenge, consistent with our results in another flock [26]. In the current study, this response was significant using ear wheal diameter measurements, and although numerically true for ear thickness measurements, the latter results were not significant. We reported previously in another flock that FR-affected ewes with WB-Se concentrations above 250 ng/mL at the time of the DTH assay had greater ear thickness and ear wheal diameter responses than FR-affected ewes with WB-Se concen-

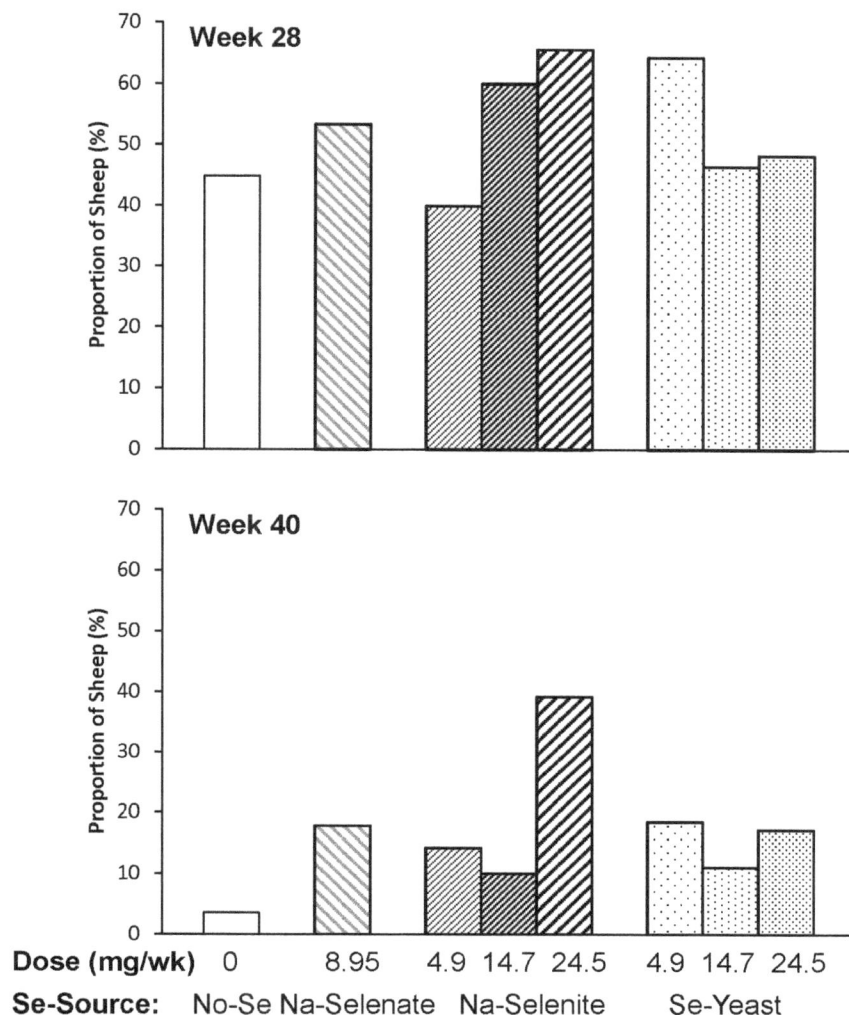

Figure 6. The effect of Se-source and dosage on sheep requiring oxytetracyline treatment. Foot rot severity was assessed after 0, 20, 28, 40, and 60 wk of Se supplementation in ewes receiving no Se treatment, Na-selenate at a dosage rate of 8.95 mg Se/wk per ewe, or Na-selenite and Se-Yeast at 4.9, 14.7, or 24.5 mg Se/wk per ewe for 62 wk. If a sheep had one foot with a score of 4, or one foot with a score of 3 and a second foot with a score of 2 or greater, or if all 4 feet had FR, then 20 mg/kg oxytetracycline was administered subcutaneously (Liquamycin LA-200; Pfizer Animal Health, Exton, PA) at the 28- and 40-wk foot trimming sessions. No significant differences in parental oxytetracycline treatment were observed for individual treatment groups at 28 wk; however; more ewes with supranutritional Na-selenite treatment were treated with oxytetracycline than ewes receiving the lowest Na-selenite dosage (*P* = 0.05). At 40 wk, more ewes receiving Na-selenite at the highest dosage had to be treated with oxytetracycline than ewes receiving no Se, Na-selenite at both lower dosages, or Se-yeast at 14.5 mg/wk (all *P* ≤ 0.05).

trations below 250 ng/mL [26]. In the current study we saw no effect of Se source and dosage on DTH responses. The DTH test may be too insensitive under field conditions to detect a difference in CMI with Se supplementation. It is well known that large variation exists in immune function measures, even among healthy animals. For example, differences in genetics, age, diet, body condition scores, stress, levels of exercise, and infectious disease history are important contributors to observed variation [8], such that demonstrating a consistent improvement in immune function with Se supplementation is challenging. In addition, differences in DTH methodology, with variable injection sites, response times, and measurement techniques may account for differences between studies. Nonetheless, finding once again an attenuated T-lymphocyte response in FR-affected sheep is important, and could be the result of decreased activation, migration, proliferation, or a combination of these effects. Se supplementation alone may or

may not (current study) be sufficient to improve the DTH response under FR-disease conditions.

We also assessed the 30-min skin test response following intradermal KLH challenge in healthy and FR-affected sheep to determine if results differed for the type I hypersensitivity reaction normally induced by histamine and inflammatory cytokines. The KLH antigen stimulates inflammatory cytokine production. We found that FR-affected ewes had attenuated 30 min ear-thickness responses to KLH intradermal injection compared with healthy ewes, consistent with results of a previous study [26]. The KLH response was not influenced by source of Se treatment. Footrot bacterial infection could suppress the Type I hypersensitivity response by affecting the release of histamine, or virulence factors such as leukotoxin, endotoxin, haemolysin, haemagglutinin, and adhesion.

A) All Sheep

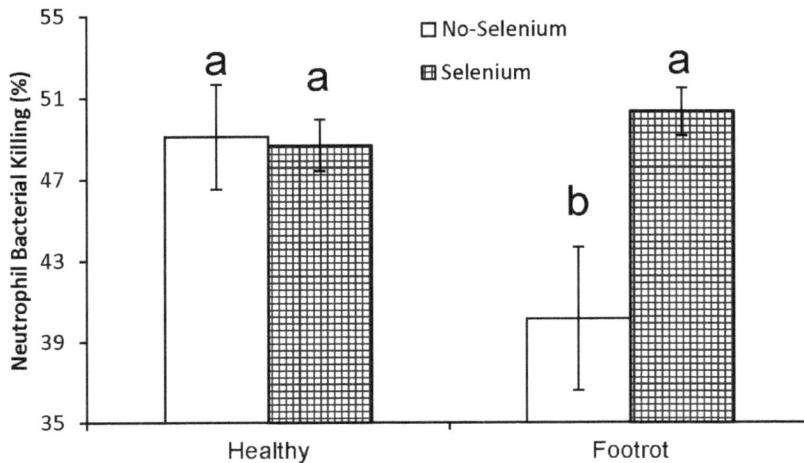

B) Footrot-affected sheep

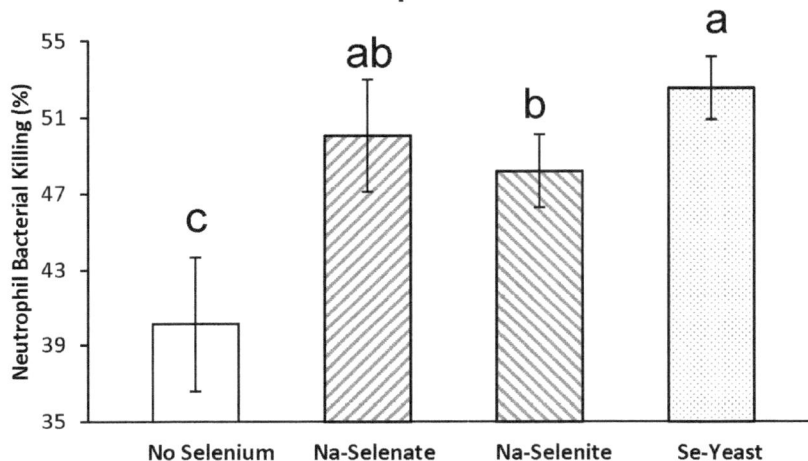

Figure 7. The effect of Se-source on ex vivo neutrophil bacterial killing in sheep. Ex vivo neutrophil bacterial killing was assessed after 60 wk of Se supplementation in healthy and footrot (FR)-affected ewes receiving no Se treatment, Na-selenate, Na-selenite, and Se-Yeast for 62 wk. **A)** Weekly oral Se drenching improved neutrophil bacterial killing in FR-affected ewes to percentages that were similar to healthy supplemented or healthy non-supplemented ewes. **B)** In FR-affected ewes, Na-selenate, Na-selenite, or Se-yeast improved neutrophil bacterial killing compared with no Se supplementation. The effect tended to be greater in ewes receiving Se-yeast compared with Na-selenite. The three dosages of Na-selenite or Se-yeast were combined because they did not differ. Different superscripts indicate group differences at $P \leq 0.10$.

Histamine normally increases capillary permeability and relaxes vascular smooth muscle, allowing edema fluid accumulation. Influx of proinflammatory cytokines triggers production of ROS. When produced in excess, ROS are important mediators of cell and tissue injury (reviewed in Murr et al. [20]). As a component of the glutathione peroxidase family of enzymes, Se contributes to the reduction of hydroperoxides in cells. Glutathione peroxidase reduces ROS to less reactive metabolites, decreasing oxidant stress. Because Se is involved in redox reactions, and immune activation is usually associated with increased production of ROS by cells of the immune system, we hypothesized that ewes receiving supranutritional Se supplementation might have suppressed 30 min skin reactions compared with ewes receiving less Se supplementation. . We observed, however, a U-shaped relationship between Se dosage and the 30-min skin test responses to KLH, similar to what we reported in a companion paper for Se-yeast dosage and neutrophil gene expression [31]. Ewes receiving

either no Se supplementation or supranutitional Se supplementation both had accentuated rather than suppressed 30-min skin test responses to KLH compared with ewes receiving the maximum FDA-allowed levels. A U-shaped relationship between Se status and human health has been postulated in a review by Rayman [13], whereby supplemental Se intake may benefit people with low Se status (or in our case under certain disease conditions such as FR), but cautions that those with adequate to high Se status might be affected adversely and should not take Se supplements. Decreased inflammation and inflammation-dependent plasma cell tumors have been reported in Se-deficient mice [45]. Both Se deficiency and high levels of Se have been reported to decrease the incidence and progression of liver tumors in transgenic mice prone to liver cancer [46,47]. A vigorous 30-min skin response to a novel antigen, induced by histamine and proinflammatory gene products, may be necessary for a successful defense against FR-causing bacteria, as FR-affected sheep had a lower 30-min skin test

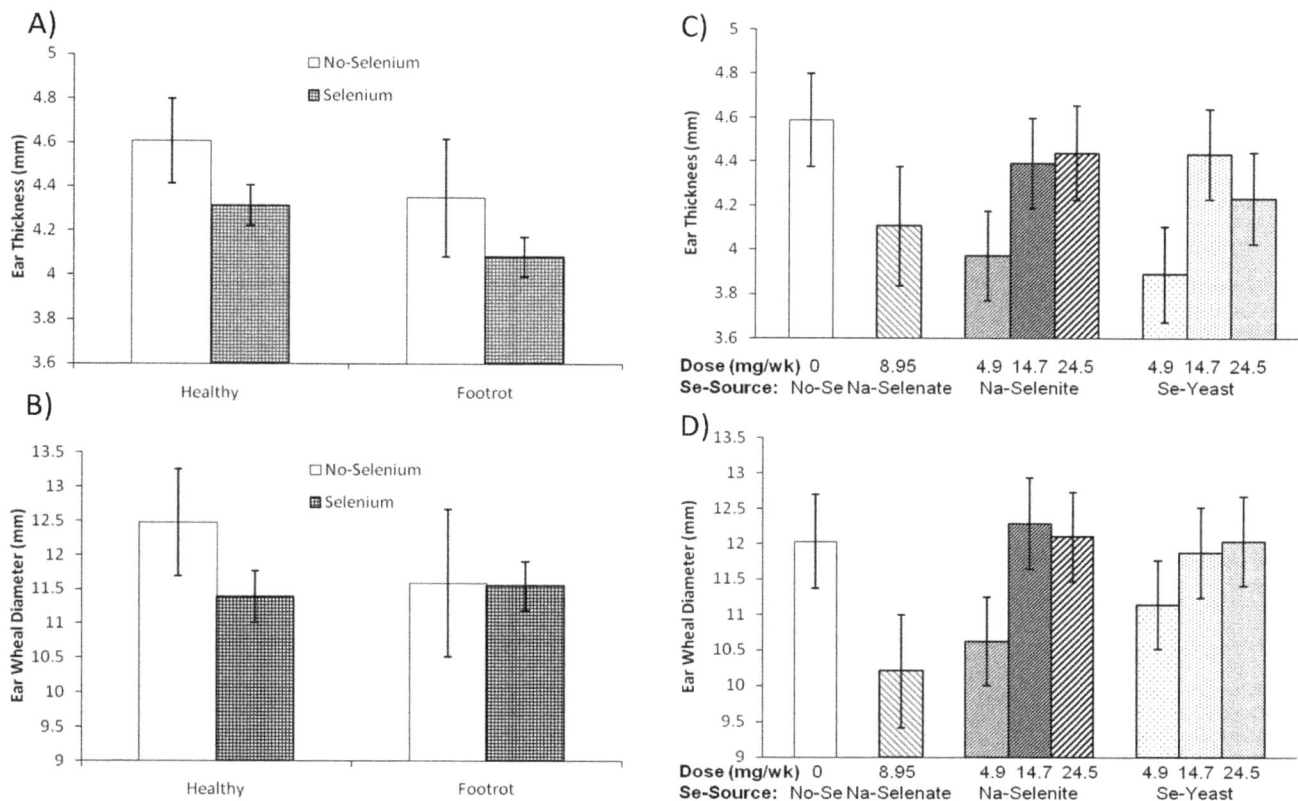

Figure 8. The effect of Se-source and dosage on 30-min skin response to keyhole limpet hemocyanin (KLH). The 30-min skin test response following KLH challenge was assessed after 52 wk of Se supplementation in healthy and footrot (FR)-affected sheep receiving no Se treatment, Na-selenate at a dosage rate of 8.95 mg Se/wk per ewe, or Na-selenite and Se-Yeast at 4.9, 14.7, or 24.5 mg Se/wk per ewe for 62 wk. **A)** Ewes affected with FR had smaller 30 min ear-thickness responses to KLH intradermal injection than healthy ewes (overall $P = 0.05$). **B)** The 30 min ear-wheal diameter response was not significantly affected by ewe FR status or Se supplementation. **C)** Selenium dosage, but not Se source affected the ear-thickness response, as Se supplementation at lower dosages (8.95 mg/wk Na-selenate, 4.9 mg/wk Na-selenite, and 4.9 mg Se-yeast/wk) had smaller ear-thickness responses than supranutritional Se supplementation (14.7 and 24.5 mg/wk Na-selenite and Se-yeast; $P = 0.01$) or no Se supplementation ($P = 0.01$). **D)** Selenium dosage, but not Se source affected the ear-thickness response, as Se supplementation at lower dosages had or tended to have smaller ear-wheal diameter responses than supranutritional Se supplementation ($P = 0.03$) or no Se supplementation ($P = 0.07$).

response than healthy sheep in this and another flock [26]. Our results suggest that optimal 30 min skin test responses vary depending on the underlying disease condition and the Se supplementation rate.

Effect of dietary Se sources and supranutritional Se-supplementation on whole-blood and serum-Se concentrations and FR morbidity in healthy and FR-affected ewes

In the current study, ewes affected with FR at baseline had lower WB- and serum-Se concentrations compared with healthy ewes, although mean concentrations were within the normal reference interval for adult sheep. [The normal reference interval for Se in WB of adult sheep > 700 days of age at the Michigan State University diagnostic laboratory is 150 to 500 ng/mL (T. Herdt, personal communication)]. This finding is consistent with our previously published observations in another flock [48]. In a companion paper, we reported that WB-Se and serum-Se concentrations increased linearly with supranutritional Se-yeast supplementation [1]. However, WB-Se concentrations in ewes receiving supranutritional Na-selenite supplementation reached a plateau similar to concentrations attained in ewes receiving 4.9 mg/wk of Se-yeast [1]. In the current analysis, we showed

that ewe FR status does not affect Se-supplementation induced increases in WB- or serum-Se concentrations irrespective of Se-source and Se-dosage. Thus, it is more likely that Se intake was lower in ewes with FR compared with healthy ewes at baseline because sheep affected with FR are less mobile and, therefore, unable or unwilling to consume as much Se-containing mineral supplement as healthy sheep [31] than the alternative hypothesis that Se requirements are higher in the presence of an infectious disease like FR because more Se is required for removal of reactive oxygen species (ROS) associated with inflammation [49].

We previously reported in another sheep flock that parenteral Se-supplementation in conjunction with routine control practices accelerated recovery from FR in sheep [48]. In the current study, even though we were able to raise WB- and serum-Se concentrations, Se supplementation did not consistently prevent FR nor accelerate recovery from FR over the 62 week treatment period compared with no Se supplementation. Selenium supplementation did decrease the percentage of healthy ewes that acquired FR during the first 20 weeks of the study; however, we did not observe a similar effect in wks 28, 40, and 60. In a companion paper, we reported that the greatest treatment success with oxytetracycline was observed in ewes receiving the highest Se-yeast dosage (24.5 mg Se/wk) compared with Se-yeast dosages of

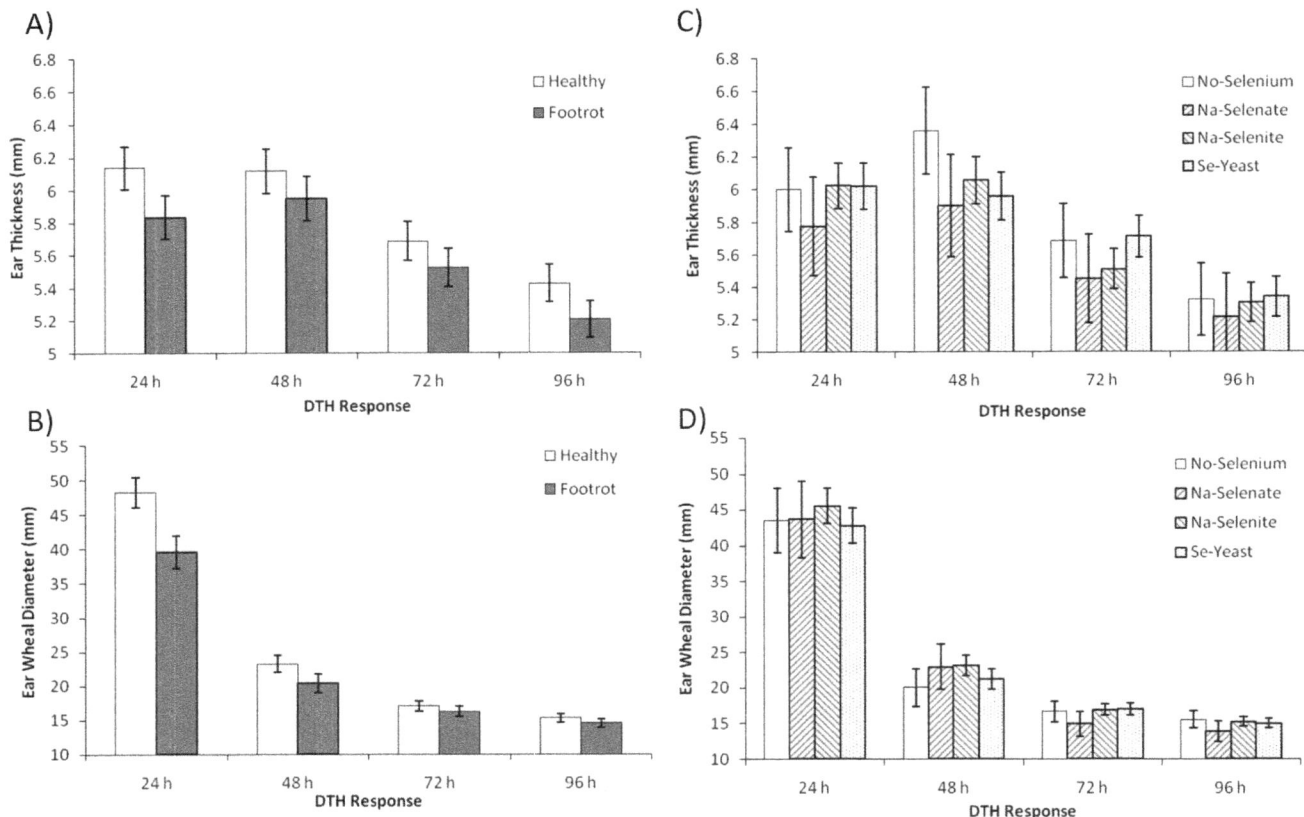

Figure 9. The effect of Se-source and dosage on delayed-type hypersensitivity response to keyhole limpet hemocyanin (KLH). Delayed-type hypersensitivity (DTH) skin test responses was assessed after 52 wk of Se supplementation in healthy and footrot (FR)- affected sheep receiving no Se treatment, Na-selenate at a dosage rate of 8.95 mg Se/wk per ewe, or Na-selenite and Se-Yeast at 4.9, 14.7, or 24.5 mg Se/wk per ewe for 62 wk. (All Se treatment groups were combined.) **A)** Ear thickness response was not significantly affected by ewe FR status ($P = 0.15$). **B)** Ear wheal diameter response was decreased in FR-affected sheep (overall $P = 0.03$), which was significant only at 24 h ($P = 0.007$; $P_{Interaction} = 0.05$). C) Ear thickness response and D) ear wheal diameter response were not significantly affected by Se source or Se dosage.

0, 4.9, or 14.5 mg Se/wk [31]. In the current analysis, we observed an improvement in FR prevalence in FR affected sheep receiving Na-selenite and Se-yeast supplementation compared with no Se and Na-selenate supplementation at wk 60, but not at earlier time points.

In general, the prevalence of FR changed according to the ewe production cycle and management practices. At baseline, ewes were selected for 50% FR prevalence, and treatment groups were stratified for FR severity and age of ewe. Ewes were kept on pastures and supplemented with grass hay during the breeding and gestation seasons. Footrot prevalence decreased to $30 \pm 2\%$ during this period. Ewes were moved from pasture into the barn for lambing around wk 20 and fed alfalfa hay and shelled corn. By wk 28 the FR prevalence had increased to $82 \pm 2\%$. Serum Se concentrations were decreasing by week 27 after increasing from week 0 to week 14. The decrease in blood Se concentrations reflected an increased Se transfer from ewes to lambs in late gestation and early lactation [1]. A corresponding decrease in WB-Se concentrations was not observed until week 40, likely because of the longer half-life of RBC.

Housing ewes in the barn at higher stocking density for the 3-month period around lambing (up to week 30) likely contributed to increased FR prevalence and severity. It is known that environmental factors play an important role in determining infection rate and progression of FR severity, with heavier infection rates

occurring under warm moist conditions [50-52]. *Dichelobacter nodosus*, one of the main organisms associated with FR is an anaerobic and fastidious bacterium that colonizes the interdigital epithelial tissue more readily during the wet seasons of spring, fall, and winter. Higher stocking density is also more conducive to heavy infection rates [51]. Crowding and moist bedding conditions could explain the higher FR prevalence in the flock in general during this time period.

Once the ewes were returned to pasture and eating grass forage at wk 30, the FR prevalence decreased to $39 \pm 3\%$ at wk 40. In part, this reflects the administration of oxytetracycline antibiotic to the more severely FR-affected ewes at wk 28 (52% of all ewes received oxytetracycline). Again, at wk 40, oxytetracycline treatment of the more severely affected ewes (16% of all ewes) was repeated. Although beneficial in helping decrease FR prevalence in these ewes from 100% to $59 \pm 9\%$, the overall FR prevalence increased between wk 40 and wk 60, to $48 \pm 3\%$.

Selenium is not recommended as the sole treatment for FR in sheep. Administration of topical or systemic antibiotics, foot paring, foot bathing in disinfectants, and vaccination with a commercially available vaccine for footrot (Footvax, MSD) containing multiple serotypes of *D. nodosus*, reviewed by Duncan et al. [53], have all been suggested for use in treating sheep with advanced FR lesions. Ewes that do not respond to treatments are often culled; this was not the case in our study, which explains why

A) All sheep

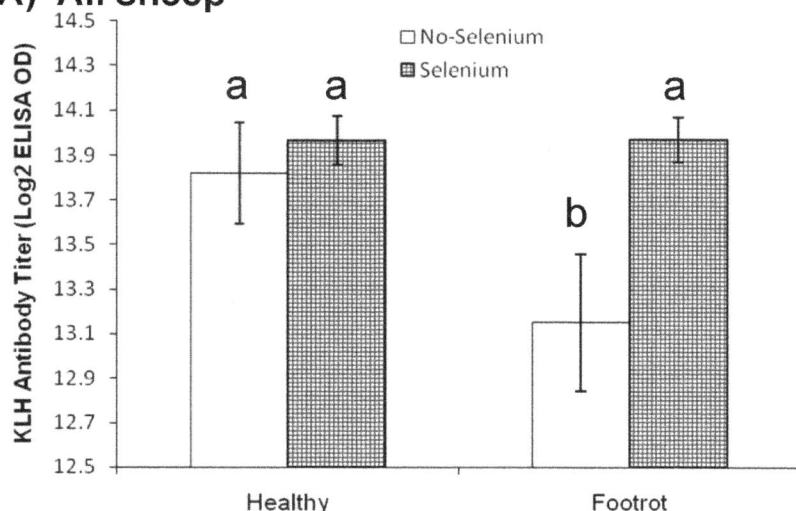

B) Footrot-affected sheep

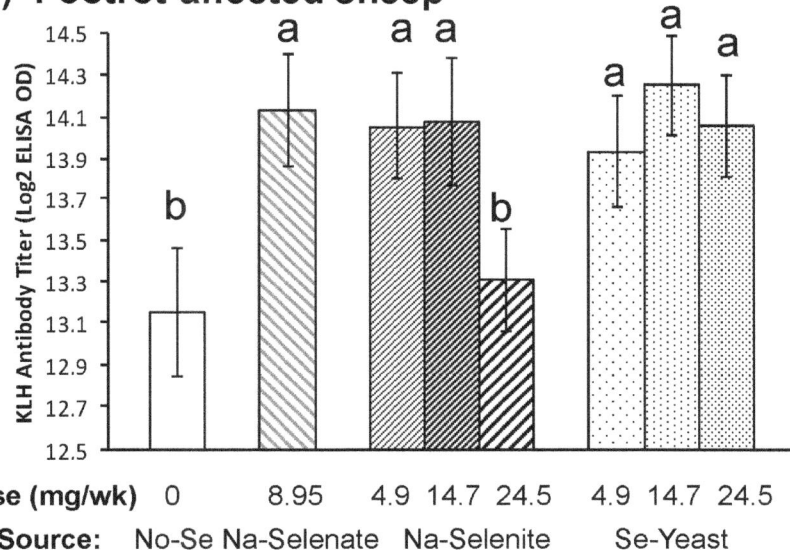

Figure 10. The effect of Se-source and dosage on antibody titers to keyhole limpet hemocyanin (KLH). Antibody titers to keyhole limpet hemocyanin (KLH) were assessed after 60 wk of Se supplementation in healthy and footrot (FR)-affected sheep receiving no Se treatment, Na-selenate at a dosage rate of 8.95 mg Se/wk per ewe, or Na-selenite and Se-Yeast at 4.9, 14.7, or 24.5 mg Se/wk per ewe for 62 wk. A) Weekly oral Se drenching improved KLH antibody titers in FR-affected ewes to titers that were similar to healthy ewes receiving no Se or healthy ewes receiving Se treatment. **B)** In FR-affected ewes, KLH antibody titers were greater in all Se-supplemented groups compared with the non supplemented group with the exception of the 24.5 mg/wk Na-selenite group. Different superscripts indicate group differences at $P \leq 0.10$.

older ewes represented the majority of cases of FR infection in our study and another [54]. Early detection of disease and prompt treatment with parenteral long-term acting oxytetracycline were control measures resulting in sheep being significantly more likely to recover from FR lesions and lameness within 5 days of treatment compared with sheep that were foot trimmed with or without parenteral administration of antibacterials [55]. In our study, ewes with FR severity scores of II (with foot scores of 4) and higher received parenteral long-term acting oxytetracycline injections at 28 and 40 wk. In hindsight, a more intensive parenteral antibiotic treatment regimen in the current study may have been beneficial in decreasing pathogen load and, thus,

infection challenge. This might have allowed us the opportunity to see more benefits associated with Se supplementation.

It is unclear why ewes receiving no Se supplementation were so resistant to FR infection. One explanation is that dietary Se status was not the most limiting factor for FR infection in this flock. Footrot is a multifactorial disease [51] and, therefore, the optimal Se supplementation dosage may vary depending on nutritional and management conditions, as well as the sheep's immune system. Another explanation is that inflammation, and thus progression of foot lesions with marginal Se deficiency, is dampened. In support, we have shown in a companion paper [31] that WB-neutrophil gene expression profiles are shifted in an anti-inflammatory direction (increased GPx4 and SEPS1) with no

Se supplementation. Others have shown that moderate Se deficiency in mice down-regulates inflammation-related genes and reduces myeloperoxidase and lysozyme activities in Se-restricted leukocytes [40]. Myeloperoxidase is expressed in neutrophils and monocytes and generates ROS that are important for antimicrobial and cytotoxic effects, as well as modulation of the immune response via nuclear factor kappa B (NF-κB) signaling. Down regulation of inflammation under marginal Se deficiency requires further study.

The goal of immunonutrition is to enhance immunity and increase resistance to disease. We are interested in supranutritional levels of Se, to determine if supplementing Se at concentrations above those currently recommended for sheep (supranutritional) can modulate the immune response in a way that reduces the severity and/or improves recovery from a disease process. Using FR as our disease model, we have shown in a companion paper that supranutritional supplementation of these ewes with Se-yeast at 24.5 mg Se/wk improved lamb growth and ewe health

compared with maximal FDA-allowed levels of Se-yeast [17]. In this study, Se supplementation did not prevent FR, but did improve innate and humoral immune functions negatively affected by FR. Future studies are warranted to evaluate whether Se supplementation enhances innate and adaptive immune responses and provides protection against other bacterial or viral pathogens.

Acknowledgments

Appreciation is expressed to B. L. Gill, W. C. Minto, K. J. Hooper, M. L. Galbraith, A. M. Harwell, Corey Stewart, C. O. Walsh, S. C. Walker, and J. M. Van Duzer for their technical assistance during the conduct of these experiments.

Author Contributions

Conceived and designed the experiments: JAH WDM GJP . Performed the experiments: JAH WRV WCS MEG WDM GJP . Analyzed the data: JAH GB. Wrote the paper: JAH GB.

References

1. Hall JA, Van Saun RJ, Bobe G, Stewart WC, Vorachek WR, et al. (2012) Organic and inorganic selenium: I. Oral bioavailability in ewes. J Anim Sci 90: 568–576.

2. Qin SY, Gao JZ, Huang KH (2007) Effects of different selenium sources on tissue selenium concentrations, blood GSH-Px activities and plasma interleukin levels in finishing lambs. Biological Trace Element Research 116: 91–102.

3. Davis PA, McDowell LR, Wilkinson NS, Buergelt CD, Van Alstyne R, et al. (2006) Tolerance of inorganic selenium by range-type ewes during gestation and lactation. Journal of animal science 84: 660–668.

4. Steen A, Strom T, Bernhoft A (2008) Organic selenium supplementation increased selenium concentrations in ewe and newborn lamb blood and in slaughter lamb meat compared to inorganic selenium supplementation. Acta Veterinaria Scandinavica 50.

5. Taylor JB, Reynolds LP, Redmer DA, Caton JS (2009) Maternal and fetal tissue selenium loads in nulliparous ewes fed supranutritional and excessive selenium during mid- to late pregnancy. Journal of Animal Science 87: 1828–1834.

6. Rooke JA, Robinson JJ, Arthur JR (2004) Effects of vitamin E and selenium on the performance and immune status of ewes and lambs. Journal of Agricultural Science 142: 253–262.

7. Kiremidjian-Schumacher L, Stotzky G (1987) Selenium and immune responses. Environ Res 42: 277–303.

8. Finch JM, Turner RJ (1996) Effects of selenium and vitamin E on the immune responses of domestic animals. Res Vet Sci 60: 97–106.

9. McClure SJ (2008) How minerals may influence the development and expression of immunity to endoparasites in livestock. Parasite Immunol 30: 89–100.

10. Hefnawy AE, Tortora-Perez JL (2010) The importance of selenium and the effects of its deficiency in animal health. Small Ruminant Research 89: 185–192.

11. Muth OH, Oldfield JE, Remmert LF, Schubert JR (1958) Effects of selenium and vitamin E on white muscle disease. Science 128: 1090.

12. FDA (2012). Code of Federal Regulations Title 21 - Food and Drugs Chapter 1 - Food and Drug Administration, Department of Health and Human Services Subchapter E - Animal drugs, feeds, and related products Part 573 - Food additive permitted in feed and drinking water of animals Subpart B - Food Additive Listing Section 573920 - Selenium

13. Rayman MP (2012) Selenium and human health. Lancet 379: 1256–1268.

14. Fairweather-Tait SJ, Collings R, Hurst R (2010) Selenium bioavailability: current knowledge and future research requirements. Am J Clin Nutr 91: 1484S–1491S.

15. Zeng H, Combs GF, Jr. (2008) Selenium as an anticancer nutrient: roles in cell proliferation and tumor cell invasion. J Nutr Biochem 19: 1–7.

16. Hall JA, Harwell AM, Van Saun RJ, Vorachek WR, Stewart WC, et al. (2011) Agronomic biofortification with selenium: Effects on whole blood selenium and humoral immunity in beef cattle. Animal Feed Science and Technology 164: 184–190.

17. Stewart WC, Bobe G, Pirelli GJ, Mosher WD, Hall JA (2012) Organic and inorganic selenium: III. Ewe and progeny performance. J Anim Sci 90: 4536–4543.

18. Stewart WC, Bobe G, Vorachek WR, Stang BV, Pirelli GJ, et al. (2013) Organic and inorganic selenium: IV. Passive transfer of immunoglobulin from ewe to lamb. J Anim Sci 91: 1791–1800.

19. Egerton JR, Roberts DS, Parsonson IM (1969) The aetiology and pathogenesis of ovine foot-rot. I. A histological study of the bacterial invasion. Journal of comparative pathology 79: 207–215.

20. Roberts DS, Egerton JR (1969) The aetiology and pathogenesis of ovine foot-rot. II. The pathogenic association of Fusiformis nodosus and F. necrophorus. Journal of comparative pathology 79: 217–227.

21. Bennett G, Hickford J, Sedcole R, Zhou H (2009) Dichelobacter nodosus, Fusobacterium necrophorum and the epidemiology of footrot. Anaerobe 15: 173–176.

22. Elleman TC, Hoyne PA, Emery DL, Stewart DJ, Clark BL (1984) Isolation of the gene encoding pilin of Bacteroides nodosus (strain 198), the causal organism of ovine footrot. FEBS letters 173: 103–107.

23. Green LE, George TR (2008) Assessment of current knowledge of footrot in sheep with particular reference to Dichelobacter nodosus and implications for elimination or control strategies for sheep in Great Britain. Veterinary journal 175: 173–180.

24. Mitchell R (2003) Footrot Eradication in Western Australia. Department of Agriculture Western Australia, Scott Print.

25. Kaler J, Daniels SL, Wright JL, Green LE (2010) Randomized clinical trial of long-acting oxytetracycline, foot trimming, and flunixine meglumine on time to recovery in sheep with footrot. Journal of veterinary internal medicine / American College of Veterinary Internal Medicine 24: 420–425.

26. Hall JA, Sendek RL, Chinn RM, Bailey DP, Thonstad KN, et al. (2011) Higher whole-blood selenium is associated with improved immune responses in footrot-affected sheep. Vet Res 42: 99.

27. Raadsma HW, O'Meara TJ, Egerton JR, Lehrbach PR, Schwartzkoff CL (1994) Protective antibody titres and antigenic competition in multivalent Dichelobacter nodosus fimbrial vaccines using characterised rDNA antigens. Veterinary immunology and immunopathology 40: 253–274.

28. Escayg AP, Hickford JGH, Bullock DW (1997) Association between alleles of the ovine major histocompatibility complex and resistance to footrot. Research in Veterinary Science 63: 283–287.

29. Egerton JR, Roberts DS (1971) Vaccination against Ovine Foot-Rot. Journal of Comparative Pathology 81: 179–&.

30. Schwartzkoff CL, Egerton JR, Stewart DJ, Lehrbach PR, Elleman TC, et al. (1993) The Effects of Antigenic-Competition on the Efficacy of Multivalent Footrot Vaccines. Australian Veterinary Journal 70: 123–126.

31. Hugejiletu H, Bobe G, Vorachek WR, Gorman ME, Mosher WD, et al. (2013) Selenium supplementation alters gene expression profiles associated with innate immunity in whole-blood neutrophils of sheep. Biol Trace Elem Res 154: 28–44.

32. NRC (2007) Nutrient Requirements of Small Ruminants: Sheep, Goats, Cervids, and New World Camelids. Washington, DC: National Academy Press.

33. Bulgin MS, Lincoln SD, Lane VM, South PJ, Dahmen JJ, et al. (1985) Evaluating an Ovine Foot-Rot Vaccine. Veterinary Medicine 80: 105–113.

34. Winter AC (2009) Footrot control and eradication (elimination) strategies. Small Ruminant Research 86: 90–93.

35. Wander RC, Hall JA, Gradin JL, Du SH, Jewell DE (1997) The ratio of dietary (n-6) to (n-3) fatty acids influences immune system function, eicosanoid metabolism, lipid peroxidation and vitamin E status in aged dogs. Journal of Nutrition 127: 1198–1205.

36. Brigelius-Flohe R, Kipp A (2009) Glutathione peroxidases in different stages of carcinogenesis. Biochim Biophys Acta 1790: 1555–1568.

37. Heirman I, Ginneberge D, Brigelius-Flohe R, Hendrickx N, Agostinis P, et al. (2006) Blocking tumor cell eicosanoid synthesis by GP x 4 impedes tumor growth and malignancy. Free Radic Biol Med 40: 285–294.

38. Kim KH, Gao Y, Walder K, Collier GR, Skelton J, et al. (2007) SEPS1 protects RAW264.7 cells from pharmacological ER stress agent-induced apoptosis. Biochem Biophys Res Commun 354: 127–132.

39. Curran JE, Jowett JB, Elliott KS, Gao Y, Gluschenko K, et al. (2005) Genetic variation in selenoprotein S influences inflammatory response. Nat Genet 37: 1234–1241.

40. Kipp AP, Banning A, van Schothorst EM, Meplan C, Coort SL, et al. (2012) Marginal selenium deficiency down-regulates inflammation-related genes in

splenic leukocytes of the mouse. The Journal of nutritional biochemistry 23: 1170–1177.

41. Hall JA, Chinn RM, Vorachek WR, Gorman ME, Jewell DE (2010) Aged Beagle dogs have decreased neutrophil phagocytosis and neutrophil-related gene expression compared to younger dogs. Vet Immunol Immunopathol 137: 130–135.

42. Gomez CR, Nomellini V, Faunce DE, Kovacs EJ (2008) Innate immunity and aging. Experimental gerontology 43: 718–728.

43. Tiwary AK, Stegelmeier BL, Panter KE, James LF, Hall JO (2006) Comparative toxicosis of sodium selenite and selenomethionine in lambs. Journal of veterinary diagnostic investigation : official publication of the American Association of Veterinary Laboratory Diagnosticians, Inc 18: 61–70.

44. Whanger PD (2002) Selenocompounds in plants and animals and their biological significance. J Am Coll Nutr 21: 223–232.

45. Felix K, Gerstmeier S, Kyriakopoulos A, Howard OM, Dong HF, et al. (2004) Selenium deficiency abrogates inflammation-dependent plasma cell tumors in mice. Cancer Res 64: 2910–2917.

46. Novoselov SV, Calvisi DF, Labunskyy VM, Factor VM, Carlson BA, et al. (2005) Selenoprotein deficiency and high levels of selenium compounds can effectively inhibit hepatocarcinogenesis in transgenic mice. Oncogene 24: 8003–8011.

47. Moustafa ME, Carlson BA, Anver MR, Bobe G, Zhong N, et al. (2013) Selenium and selenoprotein deficiencies induce widespread pyogranuloma formation in mice, while high levels of dietary selenium decrease liver tumor size driven by TGFalpha. PLoS One 8: e57389.

48. Hall JA, Bailey DP, Thonstad KN, Van Saun RJ (2009) Effect of parenteral selenium administration to sheep on prevalence and recovery from footrot. J Vet Intern Med 23: 352–358.

49. Sammalkorpi K, Valtonen V, Alfthan G, Aro A, Huttunen J (1988) Serum Selenium in Acute Infections. Infection 16: 222–224.

50. Kennan RM, Han X, Porter CJ, Rood JI (2011) The pathogenesis of ovine footrot. Vet Microbiol 153: 59–66.

51. Bennett GN, Hickford JG (2011) Ovine footrot: new approaches to an old disease. Vet Microbiol 148: 1–7.

52. Cederlof SE, Hansen T, Klaas IC, Angen O (2013) An evaluation of the ability of Dichelobacter nodosus to survive in soil. Acta Vet Scand 55: 4.

53. Duncan JS, Grove-White D, Moks E, Carroll D, Oultram JW, et al. (2012) Impact of footrot vaccination and antibiotic therapy on footrot and contagious ovine digital dermatitis. The Veterinary record 170: 462.

54. Woolaston RR (1993) Factors Affecting the Prevalence and Severity of Footrot in a Merino Flock Selected for Resistance to Haemonchus-Contortus. Australian Veterinary Journal 70: 365–369.

55. Kaler J, Daniels SL, Wright JL, Green LE (2010) Randomized clinical trial of long-acting oxytetracycline, foot trimming, and flunixine meglumine on time to recovery in sheep with footrot. J Vet Intern Med 24: 420–425.

Fine Epitope Mapping of the Central Immunodominant Region of Nucleoprotein from Crimean-Congo Hemorrhagic Fever Virus (CCHFV)

Dongliang Liu[1], Yang Li[1], Jing Zhao[1], Fei Deng[2], Xiaomei Duan[1], Chun Kou[1], Ting Wu[1], Yijie Li[1], Yongxing Wang[1], Ji Ma[1], Jianhua Yang[1,4], Zhihong Hu[2], Fuchun Zhang[1], Yujiang Zhang[3]*, Surong Sun[1]*

1 Xinjiang Key Laboratory of Biological Resources and Genetic Engineering, College of Life Science and Technology, Xinjiang University, Urumqi, Xinjiang, China, 2 State Key Laboratory of Virology, Chinese Academy of Sciences, Wuhan, Hubei, China, 3 Center for Disease Control and Prevention of the Xinjiang Uyghur Autonomous Region, Urumqi, Xinjiang, China, 4 Texas Children's Cancer Center, Department of Pediatrics, Dan L. Duncan Cancer Center, Baylor College of Medicine, Houston, Texas, United States of America

Abstract

Crimean-Congo hemorrhagic fever (CCHF), a severe viral disease known to have occurred in over 30 countries and distinct regions, is caused by the tick-borne CCHF virus (CCHFV). Nucleocapsid protein (NP), which is encoded by the S gene, is the primary antigen detectable in infected cells. The goal of the present study was to map the minimal motifs of B-cell epitopes (BCEs) on NP. Five precise BCEs (E1, ^{247}FDEAKK^{252}; E2a, ^{254}VEAL^{257}; E2b, ^{258}NGYLNKH^{264}; E3, ^{267}EVDKA^{271}; and E4, ^{274}DSMITN^{279}) identified through the use of rabbit antiserum, and one BCE (E5, ^{258}NGYL^{261}) recognized using a mouse monoclonal antibody, were confirmed to be within the central region of NP and were partially represented among the predicted epitopes. Notably, the five BCEs identified using the rabbit sera were able to react with positive serum mixtures from five sheep which had been infected naturally with CCHFV. The multiple sequence alignment (MSA) revealed high conservation of the identified BCEs among ten CCHFV strains from different areas. Interestingly, the identified BCEs with only one residue variation can apparently be recognized by the positive sera of sheep naturally infected with CCHFV. Computer-generated three-dimensional structural models indicated that all the antigenic motifs are located on the surface of the NP stalk domain. This report represents the first identification and mapping of the minimal BCEs of CCHFV-NP along with an analysis of their primary and structural properties. Our identification of the minimal linear BCEs of CCHFV-NP may provide fundamental data for developing rapid diagnostic reagents and illuminating the pathogenic mechanism of CCHFV.

Editor: Jens H. Kuhn, Division of Clinical Research, United States of America

Funding: This work was supported partly by Science and Technology Basic Work Program 2013FY113500 from Ministry of Science and Technology of China, the National Science Foundation of China (grant no. 81460303, 30860225), and the Open Research Fund Program of Xinjiang Key Laboratory of Biological Resources and Genetic Engineering (grant no. XJDX020I-2014-04). The funders had no role in study design, data collection and analysis, decision to publish, or preparation of the manuscript.

Competing Interests: The authors have declared that no competing interests exist.

* Email: xjsyzhang@163.com (YZ); sr_sun2005@163.com (SS)

Introduction

The Crimean-Congo hemorrhagic fever virus (CCHFV) is a human pathogenic agent that causes Crimean-Congo hemorrhagic fever (CCHF), a severe disease with case-fatality rates up to 30% [1–3]. CCHFV is broadly distributed across much of the Middle East, Africa, and Asia as well and has also been found in parts of Eastern Europe [4–6]. Humans are generally infected through tick bites, direct contact with blood or tissue of infected livestock, or through nosocomial infections [7–9]. In China, the first CCHF cases were reported in 1965 when the CCHFV strain BA66019 was isolated in a patient living in Bachu County of the Xinjiang Autonomous Region, which is now known to have the highest occurrences of CCHF in the country [10]. Despite the high mortality associated with CCHF, the biology and pathogenesis of the disease remain poorly understood for several reasons: CCHF outbreaks are sporadic and have been generally restricted to a relatively small number of cases, limited animal model development, and the handling of the infectious virus requires the highest

level of laboratory containment (BSL-4) [11]. Thus, early diagnosis and vaccine development are critical for both patient survival and for the prevention of potential nosocomial infection and transmission in China.

CCHFV belongs to the *Nairovirus* genus within the family Bunyaviridae [2,12]. The genome consists of three negative-stranded RNAs, designated as small (S), medium (M) and large (L) in accordance with their relative nucleotide length, and which encode the viral nucleocapsid protein (NP), the glycoprotein precursor (GP) and the putative RNA-dependent polymerase, respectively [13]. Studies have indicated that NP is the predominant protein which is present in high levels early after infection, thereby inducing a high immune response that can be detected in infected cells [14-17]. As a major protein primarily detected during the viral invasion phase, NP has been increasingly regarded as an important target of antivirus and clinical diagnosis [2]. In previous studies, complete NP expressed in bacteria has been used to detect CCHFV immunoglobulin G (IgG) and IgM antibodies; however, the instability of the protein has limited its application for routine use [18–20]. Thus there is a need to develop truncated NP or a multi-epitope peptide for CCHF diagnosis. In a prior study, Saijo et al. [21] reported that high titer sera of CCHF patients reacted only with amino acid residues 201 to 306 ($NP^{201-306}$) of the NP central fragment, a highly conserved region among various isolates. In our previous study, the NP region containing amino acid residues 237 to 305 ($NP^{237-305}$) was found to have remarkable reactivity both with a rabbit polyclonal antibody (pAb) against CCHFV-NP and with a mouse monoclonal antibody (mAb) 14B7 in Western blotting analysis [22].

Advances have made epitope mapping much easier today than it was before. Many approaches and technologies, including recombinant DNA [23], peptide synthesis [24], and peptide [25] or protein display [26] have highlighted the need for epitope mapping and raised the possibility of mapping to a sufficient level the epitopes of certain antigens of interest [27]. Biosynthetic peptide technology is often used to express several 15–25mer peptide segments covering a certain target protein to determine the presence of an antigenic region or regions for a mAb or pAb by the use of Western blotting. Epitope mapping can be subsequently performed with a set of synthetic overlapping 8mer peptides for the positive segment(s) detected by immunoblotting [28–30]. Herein, based on the findings of a previous study, we describe the fine epitope mapping of immunodominant region $NP^{237-305}$ of the CCHFV using an improved biosynthetic peptide method [28,29].

In this paper, a total of six overlapping 16–22mer peptides (Y1–Y6) and forty-one 8mer peptides (P1–P41), both fused with a truncated carrier protein, were biosynthesized and expressed for minimal epitope mapping of the antigenic properties of $NP^{237-305}$. Five potential pAb BCEs and one potential mAb BCE were identified and mapped on the stalk region of CCHFV-NP for the first time.

Materials and Methods

Ethics Statement

The study was approved by the Research Ethics Committee (Animal Ethics Committee of Xinjiang University) and the procedures that followed were in accordance with the policies and regulations of experimental animals of China. The field studies in Bachu County were permitted by Xinjiang Wildlife Conservation Association (XJWCA). The serum samples were collected using method of random sampling and this process was not involving sacrifice.

Plasmids, Antibodies and Strains

The plasmids pGEX-KG and pXXGST-1 [28] were used to express biosynthetic peptides. The prokaryotic expression plasmid pGEX-KG was maintained by the Xinjiang Key Laboratory of Biological Resources and Genetic Engineering, and pXXGST-1 was donated by Professor Wanxiang Xu of the Shanghai Institute of Planned Parenthood Research. Rabbit polyclonal antibody against CCHFV-NP (pAb) was prepared as previously described [31]. A mouse monoclonal antibody cell line (14B7) secreting IgM type monoclonal antibody 14B7 against CCHFV was obtained from Xinjiang Centers for Disease Control and Prevention (XJCDC). A high titer of mAb was separated from mice ascites [32]. A pooled sheep serum of five samples collected from Bachu County with a confirmed history of CCHFV infection was included in the study and used for reconfirming the antigenicity of identified BCEs of CCHFV-NP in Western blotting assay. Serum sample of one healthy sheep with no history of CCHFV infection was used as negative control. All the sheep sera used in

the study were collected in 2005 and kindly provided by Professor Zhang Yujiang of XJCDC [33]. The serum samples of sheep infected with CCHFV were previously identified by using indirect immunofluorescent assay (IFA) and reverse transcription polymerase chain reaction (RT-PCR) [33]. *Escherichia coli* (*E. coli*) BL21 (DE3) competent cells, used for expression of recombinant plasmids, were purchased from Beijing TransGen Biotech Co., Ltd.

Epitope Prediction

To predict the B-cell epitopes (BCEs) on the $NP^{237-305}$ fragment of CCHFV, the corresponding amino acid sequence was analyzed using the DNAStar Protean system. Secondary structure prediction of the truncated protein was performed by using the methods of Garnier and Robson [34] and Chou and Fasman [35]. The surface properties of the structural proteins, namely, hydrophilicity, flexibility, accessibility and antigenicity, were analyzed using the methods of Kyte and Doolittle [36], Karplus and Schulz [37], Emini [38] and Jameson and Wolf [39], respectively. According to the results obtained using these methods, peptides with good hydrophilicity, high accessibility, high flexibility and strong antigenicity were selected as epitope candidates. In general, peptides located in α-spiral and β-sheet regions, which do not readily form epitope regions, were excluded [40].

Biosynthesis of 8-22mer Peptides and Recombinant Plasmid Construction

Six biosynthetic 16–22mer peptides (designated Y1–Y6) spanning the $NP^{237-305}$ segment and overlapping 6 ~ 9 amino acid residues each other, which all fused with GST or a truncated GST188 carrier protein were expressed in *E. coli*, respectively [30]. The sequences of the peptides were as follows: Y1 (KLAETEGKGVFDEAKKTVEA), Y2 (AKKTVEALNGYLNKHK), Y3 (YLNKHKDEVDKASADSM), Y4 (DKASADSMITNLLKHI), Y5 (ITNLLKHIAKAQELYK) and Y6 (IAKAQELYKNSSALRAQGAQID), which correspond to $NP^{237-256}$, $NP^{250-265}$, $NP^{260-276}$, $NP^{269-284}$, $NP^{277-292}$ and $NP^{284-305}$, respectively. For the immunodominant peptides identified (Y1–Y4), four additional sets of 8mer peptides spanning the Y1 to Y4 fragments overlapping 7 residues each other were generated (totaling 41 biosynthetic peptides designated P1 to P41) for fine epitope mapping (Fig. 1c). Briefly, the synthesized DNA fragments encoding the Y1–Y6 and P1–P41 peptides (based on the S gene sequence [41] were flanked by *Bam*H I and TTA-*Sal* I sites at the 5′ and 3′ ends, respectively, then inserted into the *Bam*H I and *Sal* I sites downstream of the GST or GST188 encoding gene in the pGEX-KG or pXXGST-1 plasmid.

Expression of Fusion Proteins

The resultant recombinant plasmids expressing each 8–22mer peptide fused with GST or GST188 were transformed into *E. coli* BL21 (DE3) competent cells. Each recombinant clone was cultivated in 3 mL LB medium containing 100 μg/mL ampicillin at 30°C with continuous shaking at 200 rpm overnight. The next day, 30 μL of cell suspension was added to 3 mL fresh LB medium and grown for 4 h until reaching a bacterial density of 0.6–0.8 at OD600. The cells were grown for an additional 4 h with 0.8 mM IPTG (Y1, Y3 and Y6 fusion peptides) or without IPTG (all other fusion peptides) at 42°C to induce the expression of the recombinant proteins. For the screening of positive recombinant clones, an SDS-PAGE gel was run for each harvested cell pellet, with the pellet corresponding to GST or GST188 protein

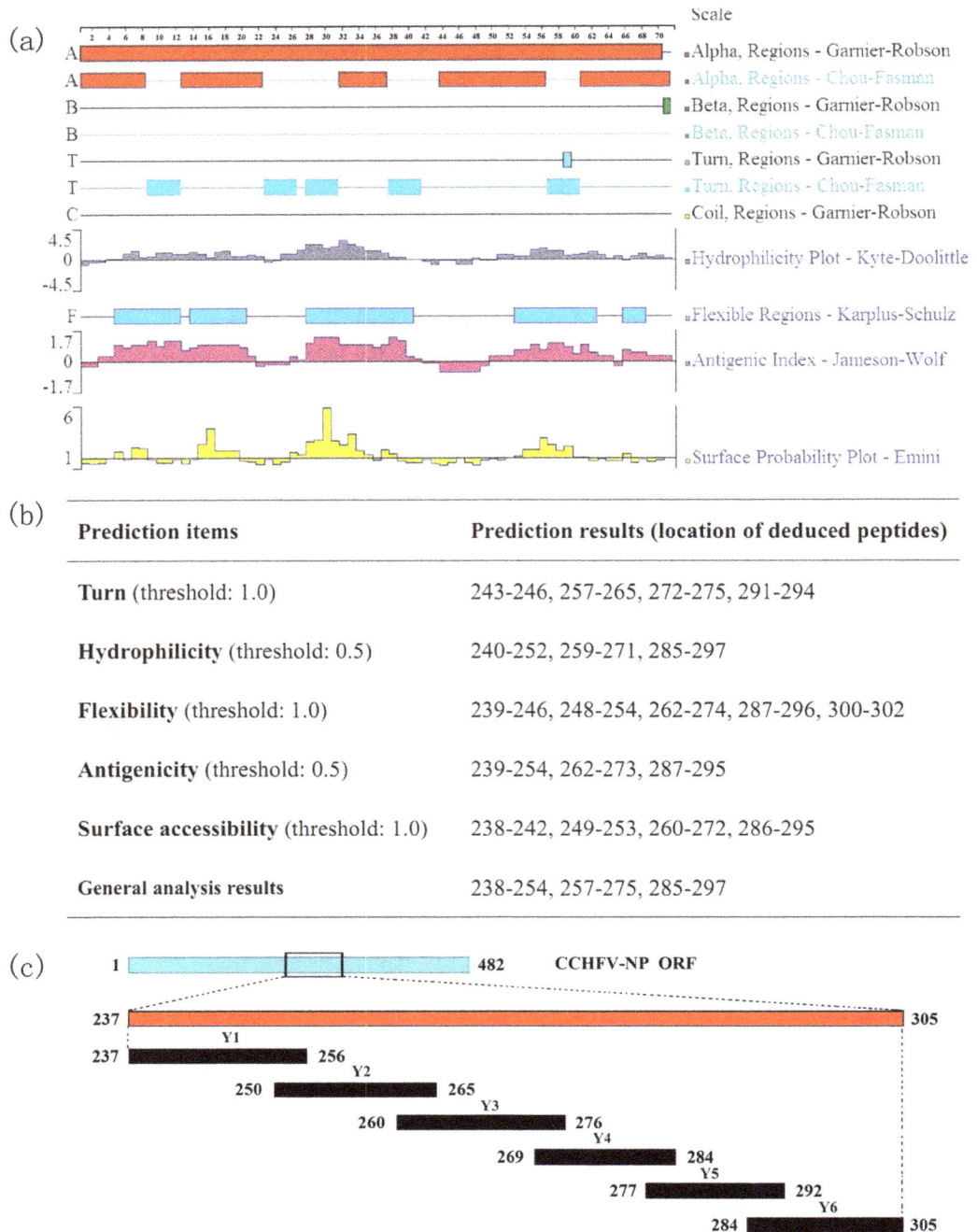

Figure 1. Prediction and mapping strategy of the epitopes in the central region of CCHFV-NP. (a) Epitope prediction for amino acid residues 237–305 of the NP sequence of the YL04057 strain using DNAStar Protean software. The secondary structure, flexibility plot, hydrophilicity, surface probability, and antigenicity index for NP$^{237-305}$ were taken into consideration. (b) Epitope predictions of the NP$^{237-305}$ fragment of the YL04057 strain based on various principles. (c) Schematic of the BCE mapping strategy with six 16–22mer overlapping biosynthetic peptides spanning the NP$^{237-305}$ fragment. The blue band represents the full length nucleoprotein. The red band represents the immunodominant fragments of YL04057 NP.

expressed by pGEX-KG or pXXGST-1 as a negative control. All recombinant clones were subsequently sent out for sequencing determination. The cell pellets containing the short peptide fusion proteins were stored at −20°C.

SDS-PAGE and Western Blotting

The cell pellets obtained from 3 mL medium were boiled in 400 μL of 1×SDS-PAGE loading buffer for 10 min, and the proteins were resolved by SDS-PAGE under reducing conditions using 15% gels [42]. Gels were either stained with Coomassie brilliant blue R-250 to analyze the bands corresponding to the fusion proteins or processed for Western blotting by electrotransfer of the proteins onto a 0.2 μm nitrocellulose membrane (Whatman GmbH, Dossel, Germany) [43]. Complete transfer of proteins was ensured by staining the nitrocellulose membrane with 0.1% (w/v) Ponceau S dye liquor. After cleaning and blocking, the nitrocellulose membrane was subsequently treated with pAb (1:1000 dilution in PBS containing 0.05% Tween 20 and 1% skim milk powder), mAb 14B7 (1:500 dilution) or a pooled serum (1:100 dilution) collected from sheep with confirmed CCHFV infection. A serum sample from a known healthy sheep with no history of CCHFV infection was used as a negative control. Specific antigen-antibody reactions on the membrane were visualized using goat anti-rabbit IgG, goat anti-mouse IgM, or rabbit anti-sheep IgG conjugated to horseradish peroxidase (HRP) (Proteintech Group, Chicago, USA) at a 1:1000 dilution. The blot was performed using ECL plus Western blotting detection reagent (GE Healthcare, Buckinghamshire, UK) according to the manufacturer's instructions.

Sequence Conservation Analysis and Three-Dimensional Modeling

To assess the sequence conservation of the identified epitopes, nine NP amino acid sequences collected from strains in different countries were obtained from GenBank. Amino acid residues 170–305 of the NP sequences from the nine virus strains were selected for multiple alignment analysis against the corresponding sequence of the YL04057 strain (GenBank code: ACM78470.1) using the ClustalW program (http://www.ebi.ac.uk/services) [44].

Three-dimensional structures of the immunodominant epitopes identified using pAb and 14B7 were simulated using PyMOL™ software [45].

Results

BCE Prediction and Mapping Strategy

Accessibility, variability, fragment mobility, charge distribution and hydrophilicity are important features of antigenic epitopes. The presence of flexible regions, such as coil and turn regions, provide further evidence for epitope identification. In this study, the secondary structure of NP$^{237-305}$ was predicted using the methods of Garnier and Robson [34] and Chou and Fasman [35] and the NP gene sequence of YL04057 CCHFV. A hydrophilicity plot, flexibility plot, surface probability plot and antigenic index for the truncated protein were obtained using the methods of Kyte and Doolittle [36], Karplus and Schulz [37], Emini [38] and Jameson and Wolf [39], respectively (Fig. 1a). The potential BCEs on NP$^{237-305}$ were predicted (Fig. 1b) based on the methods mentioned above. The finding that the secondary structure of the NP$^{237-305}$ fragment consists of five main turn motifs suggested the presence of multiple significant BCEs in this region. In our previous study, the NP$^{237-305}$ truncated fragment was found to exhibit remarkable antigen-antibody reactivity when either pAb or mAb was used in Western blotting analysis [22]. However, other unpredicted amino acids in this same region should also be considered because they may also contain BCEs, some of which may be predominant BCEs. To identify how many epitopes there are in the fragment of NP, we therefore designed a feasible strategy for BCE mapping of the NP$^{237-305}$ (Fig. 1c). Briefly, six truncated polypeptides (Y1–Y6) spanning NP$^{237-305}$ were incorporated into prokaryotic expression plasmids. Based on the results of the Western blotting analysis, sets of 8mer peptides were constructed for each of the immunodominant polypeptides identified for further BCE mapping.

Mapping Epitopes on CCHFV-NP Using pAb

All 16–22mer Peptides fused with a GST or GST188 carrier were expressed through constructing short peptide fusion expression plasmids using each synthesized encoding DNA fragments [28]. To define the fine epitopes on the NP$^{237-305}$ fragment of CCHFV, epitope mapping was performed in two steps. For the first round of antigenic peptide mapping, NP$^{237-305}$ was divided into six overlapping fragments (Y1–Y6), which were fused with GST/GST188 and expressed in E. coli, respectively. As determined by SDS-PAGE, bands corresponding to the GST-fused proteins (Lane 2, 4 and 7) were approximately 33 kD, and those corresponding to the GST188-fused proteins (Lane 3, 5 and 6) were approximately 25 kD (Fig. 2a). Western blot analysis showed that pAb reacted with polypeptides Y1–Y4 (Fig. 2b).

To further map the epitopes on NP$^{237-305}$, four sets of 8mer peptides spanning Y1 to Y4 were constructed, which have an overlap of seven amino acid residues each other in second round of fine epitope mapping. A total of 41 recombinant 8mer peptides (designated P1 to P41) were constructed and expressed in E. coli (Fig. 3). Among the 13 recombinant clones corresponding to Y1 (Fig. 3a), Western blot analysis showed that 8mer peptides P9 (GVFDEAKK), P10 (VFDEAKKT) and P11 (FDEAKKTV) were recognized by pAb against NP, suggesting that the epitope minimal motif within Y1 was the FDEAKK (named as epitope 1, E1) according to their shared residues number (Fig. 4). Three antigenic peptides Y2–Y4 were similarly identified and analyzed (Fig. 3 and 4): the fine epitopes were the VEAL (E2a) and NGYLNKH (E2b) in Y2, EVDKA (E3) in Y3 and DSMITN (E4) in Y4, respectively. Thus, five specific BCE motifs within the NP$^{237-305}$ segment were found using rabbit pAb to NP.

Epitope Mapping Using mAb 14B7 Against CCHFV-NP

In our previous study, NP$^{237-305}$ was also found to exhibit antigen-antibody reactivity with 14B7. To reveal its antibody-reactive epitope motif, using same strategy described above to map its fine epitope motif. That is, the mAb 14B7 was identified to recognize antigenic peptide Y2 in the first round of mapping (Fig. 2c) and then its epitope motif was confirm as NGYL (designated as E5, amino acid residues 258–261) in the second round of fine mapping (Fig. 5). Interestingly, its epitope motif was located in the E2b identified by the rabbit pAb, suggesting the diversity of antibody production in mouse and rabbit.

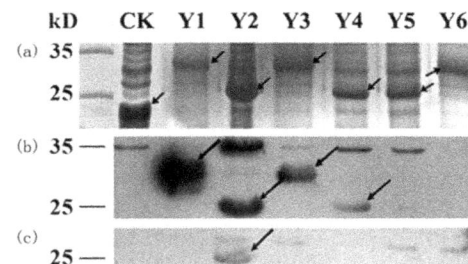

Figure 2. Prokaryotic expression and immunoblotting analysis of Y1–Y6 fused proteins. (a) SDS-PAGE analysis of expressed pXXGST-1 (CK) and Y1–Y6 peptides fused with a GST (Y1, Y3 and Y6) or GST188 tag (Y2, Y4 and Y5). **(b)** Western blotting of fusion proteins Y1–Y6 using the rabbit polyclonal antibody against CCHFV-NP. **(c)** Western blotting of fusion proteins Y1–Y6 using the mouse IgM-type monoclonal antibody 14B7 against CCHFV. The arrows represent expressed target peptides in SDS-PAGE and the corresponding positive antigenic-peptides in Western Blotting analysis.

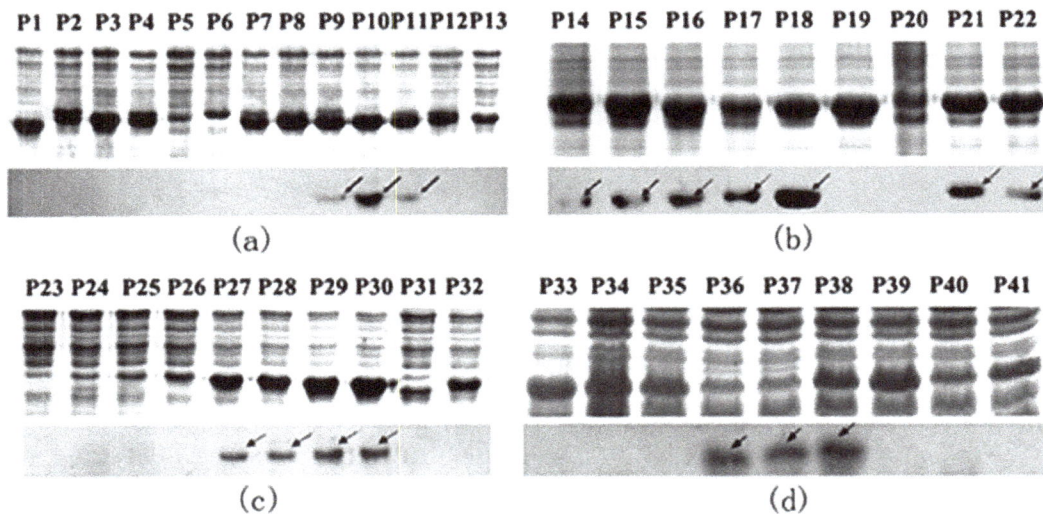

Figure 3. SDS-PAGE identification and Western blotting analysis of the minimal epitopes on NP$^{237-305}$ using pAb. (a) Thirteen 8mer peptides (P1–P13) corresponding to the Y1 protein. (b) Nine 8mer peptides (P14–P22) corresponding to the Y2 protein. (c) Ten 8mer peptides (P23–P32) corresponding to the Y3 protein. (d) Nine 8mer peptides (P33–P41) corresponding to the Y4 protein. The arrows stand for 8mer peptides which display a positive antigen-antibody reaction in Western Blotting analysis.

Determination of the Antigenicity of Identified BCEs by CCHFV Antibody-positive Sheep Sera

To determine whether the BCEs identified are rabbit/mouse specific or also recognizable by the immune systems of other host species, five randomly selected 8-mer peptides, each of which containing one of the five pAb-identified BCEs, were carried out Western blot test by using sheep sera with or without CCHFV infection (Fig. 6a). As shown in this study, the serum samples of sheep with a confirmed history of CCHFV infection could react with all the five 8-mer peptides to varying degrees, while the

Peptide items	Amino acids	Position in NP	Peptide items	Amino acids	Position in NP
P1	KLAETEGK	237-244	P23	YLNKHKDE	260-267
P2	LAETEGKG	238-245	P24	LNKHKDEV	261-268
P3	AETEGKGV	239-246	P25	NKHKDEVD	262-269
P4	ETEGKGVF	240-247	P26	KHKDEVDK	263-270
P5	TEGKGVFD	241-248	P27	HKDEVDKA	264-271
P6	EGKGVFDE	242-249	P28	KDEVDKAS	265-272
P7	GKGVFDEA	243-250	P29	DEVDKASA	266-273
P8	KGVFDEAK	244-251	P30	EVDKASAD	267-274
P9	GVFDEAKK	245-252	P31	VDKASADS	268-275
P10	VFDEAKKT	246-253	P32	DKASADSM	269-276
P11	FDEAKKTV	247-254			
P12	DEAKKTVE	248-255			
P13	EAKKTVEA	249-256			
P14	AKKTVEAL	250-257	P33	DKASADSM	269-276
P15	KKTVEALN	251-258	P34	KASADSMI	270-277
P16	KTVEALNG	252-259	P35	ASADSMIT	271-278
P17	TVEALNGY	253-260	P36	SADSMITN	272-279
P18	VEALNGYL	254-261	P37	ADSMITNL	273-280
P19	EALNGYLN	255-262	P38	DSMITNLL	274-281
P20	ALNGYLNK	256-263	P39	SMITNLLK	275-282
P21	LNGYLNKH	257-264	P40	MITNLLKH	276-283
P22	NGYLNKHK	258-265	P41	ITNLLKHI	277-284

Figure 4. The synthetic 8mer peptide sequences derived from a span of the immunodominant peptides Y1, Y2, Y3, and Y4 respectively. The yellow and magenta highlighting represents the common sequences among peptides which react with pAb or mAb using Western blotting analysis.

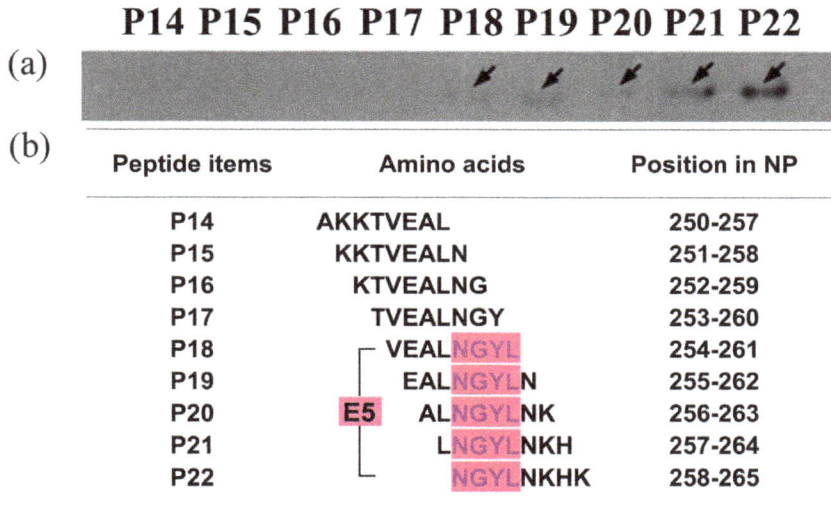

Figure 5. Minimal epitope identification on CCHFV-NP using mAb 14B7. (a) A reactivity profile of the 8mer peptides P14–P22 corresponding to Y2 using Western blotting analysis. (b) Sequences of the 8mer peptides and their positions in NP. Magenta highlighting indicates the common sequence identified as the minimal BCE (E5) on NP using 14B7. The arrows represent 8mer peptides which display positive antigen-antibody reactions using Western Blotting analysis.

CCHFV antibody-negative sera could not react with anyone of them. Of the five peptides, P10 (containing E1) and P18 (containing E2a) showed the strongest antigen-antibody reaction activities with CCHFV-infected sheep sera; Meanwhile, P29 (containing E3) and P38 (containing E4) displayed the weakest reaction intensity among the five 8-mer peptides.

Sequence Conservation Analysis and Three-Dimensional Modeling

To analyze primary structural properties of identified each BCE, the sequence corresponding to amino acid residues 170–305, which contains the identified BCEs and flanking sequences, was used to conduct multiple sequence alignment (MSA) (Fig. 7). The analysis revealed that this region (using the sequence from Chinese strain YL04057, ACM78470.1) is highly conserved when compared to the corresponding region from other CCHFV strains,

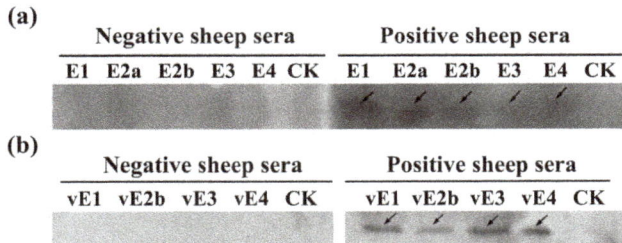

Figure 6. Western blot of five 8-mer peptides containing identified BCEs with or without one residue variation performed using positive sheep sera with a confirmed history of CCHFV infection. Five randomly selected 8-mer peptides containing identified BCEs (a) or BCEs with one residue variation (b) expressed as GST188 fusion protein in *E. coli*. A serum sample of healthy sheep with no history of CCHFV infection was used as a negative control. CK was a GST188 protein tag. The arrows represent 8mer peptides displaying positive antigen-antibody reactions based upon Western Blotting analysis.

with 90.4% sequence identity. Notably, an even higher sequence identity (92.8%) was found for $NP^{237-305}$ compared to the other nine strains. The five BCEs (E5 epitope motif was included within E2b) identified were also found to be highly conserved. Only single amino acid differences were found for four BCEs (E1, E2b, E3 and E4) compared with the other strains. To further determine whether the epitope peptides with single residue difference revealed in Fig. 7 could be used as a or all universal diagnostic reagent(s), we prepared four biosynthetic peptides (vE1, FDEAKR; vE2b, NGYLDKH; vE3, EVDRA; vE4, DNMITN) using methods mentioned above and explored their antigenic properties. Specifically, a single amino acid substitution was made within E1 (K252R), E2b (N262D), E3 (K270R), and E4 (S275N). Our investigation demonstrated that the four biosynthetic peptides with one variable residue (vE1, vE2a, vE3, vE4) can remarkably react with CCHFV antibody-positive sheep sera compared with the negative sheep sera panel (Fig. 6b), suggesting that BCEs E1–E4 derived from different CCHFV strains shared conservation in antigenicity aspect. Intriguingly, we found the E2a epitope (VEAL) and E5 epitope (NGYL) identified using mAb 14B7 were highly conserved in a majority of CCHFV isolates (Fig. 7).

Computer modeling using $PyMOL^{TM}$ software indicated that all of the antigenic motifs are located on the stalk domain of CCHFV-NP (Fig. 8a). Furthermore, the five BCEs are located in a flexible "helix-turn-helix" (HTH) structure (Fig. 8b). According to the surface representations (Fig. 8c and 8d), all of the identified BCEs are located on the surface of NP, which is consistent with the antigenic principles of surface accessibility and hydrophilicity.

Discussion

Nucleocapsid research is an important branch of viral study, as virus nucleocapsids may stimulate human immune responses, most of which are of the humoral immunity type [46]. Thus, the identification and mapping of minimal BCEs on NP represent significant steps in the development of novel diagnostic tools and multi-epitope peptide vaccines. In a previous study, a bacterially expressed recombinant NP antigen was used to detect IgG

```
ACM78470.1    (China)       LSDMIRRRNLILNRGGDENPRGPVSREHVEWCREFVKGKYIMAFNPPWGDINKSGRSGIA   229
AAL28095.2    (Yugoslavia)  ...........................................................   229
ABB72472.1    (Russia)      ...........................................................   229
AAP46054.2    (Bulgaria)    .....................S...........H.........................   229
AAZ38665.1  (South Africa)  .........Q..................................D..............   229
ADD64468.1    (China)       .......................S...................................   229
AAQ23152.2    (Tajikistan)  ................F..........................................   229
AE072054.1    (India)       ...........................................................   229
AAK40124.1    (China)       ..................S........................................   229
ABD98123.1    (Iran)        ................K..........................................   229

ACM78470.1    (China)       LVATGLAKLAETEGKGVFDEAKKTVEALNGYLNKHKDEVDKASADSMITNLLKHIAKAQE   289
AAL28095.2    (Yugoslavia)  ................FDEAKK.VEALNGYLDKHR.EVDKA..DSMITN...........   289
ABB72472.1    (Russia)      ................FDEAKK.VEALNGYLDKHR.EVDKA..DSMITN...........   289
AAP46054.2    (Bulgaria)    ................FDEAKK.VEALNGYLDKHR.EVDKA..DSMITN...........   289
AAZ38665.1  (South Africa)  ................FDEAKK.VEALNGYLDKH..EVDRA..DSMITN...........   289
ADD64468.1    (China)       ................FDEAKK.VEALNGYLNKH..EVDKA..DSMITN...........   289
AAQ23152.2    (Tajikistan)  ................FDEAKR.VEALNGYLDKH..EVDKA..DNMITN...........   289
AE072054.1    (India)       .I..............FDEAKK.VEALNGYLDKH..EVDKA..DNMITN...........   289
AAK40124.1    (China)       ............G...FDEAKK.VEALNGYLDKH..EVDKA..DNMITN...........   289
ABD98123.1    (Iran)        ................FDEAKK.VEALNGYLDKH..EVDKA..DSMITN...........   289
                                            E1       E2a  E2b      E3      E4

ACM78470.1    (China)       LYKNSSALRAQGAQID   305
AAL28095.2    (Yugoslavia)  ................   305
ABB72472.1    (Russia)      ................   305
AAP46054.2    (Bulgaria)    ................   305
AAZ38665.1  (South Africa)  ................   305
ADD64468.1    (China)       ................   305
AAQ23152.2    (Tajikistan)  ................   305
AE072054.1    (India)       ................   305
AAK40124.1    (China)       ................   305
ABD98123.1    (Iran)        ................   305
```

Figure 7. Amino acid sequence comparison of the NP$^{170-305}$ fragment from the YL04057 strain (ACM78470. 1) and other CCHFV strains using the ClustalW program. The GenBank codes and sources are shown at left. The five minimal epitopes E1, E2a, E2b, E3, and E4 recognized by pAb are highlighted in yellow, and the variable amino acids within the minimal epitopes are highlighted in red. Dots (.) indicate identical amino acids within the ten strains.

antibodies against CCHFV; the instability however, of the protein in soluble expression as well as serological diagnosis restricted the application of this protein [47,48]. The use of non-complete NP or multi-epitope peptides for CCHF diagnosis has attracted increasing attention, along with CCHF studies in general. At the same time, there have been increased efforts related to the epitope mapping of CCHFV-NP. For instance, Saijo et al. reported that in Western blotting analysis, high titer sera of CCHF patients reacted only with the highly conserved NP fragment which contained the amino acid residues 201 to 306 (NP$^{201-306}$) [21]. Similarly, Burt et al. found that NP$^{123-396}$ of CCHFV includes a highly antigenic region with application toward the development of antibody detection assays [48]. Previously, our group showed that NP$^{237-305}$ is an immunogenic region of CCHFV-NP using a polyclonal antibody and two monoclonal antibodies against CCHFV with Western blot analysis [22]. It is worth noting that the NP$^{237-305}$ region is smaller and more detailed and completely encompassed by the NP$^{201-306}$ and NP$^{123-396}$ regions. The consistent finding from these independent research groups suggests that the high antigenicity region of CCHFV-NP is located in the central region of NP rather than the N- or C-terminal regions. Although several antigenic peptides have been mapped on CCHFV-NP, to our knowledge, no minimal motifs have been previously identified, due to methodological limitations.

The biosynthetic peptide method has been successfully used by several research groups to identify the minimal epitopes on human zona pellucida protein [28–30]. In the present study, we used two prokaryotic plasmids for expressing 8–22mer peptides fused with a GST or GST188 tag to avoid the influence of different expression systems on the stability and antigenicity of the recombinant peptides generated. The simplicity, cost effectiveness, reliability, and adaptability of this approach are highly suitable for minimal motif identification [28]. Herein, we demonstrate the use of this method in mapping the minimal motifs of the BCEs of CCHFV-NP. Thus, this methodology may accelerate research requiring the minimal motif mapping of known viral antigenic epitope fragments. In the present study, we mapped six minimal BCEs on NP, five of which were identified by pAb (E1, ^{247}FDEAKK^{252}; E2a, ^{254}VEAL^{257}; E2b, ^{258}NGYLNKH^{264}; E3, ^{267}EVDKA^{271}; and E4, ^{274}DSMITN^{279}) and one by mAb (E5, ^{258}NGYL^{261}). Herein, the antigenicity of the five pAb-identified BCEs was reconfirmed by utilizing natural sera from the sheep with CCHFV infection history, indicating that the identified BCEs may have significant potential in acting as a diagnostic tool to identify whether certain wild animals or even human beings were infected by CCHFV in natural conditions. Additionally, four of the six BCEs were identified in our antigen prediction analysis (Fig. 1b), demonstrating that the epitope prediction tool combined with the biosynthetic peptide method is a reliable approach for epitope mapping and may reduce the experimental effort and expense of identifying and mapping epitopes for immunodiagnostics.

The five minimal BCEs found on NP$^{237-305}$ span amino acid residues 247 to 279 of CCHFV-NP, and all of them were found to have high sequence similarity among different CCHFV strains according to MSA analysis (88.57% for E2b, 95% for E4, 98% for E3, 98.33% for E1, and 100% for E2a and E5) (Fig. 7). To give specifics, the lysine252 was replaced by arginine within E1 (K252R) in one strain. In certain strains, a single amino acid

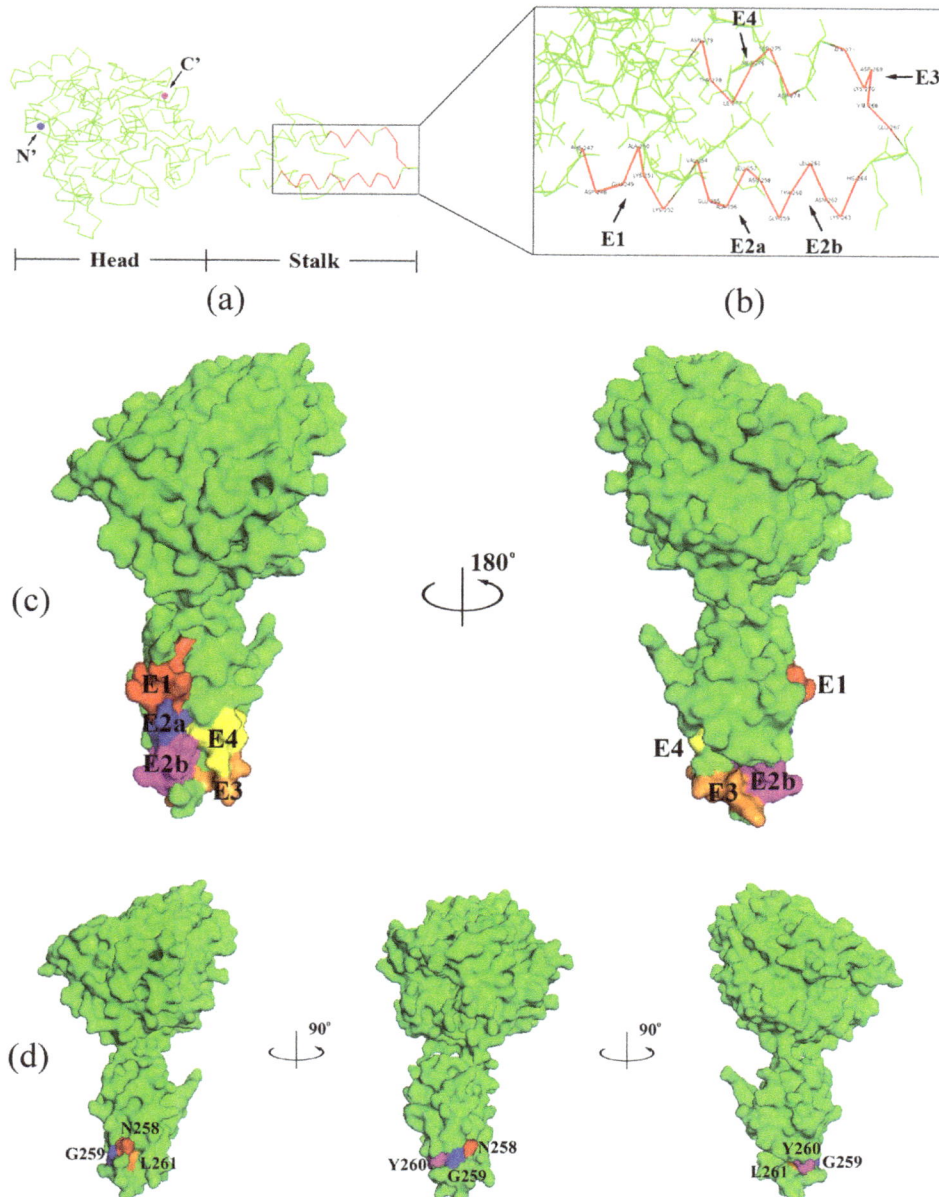

Figure 8. Location and three-dimensional structure of the epitopes identified using pAb and mAb 14B7 on CCHFV-NP. (a) The ribbon diagram shows the overall secondary structure of CCHFV-NP from strain YL04057 (PDB code: 3U3I). The motifs within the frame indicate the five minimal epitopes E1–E4. (b) E1–E4 sites on the CCHFV-NP stalk domain. (c) Surface properties of CCHFV-NP. The molecular surfaces of E1 (red), E2a (blue), E2b (magenta), E3 (orange) and E4 (yellow) are shown. (d) The structural representations show the location and spatial conformation of epitope E5 (tetrapeptide "NGYL") identified by Amb 14B7. Residues N258, G259, Y260, and L261 are shown in different colors. The figures were generated using the PyMOL molecular graphics system.

substitution was also found within E2b (N262D), E3 (K270R), and E4 (S275N). Despite one residue difference, the antigen-antibody reaction was still obvious when using positive sera of sheep naturally infected with CCHFV (Fig. 6b), reflecting highly antigenic conservation. However, ideally, the sera of CCHF patients should be utilized to verify the conservation and specificity of BCEs, which is crucial for future applications in CCHF diagnosis and prevention. In this study, we only provided the fundamental data that the antigenicity of CCHFV-NP was researched using rabbit polyclonal antiserum against the CCHFV-NP, using the mouse monoclonal antibody against the

CCHFV, and using the sera of sheep naturally infected with CCHFV. Thereby, the properties, structure, antigenicity, and immunogenicity of the NP protein, and in particular, identification of human sera infected with CCHFV will be further studied in order to be applied to CCHF diagnosis and therapy in the future. To our knowledge, the ten CCHFV strains used were isolated in countries which were directly affected by at least one of the five CCHF-epidemic areas, namely countries in Europe, Africa, Central Asia, South Asia, and the Middle East (Table S1). As depicted, epitopes E2a and E5 showed 100% conservative properties among all strains from the five CCHF-epidemic areas.

Epitope E1 was fully conserved in the countries of Europe, Africa, South Asia, the Middle East, and part of Central Asia. Similarly, epitope E3 also displayed complete homology in the different CCHF-epidemic areas, with the exception of Africa. It is worth noting that the epitope E2b from isolate YL04057, though it was not consistent with the other eight strains from different CCHF-epidemic areas, had merely one amino acid difference among ten CCHFV strains. As far as we know, eleven complete NP sequences of CCHFV strains isolated within China have been registered in the GenBank database. To further confirm whether the epitope E2b (^{258}NGYLNKH^{264}) showed high homology among strains from China, the sequences of the eleven Chinese strains corresponding to amino acid residues 241 to 300 of NP were retrieved from the GenBank for sequence alignment using the ClustalW program (Figure S1). Our study indicated that there was only a single difference, which of asparagine changing to aspartic acid (N262D) within epitope E2b in four strains of the eleven, suggesting that asparagine262 may well exist only in the Chinese CCHFV strains.

It has been previously reported that the Dugbe virus, another member of the *Nairovirus* genus, shares some antigenic and genetic properties with CCHFV [49]. Based on our findings, however, the amino acid sequences of NP$^{247-279}$ of the two viruses (GenBank codes: ACM78470.1 and AAL73396.1) display only a 9.1% similarity, despite a 57.4% sequence similarity between the two complete NP sequences (data not given). These findings suggest that the identified BCEs may be unique to CCHFV and thus highly species specific.

To further investigate the structural aspects of the minimal BCEs, the three-dimensional structure of YL04057 NP was retrieved from the Protein Databank (PDB code: 3U3I). The five pAb BCEs and one mAb BCE were all found to be located on the NP stalk domain (BCE surface properties shown in different colors in Fig. 8c and 8d). Structural analyses, particularly of the surface structure of epitopes, provide a good foundation in the search for and creation of structurally complementary drugs with clinical applications. The flexible "helix-turn-helix" structure containing the five BCEs may form a discontinuous epitope and easily react with antibodies or drugs. Human MxA protein has been shown to inhibit the CCHFV replication process [50]. In a protein-protein docking study of MxA with CCHFV-NP emphasizing epitope-based immunoinformatics, Srinivasan et al. [51] showed a complementary wrapping of the NP stalk around the MxA model. Together with the present findings, these results suggest that the CCHFV-NP stalk domain may play a critical role in immune system processes and virus interaction. Intrabodies, or intracellular antibodies, are powerful tools for cell biology studies as well as therapeutic applications [52]. They are commonly used to either block the intracellular antibody target or to image endogenous target dynamics [53,54]. It is reported that intrabodies induced cell death via activation of the caspase-3-mediated apoptotic pathway [55]. In a related vein, recent structural studies of CCHFV-NP revealed that the amino acid residues DEVD at positions 266 to 269 on the NP stalk domain comprise a putative cleavage site of caspase-3, indicating that caspase-3 cleavage of NP

may represent a host defense mechanism against lytic CCHFV infection [56–59]. Interestingly, the occurrence of the DEVD motif is within the epitope-rich region of amino acid residues 247 to 279. Our study also raises the possibility of a combination of caspase-3-dependent apoptosis and intrabody therapy in fighting a CCHFV infection in the future.

In the present study, we identified five fine linear BCEs on the stalk region of CCHFV-NP using a peptide biosynthesis strategy, thereby demonstrating the utility of this approach in peptide-based assays aimed at antibody detection. However, it remains a topic of further research whether the antigenic activities, consisting of specificity and sensitivity, can be enhanced by linearly fusing the five BCEs so that they would be more easily and more effectively applied in clinical diagnosis and epidemiological investigation.

Conclusion

In this study, the five highly conserved or 100% conserved B-cell epitopes E1, E2a, E2b (which fully overlaps E5), E3 and E4 do not only react with a prepared polyclonal antibody, but also with the positive sera of sheep naturally infected with CCHFV. More importantly, the four epitope mutants vE1, vE2b, vE3, and vE4 are distinctly recognizable through the use of naturally infected sheep sera. Our discovery has demonstrated a high antigenic-conservation of these identified minimal epitopes, which might be useable as universal epitopes in CCHF diagnosis. It is of great importance that human sera infected with CCHFV be used to test these identified epitopes in future study. Furthermore, these BCEs were determined to be located on the surface of the NP stalk region, suggesting they very well may play significant roles in the process of interaction with the host immune system, being easily recognized by antibodies. These findings would provide fundamental data for the development of novel diagnostic reagents and the illumination of the pathogenic mechanism of CCHFV.

Supporting Information

Figure S1.
(DOC)

Table S1.
(DOC)

Acknowledgments

We are very grateful to Professor Wanxiang Xu for his technical support and proofreading work.

Author Contributions

Conceived and designed the experiments: SRS DLL YJZ FCZ. Performed the experiments: DLL YL JZ XMD CK TW. Analyzed the data: SRS DLL YJL JM YXW. Contributed reagents/materials/analysis tools: DLL SRS YJZ ZHH FD. Wrote the paper: DLL SRS JHY YJZ ZHH.

References

1. Elliott RM (1990) Molecular biology of the Bunyaviridae. J Gen Virol 71: 501–522.
2. Ergönül Ö (2006) Crimean-Congo haemorrhagic fever. Lancet Infect Dis 6: 203–214.
3. World Health Organization (2001) Crimean-Congo haemorrhagic fever. Fact sheet 208.
4. Deyde VM, Khristova ML, Rollin PE, Ksiazek TG, Nichol ST (2006) Crimean-Congo hemorrhagic fever virus genomics and global diversity. J Virol 80: 8834–8842.
5. Hoogstraal H (1979) The epidemiology of tick-borne Crimean-Congo hemorrhagic fever in Asia, Europe, and Africa. J Med Entomol 15: 307–417.
6. Maltezou HC, Papa A (2010) Crimean–Congo hemorrhagic fever: Risk for emergence of new endemic foci in Europe? Travel Med Infect Di 8: 139–143.

7. Swanepoel R, Shepherd AJ, Leman PA, Shepherd SP, McGillivray GM, et al. (1987) Epidemiologic and clinical features of Crimean-Congo hemorrhagic fever in southern Africa. Am J Trop Med Hyg 36: 120–132.

8. Chinikar S (2009) An overview of Crimean-Congo hemorrhagic fever in Iran. Iran. J Microbiol 1: 7–12.

9. Gürbüz Y, Sencan I, Öztürk B, Tütüncü E (2009) A case of nosocomial transmission of Crimean–Congo hemorrhagic fever from patient to patient. Int J Infect Dis 13: e105–e107.

10. Sun SR, Dai X, Aishan M, Wang XH, Meng WW, et al. (2009). J Clin Microbiol 47: 2536–2543.

11. Bergeron É, Albariño CG, Khristova ML, Nichol ST (2010) Crimean-Congo hemorrhagic fever virus-encoded ovarian tumor protease activity is dispensable for virus RNA polymerase function. J Virol 84: 216–226.

12. Whitehouse CA (2004) Crimean–Congo hemorrhagic fever. Antivir Res 64: 145–160.

13. Walter CT, Barr JN (2011) Recent advances in the molecular and cellular biology of bunyaviruses. J Gen Virol 92: 2467–2484.

14. Magurano F, Nicoletti L (1999) Humoral response in Toscana virus acute neurologic disease investigated by viral-protein-specific immunoassays. Clin Vaccine Immunol 6: 55–60.

15. Schwarz TF, Gilch S, Pauli C, Jäger G (1996) Immunoblot detection of antibodies to Toscana virus. J Med Virol 49: 83–86.

16. Vapalahti O, Kallio-Kokko H, Närvänen A, Julkunen I, Lundkvist Å, et al. (1995) Human B-cell epitopes of puumala virus nucleocapsid protein, the major antigen in early serological response. J Med Virol 46: 293–303.

17. Dowall SD, Richards KS, Graham VA, Chamberlain J, Hewson R (2012) Development of an indirect ELISA method for the parallel measurement of IgG and IgM antibodies against Crimean-Congo haemorrhagic fever (CCHF) virus using recombinant nucleoprotein as antigen. J Virol Methods 179: 335–341.

18. Tang Q, Saijo M, Zhang Y, Asiguma M, Tianshu D, et al. (2003) A patient with Crimean-Congo hemorrhagic fever serologically diagnosed by recombinant nucleoprotein-based antibody detection systems. Clin Vaccine Immunol 10: 489–491.

19. Saijo M, Tang Q, Shimayi B, Han L, Zhang Y, et al. (2005). Recombinant nucleoprotein-based serological diagnosis of Crimean-Congo hemorrhagic fever virus infections. J Med Virol 75: 295–299.

20. Garcia S, Chinikar S, Coudrier D, Billecocq A, Hooshmand B, et al. (2006) Evaluation of a Crimean-Congo hemorrhagic fever virus recombinant antigen expressed by Semliki Forest suicide virus for IgM and IgG antibody detection in human and animal sera collected in Iran. J Clin Virol 35: 154–159.

21. Saijo M, Tang Q, Niikura M, Maeda A, Ikegami T, et al. (2002) Recombinant nucleoprotein-based enzyme-linked immunosorbent assay for detection of immunoglobulin G antibodies to Crimean-Congo hemorrhagic fever virus. J Clin Microbiol 40: 1587–1591.

22. Wei PF, Luo YJ, Li TX, Wang HL, Hu ZH, et al. (2010) Serial expression of the.truncated fragments of the nucleocapsid protein of CCHFV and identification of the epitope region. Virol Sin 25: 45–51.

23. Morrow JF, Cohen SN, Chang AC, Boyer HW, Goodman HM, et al. (1974) Replication and Transcription of Eukaryotic DNA in Esherichia coli. Proc Natl Acad Sci 71: 1743–1747.

24. Merrifield RB (1968) Solid-phase peptide synthesis. Adv Enzymol Relat Areas Mol Biol 32: 221–296.

25. Cwirla SE, Peters EA, Barrett RW, Dower WJ (1990) Peptides on phage: a vast library of peptides for identifying ligands. Proc Natl Acad Sci 87: 6378–6382.

26. Roberts BL, Markland W, Ley AC, Kent RB, White DW, et al. (1992) Directed evolution of a protein: selection of potent neutrophil elastase inhibitors displayed on M13 fusion phage. Proc Natl Acad Sci 89: 2429–2433.

27. Ladner RC (2007) Mapping the epitopes of antibodies. Biotechnol Genet Eng 24: 1–30.

28. Xu WX, He YP, Tang HP, Jia XF, Ji CN, et al. (2009) Minimal motif mapping of a known epitope on human zona pellucida protein-4 using a peptide biosynthesis strategy. J Reprod Immunol 81: 9–16.

29. Xu WX, Bhandari B, He YP, Tang HP, Chaudhary S, et al. (2012) Mapping of Epitopes Relevant for Induction of Acrosome Reaction on Human Zona Pellucida Glycoprotein-4 Using Monoclonal Antibodies. Am. J Reprod Immunol 68: 465–475.

30. Xu WX, He YP, Wang J, Tang HP, Shi HJ, et al. (2012) Mapping of Minimal Motifs of B-Cell Epitopes on Human Zona Pellucida Glycoprotein-3. Available: http://www.hindawi.com/journals/jir/2012/831010/abs/. Accepted 2 September 2011.

31. Liu DL, Li Y, Zhao J, Wu T, Sun SR (2013) Preparation of polyclonal antibody to nucleoprotein from Xinjiang hemorrhagic fever virus and its immunological evaluation. Chinese J Cell Mol Immunol 29: 838–841. (In Chinese)

32. Kints JP, Manouvriez P, Bazin H (1989) Rat monoclonal antibodies VII. Enhancement of ascites production and yield of monoclonal antibodies in rats following pretreatment with pristane and Freund's adjuvant. J Immunol Methods 119: 241–245.

33. Dai X, Muhtar, Feng CH, Sun SR, Tai XP, et al. (2006) Geography and host distribution of Crimean - Congo hemorrhagic fever in the Tarim Basin. Chin J Epidemiol 27: 1048–1052. (In Chinese)

34. Garnier J, Robson B (1989) The GOR method for predicting secondary structures in proteins. In Prediction of protein structure and the principles of protein conformation. pp. 417–465.

35. Chou PY, Fasman GD (1978) Prediction of the secondary structure of proteins from their amino acid sequence. Adv Enzymol Relat Areas Mol Biol 47: 45–148.

36. Kyte J, Doolittle RF (1982) A simple method for displaying the hydropathic character of a protein. J Mol Biol 157: 105–132.

37. Karplus PA, Schulz GE (1985) Prediction of chain flexibility in proteins. Naturwissenschaften 72: 212–213.

38. Emini EA, Hughes JV, Perlow D, Boger J (1985) Induction of hepatitis A virus-neutralizing antibody by a virus-specific synthetic peptide. J Virol 55: 836–839.

39. Jameson BA, Wolf H (1988) The antigenic index: a novel algorithm for predicting antigenic determinants. Comput Appl Biosci 4: 181–186.

40. Zhang ZW, Zhang YG, Wang YL, Pan L, Fang YZ, et al. (2010) Screening and identification of B cell epitopes of structural proteins of foot-and-mouth disease virus serotype Asia1. Vet Microbiol 140: 25–33.

41. Zhou Z, Meng W, Deng F, Xia H, Li T, et al. (2013) Complete genome sequences of two Crimean-Congo hemorrhagic fever viruses isolated in China. Genome Announc 1: e00571-13.

42. Laemmli UK (1970) Cleavage of structural proteins during the assembly of the head of bacteriophage T4. Nature 227: 680–685.

43. Towbin H, Staehelin T, Gordon J (1979) Electrophoretic transfer of proteins from polyacrylamide gels to nitrocellulose sheets: procedure and some applications. Proc Natl Acad Sci 76: 4350–4354.

44. Chenna R, Sugawara H, Koike T, Lopez R, Gibson TJ, et al. (2003) Multiple sequence alignment with the Clustal series of programs. Nucleic Acids Res 13: 3497–3500.

45. DeLano WL (2002) The PyMOL molecular graphics system.

46. Lundkvist Å, Meisel H, Koletzki D, Lankinen H, Cifire F, et al. (2002) Mapping of B-cell epitopes in the nucleocapsid protein of Puumala hantavirus. Viral Immunol 1: 177–192.

47. Samudzi RR, Leman PA, Paweska JT, Swanepoel R, Burt FJ (2012) Bacterial expression of Crimean-Congo hemorrhagic fever virus nucleoprotein and its evaluation as a diagnostic reagent in an indirect ELISA. J Virol Methods 179: 70–76.

48. Burt FJ, Samudzi RR, Randall C, Pieters D, Vermeulen J, et al. (2013) Human defined antigenic region on the nucleoprotein of Crimean-Congo haemorrhagic fever virus identified using truncated proteins and a bioinformatics approach. J Virol Methods 193: 706–712.

49. Papa A, Ma B, Kouidou S, Tang Q, Hang C, et al. (2002) Genetic characterization of the M RNA segment of Crimean Congo hemorrhagic fever virus strains, China. Emerg Infect Dis 8: 50–53.

50. Andersson I, Bladh L, Mousavi-Jazi M, Magnusson KE, Lundkvist Å, et al. (2004) Human MxA protein inhibits the replication of Crimean-Congo hemorrhagic fever virus. J Virol 78: 4323–4329.

51. Srinivasan P, Kumar SP, Karthikeyan M, Jeyakanthan J, Jasrai YT, et al. (2011) Epitope-based immunoinformatics and molecular docking studies of nucleocapsid protein and ovarian tumor domain of Crimean-Congo hemorrhagic fever virus. Front Gen 2: 72–80.

52. Lo AS, Zhu Q, Marasco WA (2008) Intracellular antibodies (intrabodies) and their therapeutic potential. Handb Exp Pharmacol 181: 343–373.

53. Moutel S, Nizak C, Perez F (2012) Selection and Use of Intracellular Antibodies (Intrabodies). Methods Mol Biol 907: 667–679.

54. Stocks M (2005) Intrabodies as drug discovery tools and therapeutics. Curr Opin Chem Biol 9: 359–365.

55. Tse E, Rabbitts TH (2000) Intracellular antibody-caspase-mediated cell killing: An approach for application in cancer therapy. Proc Natl Acad Sci USA 97: 12266–12271.

56. Karlberg H, Tan YJ, Mirazimi A (2011) Induction of caspase activation and cleavage of the viral nucleocapsid protein in different cell types during Crimean-Congo hemorrhagic fever virus infection. J Biol Chem 286: 3227–3234.

57. Wang Y, Dutta S, Karlberg H, Devignot S, Weber F, et al. (2012) Structure of Crimean-Congo hemorrhagic fever virus nucleoprotein: superhelical homo-oligomers and the role of caspase-3 cleavage. J Virol 86: 12294–12303.

58. Carter SD, Surtees R, Walter CT, Ariza A, Bergeron É, et al. (2012) Structure, function, and evolution of the Crimean-Congo hemorrhagic fever virus nucleocapsid protein. J Virol 86: 10914–10923.

59. Guo Y, Wang WM, Ji W, Deng MP, Sun YN, et al. (2012) Crimean-Congo hemorrhagic fever virus nucleoprotein reveals endonuclease activity in bunyaviruses. Proc Natl Acad Sci 109: 5046–5051.

Permissions

All chapters in this book were first published in PLOS ONE, by The Public Library of Science; hereby published with permission under the Creative Commons Attribution License or equivalent. Every chapter published in this book has been scrutinized by our experts. Their significance has been extensively debated. The topics covered herein carry significant findings which will fuel the growth of the discipline. They may even be implemented as practical applications or may be referred to as a beginning point for another development.

The contributors of this book come from diverse backgrounds, making this book a truly international effort. This book will bring forth new frontiers with its revolutionizing research information and detailed analysis of the nascent developments around the world.

We would like to thank all the contributing authors for lending their expertise to make the book truly unique. They have played a crucial role in the development of this book. Without their invaluable contributions this book wouldn't have been possible. They have made vital efforts to compile up to date information on the varied aspects of this subject to make this book a valuable addition to the collection of many professionals and students.

This book was conceptualized with the vision of imparting up-to-date information and advanced data in this field. To ensure the same, a matchless editorial board was set up. Every individual on the board went through rigorous rounds of assessment to prove their worth. After which they invested a large part of their time researching and compiling the most relevant data for our readers.

The editorial board has been involved in producing this book since its inception. They have spent rigorous hours researching and exploring the diverse topics which have resulted in the successful publishing of this book. They have passed on their knowledge of decades through this book. To expedite this challenging task, the publisher supported the team at every step. A small team of assistant editors was also appointed to further simplify the editing procedure and attain best results for the readers.

Apart from the editorial board, the designing team has also invested a significant amount of their time in understanding the subject and creating the most relevant covers. They scrutinized every image to scout for the most suitable representation of the subject and create an appropriate cover for the book.

The publishing team has been an ardent support to the editorial, designing and production team. Their endless efforts to recruit the best for this project, has resulted in the accomplishment of this book. They are a veteran in the field of academics and their pool of knowledge is as vast as their experience in printing. Their expertise and guidance has proved useful at every step. Their uncompromising quality standards have made this book an exceptional effort. Their encouragement from time to time has been an inspiration for everyone.

The publisher and the editorial board hope that this book will prove to be a valuable piece of knowledge for researchers, students, practitioners and scholars across the globe.

List of Contributors

Frauke Stock, European Cattle Genetic Diversity Consortium and Daniel G. Bradley
Smurfit Institute of Genetics, Trinity College Dublin, Dublin, Ireland

Ceiridwen J. Edwards
Smurfit Institute of Genetics, Trinity College Dublin, Dublin, Ireland
Research Laboratory for Archaeology, University of Oxford, Oxford, United Kingdom

Catarina Ginja
Veterinary Genetics Laboratory, University of California Davis, Davis, California, United States of America
Departamento de Genética, Melhoramento Animal e Reproduçāo, Instituto Nacional dos Recursos Biológicos, Fonte Boa, Vale de Santarém, Portugal

M. Cecilia T. Penedo
Veterinary Genetics Laboratory, University of California Davis, Davis, California, United States of America

Luis T. Gama
Departamento de Genética, Melhoramento Animal e Reproducāo, Instituto Nacional dos Recursos Bioló gicos, Fonte Boa, Vale de Santarém, Portugal

Juha Kantanen
Biotechnology and Food Research, MTT Agrifood Research Finland, Jokioinen, Finland

Lucía Pérez-Pardal
Area de Genética y Reproducción Animal, SERIDA, Gijón, Spain

Anne Tresset
Archéozoologie, Archéobotanique, Sociétés, Pratiques et Environnements, CNRS Muséum National d'Histoire Naturelle, Paris, France

Johannes A. Lenstra and Isaäc J. Nijman
Faculty of Veterinary Medicine, Utrecht University, Utrecht, The Netherlands

Chris N. Johnson
School of Zoology, University of Tasmania, Hobart, Australia

Linda van Bommel
School of Zoology, University of Tasmania, Hobart, Australia
Fenner School of Environment and Society, Australian National University, Canberra Australia

Xiuzhi Ma
College of Forestry, Inner Mongolia Agricultural University, Huhhot, China

Per Ambus
Department of Chemical and Biochemical Engineering, Technical University of Denmark Lyngby, Denmark

Shiping Wang
Laboratory of Alpine Ecology and Biodiversity, Institute of Tibetan Plateau Research, Chinese Academy of Sciences, Beijing, China

Yanfen Wang
Department of Life College, University of Chinese Academy of Sciences, Beijing, China

Chengjie Wang
College of Ecology and Environmental Science, Inner Mongolia Agricultural University, Huhhot, China

Abdou Razac Boukary
Department of Livestock promotion and Management of Natural Resources, ONG Karkara, Niamey, Niger
Department of Infectious and Parasitic Diseases, Research Unit in Epidemiology and Risk Analysis applied to the Veterinary Sciences (UREAR), University of Liege, Liege, Belgium
Department of Biomedical Sciences, Unit of Biostatistics and Epidemiology, Institute of Tropical Medicine, Antwerp, Belgium
Faculty of Agronomy, University of Niamey, Niamey, Niger

Claude Saegerman
Department of Infectious and Parasitic Diseases, Research Unit in Epidemiology and Risk Analysis applied to the Veterinary Sciences (UREAR), University of Liege, Liege, Belgium

Emmanuel Abatih, Reginald De Deken and Eric Thys
Department of Biomedical Sciences, Unit of Biostatistics and Epidemiology, Institute of Tropical Medicine, Antwerp, Belgium

David Fretin
Department of Bacteriology and Immunology, Veterinary and Agro-chemical Research
Centre, Uccle, Belgium

Rianatou Alambédji Bada
Service of Microbiology-Immunology-Infectious Pathology, Interstate School of Veterinary Sciences and Medicine, Dakar, Senegal

Halimatou Adamou Harouna
Ministry of Livestock, Direction of Animal Health, Maradi, Niger

Alhassane Yenikoye
Faculty of Agronomy, University of Niamey, Niamey, Niger

Verena A. Lambermont, Luc J. Zimmermann and Boris W. Kramer
Department of Pediatrics, Maastricht University Medical Center, Maastricht, the Netherlands

Marco Schlepütz, Constanze Dassow, Stefan Uhlig and Christian Martin
Institute of Pharmacology and Toxicology, University Hospital Aachen, Aachen, Germany

Peter König
Institute of Anatomy, University of Lübeck, Airway Research Center North (ARCN), Member of the German Center for Lung Research (DZL), Lübeck, Germany

Amparo M. Martínez, Juan V. Delgado and Vincenzo Landi
Departamento de Genética, Universidad de Córdoba, Córdoba, Spain

Luis T. Gama
L-INIA, Instituto Nacional dos Recursos Biológicos, Fonte Boa, Vale de Santarém, Portugal
CIISA – Faculdade de Medicina Veterinária, Universidade Técnica de Lisboa, Lisboa, Portugal

Javier Cañón, Susana Dunner and Oscar Cortés
Departamento de Producción Animal, Facultad de Veterinaria, Universidad Complutense de Madrid, Madrid, Spain

Catarina Ginja
Centre for Environmental Biology, Faculty of Sciences, University of Lisbon & Molecular Biology Group, Instituto Nacional de Recursos Biológicos, INIA, Lisbon, Portugal

Pilar Zaragoza, Inmaculada Martín-Burriel and Clementina Rodellar
Laboratorio de Genética Bioquímica, Facultad de Veterinaria, Universidad de Zaragoza Zaragoza, Spain

M. Cecilia T. Penedo
Veterinary Genetics Laboratory, University of California Davis, Davis, California, United States of America

Jose Luis Vega-Pla
Laboratorio de Investigación Aplicada, Cría Caballar de las Fuerzas Armadas, Córdoba, Spain

Atzel Acosta and Odalys Uffo
Centro Nacional de Sanidad Agropecuaria, San Joséde las Lajas, La Habana, Cuba

Jaime E. Muñoz and Luz A. Álvarez
Universidad Nacional de Colombia, Sede Palmira, Valle del Cauca, Colombia

Esperanza Camacho
IFAPA, Centro Alameda del Obispo, Córdoba, Spain

Jose R. Marques
EMBRAPA Amazônia Oriental, Belém, Pará, Brazil

Roberto Martínez
Centro Multidisciplinario de Investigaciones Tecnológicas, Dirección General de Investigación Científica y Tecnológica, Universidad Nacional de Asunción, San Lorenzo, Paraguay

Ruben D. Martínez
Genética Animal, Facultad de Ciencias Agrarias, Universidad Nacional de Lomas de Zamora, Lomas de Zamora, Argentina

Lilia Melucci
Facultad Ciencias Agrarias, Universidad Nacional de Mar del Plata, Balcarce, Argentina
Estación Experimental Agropecuaria Balcarce, Instituto Nacional de Tecnología Agropecuaria, Balcarce, Argentina

Guillermo Martínez-Velázquez and Jorge Quiroz
Instituto Nacional de Investigaciones
Forestales, Agrícolas y Pecuarias, Coyoacán, México

Alicia Postiglioni
Área Genética, Departamento de Genética y Mejora
Animal, Facultad de Veterinaria, Universidad de la
República, Montevideo, Uruguay

Philip Sponenberg
Virginia-Maryland Regional College of Veterinary
Medicine, Virginia Tech, Blacksburg Virginia,
United States of America

Axel Villalobos
Instituto de Investigación Agropecuaria, Estación
Experimental El Ejido, Los Santos, Panamá

Delsito Zambrano
Universidad Técnica Estatal de Quevedo, Quevedo,
Ecuador

**Coraline Bouet-Cararo, Corinne Bergeron, Cyril
Viarouge, Alexandra Desprat, Anthony Relmy,
Jennifer Richardson, Stephan Zientara and
Bernard Klonjkowski**
UPE, ANSES, INRA, ENVA, UMR 1161 ANSES/
INRA/ENVA, Maisons-Alfort, France

Vanessa Contreras and Isabelle Schwartz-Cornil
Virologie et Immunologie Moléculaires, UR 892
INRA, Jouy-en-Josas, France

**Stephane Bertagnoli, Gilles Foucras, Gilles Meyer,
Agathe Caruso and Sokunthea Top**
INRA,UMR1225, IHAP, Université de Toulouse,
INP, ENVT, Toulouse, France

Marion Szelechowski
Centre de Physiopathologie de Toulouse Purpan,
INSERM U1043, CNRS U5282, Université Paul-
Sabatier, Toulouse, France

**Jean-Michel Guibert, Eric Dubois and Richard
Thiery**
Unité de pathologie des petits ruminants, ANSES,
Sophia-Antipolis, France

**Veronique Dermauw and Geert Paul Jules
Janssens**
Laboratory of Animal Nutrition, Faculty of
Veterinary Medicine, Ghent University, Merelbeke,
Belgium

Marta Lopéz Alonso
Departemento de Patoloxía Animal, Universidade
de Santiago de Compostela, Lugo, Spain

Luc Duchateau
Department of Comparative Physiology and
Biometrics, Faculty of Veterinary Medicine, Ghent
University, Merelbeke, Belgium

Gijs Du Laing
Laboratory of Analytical Chemistry and Applied
Ecochemistry Ghent University, Faculty of
Bioscience Engineering, Ghent, Belgium

Tadele Tolosa
School of Veterinary Medicine, College of
Agriculture and Veterinary Medicine of Jimma
University, Jimma, Ethiopia 6 M-team and Mastitis
Quality Research Unit, Department of Reproduction,
Obstetrics, and Herd Health, Faculty of Veterinary
Medicine, Ghent University, Merelbeke, Belgium

Ellen Dierenfeld
Adjunct Faculty, Division of Animal Sciences,
University of Missouri-Columbia, Columbia,
Missouri, United States of America

Marcus Clauss
Clinic for Zoo Animals, Exotic Pets and Wildlife,
Vetsuisse Faculty, University of Zurich, Zurich,
Switzerland

**Sakthivel Dhanasekaran, Ambothi R. Vignesh, R.
Ramya, Gopal Dhinakar Raj, Krishnaswamy G.
Tirumurugaan and Angamuthu Raja**
Department of Animal Biotechnology, Madras
Veterinary College, Tamil Nadu Veterinary and
Animal Sciences University, Chennai, Tamil Nadu,
India

Elankumaran Subbiah and Moanaro Biswas
Department of Biomedical Sciences and
Pathobiology, Center for Molecular Medicine and
Infectious Diseases, Virginia-Maryland Regional
College of Veterinary Medicine, Virginia
Polytechnic Institute and State University,
Blacksburg, Virginia, United States of America

Ranjit S. Kataria
Animal Genetics Division, National Bureau of
Animal Genetic Resources, Karnal (Haryana), India

Satya Parida
Head of FMD Vaccine Differentiation Group, The
Pirbright Institute, Surrey, United Kingdom

Chellafe Ensoy, Marc Aerts and Christel Faes
Interuniversity Institute for Biostatistics and
statistical Bioinformatics, Universiteit Hasselt,
Hasselt, Belgium

Sarah Welby
Veterinary and Agrochemical Research Centre,
Brussels, Belgium

Yves Van der Stede
Veterinary and Agrochemical Research Centre,
Brussels, Belgium
University of Ghent, Laboratory of Vet. Immunology,
Merelbeke, Belgium

Lena Maria Lorenz
London School of Hygiene and Tropical Medicine,
Department of Disease Control, London, United
Kingdom

Marta Ferreira Maia
London School of Hygiene and Tropical Medicine,
Department of Disease Control, London, United
Kingdom
Ifakara Health Institute, Bagamoyo, Pwani Region,
United Republic of Tanzania

Ayimbire Abonuusum
Kumasi Centre for Collaborative Research in
Tropical Medicine, Kumasi, Ghana

Rolf Garms and Thomas Kruppa
Bernhard-Nocht Institute for Tropical Medicine,
Hamburg, Germany

Peter-Henning Clausen and Burkhard Bauer
Free University of Berlin, Faculty of Veterinary
Medicine Institute for Parasitology and Tropical
Veterinary Medicine, Berlin, Germany

**Yuanfang Fu, Zengjun Lu, Pinghua Li, Yimei Cao,
Pu Sun, Meina Tian, Na Wang, Huifang Bao,
Xingwen Bai, Dong Li, Yingli Chen and Zaixin
Liu**
State Key Laboratory of Veterinary Etiological
Biology, National Foot-and-Mouth Disease
Reference Laboratory, Lanzhou Veterinary Research
Institute, Chinese Academy of Agriculture Science,
Lanzhou, Gansu, China

**John Odden, Erlend B. Nilsen and John D. C.
Linnell**
Norwegian Institute for Nature Research,
Trondheim, Norway

Konjit Tadesse and Girmay Medhin
Animal Health and Zoonotic Research Unit, Aklilu
Lemma Institute of Pathobiology, Addis Ababa
University, Addis Ababa, Ethiopia

**Abraham Aseffa, Elena Hailu and Yohannes
Deresse**
TB Research Team, Armauer Hansen Research
Institute, Addis Ababa, Ethiopia

**Glyn Hewinson, Martin Vordermeier and Stefan
Berg**
TB Research Group, Animal Health and Veterinary
Laboratories Agency, New Haw, Addlestone,
Surrey, United Kingdom

Gobena Ameni
Animal Health and Zoonotic Research Unit, Aklilu
Lemma Institute of Pathobiology, Addis Ababa
University, Addis Ababa, Ethiopia
TB Research Team, Armauer Hansen Research
Institute, Addis Ababa, Ethiopia

**Timothy J. Snelling, Nest McKai and R. John
Wallace**
Rowett Institute of Nutrition and Health, University
of Aberdeen, Bucksburn, Aberdeen, United
Kingdom

Department of Animal Nutrition and Nutritional
Diseases, Faculty of Veterinary Medicine, Ondokuz
Mayis University, Samsun, Turke

Mick Watson
ARK Genomics, The Roslin Institute, Easter Bush,
Midlothian, United Kingdom

Sinéad M. Waters
Animal and Bioscience Research Department,
Animal and Grassland Research and Innovation
Centre, Teagasc, Grange, Dunsany, Co. Meath,
Ireland

Christopher J. Creevey
Institute of Biological Environmental and Rural
Sciences, Aberystwyth University, Aberystwyth,
Ceredigion, United Kingdom

Joanna Kemp, Michael de Veer, Jayne Sherrard and Troy Kraska
School of Biomedical Sciences, Monash University, Clayton, Victoria, Australia

School of Biomedical Sciences, Monash University, Clayton, Victoria, Australia
The ARC Centre of Excellence in Structural and Functional Microbial Genomics, Monash University, Clayton, Victoria, Australia

Martin Elhay
Veterinary Medicine Research and Development, Pfizer Animal Health, Parkville, Victoria, Australia
Abstract

Joris van Drongelen and Frederik K. Lotgering
Department of Obstetrics and Gynecology, Radboudumc, the Netherlands

Rob de Vries
Systematic Review Centre for Laboratory animal Experimentation, Radboudumc, the Netherlands

Paul Smits
Department of Pharmacology and Toxicology, Radboudumc, the Netherlands

Marc E. A. Spaanderman
Department of Obstetrics and Gynecology, Radboudumc, the Netherlands
Department of Obstetrics and Gynecology, Research School GROW, Maastricht University Medical Centre, Maastricht, the Netherlands

Bethan V. Purse and Kate R. Searle
Centre for Ecology and Hydrology, Bush Estate, Penicuik, Midlothian, United Kingdom

James Barber, Francesca Stubbins, Simon Carpenter, Eric Denison, Christopher Sanders, Philip S. Mellor, Anthony Wilson and Simon Gubbins
The Pirbright Institute, Ash Road, Pirbright, Woking, United Kingdom

Karien Labuschagne
PVVD, ARC-Onderstepoort Veterinary Institute, Private Bag X05, Onderstepoort, South Africa

Adam Butler
Biomathematics and Statistics Scotland, James Clerk Maxwell Building, The King's Buildings, Edinburgh, United Kingdom

Noel Nelson
The Met Office, Based at The Pirbright Institute, Pirbright, Woking, Surrey, United Kingdom

Miriam E. Koome, Joanne O. Davidson, Paul P. Drury, Sam Mathai, Lindsea C. Booth, Alistair Jan Gunn and Laura Bennet
Department of Physiology, the University of Auckland, Auckland New Zealand

Jean A. Hall, William R. Vorachek and M. Elena Gorman
Department of Biomedical Sciences, College of Veterinary Medicine, Oregon State University, Corvallis, Oregon, United States of America

D. Mosher and Gene J. Pirelli
Department of Animal and Rangeland Sciences, College of Agriculture, Oregon State University, Corvallis, Oregon, United States of America

Whitney C. Stewart
Department of Animal and Rangeland Sciences, College of Agriculture, Oregon State University, Corvallis, Oregon, United States of America
Current Address: Department of Animal and Range Sciences, New Mexico State University, Las Cruces, New Mexico, United States of America

Wayne and Gerd Bobe
Department of Animal and Rangeland Sciences, College of Agriculture, Oregon State University, Corvallis, Oregon, United States of America
Linus Pauling Institute, Oregon State University, Corvallis, Oregon, United States of America

Dongliang Liu, Yang Li, Jing Zhao, Xiaomei Duan, Chun Kou, Ting Wu, Yijie Li, Yongxing Wang, Ji Ma, Fuchun Zhang and Surong Sun
Xinjiang Key Laboratory of Biological Resources and Genetic Engineering, College of Life Science and Technology, Xinjiang University, Urumqi, Xinjiang, China

Fei Deng and Zhihong Hu
State Key Laboratory of Virology, Chinese Academy of Sciences, Wuhan, Hubei, China

Yujiang Zhang
Center for Disease Control and Prevention of the Xinjiang Uyghur Autonomous Region, Urumqi, Xinjiang, China

Jianhua Yang
Xinjiang Key Laboratory of Biological Resources and Genetic Engineering, College of Life Science and Technology, Xinjiang University, Urumqi, Xinjiang, China

Texas Children's Cancer Center, Department of Pediatrics, Dan L. Duncan Cancer Center, Baylor College of Medicine, Houston, Texas, United States of America Abstract

Index

A

Airway Responses, 47-51, 53-54
Ancestral Genetic Contributions, 58
Anopheles Gambiae, 108, 110-113
Antenatal Dexamethasone, 185, 187, 189, 191
Antigenicity, 213-214, 216-219
Asphyxia, 185-192

B

Bacteriological Findings, 134
Bacteriological Testing, 38
Biogenic Amines, 47-49
Bluetongue Virus, 68-69, 71, 73, 75-76, 96, 106-107, 172, 183-184
Breed Differentiation, 56
Bronchoconstriction, 47-54
Brucella Spp., 35, 37, 39, 41, 43-45

C

Cattle Haplogroups Defined, 2
Cattle Haplotypes, 2
Cattle Movements, 96-97, 99, 101, 103-105, 107
Central Immunodominant Region, 212-213, 215, 217, 219, 221
Community Composition, 140-141, 143, 145, 147
Comparative Intradermal Tuberculin (cidt), 130-131
Comprehensive Survey, 1, 3, 5, 7, 9, 11, 13, 66
Crimean-congo Hemorrhagic Fever Virus (cchfv), 212
Culicoides Phenology, 172-173, 175-177, 179, 181, 183
Culturing Of Sputum, 131

D

Dairy Cattle Farming, 1, 3, 5, 7, 9, 11, 13, 66
Deltamethrin, 108-109, 111-113
Different Adjuvants, 149, 151, 153, 155
Disease Policy, 172, 182
Domestic Sheep, 24, 122-125, 127, 129

E

Edible Tissues, 77-81, 83
Electric Field Stimulation (efs), 48, 52
Element Distribution, 77, 79, 81, 83
Environmental Covariates, 175

F

Faecal Egg Counts (fec), 150
Fine Epitope Mapping, 212-213, 215, 217, 219, 221
Foot-and-mouth Disease (fmd), 96, 114
Footrot-affected Sheep, 193, 195, 197-199, 201, 203, 205, 207, 209, 211
Functional Vascular Changes, 157, 159, 161, 163, 165, 167, 169, 171

G

Genetic Diversity, 1, 10-13, 56, 60, 64-66
Genetic Footprints, 55, 57, 59, 61, 63, 65, 67
Genetic Polymorphisms, 85, 89, 92

H

Haemonchus Contortus, 149, 156
Hippocampal Neuronal Density, 185
Histological Procedures, 186
Humoral Immune Responses, 193, 195, 197, 199, 201, 203, 205, 207, 209, 211

I

Iberian Cattle, 9, 55-67
Immune Parameters, 149-151, 153, 155
Induced Immune Responses, 68, 156
Inner Mongolia Grassland Soils, 26-27, 29, 31, 33
Intradermal Injections, 151, 153
Intrinsic Activation, 49
Isotopic Characteristics, 29

L

Leymus Chinensis Vegetation, 27
Livestock Guardian Dogs, 14-15, 17, 19, 21, 23-25

M

Maremma Sheepdogs, 14-15, 17, 19, 21, 23-25
Mast Cell Staining, 49
Methanogenic Archaea, 140-147
Mineral Analyses, 78, 83
Molecular Typing Of Mycobacteria, 133
Monoclonal Antibody, 69, 114-115, 117, 119, 121, 212-213, 215, 219
Mycobacterium Tuberculosis, 130-131, 133, 135, 137-139

N

Net Fencing, 108-109, 111-113

Neural Injury, 185-187, 189-191

Nucleoprotein, 212-215, 217, 219, 221

O

Outdoor-biting Mosquitoes, 108-109, 111, 113

P

Parasitological, 150, 152, 155

Pathological Examination, 131

Peripheral Blood Mononuclear Cells (pbmc), 85-86

Periurban And Rural Areas, 35, 39

Peste Des Petits Ruminants, 85, 87, 89, 91, 93-95

Phylogenetic Analysis, 95, 144-147

Phylogeography Of Y-haplotypes, 3

Population Genetic Structure, 58

Precision-cut Lung Slices (pcls), 47, 49, 51, 53-54

Predominant Epitope, 114-115, 117, 119, 121

Pregnancy In Animals, 157, 159, 161, 163, 165, 167, 169, 171

Preterm Fetal Sheep, 185, 187-189, 191-192

Protective Potential, 68

R

Risk Factor Analysis, 38

S

Sanger Amplicon Sequencing, 142

Scottish Upland Sheep, 140-141, 143, 145, 147

Selenium Supplementation, 193, 195, 197, 199, 201, 203, 205, 207, 209-211

Seroprevalence, 35-37, 39-43, 45

Socio-economic System, 124

Soil Carbon Decomposition, 26-27, 29, 31, 33

Soil Organic Matter (som), 26, 28

Spatio-temporal Model, 96-97, 99-101, 103, 105-107

T

Taurine Haplotypes, 1, 3, 9

Timing Metrics, 173, 175

Toll-like Receptor, 85, 87, 89, 91, 93-95, 203

Traditional Cattle, 35, 37, 39, 41, 43, 45

Transport Of Animals, 96-97

V

Vaccinated Animals, 68, 71, 114, 116-117, 119, 121

Viral Vectors, 68-69, 71, 73, 75

W

Water Buffalo, 84-93, 95, 141

Y

Y-chromosomal Variation, 1, 3, 5, 7, 9, 11, 13, 66

www.ingramcontent.com/pod-product-compliance
Lightning Source LLC
Chambersburg PA
CBHW061249190326

41458CB00011B/3625

* 9 7 8 1 6 3 2 3 9 7 8 0 5 *